1Mπ

One Million Digits of Pi

<u>Data</u>

Computation of 1000000 Digits of Pi
Method used : Chudnovsky
Size of FFT : 64 K
Physical memory used : ~ 3072 K
Disk memory used : ~ 1.91 Meg

Computation run information :

Duration : 7.11 seconds

ISBN-13: 978-1523410569
ISBN-10: 1523410566
Author: Alberto Sousa © 2016

Simply Enjoy Pi

Pi = 3.

1415926535 8979323846 2643383279 5028841971 6939937510 5820974944 5923078164
0628620899 8628034825 3421170679 : 1
8214808651 3282306647 0938446095 5058223172 5359408128 4811174502 8410270193
8521105559 6446229489 5493038196 : 2
4428810975 6659334461 2847564823 3786783165 2712019091 4564856692 3460348610
4543266482 1339360726 0249141273 : 3
7245870066 0631558817 4881520920 9628292540 9171536436 7892590360 0113305305
4882046652 1384146951 9415116094 : 4
3305727036 5759591953 0921861173 8193261179 3105118548 0744623799 6274956735
1885752724 8912279381 8301194912 : 5
9833673362 4406566430 8602139494 6395224737 1907021798 6094370277 0539217176
2931767523 8467481846 7669405132 : 6
0005681271 4526356082 7785771342 7577896091 7363717872 1468440901 2249534301
4654958537 1050792279 6892589235 : 7
4201995611 2129021960 8640344181 5981362977 4771309960 5187072113 4999999837
2978049951 0597317328 1609631859 : 8
5024459455 3469083026 4252230825 3344685035 2619311881 7101000313 7838752886
5875332083 8142061717 7669147303 : 9
5982534904 2875546873 1159562863 8823537875 9375195778 1857780532 1712268066
1300192787 6611195909 2164201989 : 10
3809525720 1065485863 2788659361 5338182796 8230301952 0353018529 6899577362
2599413891 2497217752 8347913151 : 11
5574857242 4541506959 5082953311 6861727855 8890750983 8175463746 4939319255
0604009277 0167113900 9848824012 : 12
8583616035 6370766010 4710181942 9555961989 4676783744 9448255379 7747268471
0404753464 6208046684 2590694912 : 13
9331367702 8989152104 7521620569 6602405803 8150193511 2533824300 3558764024
7496473263 9141992726 0426992279 : 14
6782354781 6360093417 2164121992 4586315030 2861829745 5570674983 8505494588
5869269956 9092721079 7509302955 : 15
3211653449 8720275596 0236480665 4991198818 3479775356 6369807426 5425278625
5181841757 4672890977 7727938000 : 16
8164706001 6145249192 1732172147 7235014144 1973568548 1613611573 5255213347
5741849468 4385233239 0739414333 : 17
4547762416 8625189835 6948556209 9219222184 2725502542 5688767179 0494601653
4668049886 2723279178 6085784383 : 18
8279679766 8145410095 3883786360 9506800642 2512520511 7392984896 0841284886
2694560424 1965285022 2106611863 : 19
0674427862 2039194945 0471237137 8696095636 4371917287 4677646575 7396241389
0865832645 9958133904 7802759009 : 20
9465764078 9512694683 9835259570 9825822620 5224894077 2671947826 8482601476
9909026401 3639443745 5305068203 : 21
4962524517 4939965143 1429809190 6592509372 2169646151 5709858387 4105978859
5977297549 8930161753 9284681382 : 22
6868386894 2774155991 8559252459 5395943104 9972524680 8459872736 4469584865
3836736222 6260991246 0805124388 : 23
4390451244 1365497627 8079771569 1435997700 1296160894 4169486855 5848406353
4220722258 2848864815 8456028506 : 24
0168427394 5226746767 8895252138 5225499546 6672782398 6456596116 3548862305
7745649803 5593634568 1743241125 : 25
1507606947 9451096596 0940252288 7971089314 5669136867 2287489405 6010150330
8617928680 9208747609 1782493858 : 26
9009714909 6759852613 6554978189 3129784821 6829989487 2265880485 7564014270
4775551323 7964145152 3746234364 : 27

5428584447 9526586782 1051141354 7357395231 1342716610 2135969536 2314429524
8493718711 0145765403 5902799344 : 28
0374200731 0578539062 1983874478 0847848968 3321445713 8687519435 0643021845
3191048481 0053706146 8067491927 : 29
8191197939 9520614196 6342875444 0643745123 7181921799 9839101591 9561814675
1426912397 4894090718 6494231961 : 30
5679452080 9514655022 5231603881 9301420937 6213785595 6638937787 0830390697
9207734672 2182562599 6615014215 : 31
0306803844 7734549202 6054146659 2520149744 2850732518 6660021324 3408819071
0486331734 6496514539 0579626856 : 32
1005508106 6587969981 6357473638 4052571459 1028970641 4011097120 6280439039
7595156771 5770042033 7869936007 : 33
2305587631 7635942187 3125147120 5329281918 2618612586 7321579198 4148488291
6447060957 5270695722 0917567116 : 34
7229109816 9091528017 3506712748 5832228718 3520935396 5725121083 5791513698
8209144421 0067510334 6711031412 : 35
6711136990 8658516398 3150197016 5151168517 1437657618 3515565088 4909989859
9823873455 2833163550 7647918535 : 36
8932261854 8963213293 3089857064 2046752590 7091548141 6549859461 6371802709
8199430992 4488957571 2828905923 : 37
2332609729 9712084433 5732654893 8239119325 9746366730 5836041428 1388303203
8249037589 8524374417 0291327656 : 38
1809377344 4030707469 2112019130 2033038019 7621101100 4492932151 6084244485
9637669838 9522868478 3123552658 : 39
2131449576 8572624334 4189303968 6426243410 7732269780 2807318915 4411010446
8232527162 0105265227 2111660396 : 40
6655730925 4711055785 3763466820 6531098965 2691862056 4769312570 5863566201
8558100729 3606598764 8611791045 : 41
3348850346 1136576867 5324944166 8039626579 7877185560 8455296541 2665408530
6143444318 5867697514 5661406800 : 42
7002378776 5913440171 2749470420 5622305389 9456131407 1127000407 8547332699
3908145466 4645880797 2708266830 : 43
6343285878 5698305235 8089330657 5740679545 7163775254 2021149557 6158140025
0126228594 1302164715 5097925923 : 44
0990796547 3761255176 5675135751 7829666454 7791745011 2996148903 0463994713
2962107340 4375189573 5961458901 : 45
9389713111 7904297828 5647503203 1986915140 2870808599 0480109412 1472213179
4764777262 2414254854 5403321571 : 46
8530614228 8137585043 0633217518 2979866223 7172159160 7716692547 4873898665
4949450114 6540628433 6639379003 : 47
9769265672 1463853067 3609657120 9180763832 7166416274 8888007869 2560290228
4721040317 2118608204 1900042296 : 48
6171196377 9213375751 1495950156 6049631862 9472654736 4252308177 0367515906
7350235072 8354056704 0386743513 : 49
6222247715 8915049530 9844489333 0963408780 7693259939 7805419341 4473774418
4263129860 8099888687 4132604721 : 50
5695162396 5864573021 6315981931 9516735381 2974167729 4786724229 2465436680
0980676928 2382806899 6400482435 : 51
4037014163 1496589794 0924323789 6907069779 4223625082 2168895738 3798623001
5937764716 5122893578 6015881617 : 52
5578297352 3344604281 5126272037 3431465319 7777416031 9906655418 7639792933
4419521541 3418994854 4473456738 : 53
3162499341 9131814809 2777710386 3877343177 2075456545 3220777092 1201905166
0962804909 2636019759 8828161332 : 54
3166636528 6193266863 3606273567 6303544776 2803504507 7723554710 5859548702
7908143562 4014517180 6246436267 : 55

9456127531 8134078330 3362542327 8394497538 2437205835 3114771199 2606381334
6776879695 9703098339 1307710987 : 56
0408591337 4641442822 7726346594 7047458784 7787201927 7152807317 6790770715
7213444730 6057007334 9243693113 : 57
8350493163 1284042512 1925651798 0694113528 0131470130 4781643788 5185290928
5452011658 3934196562 1349143415 : 58
9562586586 5570552690 4965209858 0338507224 2648293972 8584783163 0577775606
8887644624 8246857926 0395352773 : 59
4803048029 0058760758 2510474709 1643961362 6760449256 2742042083 2085661190
6254543372 1315359584 5068772460 : 60
2901618766 7952406163 4252257719 5429162991 9306455377 9914037340 4328752628
8896399587 9475729174 6426357455 : 61
2540790914 5135711136 9410911939 3251910760 2082520261 8798531887 7058429725
9167781314 9699009019 2116971737 : 62
2784768472 6860849003 3770242429 1651300500 5168323364 3503895170 2989392233
4517220138 1280696501 1784408745 : 63
1960121228 5993716231 3017114448 4640903890 6449544400 6198690754 8516026327
5052983491 8740786680 8818338510 : 64
2283345085 0486082503 9302133219 7155184306 3545500766 8282949304 1377655279
3975175461 3953984683 3936383047 : 65
4611996653 8581538420 5685338621 8672523340 2830871123 2827892125 0771262946
3229563989 8989358211 6745627010 : 66
2183564622 0134967151 8819097303 8119800497 3407239610 3685406643 1939509790
1906996395 5245300545 0580685501 : 67
9567302292 1913933918 5680344903 9820595510 0226353536 1920419947 4553859381
0234395544 9597783779 0237421617 : 68
2711172364 3435439478 2218185286 2408514006 6604433258 8856986705 4315470696
5747458550 3323233421 0730154594 : 69
0516553790 6866273337 9958511562 5784322988 2737231989 8757141595 7811196358
3300594087 3068121602 8764962867 : 70
4460477464 9159950549 7374256269 0104903778 1986835938 1465741268 0492564879
8556145372 3478673303 9046883834 : 71
3634655379 4986419270 5638729317 4872332083 7601123029 9113679386 2708943879
9362016295 1541337142 4892830722 : 72
0126901475 4668476535 7616477379 4675200490 7571555278 1965362132 3926406160
1363581559 0742202020 3187277605 : 73
2772190055 6148425551 8792530343 5139844253 2234157623 3610642506 3904975008
6562710953 5919465897 5141310348 : 74
2276930624 7435363256 9160781547 8181152843 6679570611 0861533150 4452127473
9245449454 2368288606 1340841486 : 75
3776700961 2071512491 4043027253 8607648236 3414334623 5189757664 5216413767
9690314950 1910857598 4423919862 : 76
9164219399 4907236234 6468441173 9403265918 4044378051 3338945257 4239950829
6591228508 5558215725 0310712570 : 77
1266830240 2929525220 1187267675 6220415420 5161841634 8475651699 9811614101
0029960783 8690929160 3028840026 : 78
9104140792 8862150784 2451670908 7000699282 1206604183 7180653556 7252532567
5328612910 4248776182 5829765157 : 79
9598470356 2226293486 0034158722 9805349896 5022629174 8788202734 2092222453
3985626476 6914905562 8425039127 : 80
5771028402 7998066365 8254889264 8802545661 0172967026 6407655904 2909945681
5065265305 3718294127 0336931378 : 81
5178609040 7086671149 6558343434 7693385781 7113864558 7367812301 4587687126
6034891390 9562009939 3610310291 : 82
6161528813 8437909904 2317473363 9480457593 1493140529 7634757481 1935670911
0137751721 0080315590 2485309066 : 83

9203767192 2033229094 3346768514 2214477379 3937517034 4366199104 0337511173
5471918550 4644902636 5512816228 : 84
8244625759 1633303910 7225383742 1821408835 0865739177 1509682887 4782656995
9957449066 1758344137 5223970968 : 85
3408005355 9849175417 3818839994 4697486762 6551658276 5848358845 3142775687
9002909517 0283529716 3445621296 : 86
4043523117 6006651012 4120065975 5851276178 5838292041 9748442360 8007193045
7618932349 2292796501 9875187212 : 87
7267507981 2554709589 0455635792 1221033346 6974992356 3025494780 2490114195
2123828153 0911407907 3860251522 : 88
7429958180 7247162591 6685451333 1239480494 7079119153 2673430282 4418604142
6363954800 0448002670 4962482017 : 89
9289647669 7583183271 3142517029 6923488962 7668440323 2609275249 6035799646
9256504936 8183609003 2380929345 : 90
9588970695 3653494060 3402166544 3755890045 6328822505 4525564056 4482465151
8754711962 1844396582 5337543885 : 91
6909411303 1509526179 3780029741 2076651479 3942590298 9695946995 5657612186
5619673378 6236256125 2163208628 : 92
6922210327 4889218654 3648022967 8070576561 5144632046 9279068212 0738837781
4233562823 6089632080 6822246801 : 93
2248261177 1858963814 0918390367 3672220888 3215137556 0037279839 4004152970
0287830766 7094447456 0134556417 : 94
2543709069 7939612257 1429894671 5435784687 8861444581 2314593571 9849225284
7160504922 1242470141 2147805734 : 95
5510500801 9086996033 0276347870 8108175450 1193071412 2339086639 3833952942
5786905076 4310063835 1983438934 : 96
1596131854 3475464955 6978103829 3097164651 4384070070 7360411237 3599843452
2516105070 2705623526 6012764848 : 97
3084076118 3013052793 2054274628 6540360367 4532865105 7065874882 2569815793
6789766974 2205750596 8344086973 : 98
5020141020 6723585020 0724522563 2651341055 9240190274 2162484391 4035998953
5394590944 0704691209 1409387001 : 99
2645600162 3742880210 9276457931 0657922955 2498872758 4610126483 6999892256
9596881592 0560010165 5256375678 : 100
5667227966 1988578279 4848855834 3975187445 4551296563 4434803966 4205579829
3680435220 2770984294 2325330225 : 101
7634180703 9476994159 7915945300 6975214829 3366555661 5678736400 5366656416
5473217043 9035213295 4352916941 : 102
4599041608 7532018683 7937023488 8689479151 0716378529 0234529244 0773659495
6305100742 1087142613 4974595615 : 103
1384987137 5704710178 7957310422 9690666702 1449863746 4595280824 3694457897
7233004876 4765241339 0759204340 : 104
1963403911 4732023380 7150952220 1068256342 7471646024 3354400515 2126693249
3419673977 0415956837 5355516673 : 105
0273900749 7297363549 6453328886 9844061196 4961627734 4951827369 5588220757
3551766515 8985519098 6665393549 : 106
4810688732 0685990754 0792342402 3009259007 0173196036 2254756478 9406475483
4664776041 1463233905 6513433068 : 107
4495397907 0903023460 4614709616 9688688501 4083470405 4607429586 9913829668
2468185710 3188790652 8703665083 : 108
2431974404 7718556789 3482308943 1068287027 2280973624 8093996270 6074726455
3992539944 2808113736 9433887294 : 109
0630792615 9599546262 4629707062 5948455690 3471197299 6409089418 0595343932
5123623550 8134949004 3642785271 : 110
3831591256 8989295196 4272875739 4691427253 4366941532 3610045373 0488198551
7065941217 3524625895 4873016760 : 111

0298865925 7866285612 4966552353 3829428785 4253404830 8330701653 7228563559
1525347844 5981831341 1290019992 : 112
0598135220 5117336585 6407826484 9427644113 7639386692 4803118364 4536985891
7544264739 9882284621 8449008777 : 113
6977631279 5722672655 5625962825 4276531830 0134070922 3343657791 6012809317
9401718598 5999338492 3549564005 : 114
7099558561 1349802524 9906698423 3017350358 0440811685 5265311709 9570899427
3287092584 8789443646 0050410892 : 115
2669178352 5870785951 2983441729 5351953788 5534573742 6085902908 1765155780
3905946408 7350612322 6112009373 : 116
1080485485 2635722825 7682034160 5048466277 5045003126 2008007998 0492548534
6941469775 1649327095 0493463938 : 117
2432227188 5159740547 0214828971 1177792376 1225788734 7718819682 5462981268
6858170507 4027255026 3329044976 : 118
2778944236 2167411918 6269439650 6715157795 8675648239 9391760426 0176338704
5499017614 3641204692 1823707648 : 119
8783419689 6861181558 1587360629 3860381017 1215855272 6683008238 3404656475
8804051380 8016336388 7421637140 : 120
6435495561 8689641122 8214075330 2655100424 1048967835 2858829024 3670904887
1181909094 9453314421 8287661810 : 121
3100735477 0549815968 0772009474 6961343609 2861484941 7850171807 7930681085
4690009445 8995279424 3981392135 : 122
0558642219 6483491512 6390128038 3200109773 8680662877 9239718014 6134324457
2640097374 2570073592 1003154150 : 123
8936793008 1699805365 2027600727 7496745840 0283624053 4603726341 6554259027
6018348403 0681138185 5105979705 : 124
6640075094 2608788573 5796037324 5141467867 0368809880 6097164258 4975951380
6930944940 1515422221 9432913021 : 125
7391253835 5915031003 3303251117 4915696917 4502714943 3151558854 0392216409
7229101129 0355218157 6282328318 : 126
2342548326 1119128009 2825256190 2052630163 9114772473 3148573910 7775874425
3876117465 7867116941 4776421441 : 127
1112635835 5387136101 1023267987 7564102468 2403226483 4641766369 8066378576
8134920453 0224081972 7856471983 : 128
9630878154 3221166912 2464159117 7673225326 4335686146 1865452226 8126887268
4459684424 1610785401 6768142080 : 129
8850280054 1436131462 3082102594 1737562389 9420757136 2751674573 1891894562
8352570441 3354375857 5342698699 : 130
4725470316 5661399199 9682628247 2706413362 2217892390 3176085428 9437339356
1889165125 0424404008 9527198378 : 131
7386480584 7268954624 3882343751 7885201439 5600571048 1194988423 9060613695
7342315590 7967034614 9143447886 : 132
3604103182 3507365027 7859089757 8272731305 0488939890 0992391350 3373250855
9826558670 8924261242 9473670193 : 133
9077271307 0686917092 6462548423 2407485503 6608013604 6689511840 0936686095
4632500214 5852930950 0009071510 : 134
5823626729 3264537382 1049387249 9669933942 4685516483 2611341461 1068026744
6637334375 3407642940 2668297386 : 135
5220935701 6263846485 2851490362 9320199199 6882851718 3953669134 5222444708
0459239660 2817156551 5656661113 : 136
5982311225 0628905854 9145097157 5539002439 3153519090 2107119457 3002438801
7661503527 0862602537 8817975194 : 137
7806101371 5004489917 2100222013 3501310601 6391541589 5780371177 9277522597
8742891917 9155224171 8958536168 : 138
0594741234 1933984202 1874564925 6443462392 5319531351 0331147639 4911995072
8584306583 6193536932 9699289837 : 139

9149419394 0608572486 3968836903 2655643642 1664425760 7914710869 9843157337
4964883529 2769328220 7629472823 : 140
8153740996 1545598798 2598910937 1712621828 3025848112 3890119682 2142945766
7580718653 8065064870 2613389282 : 141
2994972574 5303328389 6381843944 7707794022 8435988341 0035838542 3897354243
9564755568 4095224844 5541392394 : 142
1000162076 9363684677 6413017819 6593799715 5746854194 6334893748 4391297423
9143365936 0410035234 3777065888 : 143
6778113949 8616478747 1407932638 5873862473 2889645643 5987746676 3847946650
4074111825 6583788784 5485814896 : 144
2961273998 4134427260 8606187245 5452360643 1537101127 4680977870 4464094758
2803487697 5894832824 1239292960 : 145
5829486191 9667091895 8089833201 2103184303 4012849511 6203534280 1441276172
8583024355 9830032042 0245120728 : 146
7253558119 5840149180 9692533950 7577840006 7465526031 4461670508 2768277222
3534191102 6341631571 4740612385 : 147
0425845988 4199076112 8725805911 3935689601 4316682831 7632356732 5417073420
8173322304 6298799280 4908514094 : 148
7903688786 8789493054 6955703072 6190095020 7643349335 9106024545 0864536289
3545686295 8531315337 1838682695 : 149
1786227363 7169757741 8302398600 6591481616 4049449650 1173213138 9574706208
8474802365 3710311508 9842799275 : 150
4426853277 9743113951 4357417221 9759799359 6852522857 4526379628 9612691572
3579866205 7340837576 6873884266 : 151
4059909935 0500081337 5432454635 9675048442 3528487470 1443545419 5762584735
6421619813 4073468541 1176688311 : 152
8654489377 6979566517 2796623267 1481033864 3913751865 9467300244 3450054499
5399742372 3287124948 3470604406 : 153
3471606325 8306498297 9551010954 1836235030 3094530973 3583446283 9476304775
6450150085 0757894954 8931393944 : 154
8992161255 2559770143 6858943585 8775263796 2559708167 7643800125 4365023714
1278346792 6101995585 2247172201 : 155
7772370041 7808419423 9487254068 0155603599 8390548985 7235467456 4239058585
0216719031 3952629445 5439131663 : 156
1345308939 0620467843 8778505423 9390524731 3620129476 9187497519 1011472315
2893267725 3391814660 7300089027 : 157
7689631148 1090220972 4520759167 2970078505 8071718638 1054967973 1001678708
5069420709 2232908070 3832634534 : 158
5203802786 0990556900 1341371823 6837099194 9516489600 7550493412 6787643674
6384902063 9640197666 8559233565 : 159
4639138363 1857456981 4719621084 1080961884 6054560390 3845534372 9141446513
4749407848 8442377217 5154334260 : 160
3066988317 6833100113 3108690421 9390310801 4378433415 1370924353 0136776310
8491351615 6422698475 0743032971 : 161
6746964066 6531527035 3254671126 6752246055 1199581831 9637637076 1799191920
3579582007 5956053023 4626775794 : 162
3936307463 0569010801 1494271410 0939136913 8107258137 8135789400 5599500183
5425118417 2136055727 5221035268 : 163
0373572652 7922417373 6057511278 8721819084 4900617801 3889710770 8229310027
9766593583 8758909395 6881485602 : 164
6322439372 6562472776 0378908144 5883785501 9702843779 3624078250 5270487581
6470324581 2908783952 3245323789 : 165
6029841669 2254896497 1560698119 2186584926 7704039564 8127810217 9913217416
3058105545 9880130048 4562997651 : 166
1212415363 7451500563 5070127815 9267142413 4210330156 6165356024 7338078430
2865525722 2753049998 8370153487 : 167

9300806260 1809623815 1613669033 4111138653 8510919367 3938352293 4588832255
0887064507 5394739520 4396807906 : 168
7086806445 0969865488 0168287434 3786126453 8158342807 5306184548 5903798217
9945996811 5441974253 6344399602 : 169
9025100158 8827216474 5006820704 1937615845 4712318346 0072629339 5505482395
5713725684 0232268213 0124767945 : 170
2264482091 0235647752 7230820810 6351889915 2692889108 4555711266 0396503439
7896278250 0161101532 3516051965 : 171
5904211844 9499077899 9200732947 6905868577 8787209829 0135295661 3978884860
5097860859 5701773129 8155314951 : 172
6814671769 5976099421 0036183559 1387778176 9845875810 4466283998 8060061622
9848616935 3373865787 7359833616 : 173
1338413385 3684211978 9389001852 9569196780 4554482858 4837011709 6721253533
8758621582 3101331038 7766827211 : 174
5726949518 1795897546 9399264219 7915523385 7662316762 7547570354 6994148929
0413018638 6119439196 2838870543 : 175
6777432242 7680913236 5449485366 7680000010 6526248547 3055861598 9991401707
6983854831 8875014293 8908995068 : 176
5453076511 6803337322 2651756622 0752695179 1442252808 1651716677 6672793035
4851542040 2381746089 2328391703 : 177
2754257508 6765511785 9395002793 3895920576 6827896776 4453184040 4185540104
3513483895 3120132637 8369283580 : 178
8271937831 2654961745 9970567450 7183320650 3455664403 4490453627 5600112501
8433560736 1222765949 2783937064 : 179
7842645676 3388188075 6561216896 0504161139 0390639601 6202215368 4941092605
3876887148 3798955999 9112099164 : 180
6464411918 5682770045 7424343402 1672276445 5893301277 8158686952 5069499364
6101756850 6016714535 4315814801 : 181
0545886056 4550133203 7586454858 4032402987 1709348091 0556211671 5468484778
0394475697 9804263180 9917564228 : 182
0987399876 6973237695 7370158080 6822904599 2123661689 0259627304 3067931653
1149401764 7376938735 1409336183 : 183
3216142802 1497633991 8983548487 5625298752 4238730775 5955595546 5196394401
8218409984 1248982623 6737714672 : 184
2606163364 3296406335 7281070788 7581640438 1485018841 1431885988 2769449011
9321296827 1588841338 6943468285 : 185
9006664080 6314077757 7257056307 2940049294 0302420498 4165654797 3670548558
0445865720 2276378404 6682337985 : 186
2827105784 3197535417 9501134727 3625774080 2134768260 4502285157 9795797647
4670228409 9956160156 9108903845 : 187
8245026792 6594205550 3958792298 1852648007 0683765041 8365620945 5543461351
3415257006 5974881916 3413595567 : 188
1964965403 2187271602 6485930490 3978748958 9066127250 7948282769 3895352175
3621850796 2977851461 8843271922 : 189
3223810158 7444505286 6523802253 2843891375 2738458923 8442253547 2653098171
5784478342 1582232702 0690287232 : 190
3300538621 6347988509 4695472004 7952311201 5043293226 6282727632 1779088400
8786148022 1475376578 1058197022 : 191
2630971749 5072127248 4794781695 7296142365 8595782090 8307332335 6034846531
8730293026 6596450137 1837542889 : 192
7557971449 9246540386 8179921389 3469244741 9850973346 2679332107 2686870768
0626399193 6196504409 9542167627 : 193
8409146698 5692571507 4315740793 8053239252 3947755744 1591845821 5625181921
5523370960 7483329234 9210345146 : 194
2643744980 5596103307 9941453477 8457469999 2128599999 3996122816 1521931488
8769388022 2810830019 8601654941 : 195

6542616968 5867883726 0958774567 6182507275 9929508931 8052187292 4610867639
9589161458 5505839727 4209809097 : 196
8172932393 0106766386 8240401113 0402470073 5085782872 4627134946 3685318154
6969046696 8693925472 5194139929 : 197
1465242385 7762550047 4852954768 1479546700 7050347999 5888676950 1612497228
2040303995 4632788306 9597624936 : 198
1510102436 5553522306 9061294938 8599015734 6610237122 3547891129 2547696176
0050479749 2806072126 8039226911 : 199
0277722610 2544149221 5765045081 2067717357 1202718024 2968106203 7765788371
6690910941 8074487814 0490755178 : 200
2038565390 9910477594 1413215432 8440625030 1802757169 6508209642 7348414695
7263978842 5600845312 1406593580 : 201
9041271135 9200419759 8513625479 6160632288 7361813673 7324450607 9244117639
9759746193 8358457491 5988097667 : 202
4470930065 4634242346 0634237474 6660804317 0126005205 5928493695 9414340814
6852981505 3947178900 4518357551 : 203
5412522359 0590687264 8786357525 4191128887 7371766374 8602766063 4960353679
4702692322 9718683277 1739323619 : 204
2007774522 1262475186 9833495151 0198642698 8784717193 9664976907 0825217423
3656627259 2844062043 0214113719 : 205
9227852699 8469884770 2323823840 0556555178 8908766136 0130477098 4386116870
5231055314 9162517283 7327286760 : 206
0724817298 7637569816 3354150746 0883866364 0693470437 2066886512 7568826614
9730788657 0156850169 1864748854 : 207
1679154596 5072342877 3069985371 3904300266 5307839877 6385032381 8215535597
3235306860 4301067576 0838908627 : 208
0498418885 9513809103 0423595782 4951439885 9011318583 5840667472 3702971497
8508414585 3085781339 1562707603 : 209
5639076394 7311455495 8322669457 0249413983 1634332378 9759556808 5683629725
3867913275 0555425244 9194358912 : 210
8405045226 9538121791 3191451350 0993846311 7740179715 1228378546 0116035955
4028644059 0249646693 0707769055 : 211
4810288502 0808580087 8115773817 1917417760 1733073855 4758006056 0143377432
9901272867 7253043182 5197579167 : 212
9296996504 1460706645 7125888346 9797964293 1622965520 1687973000 3564630457
9308840327 4807718115 5533090988 : 213
7025505207 6804630346 0865816539 4876951960 0440848206 5967379473 1680864156
4565053004 9881616490 5788311543 : 214
4548505266 0069823093 1577765003 7807046612 6470602145 7505793270 9620478256
1524714591 8965223608 3966456241 : 215
0519551052 2357239739 5128818164 0597859142 7914816542 6328920042 8160913693
7773722299 9833270820 8296995573 : 216
7727375667 6155271139 2258805520 1898876201 1416800546 8736558063 3471603734
2917039079 8639652296 1312801782 : 217
6797172898 2293607028 8069087768 6605932527 4637840539 7691848082 0410219447
1971386925 6084162451 1239806201 : 218
1318454124 4782050110 7987607171 5568315407 8865439041 2108730324 0201068534
1947230476 6667217498 6986854707 : 219
6781205124 7367924791 9315085644 4775379853 7997322344 5612278584 3296846647
5133365736 9238720146 4723679427 : 220
8700425032 5558992688 4349592876 1240075587 5694641370 5625140011 7971331662
0715371543 6006876477 3186755871 : 221
4878398908 1074295309 4106059694 4315847753 9700943988 3949144323 5366853920
9946879645 0665339857 3888786614 : 222
7629443414 0104988899 3160051207 6781035886 1166020296 1193639682 1349607501
1164983278 5635316145 1684576956 : 223

8710900299 9769841263 2665023477 1672865737 8579085746 6460772283 4154031144
1529418804 7825438761 7707904300 : 224
0156698677 6795760909 9669360755 9496515273 6349811896 4130433116 6277471233
8817406037 3174397054 0670310967 : 225
6765748695 3587896700 3192586625 9410510533 5843846560 2339179674 9267844763
7084749783 3365557900 7384191473 : 226
1988627135 2595462518 1604342253 7299628632 6749682405 8060296421 1463864368
6422472488 7283434170 4415734824 : 227
8183330164 0566959668 8667695634 9141632842 6414974533 3499994800 0266998758
8815935073 5781519588 9900539512 : 228
0853510357 2613736403 4367534714 1048360175 4648830040 7846416745 2167371904
8310967671 1344349481 9262681110 : 229
7399482506 0739495073 5031690197 3185211955 2635632584 3390998224 9862406703
1076831844 6607291248 7475403161 : 230
7969941139 7387765899 8685541703 1884778867 5929026070 0432126661 7919223520
9382278788 8098863359 9116081923 : 231
5355570464 6349113208 5918979613 2791319756 4909760001 3996234445 5350143464
2686046449 5862476909 4347048293 : 232
2941404111 4654092398 8344435159 1332010773 9441118407 4107684981 0663472410
4823935827 4019449356 6516108846 : 233
3125678529 7769734684 3030614624 1803585293 3159734583 0384554103 3701091676
7763742762 1021370135 4854450926 : 234
3071901147 3184857492 3318167207 2137279355 6795284439 2548156091 3728128406
3330393735 6242001604 5664557414 : 235
5881660521 6660873874 8047243391 2129558777 6390696903 7078828527 7538940524
6075849623 1574369171 1317613478 : 236
3882719416 8606625721 0368513215 6647800147 6752310393 5786068961 1125996028
1839309548 7090590738 6135191459 : 237
1819510297 3278755710 4972901148 7171897180 0469616977 7001791391 9613791417
1627070189 5846921434 3696762927 : 238
4591099400 6008498356 8425201915 5937037010 1104974733 9493877885 9894174330
3178534870 7603221982 9705797511 : 239
9144051099 4235883034 5463534923 4982688362 4043327267 4155403016 1950568065
4180939409 9820206099 9414021689 : 240
0900708213 3072308966 2119775530 6659188141 1915778362 7292746156 1857103721
7247100952 1423696483 0864102592 : 241
8874579993 2237495519 1221951903 4244523075 3513380685 6807354464 9951272031
7448719540 3976107308 0602699062 : 242
5807602029 2731455252 0780799141 8429063884 4373499681 4582733720 7266391767
0201183004 6481900024 1308350884 : 243
6584152148 9912761065 1374153943 5657211390 3285749187 6909441370 2090517031
4877734616 5287984823 5338297260 : 244
1361109845 1484182380 8120540996 1252745808 8109948697 2216128524 8974255555
1607637167 5054896173 0168096138 : 245
0381191436 1143992106 3800508321 4098760459 9309324851 0251682944 6726066613
8151745712 5597549535 8023998314 : 246
6982203613 3808284993 5670557552 4712902745 3977621404 9318201465 8008021566
5360677655 0878380430 4134310591 : 247
8046068008 3459113664 0834887408 0057412725 8670479225 8319127415 7390809143
8313845642 4150940849 1339180968 : 248
4025116399 1936853225 5573389669 5374902662 0923261318 8558915808 3245557194
8453875628 7861288590 0410600607 : 249
3746501402 6278240273 4696252821 7174941582 3317492396 8353013617 8653673760
6421667781 3773995100 6589528877 : 250
4276626368 4183068019 0804609849 8094697636 6733566228 2915132352 7888061577
6827815958 8669180238 9403330764 : 251

4191240341 2022316368 5778603572 7694154177 8826435238 1319050280 8701857504
7046312933 3537572853 8660588890 : 252
4583111450 7739429352 0199432197 1171642235 0056440429 7989208159 4307167019
8574692738 4865383343 6145794634 : 253
1759225738 9858800169 8014757420 5429958012 4295810545 6510831046 2972829375
8416116253 2562516572 4980784920 : 254
9989799062 0035936509 9347215829 6517413579 8491047111 6607915874 3698654122
2348341887 7229294463 3517865385 : 255
6731962559 8520260729 4767407261 6767145573 6498121056 7771689348 4917660771
7052771876 0119990814 4113058645 : 256
5779105256 8430481144 0261938402 3224709392 4980293355 0731845890 3553971330
8844617410 7959162511 7148648744 : 257
6861124760 5428673436 7090466784 6867027409 1881014249 7111496578 1772427934
7070216688 2956108777 9440504843 : 258
7528443375 1088282647 7197854000 6509704033 0218625561 4733211777 1174413350
2816088403 5178145254 1964320309 : 259
5760186946 4908868154 5285621346 9883554445 6024955666 8436602922 1951248309
1060537720 1980218310 1032704178 : 260
3866544718 1260397190 6884623708 5751808003 5327047185 6594994761 2424811099
9288679158 9690495639 4762460842 : 261
4065930948 6215076903 1498702067 3533848349 5508363660 1784877106 0809804269
2471324100 0946401437 3603265645 : 262
1845667924 5666955100 1502298330 7984960799 4988249706 1723674493 6122622296
1790814311 4146609412 3415935930 : 263
9585407913 9087208322 7335495720 8075716517 1876599449 8569379562 3875551617
5754380917 8052802946 4200447215 : 264
3962807463 6021132942 5591600257 0735628126 3873310600 5891065245 7080244749
3754318414 9401482119 9962764531 : 265
0680066311 8382376163 9663180931 4446712986 1552759820 1451410275 6006892975
0246304017 3514891945 7636078935 : 266
2855505317 3314164570 5049964438 9093630843 8744847839 6168405184 5273288403
2345202470 5685164657 1647713932 : 267
3775517294 7951261323 9822960239 4548579754 5865174587 8771331813 8752959809
4121742273 0035204650 8089177705 : 268
0682592488 2232215493 8048371454 7816472139 7682096332 0508305647 9204820859
2047549985 7320388876 3916019952 : 269
4091893894 5576768749 7308569559 5801065952 6503036266 1597506622 2508406742
8898265907 5106375635 6996821151 : 270
0949669744 5805472886 9363102036 7823250182 3237084597 9011154847 2087618212
4778132663 3041207621 6587312970 : 271
8112307581 5982124863 9807212407 8688781145 0165582513 6178903070 8608701989
7588980745 6643955157 4153631931 : 272
9198107057 5336633738 0382721527 9884935039 7480015890 5194208797 1130805123
3933221903 4662499171 6915094854 : 273
1401871060 3546037946 4337900589 0957721180 8044657439 6280618671 7861017156
7409676620 8029576657 7051291209 : 274
9079443046 3289294730 6159510430 9022214393 7184956063 4056189342 5130572682
9146578329 3340524635 0289291754 : 275
7087256484 2600349629 6116541382 3007731332 7298305001 6025672401 4185152041
8907011542 8857992081 2198449315 : 276
6999059182 0118197335 0012618772 8036812481 9958770702 0753240636 1259313438
5955425477 8196114293 5163561223 : 277
4966615226 1473539967 4051584998 6035529533 2924575238 8810136202 3476246690
5581643896 7863097627 3655047243 : 278
4864307121 8494373485 3006063876 4456627218 6661701238 1277156213 7974614986
1328744117 7145524447 0899714452 : 279

2885662942 4402301847 9120547849 8574521634 6964489738 9206240194 3518310088
2834802492 4908540307 7863875165 : 280
9113028739 5878709810 0772718271 8745290139 7283661484 2142871705 5317965430
7650453432 4600536361 4726181809 : 281
6997693348 6264077435 1999286863 2383508875 6683595097 2655748154 3194019557
6850437248 0010204137 4983187225 : 282
9677387154 9583997184 4490727914 1965845930 0839426370 2087563539 8216962055
3248032122 6749891140 2678528599 : 283
6734052420 3109179789 9905718821 9493913207 5343170798 0023736590 9853755202
3891164346 7185582906 8537118979 : 284
5262623449 2483392496 3424497146 5684659124 8918556629 5893299090 3523923333
3647435203 7077010108 4388003290 : 285
7598342170 1855422838 6161721041 7603011645 9187805393 6744747205 9985023582
8918336929 2233732399 9480437108 : 286
4196594731 6265482574 8099482509 9918330069 7656936715 9689364493 3488647442
1350084070 0660883597 2350395323 : 287
4017958255 7036016936 9909886711 3210979889 7070517280 7558551912 6993067309
9250704070 2455685077 8679069476 : 288
6126298082 2516331363 9952117098 4528092630 3759224267 4257559989 2892783704
7444521893 6320348941 5521044597 : 289
2618838003 0067761793 1381399162 0580627016 5102445886 9247649246 8919246121
2531027573 1390840470 0071435613 : 290
6231699237 1694848132 5542009145 3041037135 4532966206 3921054798 2439212517
2540132314 9027405858 9206321758 : 291
9494345489 0684639931 3757091034 6332714153 1622328055 2297297953 8018801628
5907357295 5416278867 6498274186 : 292
1642187898 8574107164 9069191851 1628152854 8679417363 8906653885 7642291583
4250067361 2453849160 6741373401 : 293
7357277995 6341043326 8835695078 1493137800 7362354180 0706191802 6732855119
1942676091 2210359874 6924117283 : 294
7493126163 3950012395 9924050845 4375698507 9570462226 6461900010 3500490183
0341535458 4283376437 8111988556 : 295
3187777925 3720116671 8539541835 9844383052 0376281944 0761594106 8207169703
0228515225 0573126093 0468984234 : 296
3315273213 1361216582 8080752126 3154773060 4423774753 5059522871 7440266638
9148817173 0864361113 8906942027 : 297
9088143119 4487994171 5404210341 2190847094 0802540239 3294294549 3878640230
5129271190 9751353600 0921971105 : 298
4120966831 1151632870 5423028470 0731206580 3262641711 6165957613 2723515666
6253667271 8998534199 8952368848 : 299
3099930275 7419916463 8414270779 8870887422 9277053891 2271724863 2202889842
5125287217 8260305009 9451082478 : 300
3572905691 9885554678 8607946280 5371227042 4665431921 4528176074 1482403827
8358297193 0101788834 5674167811 : 301
3989547504 4833931468 9630763396 6572267270 4339321674 5421824557 0625247972
1997866854 2798977992 3395790575 : 302
8189062252 5473582205 2364248507 8340711014 4980478726 6919901864 3882293230
5382318559 7328697809 2225352959 : 303
1017341407 3348847610 0556401824 2392192695 0620831838 1454698392 3664613639
8910121021 7709597670 4908305081 : 304
8547041946 6437131229 9692358895 3849301363 5657618610 6062228705 5994233716
3102127845 7446463989 7381885667 : 305
4626087948 2018647487 6727272220 6267646533 8099801966 8836809941 5907577685
2639865146 2533363124 5053640261 : 306
0569605513 1838131742 6118442018 9088853196 3569869627 9503673842 4313011331
7533053298 0201668881 7481342988 : 307

6815855778 1034323175 3064784983 2106297184 2518438553 4427620128 2345707169 8853051832 6179641178 5796088881 : 308
5032960229 0705614476 2209150947 3903594664 6916235396 8092013945 7817589108 8931992112 2600739281 4916948161 : 309
5273842736 2642980982 3406320024 4024495894 4561291670 4950823581 2487391799 6486411334 8032475777 5219708932 : 310
7722623494 8601504665 2681439877 0516153170 2669692970 4928316285 5042128981 4670619533 1970269507 2143782304 : 311
7687528028 7354126166 3917082459 2517001071 4180854800 6369232594 6201900227 8087409859 7719218051 5853214739 : 312
2653251559 0354102092 8466592529 9914353791 8253145452 9059841581 7637058927 9069098969 1116438118 7809435371 : 313
5213322614 4362531449 0127454772 6957393934 8154691631 1624928873 5747188240 7150399500 9446731954 3161938554 : 314
8520766573 8825139639 1635767231 5100555603 7263394867 2082078086 5373494244 0115799667 5073607111 5935133195 : 315
9197120948 9647175530 2453136477 0942094635 6969822266 7377520994 5168450643 6238242118 5353488798 9395673187 : 316
8066061078 8544000550 8276570305 5874485418 0577889171 9207881423 3511386629 2966717964 3468760077 0479995378 : 317
8338787034 8718021842 4373421122 7394025571 7690819603 0920182401 8842705704 6092622564 1783752652 6335832424 : 318
0661253311 5294234579 6556950250 6810018310 9004112453 7901533296 6156970522 3792103257 0693705109 0830789479 : 319
9990049993 9532215362 2748476603 6136776979 7856738658 4670936679 5885837887 9562594646 4891376652 1995882869 : 320
3380183601 1932368578 5585581955 5604215625 0883650203 3220245137 6215820461 8106705195 3306530606 0650105488 : 321
7167245377 9428313388 7163139559 6905832083 4168984760 6560711834 7136218123 2462272588 4199028614 2087284956 : 322
8796393254 6428534307 5301105285 7138296437 0999035694 8885285190 4029560473 4613113826 3878897551 7885604249 : 323
9874831638 2804046848 6189381895 9054203988 9872650697 6202019955 4841265000 5394428203 9301274816 3815853039 : 324
6439925470 2016727593 2857436666 1644110962 5663373054 0921951967 5148328734 8089574777 7527834422 1091073111 : 325
3518280460 3634719818 5655572957 1447476825 5285786334 9342858423 1187494400 0322969069 7758315903 8580393535 : 326
2135886007 9600342097 5473922967 3331064939 5601812237 8128545843 1760556173 3861126734 7807458506 7606304822 : 327
9409653041 1183066710 8189303110 8871728167 5195796753 4718853722 9309616143 2040063813 2246584111 1157758358 : 328
5811350185 6904781536 8938137718 4728147519 9835050478 1297718599 0847076219 7460588742 3256995828 8925350419 : 329
3795826061 6211842368 7685114183 1606831586 7994601652 0577405294 2305360178 0313357263 2670547903 3840125730 : 330
5912339601 8801378254 2192709476 7337191987 2873852480 5742124892 1183470876 6296672072 7232565056 5129333126 : 331
0595057777 2754247124 1648312832 9820723617 5057467387 0128209575 5443059683 9555568686 1188397135 5220844528 : 332
5264008125 2027665557 6774959696 2661260456 5245684086 1392382657 6858338469 8499778726 7065551918 5446869846 : 333
9478495734 6226062942 1962455708 5371272776 5230989554 5019303773 2166649182 5781546772 9200521266 7143463209 : 334
6378918523 2321501897 6126034373 6840671941 9303774688 0999296877 5824410478 7812326625 3181845960 4538535438 : 335

3911449677 5312864260 9252115376 7325886672 2604042523 4910870269 5809964759
5805794663 9734190640 1003636190 : 336
4042033113 5793365424 2630356145 7009011244 8008900208 0147805660 3710154122
3288914657 2239314507 6071670643 : 337
5568274377 4396578906 7972687438 4730763464 5167756210 3098604092 7170909512
8086309029 7385044527 1828927496 : 338
8921210667 0081648583 3955377359 1913695015 3162018908 8874842107 9870689911
4804669270 6509407620 4650277252 : 339
8650728905 3285485614 3316081269 3005693785 4178610969 6920253886 5034577183
1766868859 2368148847 5276498468 : 340
8219497397 2970773718 7188400414 3231276365 0481453112 2850990020 7424092558
5925292610 3021067368 1543470152 : 341
5234878635 1643976235 8604191941 2969769040 5264832347 0099111542 4260127343
8022089331 0966863678 9869497799 : 342
4001260164 2276092608 2349304118 0643829138 3473546797 2539926233 8791582998
4864592717 3405922562 0749105308 : 343
5315371829 1168163721 9395188700 9577881815 8685046450 7699343940 9874335144
3162633031 7247747486 8979182092 : 344
3948083314 3970840673 0840795893 5810896656 4775859905 5637695252 3265361442
4780230826 8118310377 3588708924 : 345
0613031336 4773710116 2821461466 1679404090 5186152603 6009252194 7218890918
1073358719 6414214447 8654899528 : 346
5823439470 5007983038 8538860831 0357193060 0277119455 8021911942 8999227223
5334870756 6246926177 6631788551 : 347
4435021828 7026685610 6650035310 5021631820 6017609217 9846849368 6316129372
7951873078 9726373537 1715025637 : 348
8733579771 8081848784 5886650433 5824377004 1477104149 3492743845 7587107159
7315594394 2641257027 0965125108 : 349
1155482479 3940359768 1188117282 4721582501 0949609662 5393395380 9221955919
1818855267 8062149923 1727631632 : 350
1833989693 8075616855 9117529984 5013206712 9392404144 5938623988 0938124045
2191484831 6462101473 8918251010 : 351
9096773869 0664041589 7361047643 6500068077 1056567184 8628149637 1118832192
4456639458 1449148616 5500495676 : 352
9826903089 1118568798 6929470513 5248160917 4324301538 3684707292 8989828460
2223730145 2655679898 6277679680 : 353
9146979837 8268764311 5988321090 4371561129 9766521539 6354644208 6919756737
0005738764 9784376862 8768179249 : 354
7469438427 4652563163 2300555130 4174227341 6464551278 1278457777 2457520386
5437542828 2567141288 5834544435 : 355
1325620544 6424101103 7955464190 5811686230 5964476958 7054072141 9852121067
3433241075 6767575818 4569906930 : 356
4604752277 0167005684 5439692340 4171108988 8993416350 5851578873 5343081552
0811772071 8803791040 4698306957 : 357
8685473937 6564336319 7978680367 1873079693 9242363214 4845035477 6315670255
3900654231 1792015346 4977929066 : 358
2415083288 5839529054 2637687668 9688050333 1722780018 5885069736 2324038947
0047189761 9347344308 4374437599 : 359
2503417880 7972235859 1342458131 4404984770 1732361694 7197657153 5319775499
7162785663 1190469126 0918259124 : 360
9890367654 1769799036 2375528652 6375733763 5269693443 5440047306 7198868901
9681474287 6779086697 9688522501 : 361
6369498567 3021752313 2529265375 8964151714 7955953878 4278499866 4563028788
3196209983 0494519874 3963690706 : 362
8276265748 5810439112 2326187940 5994155406 3270131989 8957037611 0532360629
8674803779 1537675115 8304320849 : 363

8720920280 9297526498 1256916342 5000522908 8726469252 8466610466 5392171482 0801305022 9805263783 6426959733 : 364

7070539227 8915351056 8883938113 2497570713 3102950443 0346715989 4487868471 1643832805 0692507766 2745001220 : 365

0352620370 9466023414 6489983902 5258883014 8678162196 7751945831 6771876275 7200505439 7944124599 0077115205 : 366

1546199305 0983869825 4284640725 5540927403 1325716326 4079293418 3342147090 4125425335 2324802193 2277075355 : 367

5467958716 3835875018 1593387174 2360615511 7101312352 5633485820 3651461418 7004920570 4372018261 7331947157 : 368

0086757853 9336078622 7395581857 9758725874 4102542077 1054753612 9404746010 0094095444 9596628814 8691590389 : 369

9071865980 5636171376 9222729076 4197755177 7201042764 9694961105 6220592502 4202177042 6962215495 8726453989 : 370

2276976603 1052498085 5759471631 0758701332 0886146326 6412591148 6338812202 8444069416 9488261529 5776253250 : 371

1987035987 0674380469 8219420563 8125583343 6421949232 2759372212 8905642094 3082352544 0841108645 4536940496 : 372

9271494003 3197828613 1818618881 1118408257 8659287574 2638445005 9944229568 5864604810 3301538891 1499486935 : 373

4360302218 1094346676 4000022362 5505736312 9462629609 6198760564 2599639461 3869233083 7196265954 7392346241 : 374

3459779574 8524647837 9807956931 9865081597 7675350553 9189911513 3525229873 6112779182 7485420086 8953965835 : 375

9421963331 5028695611 9201229888 9887006079 9927954111 8826902307 8913107603 6176347794 8943203210 2773359416 : 376

9086500719 3280401716 3840644987 8717537567 8118532132 8408216571 1075495282 9497493621 4608215583 2056872321 : 377

8557406516 1096274874 3750980922 3021160998 2633033915 4694946444 9100451528 0925089745 0748967603 2409076898 : 378

3652940657 9201983152 6541065813 6823791984 0906457124 6894847020 9357761193 1399802468 1340520039 4781949866 : 379

2026240089 0215016616 3813538381 5150377350 2296607462 7952910384 0686855690 7015751662 4192987244 4827194293 : 380

3100485482 4454580718 8976330032 3252582158 1280327467 9620028147 6243182862 2171054352 8983482082 7345168018 : 381

6131719593 3247110746 6222850871 0666117703 4653528395 7762599774 4672185715 8161264111 4327179434 7885990892 : 382

8084866949 1413909771 6736900277 7585026866 4654056595 0394867841 1107901161 0400857274 4562938425 4941675946 : 383

0548711723 5946429105 8509099502 1495879311 2196135908 3158826206 8233215615 3086833730 8381732793 2819698387 : 384

5087083483 8804638847 8441884003 1847126974 5437093732 9836240287 5197920802 3218787448 8287284372 7378017827 : 385

0080587824 1074935751 4889978911 7397461293 2035108143 2703251409 0304874622 6294234432 7571260086 6425083331 : 386

8768865075 6429271605 5252895449 2153765175 1492196367 1810494353 1785838345 3865255656 6406572513 6357506435 : 387

3236508936 7904317025 9787817719 0314867963 8408288102 0946149007 9715137717 0990619549 6964007086 7667102330 : 388

0486726314 7551053723 1757114322 3174114116 8062286420 6388906210 1923552235 4671166213 7499693269 3217370431 : 389

0598722503 9456574924 6169782609 7025335947 5020913836 6737728944 3869640002 8110344026 0847128990 0074680776 : 390

4844088711 3413525033 6787731679 7709372778 6821661178 6534423173 2264637847 6978751443 3209534000 1650692130 : 391

5464768909 8505020301 5044880834 2618452087 3053097318 9492916425 3229336124
3151430657 8264070283 8984098416 : 392
0295030924 1897120971 6016492656 1341343342 2298827909 9217860426 7981245728
5345801338 2609958771 7811310216 : 393
7340256562 7440072968 3406619848 0676615805 0216918337 2368039902 7931606420
4368120799 0031626444 9146190219 : 394
4582296909 9212278855 3948783538 3056468648 8165556229 4315673128 2743908264
5061162894 2803501661 3366978240 : 395
5177015521 9626522725 4558507386 4058529983 0379180350 4328767038 0925216790
7571204061 2375963276 8567484507 : 396
9151147313 4400018325 7034492090 9712435809 4479004624 9431345502 8900680648
7042935340 3743603262 5820535790 : 397
1183956490 8935434510 1342969617 5452495739 6062149028 8728932792 5206965353
8639644322 5388327522 4996059869 : 398
7475988232 9916263545 9733244451 6375533437 7492928990 5811757863 5555562693
7426910947 1170021654 1171821975 : 399
0519831787 1371060510 6379555858 8905568852 8879890847 5091576463 9074693619
8815078146 8526213325 2473837651 : 400
1929901561 0918977792 2008705793 3964638274 9068069876 9168197492 3656242260
8715417610 0430608904 3779766785 : 401
1966189140 4144925270 4808819714 9880154205 7787006521 5940092897 7760133075
6847966992 9554336561 3984773806 : 402
0394368895 8876460549 8387147896 8482805384 7017308711 1776115966 3505039979
3438693391 1978988710 9156541709 : 403
1330826076 4740630571 1411098839 3880954814 3782847452 8838368079 4188843426
6622207043 8722887413 9478010177 : 404
2139228191 1992365405 5163958934 7426395382 4829609036 9002883593 2774585506
0801317988 4071624465 6399794827 : 405
5783650195 5142215513 3928197822 6984278638 3916797150 9126241054 8725700924
0700454884 8569295044 8110738087 : 406
9965474815 6891393538 0943474556 9721289198 2717702076 6613602489 5814681191
3361412125 8783895577 3571949863 : 407
1721084439 8901423948 4966592517 3138817160 2663261931 0653665350 4147307080
4414939169 3632623737 6777709585 : 408
0313255990 0957627319 5730864804 2467701212 3270205337 4266705314 2448208168
1303063973 7873664248 3672539837 : 409
4876909806 0218278578 6216512738 5635132901 4890350988 3270617258 9325753639
9397905572 9175160097 6154590447 : 410
7169226580 6315111028 0384360173 7474215247 6085152099 0161585823 1257159073
3421736576 2671423904 7827958728 : 411
1505095633 0928026684 5893764964 9770232973 6413190609 8274063353 1089792464
2421345837 4090116939 1964250459 : 412
1288134034 9881063540 0887596820 0544083643 8651661788 0557608956 8967275315
3808194207 7332597917 2784376256 : 413
6118431989 1025007491 8290864751 4979400316 0703845549 4653859460 2745244746
6812314687 9434416109 9333890899 : 414
2638411847 4252570445 7251745932 5738989565 1857165759 6148126602 0310797628
2541655905 0604247911 4016957900 : 415
3383565748 6925280074 3025623419 4982864679 1447632277 4005529460 9039401775
3633565547 1931000175 4300475047 : 416
1914489984 1040015867 9461792416 1001645471 6551337074 0739502604 4276953855
3834397550 5488710997 8520540117 : 417
5169747581 3449260794 3368954378 3221172450 6873442319 8987884412 8542064742
8097356258 0706698310 6979935260 : 418
6933921356 8588139121 4807354728 4632277849 0808700246 7776303605 5512323866
5629517885 3719673034 6347012229 : 419

3958160679 2509153217 4890308408 8651606111 9011498443 4123501246 4692802880
5996134283 5118847154 4977127847 : 420
3361766285 0621697787 1774382436 2565711779 4500644777 1837022199 9106695021
6567576440 4499794076 5037999954 : 421
8450027106 6598781360 3802314126 8369057831 9046079276 5297277694 0436130230
5178708054 6511542469 3952651271 : 422
0105292707 0306673024 4471259739 3995051462 8404767431 3637399782 5918454117
6413327906 4606365841 5292701903 : 423
0276017339 4748669603 4869497654 1752429306 0407270050 5903950314 8522921392
5755948450 7886797792 5253931765 : 424
1564161971 6844352436 9794447355 9642606333 9105512682 6061595726 2170366985
0647328126 6724521989 0605498802 : 425
8078288142 9796336696 7441248059 8219214633 9565745722 1022986775 9974673812
6069367069 1340815594 1201611596 : 426
0190237753 5255563006 0624798326 1249881288 1929373434 7686268921 9239777833
9107331065 8825681377 7172328315 : 427
3290825250 9273304785 0724977139 4483338925 5208117560 8452966590 5539409655
6854170600 1179857293 8139982583 : 428
1929367910 0391844099 2865756059 9359891000 2969864460 9747147184 7010153128
3762631146 7742091455 7404181591 : 429
8800064943 2378558393 0853082830 5476076799 5243573916 3122188605 7549673832
2431956506 5546085288 1201902363 : 430
6447127037 4863442172 7257879503 4284863129 4491631847 5347531435 0413920961
0879605773 0987201352 4840750576 : 431
3719925365 0470908582 5139368634 6386336804 2891767107 6021111598 2887553994
0120076013 9470336617 9371539630 : 432
6139863655 4922137415 9790511908 3588290097 6566473007 3387931467 8913181465
1093167615 7582135142 4860442292 : 433
4453041131 6065270097 4330088499 0346754055 1864067734 2603583409 6086055337
4736276093 5658853109 7609942383 : 434
4738222208 7292464497 6845605795 6251676557 4088410321 7313456277 3585605235
8236389532 0385340248 4227337163 : 435
9123973215 9954408284 2166663602 3296545694 7035771848 7344203422 7706653837
3875061692 1276801576 6181095420 : 436
0977083636 0436111059 2409117889 5403380214 2652394892 9686439808 9261146354
1457153519 4342850721 3534530183 : 437
1587562827 5733898268 8985235577 9929572764 5229391567 4775666760 5108788764
8453493636 0682780505 6462281359 : 438
8885879259 9409464460 4170520447 0046315137 9754317371 8775603981 5962647501
4109066588 6616218003 8266989961 : 439
9655805872 0863972117 6995219466 7898570117 9833244060 1811575658 0742841829
1061519391 7630059194 3144346051 : 440
5404771057 0054339000 1824531177 3371895585 7603607182 8605063564 7997900413
9761808955 3636696031 6219311325 : 441
0223851791 6720551806 5926351803 6251214575 9262383693 4822266589 5576994660
4919381124 8660909979 8128571823 : 442
4940066155 5219611220 7203092277 6462009993 1524427358 9488710576 6238946938
8944649509 3960330454 3408421024 : 443
6240104872 3328750081 7491798755 4387938738 1439894238 0117627008 3719605309
4383940063 7561164585 6094312951 : 444
7597713935 3960743227 9248922126 7045808183 3137641658 1826956210 5872892447
7400359470 0926866265 9651422050 : 445
6300785920 0248829186 0839743732 3538490839 6432614700 0532423540 6470420894
9921025040 4726781059 0836440074 : 446
6638002087 0126664209 4571817029 4675227854 0074508552 3777208905 8168391844
6592829417 0182882330 1497155423 : 447

5235911774 8186285929 6760504820 3864343108 7795628929 2540563894 6621948268
7110428281 6389397571 1757786915 : 448
4301650586 0296521745 9581988878 6804081103 2843273986 7198621306 2055598552
6603640504 6282152306 1545944744 : 449
8990883908 1999738747 4529698107 7620148713 4000122535 5222466954 0931521311
5337915798 0269795557 1050850747 : 450
3874750758 0687653764 4578252443 2638046143 0428892359 3485296105 8269382103
4980004052 4840708440 3561167817 : 451
1705128133 7880570564 3450616119 3304244407 9826037795 1198548694 5591520519
6009304127 1007277849 3015550388 : 452
9536033826 1929343797 0818743209 4991415959 3396368110 6275572952 7800425486
3060054523 8391510689 9891357882 : 453
0019411786 5356821491 1852820785 2130125518 5184937115 0342215954 2244511900
2073935396 2740020811 0465530207 : 454
9328672547 4054365271 7595893500 7163360763 2161472581 5407642053 0200453401
8357233829 2661915308 3540951202 : 455
2632916505 4426123619 1970516138 3935732669 3760156914 4299449437 4485680977
5696303129 5887191611 2929468188 : 456
4936338647 3927476012 2696415884 8900965717 0861605981 4720446742 8664208765
3347998582 2209061980 2173211614 : 457
2304194777 5499073873 8567941189 8246609130 9169177227 4207233367 6350326783
4058630193 0193242996 3972044451 : 458
7928812285 4478211953 5308989101 2534297552 4727635730 2262813820 9180743974
8671453590 7786335301 6082155991 : 459
1314144205 0914472935 3502223081 7193663509 3468658586 5631485557 5862447818
6201087118 8976065296 9899269328 : 460
1787055764 3514338206 0141077329 2610634315 2533718224 3385263520 2177354407
1528189813 7698755157 5745469397 : 461
2715048846 9793619500 4777209705 6179391382 8989845327 4262272886 4710888327
0173723258 8182446584 3624958059 : 462
2560338105 2156062061 5571329915 6084892064 3403033952 6226345145 4283678698
2880742514 2256745180 6184149564 : 463
6861116354 0497189768 2154227722 4794740335 7152743681 9409892050 1136534001
2384671429 6551867344 1537416150 : 464
4256325671 3430247655 1252192180 3578016924 0326699541 7460875924 0920700466
9340396510 1781348578 3569444076 : 465
0470232540 7555577647 2845075182 6890418293 9661133101 6013111907 7398632462
7782190236 5066037404 1606724962 : 466
4901374332 1724645409 7412995570 5291424382 0807609836 4823465973 8866913499
1978401310 8015581343 9791948528 : 467
3043673901 2482082444 8141280954 4377389832 0059864909 1595053228 5791457688
4962578665 8859991798 6752055455 : 468
8099004556 4611787552 4937012455 3217170194 2828846174 0273664997 8475508294
2280202329 0122163010 2309772151 : 469
5694464279 0980219082 6689868834 2630716092 0791408519 7695235553 4886577434
2527753119 7247430873 0436195113 : 470
9611908003 0255878387 6442060850 4473063129 9277888942 7291897271 6989057592
5244679660 1897074829 6094919064 : 471
8764693702 7507738664 3239191904 2254290235 3189233772 9316673608 6996228032
5571853089 1928440380 5071030064 : 472
7768478632 4319100022 3929785255 3723755662 1364474009 6760539439 8382357646
0699246526 0089090624 1059042154 : 473
5392790441 1529580345 3345002562 4410100635 9530039598 8644661695 9562635187
8060688513 7234627079 9732723313 : 474
4693971456 2855426154 6765063246 5676620279 2452085813 4771760852 1691340946
5203076733 9184114750 4140168924 : 475

1213198268 8156866456 1485380287 5393311602 3229255561 8941042995 3356400957 8649534093 5115266454 0244187759 : 476
4931693056 0448686420 8627572011 7231952640 5023099774 5676478384 8897346431 7215980626 7876718380 0524769688 : 477
4084989185 0861490034 3240347674 2686245952 3958903585 8213500645 0998178244 6360873177 5437885967 7672919526 : 478
1112138591 9472545140 0301180503 4378752776 6440276261 8941017576 8726804281 7662386068 0477885242 8874302591 : 479
4524707395 0546525135 3394595987 8961977891 1041890292 9438185672 0507096460 6263541732 9446495766 1265195349 : 480
5701860015 4126239622 8641389779 6733329070 5673769621 5649818450 6842263690 3678495559 7002607986 7996261019 : 481
0393312637 6855696876 7029295371 1625280055 4310078640 8728939225 7145124811 3577862766 4902425161 9902774710 : 482
9033593330 9304948380 5978566288 4478744146 9841499067 1237647895 8226329490 4679812089 9848571635 7108783119 : 483
1848630254 5016209298 0582920833 4813638405 4217200561 2198935366 9371336733 3924644161 2522319694 3471206417 : 484
3754912163 5700857369 4397305979 7097197266 6664226743 1117762176 4030686813 1035189911 2271339724 0368870009 : 485
9686292254 6465006385 2886203938 0050477827 6912835603 3725482557 9391298525 1506829969 1077542576 4748832534 : 486
1412132800 6267170940 0909822352 9657957997 8030182824 2849022147 0748111124 0186076134 1515038756 9830918652 : 487
7806588966 8236252393 7845272634 5304204188 0250844236 3190383318 3845505223 6799235775 2929106925 0432614469 : 488
5010986108 8899914658 5518818735 8252816430 2520939285 2580779697 3762084563 7482114433 9881627100 3170315133 : 489
4402309526 3519295886 8069082135 5853680161 0002137408 5115448491 2685841268 6958991741 4913382057 8492800698 : 490
2551957402 0181810564 1297250836 0703568510 5533178784 0829000041 5525118657 7945396331 7538532092 1497205266 : 491
0783126028 1961164858 0986845875 2512999740 4092797683 1766399146 5538610893 7587952214 9717317281 3151793290 : 492
4431121815 8710235187 4075722210 0123768721 9447472093 4931232410 7065080618 5623725267 3254073332 4875754482 : 493
9675734500 1932190219 9119960797 9893733836 7324257610 3938985349 2787774739 8050808001 5544764061 0535222023 : 494
2540944356 7718794565 4304067358 9649101761 0775948364 5408234861 3025471847 6485189575 8366743997 9150851285 : 495
8020607820 5544629917 2320202822 2914886959 3997299742 9747115537 1858924238 4938558585 9540743810 4882624648 : 496
7880533042 7146301194 1589896328 7926783273 2245610385 2197011130 4665871005 0008328517 7311776489 7352309266 : 497
6123458887 3102883515 6264460236 7199664455 4727608310 1187883891 5114934093 9344750073 0258558147 5619088139 : 498
8752357812 3313422798 6650352272 5367171230 7568610450 0454897036 0079569827 6263923441 0714658489 5780241408 : 499
1584052295 3693749971 0665594894 4592462866 1996355635 0652623405 3394391421 1127181069 1052290024 6574236041 : 500
3009369188 9255865784 6684612156 7955425660 5416005071 2766417660 5687427420 0329577160 6434486062 0123982169 : 501
8271723197 8268166282 4993871499 5449137302 0518436690 7672357740 0053932662 6227603236 5975171892 5901801104 : 502
2903842741 8550789488 7438832703 0632832799 6300720069 8012244365 1163940869 2222074532 0244624121 1558043545 : 503

4206421512 1585056896 1573564143 1306888344 3185280853 9759277344 3365538418 8340303517 8229462537 0201578215 : 504
7373265523 1857635540 9895403323 6382319219 8921711774 4946940367 8296185920 8034038675 7583411151 8824177439 : 505
1450773663 8407188048 9358256868 5420116450 3135763335 5509440319 2367203486 5101056104 9872726472 1319865434 : 506
3545040913 1859513145 1812764373 1043897250 7004981987 0521762724 9406521461 9959232142 3144397765 4670835171 : 507
4749367986 1865527917 1582408065 1063799500 1842959387 9915835017 1580759883 7849622573 9851212981 0326379376 : 508
2183224565 9423668537 6799113140 1080431397 3233544909 0824910499 1433258432 9882103398 4698141715 7560108297 : 509
0658306521 1347076803 6806953229 7199059990 4451209087 2757762253 5104090239 2888779424 6304832803 1913271049 : 510
5478599180 1969678353 2146444118 9260631526 6181674431 9355081708 1875477050 8026540252 9410921826 4858213857 : 511
5266881555 8411319856 0022135158 8872103656 9608751506 3187533002 9421186822 2189377554 6027227291 2905042922 : 512
5978771066 7873840000 6167721546 3844129237 1193521828 4998243509 2089180168 5572798156 4218581911 9749098573 : 513
0570332667 6464607287 5743056537 2602768982 3732597450 8447964954 5648030771 5981539558 2777913937 3601717422 : 514
9960273531 0276871944 9444917939 7851446315 9731443535 1850491413 9415573293 8204854212 3508173912 5497498193 : 515
0871439661 5132942045 9193801062 3142177419 9184060180 3479498876 9105155790 5554806953 8785400664 5337598186 : 516
2846419905 2204528033 0626369562 6490910827 6271159038 5699505124 6529996062 8554438383 3032763859 9800792922 : 517
8466595035 5121124528 4087516229 0602620118 5777531374 7949362055 4964010730 0134885315 0735487353 9056029089 : 518
3352640071 3274732621 9603117734 3394367338 5759124508 1493357369 1166454128 1788171454 0230547506 6713651825 : 519
8284898099 5121391939 9563324133 6556777098 0030819102 7204099714 8687418134 6670060940 5102146269 0280449159 : 520
6465453301 0775469541 3088714165 3125448130 6119240782 1188690056 0277818242 3502269618 9344352547 6335735364 : 521
8561936325 4417756613 9817039306 3287216690 5722259745 2091929172 6219984440 9646158269 4563802395 0283712168 : 522
6446561785 2355651641 2771282691 8688615572 7162014749 3405227694 6595712198 3149433816 2211400693 6307430444 : 523
1732847861 0177774383 7977037231 7952554341 0722344551 2555589998 6461838767 6490397246 1167959018 1000350989 : 524
2864120419 5163551108 7632042676 1297982652 9425882951 1412758412 6273279079 8807559751 8515768412 6474220947 : 525
9721843309 3529726652 1001566251 4552994745 1276315509 1763673025 9462132930 1904028379 5424632325 8550301096 : 526
7069227202 270 486341 9005438302 6506812141 4213505715 4175057508 6399076739 4633514620 9082888934 9383764393 : 527
9925690060 4067311422 0933121959 3620298297 2351163259 3867722414 7791162957 2780752395 0562515816 0313335938 : 528
2311500518 6268905306 5836812998 8108663263 2719806112 7154885879 8093487912 9137074982 3057592909 1862939195 : 529
0147211975 8606727009 2547718025 7503377307 9939713453 9532646195 2699965963 8565491759 0458333585 7991020127 : 530
1320458390 3200853878 8816336376 8518208372 7885131175 2277696097 8796214237 2162545214 5912818317 9821604411 : 531

1311671406 9148271709 8101545778 1939202311 5638719508 0502467972 5792497605
7726259133 2855972637 1211201905 : 532
7207714091 4864507409 4926718035 8151575715 1405039761 0963846755 5692989703
8354731410 0223802583 4687673501 : 533
2977541327 9532060971 1545064842 1218593649 0997917766 8747744818 8287063231
5515865032 8981642282 8823274686 : 534
6106592732 1979071623 8464215348 9852476216 7890502609 9804526648 3929542357
2873439776 8049577409 1449538391 : 535
5755654854 5905897649 5198513801 0079580107 8375994577 5299196700 5476022525
5203445398 8712538780 1719607181 : 536
6407812484 7847257912 4078245443 6168234523 9570689514 2722697504 3187363326
3011103053 4233358216 0933319121 : 537
8806608268 3414289104 1517324721 6053355849 9932245487 3077882290 5252324234
8615315209 7693846104 2582849714 : 538
9634753418 3756200301 4915703279 6853018686 3157248840 1526639835 6895636346
5743532178 3493199825 5421173084 : 539
6774529708 5839507616 4582296303 2442432823 7737450517 0285606980 6788952176
8198156710 7816334052 6675953942 : 540
4926280756 9683261074 9532339053 6223090807 0814559198 3735537774 8742029039
0181429373 1152933464 4468151212 : 541
9450975965 3430628421 5319445727 1186149000 1765055817 7095302468 8752632501
1970520947 6159416768 7277844720 : 542
0019278913 7251841622 8577837922 8443908430 1181121496 3664246590 3363419454
0657183544 7719124466 2125939265 : 543
6620306888 5200555991 2123536371 8226922531 7814587925 9375044144 8933981608
6579008761 6502463519 7045828895 : 544
4817937566 8104647461 4105142498 8702521399 3687050937 2305447734 1126413548
9280684105 9107716677 8212383328 : 545
1026218558 7751312721 1793444482 0144042574 5083063944 7383637939 0628300897
3306241380 6145894142 2769474793 : 546
1665717623 1824721683 5067807648 7573420491 5576282175 8397297513 4478990696
5895325489 4033561561 3167403276 : 547
4724692125 0575911625 1529654568 5446334981 1431767025 7295661844 7754874693
7846423373 7238981920 6620485118 : 548
9437886822 4807279352 0225017965 4534375727 4163910791 9729529508 1294292220
5347717304 1844779156 7399173841 : 549
8311710362 5243957161 5271466900 5814700002 6330104526 4354786590 3290733205
4683388720 7873544476 2647925297 : 550
6901709120 0787418373 6735087713 3769776834 9634425241 9949951388 3150748775
3743384945 8259765560 9965559543 : 551
1804092017 8497184685 4973706962 1208852437 7013853757 6814166327 2241263442
3982152941 6453780004 9250726276 : 552
5150789085 0712659970 3670872669 2764308377 2296859851 6912230503 7462744310
8529343052 7307886528 3977335246 : 553
0174635277 0320593817 9125396915 6210636376 2588293757 1373840754 4064689647
8310070458 0613446731 2715911946 : 554
0843593582 5987782835 2665311510 6504162329 5329047772 1740835593 4972375855
2138048305 0900096466 7608830154 : 555
0612824308 7406455944 3185341375 5220166305 8121110334 5312074508 6824339432
1590435944 3031243122 7471385842 : 556
0303901060 7094031523 5556172767 9941600203 9397509989 7629335325 8555756248
0899669182 9864222677 5023601932 : 557
5797472674 2578211119 7347094023 5745722227 1212526852 3842958742 7350156366
0093188045 4933389897 4157149054 : 558
4182559738 0808715652 8143010267 0460284316 8192303925 3529779576 5862414392
7015497408 7927313105 1636119137 : 559

5770089295 6482332364 8298263024 6079758757 6774537716 0102490804 6243018565
2416175665 5600160859 1215345562 : 560
6760219268 9982855377 8725831451 4408265458 3484409478 4631787773 7479465358
0169960779 4055687011 9232860804 : 561
1130904629 3508718271 2593466871 2766694873 8998245985 2778649956 9165464029
4589350649 6433580982 4765965165 : 562
1420909867 5520380830 9203230487 3427034682 8875160407 1546653834 6196112230
1375945157 9252696743 6425319273 : 563
9003603860 8236450762 6988274976 1872357547 6762889950 7521148048 5252795084
5033958570 8381304769 3788132112 : 564
3674281319 4879502280 6632017002 2460331989 6719706491 6374117585 4851878484
0120548446 7258885140 1562725019 : 565
8217190669 6081262778 5485964818 3696214107 2171421498 6361918774 7545096503
0895709947 0934337856 9816744658 : 566
2826791194 0611956037 8453978558 3924076127 6344105766 7510243075 5981455278
6167815949 6570625597 5507430652 : 567
1085301597 9080733437 3607943286 6757890533 4836695554 8680391343 3720156498
8342208933 9997164147 9746938696 : 568
9054800891 9306713805 7171505857 3071488156 4992071408 6758259602 8760564597
8242377024 2469805328 0566327870 : 569
4192676846 7116266879 4634869504 6450742021 9373945259 2626686135 5294062478
1361206202 6364981999 9949840514 : 570
3868285258 9563422643 2870766329 9304891723 4007254717 6418868535 1372332667
8779217383 4754148002 2803392997 : 571
3579361524 1275582956 9276837231 2347989894 4627433045 4566790062 0324205163
9628258844 3085438307 2014956721 : 572
0646053323 8537203143 2421126074 2448584509 4580494081 8209276391 4000854042
2023556260 2185643489 9414543995 : 573
0410980591 8179488826 2805206644 1086319001 6885681551 6922948620 3010738897
1810077092 9059048074 9092427141 : 574
0189335428 1842999598 8169660993 8369616443 8152887721 4085268088 7574882932
5873580990 5670755817 0179491619 : 575
0611400190 8553744882 7262009366 8560447559 6557476485 6740081773 8170330738
0305476973 6097865438 5938218722 : 576
0583902344 4435088674 9986650604 0645874346 0053318274 3629617786 2518081893
1443632512 0510709469 0813586440 : 577
5192295129 3245007883 3398788429 3393424351 2634336520 4385812912 8343452973
0865290978 3300671261 7981303167 : 578
9438553572 6296998740 3595704584 5223085639 0098913179 4759487521 2639707837
5944861139 4519602867 5121056163 : 579
8976008880 0927461158 6080020780 3341591451 7970730368 3519697776 6076373785
3330120241 2011204698 8609209339 : 580
0853657732 2239241244 9051532780 9509558664 5947763448 2269986074 8132973026
3097502881 2103517723 1244650953 : 581
4965369309 0018637764 0940943498 3731325132 1862080214 8099226855 0294845466
1814715557 4447096695 3017769043 : 582
4272031892 7706047177 8452793916 0472281534 3798035396 7986142437 0956683221
4914654380 1459382927 7393396032 : 583
7540480095 5223181666 7380357183 9327570771 4204672383 8624617803 9762923771
3120958078 9363841447 9298025880 : 584
6552212926 2093623930 6373134966 4018661951 0811583471 1733120258 0586672763
9992763579 0780638188 1306915636 : 585
6274125431 2595899361 1964762610 1405563503 3995231403 2311381965 6236327198
9618372548 4533370206 2563464223 : 586
9527669435 6837676136 8711962921 8187545760 8161705303 1590728828 7007123136
6630872275 4918661395 7737305460 : 587

6599743781 0987649802 4140112421 4277366808 2751390959 3134041558 2626678951
0846776118 6659576601 6599817808 : 588
9414985754 9762843878 5610026379 6543178313 6340251358 1416115190 2096499133
5487331311 1502270068 1930135929 : 589
5959716401 9719605362 5033558479 9809634887 1803911161 2813595968 5654788683
2585643789 6173159762 0024196215 : 590
5289629790 4819822199 4622694871 3746244472 9093456470 0285376949 5885959160
6789282491 0544125159 9630078136 : 591
8367490209 3749157328 9627002865 6829344431 3423473512 3929825916 6739503425
9958689706 9726733258 2735903121 : 592
2887466604 5146148785 0346142827 7659916080 9039865257 5717263081 8334944418
2019353338 5071292345 7743755793 : 593
4406217871 1330063106 0033240539 9169368260 3746176638 5657588775 8020122936
6353270267 1006812618 2517291460 : 594
8202541892 8859352444 9107013820 6211553827 7935652969 1457650204 8643282865
5579347072 0963480737 2692141186 : 595
8954673227 6775133569 0190153723 6690368653 8916129168 8887876407 5254934942
4973342718 1178892759 9315967193 : 596
5475898809 7924525262 3636590363 2007085444 0784544797 3482918020 8204492667
0634420437 5553250505 2752283377 : 597
8887040804 0335319234 0768563010 9347772125 6390886404 1310107381 7853338316
0381352808 2811904083 2564401842 : 598
0537467929 9262203769 8718018061 1226244909 0924264198 5820861751 1771137890
5160914038 1575003366 4241560952 : 599
1632819712 2335023167 4226005679 4128140621 7219641842 7057843289 5980288233
5059828208 1966662490 3585778994 : 600
0333152274 8177769528 4368163008 8531769694 7836905806 7106482808 3598046698
8410981351 5865490693 3319522394 : 601
3632879239 9053481098 7830274500 1720654336 9906611778 4554364687 7236318444
6476806914 2828004551 0746866453 : 602
9280539940 9108754939 1660957316 1971503316 6968309929 4663491427 9878084225
7220697148 8755806374 8030886299 : 603
5118473187 1247772919 1007022758 8893486939 4562895158 0296537215 0409603107
7612898312 6358996489 3410247036 : 604
0366450586 8728758905 1406841238 1242473863 8542790828 2733827973 3268855049
3587430316 0274749063 1295723497 : 605
4261122151 7417153133 6186224109 1386950068 8835898962 3492763173 1647834007
7460886655 5987333821 1382992877 : 606
6911495492 1841920877 7160606847 2874673681 8861675072 2101726110 3830671787
8566948129 4878504894 3063086169 : 607
9487987031 6051588410 8282351274 1535385133 6589533294 8629494495 0618685147
7910580469 6039069372 6626703865 : 608
1290520113 7810858616 1888869479 5760741358 5534585151 7680519733 3443349523
0120395770 7396237713 1603024288 : 609
7200537320 9982530089 7761897312 9817881944 6717311606 4723147624 8457551928
7327828251 2718244680 7824215216 : 610
4695678192 9409823892 6284943760 2488522790 0362021938 6696482215 6280936053
7317804086 3727268426 6964219299 : 611
4681921490 8701707533 3610947913 8180406328 7387593848 2695355830 7739576144
7997270003 4728801827 8528138950 : 612
3217986345 2161110666 0883931405 3226944905 4555278678 9441757920 2440021450
7801920998 0446138254 7805858048 : 613
4424164047 7503153605 4906591430 0781583724 3012313751 1562284015 8386442708
9071828481 6757527123 8467824595 : 614
3433444962 2010096071 0513706084 6180118754 3120725491 3349942476 1711563332
1408934609 1565615506 0031738421 : 615

8701570226 1031019166 0388706466 1438897736 3187809407 1152752817 4689576401
5810470169 6524755774 0891644568 : 616
6777171585 0058326994 3401677202 1567677240 6812836656 5264122982 4394651331
9735919970 9403275938 5026695574 : 617
7023181320 3243716420 5861410336 0652453693 9160050644 9530601612 6782264894
2437397166 7176612310 4897503188 : 618
5732165554 9883421218 0284691252 9086101485 5278152776 2562375045 6375769497
7343368460 1560772703 5509629049 : 619
3924870884 0628106794 3622418704 7470083688 4267102255 8302403599 8416459511
2248527263 3632645114 0173952480 : 620
8619463584 0783753556 8856223171 1552094722 3065437092 6067973510 0056554938
1224575483 7285457117 9739361575 : 621
6167641692 8958052572 9752233855 8611388322 1711073622 6581621884 2443178857
4887981090 2665379342 6664216990 : 622
9140565364 3224930133 4867988154 8866286650 5234699723 5574738424 8305904236
7714327879 2316422403 8777643301 : 623
9260019228 4778313837 6325361210 2533693581 2624086866 6997382759 7736568222
7907215832 4788886423 6934639616 : 624
4363308730 1398142114 3030600873 0666164803 6789840913 3592629340 2304324974
9268878316 4360268101 1309570716 : 625
1419128306 8657732353 2639653677 3903176613 6131596555 3584999398 6005651559
2193675997 7717933019 7446881483 : 626
7110320650 3693192894 5214026509 1546518430 9936553493 3371834252 9843367991
5939417466 2239003895 2767381333 : 627
0617747629 5749438687 1697845376 7219493506 5908757119 1772087547 7107189937
9608947745 1265475750 1871194870 : 628
7387367858 9020061737 3321075693 3022163206 2843206567 1192096950 5857611739
6163232621 7708945426 2146098584 : 629
1023781321 5817727602 2227381334 9541048100 3073275107 7999489919 7796388353
0734443457 5329759142 6376840544 : 630
2264784216 0631227696 4696715647 3999043715 9033239065 6072664411 6438605404
8388471619 1210900870 1019130726 : 631
0710441141 4324197679 6828547885 5247794764 8180295973 6049439700 4795960402
9274629920 3572099761 9501403483 : 632
1538094771 4601056333 4469988208 2212058728 1510729182 9712119178 7642488035
4672316916 5418522567 2923442918 : 633
7128163232 5969654135 4858957713 3208339911 2887759172 2611527337 9010341362
0856145779 9239877832 5083550730 : 634
1998184590 2595835598 9260553299 6737704917 2245493532 9683300002 2301815172
2657578752 4058832249 0858212800 : 635
8974790932 6100762578 7704286560 0699617621 2176845478 9964407050 6624171021
3327486796 2374302291 5535820078 : 636
0141165348 0656474882 3061500339 2068983794 7662550365 4982280532 9662862117
9306284301 7049240230 1985719978 : 637
9488368971 8304380518 2174419147 6604297524 3725168343 5411217038 6313794114
2209529588 5798060152 9387527537 : 638
9903093887 1683572095 7607152219 0027937929 2786303637 2687658226 8124199338
4808166021 6037221547 1014300737 : 639
7537792699 0695871212 8928801905 2031601285 8618254944 1335382078 4883465311
6326504076 4242839087 0121015194 : 640
2319616522 6842200371 1230464300 6734420647 4771802135 3070124098 8603533991
5266792387 1101706221 8658835737 : 641
8121093517 9775604425 6346949997 8725112544 0854522274 8109148743 0725986960
2040275941 1789425812 8188215995 : 642
2359658979 1811440776 5335432175 7595255536 1581280011 6384672031 9346507296
8079907939 6371496177 4312119402 : 643

0212975731 2516525376 8017359101 5573381537 7200195244 4543620071 8484756634
1540744232 8621060997 6132434875 : 644
4884743453 9665981338 7174660930 2053507027 1952983943 2714253711 5576660002
5784423031 0734295515 3394506048 : 645
6222764966 6876240793 2435319299 2639253731 0768921353 5257232108 0889819339
1686682789 4828117047 2624501948 : 646
4097009757 6092098372 4090074717 9733407881 4182519584 2598096241 7476101382
5264395513 5259311885 0456362641 : 647
8830033853 9652435997 4169313228 9471987830 8427600401 3680747039 0409723847
3945834896 1865397905 9411859931 : 648
0356168436 8692194853 8205578039 5773881360 6795499000 8512325944 2529724486
6667668346 4140218991 5944565309 : 649
4234406506 6785194841 7766779470 4720419588 2204329538 0326310537 4948831221
8039127967 8446100139 7267538921 : 650
9511911783 6587662528 0836900532 4900459741 0947068772 9123282143 0463533728
3519953648 2743258331 1914445901 : 651
7809607782 8835837301 1185754365 9958982724 5319253105 8811502630 7542571493
9430244539 3187017992 3608166611 : 652
3054262539 9583389794 2971602070 3387678150 3301028012 0095997252 2222808014
2357109476 0351925544 4349299867 : 653
6781789104 5559063015 9538097618 7592035893 7341978962 3589311259 8390259831
0267193304 1892151096 8915622506 : 654
9659119828 3234555030 5908173073 5195503721 6658702880 5399213857 6037035377
1051780212 8012956684 1984140362 : 655
8727256232 1442875430 2210909472 7210734741 3497551419 0737043318 2766261772
7599688882 6027225247 1336833534 : 656
5281669277 9591328861 3817663498 5772893690 0965749562 2871030243 6259077241
2219094300 8717556926 2575806570 : 657
9912016659 6224360802 4287002454 7362036394 8412559548 8172727247 3653467783
6472019183 0399871762 7037515724 : 658
6499222894 6793232269 3619177641 6146187956 1395669956 7783068290 3165896994
3076733350 8234990790 6241002025 : 659
0613405734 4300695745 4746821756 9044165154 0636584680 4636926212 7421107539
9042188716 1276177870 1425886482 : 660
5775223889 1845995233 7629237791 5585744549 4773612955 2595222657 8636462118
3775984737 0034797140 8206994145 : 661
5807190802 1359073226 9233100831 7595106590 1912129479 5408603640 7573587502
0589020870 4579670007 0552625058 : 662
1142066390 7459215273 3094068236 4944159089 1009220296 6805233252 6619891131
1842016291 6310768940 8472356436 : 663
6808182168 6572196882 6835840278 5500782804 0434537101 8365109695 1782335743
0305048526 5373807353 1074185917 : 664
7056103973 9506264035 5442275156 1011072617 7937063472 3804990666 9221619711
9425912044 5084641746 3835899382 : 665
3994651739 5509000859 4799901360 2667426149 4290066467 1150671754 2217703877
4507673563 7421547829 0591101261 : 666
9157555870 2389570014 0511782264 6989944917 9083017954 7587676016 8094100135
8376135785 9135692445 5647764464 : 667
1786671153 9195135769 6104864922 4900834467 1548638305 4477914330 0976804868
7834818467 2733758436 8927243104 : 668
4740680768 5278625585 1650920882 6381323362 3148733336 7147645204 5087662761
4950389949 5048095604 6098960432 : 669
9123358348 8599902945 2640028499 4280878624 0398118148 8476730121 6754161106
6299955536 6819312328 7425702063 : 670
7383520200 8686369131 1733469731 7412191536 3324674532 5630871347 3027921749
5622701468 7325867891 7345583799 : 671

6435135880 0959350877 5563562488 1049385299 9007675135 5135277924 1242927748
8565888566 5132473025 1471021057 : 672
5352516511 8148509027 5047684551 8252096331 8990685276 1443513821 3662152368
8905787866 9943228881 6028377482 : 673
0355060160 2989400911 9713850179 8716836337 4413927597 3644017007 0147637066
5570350433 8121113576 4150184518 : 674
2141361982 3495159601 0647527125 7593518530 4332875537 7830575095 6742544268
4712219618 7091785607 8393614451 : 675
1383335649 1032564057 3389866717 8123972237 5193164306 1701385953 9474367843
3926709867 1245221118 9690840236 : 676
3274114966 0124348309 8929941738 0305884171 6661307304 0067588380 4321115553
7944060549 7721705942 8215148861 : 677
6567277124 0903387727 7456290971 1013488518 4374118695 6554497457 3684521806
6982911045 0580042998 8795389902 : 678
7804383596 2824094218 6055628778 8428802127 5538848037 2864001944 1614257499
9042720095 9520465417 0598104989 : 679
9675045119 3647117277 2220436102 6140797508 0968697517 6600237187 7483480161
2031023468 0567112644 7661237476 : 680
2785219024 1202569943 5347162266 6089367521 9833111813 5111465038 5489502512
0655772636 1454736044 2685949807 : 681
4396932331 2971273771 5734709971 3952291182 6534851555 8713733662 9120242714
3025037632 6950135091 1612952993 : 682
7858646813 0722648600 8270881333 5381937036 8259886789 3321238327 0532976258
5738279009 7826460545 5985551318 : 683
3668884462 8265133798 4916678394 0976135376 6251798258 2496634587 7195012438
4040359140 8492097337 5464247448 : 684
8176184070 0235695801 7741017769 6925077814 8933866725 5789856458 9851056891
9609243988 4156928069 6983352240 : 685
2256345704 9731224526 9354193837 0048431833 5719651662 6721575524 1934019330
9901831930 9196582920 9696562476 : 686
6768365964 7019595754 7393455143 3741370876 1517323677 2042273856 7427917069
8204549953 0959188724 3493952409 : 687
4441678998 8463198455 0485239366 2972079777 4528143994 1825678945 7795712552
4268260899 4086331737 1538896262 : 688
8896294021 1210888442 7376568624 5276121303 7101730078 5135715404 5330415079
5944777614 3597437803 7424366469 : 689
7324713841 0492124314 1389035790 9241603640 6314038149 8314819052 5172093710
3964026808 9948325722 9795456404 : 690
2701757722 9041732347 9607361878 7889913318 3058430693 9482596131 8713816423
4672187308 4513387721 9086975104 : 691
9428437693 2502498165 6673816260 6159417682 5250999374 1672883951 7440669325
4965340310 1452225316 1890092353 : 692
7648637848 2881344209 8700480962 2717122640 7489571939 0029185733 0746010436
0729190945 7679946149 2929042798 : 693
1687729426 4877299528 5843464777 5386906950 1489841339 2454039414 4680263625
4021186143 1703125111 7577642829 : 694
9146445334 0892097696 1699098372 6523617687 4560589470 4968170136 9749095230
7208268288 7890730190 0182534258 : 695
0534342170 5928713931 7379931424 1085264739 0948284596 4180936141 3847583113
6130576108 4623668372 3769591349 : 696
2615824516 2215521348 7924414504 1756848064 1206365201 7038633012 9532777699
0231186480 2006755690 5682295016 : 697
3549319923 0591424639 6217025329 7475731140 9422018019 9368035026 4956369558
6642590676 2685687372 1103391567 : 698
9383989576 5565193177 8830002416 1353956243 7777840801 7488193730 9502069990
0890899328 0883974303 6773659552 : 699

4891300156 6332940779 0713961546 4534088791 5103006513 2193448667 3248275907
9468078798 1942501958 2622320395 : 700
1312520141 0996053126 0696555404 2486705499 8678692302 1746989009 5478507256
7297879476 9888831093 4874644264 : 701
0071818316 0331655511 5342761556 2240547447 3378049246 2149521332 5852769884
7336269182 6491743389 8782478927 : 702
8468918828 0546699823 0368993978 3413747587 0258057163 4941356843 3929396068
1920617733 3179173820 8562436433 : 703
6353598634 9449689078 1064019674 0744365836 6707158692 4521182997 8938040771
3750129085 8646578905 7714268335 : 704
8276897855 4717687184 4277261205 0926648610 2051535642 8406323684 8180728794
0717127966 8200607275 5955590404 : 705
0233178749 4473464547 6062818954 1512139162 9184442976 5106694796 9354016866
0100551960 7768733539 6511614930 : 706
9375709685 5455938151 3789569039 2510149532 6562814701 1998326992 2000663928
7537471313 5236421589 2651262040 : 707
7288771657 8358405219 6460541054 3544364216 6562244565 0429990102 5658692727
9142752931 1720827939 3775132610 : 708
6052881235 3734510683 7293989358 0871243869 3859343891 7571337630 0720319760
8166044646 8393772580 6909237297 : 709
5234867029 1691042636 9262090199 6052041210 2407764819 0316014085 8635584276
0953708655 8164273995 3493465463 : 710
1450404019 9528537252 0049578052 5465625115 4109252437 9913262627 1360909940
2902262062 8367521323 0506518393 : 711
4057450112 0993414649 1843332364 6569371725 9144893241 5900624202 0612885732
9261335968 0872650004 5628284557 : 712
5745965921 2053034131 0111827501 3069615098 3551563200 4310784601 9065654938
0654252522 9161991819 9596027523 : 713
2770224985 5738824899 8827074659 3635576858 2560518068 9642853768 5077201222
0347920993 9361792682 0659014216 : 714
5615925306 7379445689 4907085326 3568196831 8617722682 4991147261 5732035807
6462981162 4401331673 7892788689 : 715
2290325933 4986179702 1994981925 7396176730 7583441709 8559222170 1718257127
7753449150 8205278430 9046194608 : 716
3521740200 5838672849 7094110232 6695392144 5461066215 0064106747 4020700918
9911951376 4669044812 6725369153 : 717
7162290791 3854039375 6007783515 3374167747 9421003840 0230895185 0994548779
0393461222 2086506016 0500351776 : 718
2648316111 5332558770 5073541279 2499098593 7347378708 1194253055 1214369797
4991495186 0535920403 8302357163 : 719
5272763087 4693219622 1900642608 8618367610 3346002255 4774778136 4101269190
6569686495 0126883762 9690723396 : 720
1276287223 0411418136 1006026404 4030035996 9889199458 2739762411 4613744804
0596970625 7676472376 6065541618 : 721
5746905272 2923822827 5186799156 9833907476 7114610302 2776606020 0612468764
7772881909 6791613354 0198814027 : 722
5799217416 7678799231 6039635694 9285151363 3647219540 6111717673 8737255572
8522940054 3617851765 0230754469 : 723
3869307873 4991103521 8253292972 6044553210 7978877114 4989887091 1511237250
6042387537 3484125708 6064069052 : 724
0584521227 5453384800 8205302450 4565176695 1857691320 0042816758 0549248117
8051983264 6032445792 8297301291 : 725
0531838563 6821206215 5312886685 6495651261 3892261367 0640939533 3457052698
6959692350 3530942245 4386527867 : 726
7673027540 4027022463 8448355323 9914751363 4410440500 9233036127 1496081355
4905315390 2100229959 5756583705 : 727

3812619656 8314428605 7956696622 1547216956 2087001372 7768536960 8407048333
2513279311 2232507148 6302069512 : 728
4539500373 5723346807 0946564830 8920980153 4878705633 4910923660 5755405086
4111521441 4814346304 3727327104 : 729
5027768661 9531078583 2333485784 0297160925 2153260925 5893265560 0672124359
4642550659 9677177038 8445396181 : 730
6328796144 6081778927 2171836908 8801267782 0743010642 2524634807 4543004764
9288555340 9062185153 6543554741 : 731
2547615276 9772667769 7727770583 1580141218 5688011705 0283652755 4321480348
8004442979 9980621579 0456416195 : 732
7212784508 9284898064 2649742709 0579129069 2178072987 6947797511 2447305991
4060506299 4689428093 1034216416 : 733
6299356148 2813099887 0745292716 0484336308 1840412646 9637925843 0941854422
1635908457 6146078558 5624738149 : 734
3142707826 6215185541 6038702068 7698046174 7400808324 3436653823 5455510944
9498431093 4947599446 7267366535 : 735
2517662706 7721941831 9197719637 8015702169 9336750837 6005716345 4643671776
7233875886 4340564487 1566964321 : 736
0412825956 4534984138 8412890420 6820470076 1559691684 3038999348 3667935425
4921032811 3363184722 5923055543 : 737
8305820694 1675629992 0133731754 8912203723 0349072681 0685344540 3599356182
3576312837 7676406310 1312533521 : 738
2141994611 8693508331 7658785204 7112364331 2267651299 6417132521 7513553261
8676819423 3879036546 8908001827 : 739
1352835848 8844411176 1234101179 9187092365 0718485785 6221021104 0097769944
5312179502 2479578069 5065329659 : 740
4038398736 9907240797 6790408267 9400761872 9547835963 4927939045 7697366164
3405359792 2192858705 7495748169 : 741
6694062334 2726197335 1813662606 3735982575 5524965098 0726012366 8283605928
3418558480 2695841377 2558970883 : 742
7899429105 4980033111 3884603401 9391661221 8669605849 1571485733 5682861495
0001909759 1125218800 3964197621 : 743
6355937574 3718011480 5594422987 3041819680 8085647265 7135476128 3162920044
9880315402 1055305970 7666636274 : 744
9328308916 8809323592 9008178741 1985738317 1926167288 3491840242 9721290434
9655269427 2640255964 1463525914 : 745
3484006758 6769035038 2320572934 1329815935 3304444649 6829441367 3234421583
8076169483 1219333119 8190610961 : 746
4295220153 6170298575 1055943264 6146850545 2684975764 8078080092 2133581137
8197749271 7685450755 3832876887 : 747
4474591593 7311624706 0109124460 9829424841 2875202244 6259447763 8749491997
8404468292 5736096853 4549843266 : 748
5368628444 8936570411 1817793806 4416165312 2360021491 8768769467 3984075171
7630751684 9856359201 4868929431 : 749
0594020245 7969622924 5666448819 6757629434 9535326382 1716133957 5779076637
0764569570 2597388004 3841580589 : 750
4336137106 5518599876 0075492418 7211714889 2952217377 2114608115 4344982665
4798725800 5667472405 1122007383 : 751
4592715757 2771521858 9946948117 9406444663 9943237004 4291140747 2181802248
2583773601 7346685300 7449855647 : 752
1542003612 3593397312 9144585915 2288740871 9508708632 2188372882 6282288463
1843717261 9033057771 4765156414 : 753
3822306791 8473860391 4768310814 1358275755 8536435977 2165002827 7803713422
8696887873 4979509603 1108899196 : 754
1433866640 6845069742 0787700280 5093672033 8723262963 7856038653 2164323488
1555755701 8469089074 6478791224 : 755

3637555666 8678067610 5449550172 6079114293 0831285761 2544819444 4947324481
9093795369 0082063846 3167822506 : 756
4809531810 4065702543 2760438570 3505922818 9198780658 6541218429 9217273720
9551032422 5107971807 7833042609 : 757
0867942734 2895573555 9252723805 5114404380 0123904168 7716445180 2264916816
4192740110 6451622431 1017000566 : 758
9112173318 9423400547 9596846698 0429801736 2570406733 2821299621 5368488140
4102194463 4246462207 4557564396 : 759
0452985313 0714090846 0849965376 7803793201 8991408658 1466217531 9337665970
1143306086 2500982956 6917638846 : 760
0567629729 3146491149 3704624469 3519840395 3444913514 1193667933 3019366176
6365255514 9174982307 9870722808 : 761
6085962611 2660504289 2969665356 5251668888 5572112276 8027727437 0891738963
9772257564 8905334010 3885593112 : 762
5679991516 5890250164 8696142720 7005916056 1661597024 5198905183 2969278935
5503039346 8121976158 2183980483 : 763
9605625230 9146263844 7386296039 8489243861 8729850777 5928792722 0685548072
1049781765 3286210187 4767668972 : 764
4884113956 0349480376 7270363169 2100735083 4073865261 6845074824 9644859742
8134936480 3724261167 0426687083 : 765
1925040997 6153190768 5577032742 1785010006 4419841242 0739640013 9603601583
8105659284 1368457411 9102736420 : 766
2741637234 8821452410 1347716529 6031284086 5841978795 1116511529 8278146203
7913985500 6399960326 5912485253 : 767
0849369031 3130100799 9771913622 3086601109 9929142871 2493885416 1203802041
1340188887 2196934779 0449752745 : 768
4288072803 5093058287 5442075513 4816660927 8793535665 2125562013 9988249628
4787262144 3236285367 6502591450 : 769
4683776352 8258765213 9156480972 1419296755 4938437558 2600253168 5363567313
7926247587 8049445944 1834291727 : 770
5698837622 6261846365 4527434976 6241113845 1305481449 8363117897 8448973207
6719508784 1586188796 9295581973 : 771
3250699951 4026015116 7552975057 5437810242 2389579257 8656212843 2731202200
7167305740 6928686936 3930186765 : 772
9582513264 9914595026 0917069347 5194089753 5746401683 0811798846 4524736189
5605647942 6358070562 5632811892 : 773
6966302647 9535951097 1276591362 3318086692 1535788607 8127599105 3717140220
4506186075 3748663063 5059148391 : 774
6467656723 2057145168 8617079098 4695932236 7249467375 8309960704 2589220481
5507991327 5208858378 1117685214 : 775
2693347869 2189524062 2657921043 6203488529 2626798401 3953216458 7911515790
5046057971 0838983371 8640380244 : 776
1751134722 6472547010 7947939969 5355466961 9726763255 2299146549 3349966323
4185951450 3609803440 9221220671 : 777
2567698723 4279407088 5707047429 3173329188 5238967219 7135392449 2426178641
1886377909 6281448691 7869468177 : 778
5917171506 6911148002 0759432012 0619696377 9510322708 9029566085 5622254526
0261046073 6131368869 0092817210 : 779
6819861855 3780982018 4711541636 3032626569 9283424155 0236009780 4641710852
5537612728 9053350455 0613568414 : 780
3775854429 6779770146 6029438768 7225115363 8011917581 5402812081 8255606485
4107879335 9892106442 7244898618 : 781
9616294134 1800129513 0683638609 2941000831 3667337215 3008352696 2357371753
3073865333 8204842190 3081864491 : 782
8409372394 4033405244 9095545580 1640646076 1581010301 7674884750 1766190869
2946098769 2016912021 8168829104 : 783

0870709560 9514704169 2114702741 3390052253 3408348128 7035303102 3919699978
5974139085 9360543359 9697075604 : 784
4601342424 5368249609 8772581311 0247327985 6207212657 2499003468 2938868723
0489556225 3204463602 6398542252 : 785
5841646432 4271611419 8178024825 9556354490 7219226583 8636626637 5083594431
4877635156 1457107455 2801615967 : 786
7048442714 1944351832 7569840755 2677926411 2617652506 1596523545 7187956673
1709133193 5876162825 5920783080 : 787
1852068901 5150471334 0386100310 0559148178 5211038475 4542933389 1884441205
1794396997 0194112695 1195265649 : 788
1959418997 5418393234 6474242907 0271887522 3534393673 6336632003 0723274703
7407123982 5620246626 5197409019 : 789
9762452056 1985576257 6000870817 3083288344 3818310700 5451449354 5885422678
5785519153 7229237955 5494333410 : 790
1744201696 0009069641 5612732297 7702212179 5186837635 9082255128 8164700219
9234886404 3959153018 4640047143 : 791
2118636062 2527011541 1222838027 7853891109 8490201342 7410141215 5976996543
8877197485 3764311582 2983853312 : 792
3071751132 9619045590 0793806427 6695819014 8426279912 2179294798 7348901868
4716765038 2732855205 9082984529 : 793
8062592503 5212845192 5927986593 5061329619 4679625237 3972565584 1578537445
6755899803 2405492186 9628884903 : 794
3256085145 5344391660 2262577755 1291620077 2796852629 3879375304 5418108072
9285891989 7153817973 4349618723 : 795
2927614747 8501926114 5041327487 3242970583 4084711123 3374627461 7274626582
4153242710 5932250625 5302314738 : 796
7592517247 8732288149 1455915605 0363345754 2423377916 0374952502 4930223514
8196138116 2563911415 6103268449 : 797
5807250827 3431765944 0540982697 6526934457 9863479709 7431244982 7193311386
3873159636 3612186234 9726140955 : 798
6079920628 3169994200 7205481152 5353393946 0768500199 0988655386 1433495781
6500899616 4907967814 2901148387 : 799
6456821749 1407562376 7618453775 1440314754 1120676016 0726460556 8592577993
2207033733 3398916369 5043466906 : 800
9482843662 9980037414 5276277165 4762382554 6170883189 8108688068 4785370553
6480469350 9588180253 6052974079 : 801
3538676511 1950793732 8208314626 8960071075 1755206144 3378411454 9950136432
4463281933 4638905093 6545714506 : 802
9008644834 4018042836 3390513578 1572739733 3453728426 3372174065 7757710798
3051755572 1036795976 9018899584 : 803
9413019599 9573017901 2401939086 8135658553 9661941371 7944876320 7986880037
1607303220 5474235722 6689680188 : 804
2123424391 8859841689 7227765219 4032493227 3147936692 3400484897 6059037958
0946960417 5427961378 2553781223 : 805
9476461478 3292697654 5162290281 7011004378 4603875654 4151739433 9600489153
1881757665 0500951697 4024156447 : 806
7129365661 4253949368 8842305174 0012992055 6854289853 8979426699 5677702708
9146513736 8922061044 1548166215 : 807
6804219838 4767308717 8759027920 9175900695 2734566820 2651337311 1518000181
4341209626 0165862982 1076663523 : 808
3617740078 3778342370 9152644063 0540718078 4335806107 2961105550 0204151316
9637304684 9213356837 2654003075 : 809
0982908936 4612047891 1147530370 4989395283 3457824082 8173864413 2271000296
8311940203 3234564208 2647327623 : 810
3830294639 3789983758 3655455991 9340866235 0909679611 3400486702 7123176526
6637107787 2511186035 4037554487 : 811

4186935197 3365662177 2359229396 7764632515 6202348757 0113795712 0962377234
3137021203 1004965152 1119760131 : 812
7641940820 3437348512 8526029133 3491512508 3119802850 1778557107 2537314913
9215709105 1309650598 8599993156 : 813
0863655477 4035518981 6673353588 0048214665 0997414337 6118277772 3351910741
2175728415 9258087259 1315074606 : 814
0256349037 7726337391 4461377038 0213183474 4730111303 2670296917 3350477016
3210661622 7830027269 2833655840 : 815
1179141944 7808748253 3607144032 9625228577 5009808599 6090409363 1263562132
8162071453 4061042241 1208301000 : 816
8587264252 1122624801 4264751942 6184325853 3867538740 5474349107 2710049754
2811594660 1713612259 0440158991 : 817
6002298278 0179603519 4080046513 5347526987 7760952783 9984368086 9089891978
3969353217 9980139135 4425527179 : 818
1022539701 0810632143 0485113782 9149851138 1969143043 4975001899 8068164441
2123273328 3071928243 6240673319 : 819
6554692677 8511931527 7511344646 8905504248 1133614349 8460484905 1258345683
2664415284 8971397237 6040328212 : 820
6602535166 9391408204 9947320486 0216277597 9177123475 1097502403 0789357599
3771509502 1751693555 8270725339 : 821
1189233407 0223832077 5858021371 7477837877 8391015234 1320984894 2345961369
2340497998 2793041444 6316270721 : 822
4796117456 9757196812 3929191374 0982925805 5619552074 3424329598 2898980529
2333664154 1925636738 0689494201 : 823
4712413405 2507220406 1794355252 5552250087 4879008656 8314542835 1677505422
9480327478 3044056438 5815919526 : 824
6675828292 9705226127 6287110401 3480178722 4801789684 0524079243 6058274246
7443076721 6452703134 5135416764 : 825
9668901274 7868010102 9513386269 8649748212 1186290403 3769156857 6240699296
3724930972 0162870720 0189835423 : 826
6903641492 7023696193 8547372480 3298550451 1208919287 9829874467 8641291594
1753167560 2533435310 6267452545 : 827
0711418148 3239880607 2971402347 2552071349 0798398982 3552687239 5090936566
7878992383 7125789762 4875599044 : 828
3228895388 3773173489 4112275707 1410959790 0479193010 4674075041 1435381782
4646307959 8955563899 1884773781 : 829
3413470702 4674736211 2048986226 9918885174 5625173251 9341352038 1158633501
2391305444 1910073628 4475675141 : 830
6105041097 3505852762 0444891909 7890198431 5485280533 9857778443 1393388399
4310444465 6692445508 8594631408 : 831
1751220331 3906815965 9251054685 8013133838 1521764182 1043342978 8826119630
4431113887 9625874609 0226130900 : 832
8499754303 9577124323 0616906262 9194039214 3974027089 4777663702 4881554993
2245882597 9020631257 4369109463 : 833
9325280624 1642476868 4954553249 3801763937 1615636847 8598237159 0238542126
5840615367 2286071317 0267474013 : 834
1145261063 7653833903 1592194346 9817605358 3803106128 8785205154 6933639241
0884676320 0956708971 8367490578 : 835
1630851581 3816196688 2222047570 4375906143 3804072585 3862083565 1769984267
7452319582 4182683698 2701602374 : 836
1493836349 6629351576 8540613973 4274647089 9685618170 1605511048 8097155485
9118617189 6680259735 4170542398 : 837
5135560018 7203350790 6094642127 1143993196 0465274240 5088222535 9773481519
1354385712 5325854049 3946010865 : 838
7937980586 2014336607 8825219717 8090258173 7087091646 0452727977 1535099103
4073642502 0386386718 2205228796 : 839

9445838765 2947951048 6607173902 2932745542 6785669776 8659399234 1683412227
4663015062 1553205026 5534146099 : 840
5249356050 8549217565 4913483095 8906536175 6938176374 7364418337 8974229700
7035452066 6317092960 7591989627 : 841
7324230902 5239744386 1014263098 6877339138 8251868431 6501027964 9114977375
8288891345 0341148865 9486702154 : 842
9210108432 8080783428 0894172980 0898329753 6940644969 9031253998 6391958160
1468995220 8806622854 0841486427 : 843
4786281975 5466292788 1462160717 1381880180 8405720847 1586890683 6919393381
8642784545 3795671927 2397972364 : 844
6516675920 1105799566 3962598535 5127635587 6814021340 9829016296 8734298507
9247184605 6874828331 3812591619 : 845
6247615690 2875901072 7331032991 4062386460 8333378638 2579263023 9159000355
7609032477 2813388873 3917809696 : 846
6601469615 0317542267 5112599331 5529674213 3363002229 6490648093 4582008181
0618021002 2766458040 0278213336 : 847
7585730190 1137175467 2763059044 3531313190 3609248909 7246427928 4555499134
9000518029 5707082919 0525567818 : 848
8991389962 5138662319 3800536113 4622429461 0248954072 4048571232 5662888893
1722116432 9478161905 5486805494 : 849
3441034090 6807160880 2822795968 6950133643 8142682521 7047287086 3010137301
1552368614 1690837567 5747637239 : 850
7631857570 3810944339 0564564468 5241830281 4810799837 6918512127 2019350440
4180460472 1626939445 7883770901 : 851
0597469321 9720558114 0787759897 7207200968 9382249303 2368305158 6265728111
4637996983 1375179376 2321511125 : 852
2349734305 2406221052 4423435373 2905655163 4066695061 6589287821 8707756794
1760807129 7378133518 7117931650 : 853
0331555238 2248773065 3444179453 4153952024 2444970341 0120874072 1881093882
6816751204 2299404948 1794494727 : 854
3289477011 1574139441 2284555218 2842492224 0658752689 1722727806 0711675404
6973008037 0396187877 9669488255 : 855
5614674384 3925701158 2954666135 8678671897 6612973112 6720007297 1553613027
5035561678 1776544228 7442114729 : 856
8816148027 0524380681 7653573275 5786025058 4708401320 8837932816 0087690813
0049249147 3682517035 3822196190 : 857
3901499952 3495387105 9973511434 7829233949 9187936608 6923013755 9636853237
3806703591 1442432685 6151210940 : 858
4259582639 3016780171 2866923928 3231057658 8517140202 1119695706 4799814031
5056330451 4156441462 3163763809 : 859
9044028162 5691757648 9142569714 1635984393 1743327023 7812336938 0430128926
2637538266 7795034169 3343236075 : 860
0024817574 1808750388 4750949394 5489620974 0485442635 6371649959 4992098088
4294790363 6662975260 0324385635 : 861
2945844728 9445471662 0929749549 6616877414 1208821304 7702281611 6456044007
2363515811 4972973921 8966737382 : 862
6472047226 4222124201 6560150284 9713063327 9581430251 6013694825 5670147809
3579088965 7134926158 1613469018 : 863
0696508955 6310121218 4918058479 2272069187 1696316330 0448580201 0286065785
8591269974 6376617414 6393415956 : 864
9539554203 3146280265 1895110793 8074573315 7598460861 7370268786 7602943677
7805002446 7339133243 1669880354 : 865
0732323882 8184750105 1641331189 5370364884 2269027047 8052742490 6034920829
5475505400 3457160184 0725745369 : 866
3814553117 5354210726 5578356154 9987444748 0427323457 8800618731 4934156604
6352979779 4550753593 0479568720 : 867

9316724536 5472083816 8585560604 3801977030 7642460834 8987610134 5709394877
0029461757 9206195254 9255757109 : 868
0385251714 8852526567 1045349813 4198033906 4152987634 3695420256 0802776144
2191431892 1393908834 5431317696 : 869
8510184010 3844472348 9488695209 8194353190 6506555354 6173358140 4554483788
4752526253 9496658699 9205841765 : 870
2780125341 0338964698 1864243003 4146791380 6190280596 0785488801 0789705516
9462152287 7309010446 7462497979 : 871
9926271209 5168477956 8482583341 4022664772 1084336243 7593741610 5367340419
5473896419 7895425335 0363018614 : 872
0095153476 6961476255 6518738232 9246854735 6935802896 0115367917 8730355315
9378363082 2486151777 7054157757 : 873
6561759358 5120166929 4311113886 3582159667 6188303261 0416465171 4846979385
4226216871 6140012237 8213779774 : 874
1312689772 6671299202 5922017408 7700769562 8347393220 1088159356 2862819285
6357189338 4958850603 8531581797 : 875
6067947984 0878360975 9601497334 2057270460 3521790605 6476032855 6927627349
5182203236 1441125841 8242624771 : 876
2012035776 3888959743 1823282787 1314608053 5335744942 9762179678 9034568169
8895535185 0447832561 6380709476 : 877
9516990862 4710001974 8809205009 5219436323 7871976487 0339223811 5403634754
8862684595 6159755193 7654101150 : 878
1406700122 6927474393 8885899438 5973024541 4801061235 9080362745 8528849356
3251585384 3832424932 5266608758 : 879
8908318700 7091002373 7710657698 5056433928 8543376583 4259675065 3715005333
5144899082 9388773735 2051459333 : 880
0496265314 1514138612 4437935885 0709446880 4548697535 8170212908 4907873478
0681436632 3322819415 8273456713 : 881
5644317153 7967818058 1958524648 4008403290 9981943781 7181773023 1700398973
3050495387 3561162610 2399943325 : 882
9780126893 4326055847 1027876490 1070923443 8846340117 3555686590 3585244919
3701810416 2620850429 9258697435 : 883
8170981338 9404593447 1937493877 6242324098 5283276226 6604942385 1297094532
4558625210 3600829286 6497241749 : 884
1914198896 6129558076 7709795947 9530601311 9159011773 9431042090 4907942444
8868513086 8444937059 0902600612 : 885
0649425744 7103535476 5785924270 8130410618 5462198818 3009063458 8187038755
8562749115 8737542106 4667951346 : 886
4875867715 4383801852 1348281915 8124625993 3516019893 5595167968 9328522058
2479942103 4512715877 1633452229 : 887
9541883968 0448835529 7533612868 3722593539 0079201666 9413390911 6875880398
8828869216 0023732573 6158820716 : 888
3516271332 8105181876 0210485218 0675526648 6739089009 0719513805 8626735124
3122156916 3790227732 8705410842 : 889
0378415256 8328871804 6987952513 0732663402 7851905941 7338920358 5403956770
3561132935 4482585628 2876106106 : 890
9822972142 0961993509 3313121711 8789107876 6872044548 8760894101 7479864713
7882462153 9559333332 7556200943 : 891
9580434537 9197822805 9039595992 7436913793 7786649409 6404877784 1748336432
6840262829 3240626008 1908081804 : 892
3909145563 5193685606 3045089142 2896452199 8779884934 7477729132 7972660276
5840166789 0136490508 7411421268 : 893
6196986204 4126965282 9810870454 7986155954 5338021201 1556469799 7678573892
0186243599 3267776894 5406050821 : 894
8838227909 8336271671 2449002676 1178498264 3770330020 8184459000 9717235204
3319947082 4209877151 4449751017 : 895

0556430295 4282181967 0009202515 6158441742 0593365814 8134902693 1115170938
7226002645 8630561325 6057925609 : 896
2733226557 9346280805 6834439213 7368840565 0434307396 5740610177 7937014142
4615493070 7413608054 4210029560 : 897
0095663588 9778992676 3051771878 1943706761 4982175641 8659011616 0865408635
3915130392 0131680576 9034172596 : 898
4536923508 0641744656 2351523929 0504094799 5318407486 2151210561 8338545661
7665260639 3713658802 5216662235 : 899
7613220194 1701372664 9660732520 1077194793 1265282763 3024138051 6490717456
5964853748 3546691945 2358031530 : 900
1969160480 9946068149 0403781982 9732360930 0871357607 9862142542 2096419004
3679054790 4993007837 2421581954 : 901
5354183711 2936865843 0553842717 6280352791 2882112930 8351575656 5999447417
8843838156 5148434229 8587042455 : 902
9243469329 5232821803 5083337262 8379183021 6591836181 5542171574 4846577842
0134329982 5945668845 5826617197 : 903
9012180849 4803324487 8725818377 4805522268 1510113717 4536841787 0280274452
4429054745 1823467491 9564188551 : 904
2444213377 8352142386 5979925988 2032870851 0933838682 9906571994 6149062902
5742768603 8850511032 6385445404 : 905
1918495886 6538545040 5713236296 8106914681 4847869659 1668618427 5679846004
1868762298 0555629630 4595322792 : 906
3051616721 5919686758 4952363529 8935788507 7460815373 2145464298 4792310511
6763577494 9462295256 9497660359 : 907
4739624309 9534331040 4994209677 8838270027 1447849406 9037073249 1064441516
9605325656 0586778757 4174721108 : 908
2743577431 5194060757 9835636291 4332639781 2218946287 4477981198 0722564671
4664054850 1310096567 8631488009 : 909
0303749338 8753641831 6513498254 6694673316 1181233648 5439764932 5026179549
3572043054 0218297487 1251107404 : 910
0116114058 9991109306 2492312813 1163405492 6257135672 1818628932 7861388337
1802853505 6503591952 7414008695 : 911
1092616754 1476792668 0321092374 6708721360 6278332922 3864136195 9412133927
8036118276 3241060047 4097111104 : 912
8140003623 3427145144 8333464167 5466354699 7314947566 4342365949 3496845884
5515241507 5637660508 6632827424 : 913
7941360628 7604129064 4913828519 4564026431 5322585862 4043141838 6695906332
4506300039 2213192647 6259626915 : 914
1090445769 5301444054 6180378575 0303668621 2462278639 7527466678 7012100339
2984873375 0144756003 2210062235 : 915
8029343774 9550320370 1273846816 3061026570 3008722754 6296679688 0890587127
6763610662 2572235222 9739206443 : 916
0935243272 2810085997 3095132528 6306011054 9791564479 1845004618 0467624089
2892568091 2930592960 6423570210 : 917
6152464620 5023248966 5939873249 3396737695 2023991760 8984745718 4353193664
6529125848 0644801965 2016283879 : 918
5189499336 7592414856 2613699594 5307287254 5324632915 2911012876 3770605570
6095313775 2775186792 3292134955 : 919
2451330898 6796916512 9073841302 1675732386 3757582008 0363575728 0027544903
2795307990 0799442541 1087256931 : 920
8801466793 5595834676 4328688769 6661009739 5749967836 5933978463 4695994895
0610490383 6474095046 9522606385 : 921
8046758073 0699122904 7408987916 6872117147 5276447116 0440195271 8169508289
7335371485 3092893704 6384420893 : 922
2997711258 5684084660 8339934045 6890267875 1600877546 1267988015 4658565220
6121095349 0796707365 5397025761 : 923

9943137663 9960606061 1064069593 3082817187 6426043573 4253617569 4378484849
5250108266 4883951597 0049059838 : 924
0812105221 1110919433 2395113605 1446459834 2107990580 8209371646 4523127704
0231600721 3854372346 1267260997 : 925
8703856570 9199850759 5634613248 4601884098 5019428768 7902268734 5565005191
2154654406 3829253851 2763176639 : 926
2205093834 5204300773 0170299403 6261543400 1322763910 9129883278 6392041230
0445551684 0548898090 8077917463 : 927
6092439334 9126411642 4009388074 6356607262 3366958427 6458369826 8734815881
9610585718 3576746200 9650526065 : 928
9292635482 9149904576 8307210893 2458570737 0166071739 8194485028 8426039636
6074603118 4786225831 0565808708 : 929
7030556759 5861341700 7454029656 8763477417 6431051751 0367328692 4555858208
2372038601 7817394051 7513043799 : 930
4868822320 0443780431 0317092103 4261674998 0000730160 9481458637 4488778522
2730763304 9538394434 5382770608 : 931
7607635420 9844500830 6247630253 5727810327 8346176697 0544287155 3153400164
9707665719 5985041748 1990872014 : 932
9087568603 7783591994 7193433527 7294728553 7925787684 8323011018 5936580071
7291186967 6176550537 7503029303 : 933
3830706448 9128114120 2550615089 6411007623 8245744886 5518258105 8140345320
1247547232 6908754750 7078577659 : 934
7325428444 5935304499 2070014538 7489482265 5644222369 6365544194 2254413382
1222547749 7535494624 8276805333 : 935
3698328415 6138692363 4433585538 6847111143 0498248398 9918031654 5863828935
3799130535 2228334301 3795337295 : 936
4016257623 2280811384 9949187614 4141322933 7671065634 9252881452 8239506209
0223578766 8465011666 0097382753 : 937
6604054469 4165342223 9052108314 5858470355 2935221992 8272760574 8212660652
9138553034 5549744551 4703449394 : 938
8686342945 9658431024 1907859236 8022456076 3936784166 2705185551 7870290407
3557304620 6396924533 0779578224 : 939
5949710420 1880430001 8388142900 8173039450 5073427870 1312446686 0092778581
8110409115 1172937487 3627887874 : 940
9074652855 6543474888 6831064110 0510230208 7510776891 8781525622 7352515503
7953244485 7787277617 0019648537 : 941
0355516765 5209119339 3437628662 8461984402 6295252183 6785223674 7510880978
1507098978 4130862458 8152266096 : 942
3551401874 4958369269 1779904712 0726494905 7372642860 0521140358 1231076006
6995185361 2486274675 6375896225 : 943
2991164960 6687650826 1734178484 7893372950 5673900787 8617925351 4406210453
6625064046 3728815698 2323175005 : 944
9626108092 1955211150 8593029556 5496753886 2612972339 9146283584 7604862762
7027309739 2020014322 4870758233 : 945
7354915246 0856082103 2888297418 3906478869 9232736913 6004883743 6615223517
0584377055 4521081551 3361262142 : 946
9118156153 0175888257 3594892507 1088792621 2864139244 3309383797 3338678061
3179523731 5266773820 8580247014 : 947
3352700924 3803266951 7421195076 7088432634 6442749127 5589077468 6358216216
6042741315 1702124585 8605623363 : 948
1493164646 9139465624 9747174195 8354218607 7487110573 3845843368 9939645913
7406033821 5935224359 4751626239 : 949
1886853078 2282176398 3237306180 2042465604 7752794310 4796189724 2995330297
9249748168 4052893791 0449470045 : 950
9086499187 2727345413 5081019838 8186467360 9392571930 5119686456 0185578245
0218231065 8894379865 2243205067 : 951

7379966196 9554724405 8592241795 3006820451 7953700434 7245176289 3566770508
4902131077 3662575169 7335527462 : 952
3029430312 0359626095 3423574397 2496592110 1065781782 6108745318 8748031874
3082357369 9195156340 9571627009 : 953
9244492974 9105489851 5196586647 4014822510 6335367949 7371425102 2934188258
5117371994 4991150975 8374613010 : 954
5505064197 7215319293 5487537119 1630262030 3285886585 2848019350 9225875775
5974252765 8401172134 2323648084 : 955
0271433563 6754204637 5182552524 9443296570 4386138786 5901965738 8028684018
9408767281 6714137033 6617326501 : 956
2057865391 5780703088 7142615190 7500149257 6112927675 1930967284 5397116021
3606303090 5422439663 2067432358 : 957
2797889332 3244057791 9927848463 3339777737 6559018705 7480682867 8347965624
1461028995 0848739969 2970750432 : 958
7530299728 7229732793 4442988646 4127253481 6060377970 7298299173 0292963086
9580199631 2413304939 3504933254 : 959
1235507105 4461182591 1411164545 3471032988 1047844067 7801380771 3146540009
9386306481 2666143308 5820681139 : 960
5838319169 5455582594 2689576984 1428893743 4670841079 4631893253 9106963955
7807060212 4597489829 3564613560 : 961
7889834724 1997947856 4362042094 6134123876 1319886535 2358312996 8622689486
0840845665 5606876954 5012744866 : 962
3140505473 5351746873 0098063227 8046891224 6821460806 7276277084 0240226615
5485024008 9528916571 1761743902 : 963
0337584877 8429112896 2324705919 1874691042 0058483261 4067733375 1027195653
9946971625 1724831223 0633919328 : 964
7079838007 4848572651 6123434933 2733566644 7335855643 0235280883 9243482787
6088616494 3289399166 3992104883 : 965
0784777704 8045728491 4563033532 6507002958 8906265915 4985094079 7276756712
9795010098 2294762289 6189159144 : 966
1520032283 8787734851 3097908101 9129267227 1037788980 5396415636 2364169154
9857684083 9846886168 4375407065 : 967
1210390625 0612810766 3799047908 8796747780 6973847317 0475253442 1563903872
0123880632 3688037017 9493089549 : 968
0077633152 3063548374 2568166533 6160664198 0030188287 1237674818 9833024683
6371488309 2592833759 0227894258 : 969
8060087286 0388591688 4973069394 8020511221 7663591382 5152427867 0094406942
3551202015 6837777885 1824670025 : 970
6517085092 4962374772 6813694284 3500629388 1442998790 5301056217 3754591826
7997321773 5029368928 0652100253 : 971
9626880749 8092643458 0116557158 8670044350 3976505323 4782873273 6884086354
0002740676 7838219635 2222653929 : 972
0939807367 3913640828 9872201777 6747168118 1958561337 2158311905 4682936083
2369761134 5028175783 0202934845 : 973
9829250008 9568263027 1263295866 2921476531 4223335179 3093387951 3570953463
7718368409 2444422096 3193312956 : 974
2030557551 7340067973 7406141621 0792363342 3805646850 0920371671 5264255637
1853889571 4164197723 8742261059 : 975
6667396997 1731681694 1543509528 3193556417 7056686222 1521799115 1355639707
1433128936 5755384464 8326201206 : 976
4243380169 5586269856 1022460646 0693307938 4785881436 7407000599 7697036490
1927332882 6135329363 1124036506 : 977
9865216063 8987250267 2380874033 9674439783 0258296894 2568967418 6433613497
9475245526 2914265228 4241924308 : 978
3388103580 0537870239 9954217211 3686550275 3413622116 9314069466 9513186928
1025747959 8560514500 5021715913 : 979

3177516099 5786555198 1886193211 2821107094 4228724044 2481153406 0558959583
5581523201 2184605820 5635926993 : 980
0347885113 2068626627 5887714460 3599665610 8430725696 5005630644 8918759946
6596772847 1715395736 1210818084 : 981
1547273142 6617489331 3417463266 2354222072 6001460127 0120693463 9520564445
5432916629 8666078308 9068118790 : 982
0908152950 6362678207 5614388815 7813511346 9536630387 8412092346 9428687308
3932043233 3872775496 8052103028 : 983
2154432472 3388845215 3437272501 2858974769 1460808314 4041258681 8154004918
7772287869 8018534545 3700652665 : 984
5649170915 4295227567 0922221747 4112062720 6566229898 0603289167 2068743654
9482461086 9736722554 7404812889 : 985
2424718543 2360575341 1672850757 5520571311 5669795458 4887398742 2281358879
8584078313 5060548290 5514827852 : 986
9489112190 5383195624 2287194847 5940785939 8047901094 1940706717 6443903273
0712135887 3850499936 3883820550 : 987
1683402777 4960702768 4488028191 2220636888 6368110435 6952930065 2195528261
5269912716 3727738841 8993287130 : 988
5634646882 2739828876 3198645709 8363089177 8648708667 6185485680 0476725526
7541474285 1028145807 4031529921 : 989
9781455775 6843681110 1853174981 6701642664 7884090262 6828244482 5802753209
4549915104 5185177165 4631180490 : 990
4567985713 2575281179 1365627815 8111288816 5622858760 3087597496 3849435275
6766121689 5926148503 0785362045 : 991
2745077529 5063101248 0341804584 0594329260 7985443562 0093708091 8215239203
7179067812 1992280496 0697382387 : 992
4331262673 0306795943 9609549571 8957721791 5597300588 6936468455 7667609245
0906088202 2122357192 5453671519 : 993
1834872587 4239194108 9044411595 9932760044 5065562064 6116465566 5487594247
3692523369 5599303035 5095817626 : 994
1762318495 6190649483 9673002037 7638743693 4399982943 0209147073 6189479326
9276244518 6560239559 0537051289 : 995
7816345542 3320114975 9948962784 2432748378 8032701418 6769526211 8097500640
5149755889 6502930048 6760520801 : 996
0491537885 4139094245 3169171998 7628941277 2211294645 6829486028 1493181560
2496778879 4981377721 6229359437 : 997
8110044480 6079767242 9276249510 7841534464 2915084276 4520002042 7694706980
4177583220 9097020291 6573472515 : 998
8290463091 0359037842 9775726517 2087724474 0952267166 3060054697 1638794317
1196873484 6887381866 5675127929 : 999
8575016363 4113146275 3049901913 5646823804 3299706957 7015078933 7728658035
7127909137 6742080565 5493624646 : 1000
4126002437 9684543777 3390264725 1281941632 0076848736 2517640659 6754069362
1758879307 8559164787 7727473927 : 1001
2002910342 9495624476 6130820072 9250734529 1707642266 2104767303 7863169954
2374551174 5652202278 3324096803 : 1002
5246676631 9086101120 6745856287 3174135111 6229207886 5132941244 8154716281
8207987716 8346341322 3622341177 : 1003
8823102765 9825109358 8923591620 5510876329 8087993165 1725289380 0123781743
4896832151 5905624933 4737020683 : 1004
2232100118 6373957705 6747386710 2173212375 2243252416 2635803437 6253606808
6691635715 9455152781 7803921774 : 1005
3228234366 3377281118 6390511893 0759016666 5074295275 8384008544 6354193171
9053136365 9724905158 4091065822 : 1006
0181473479 9022359067 1381469051 1605192230 1269482316 1134174399 4471483304
0862484269 1395023367 1341242512 : 1007

3864026657 2581309439 6762193965 5407386524 2298978797 8219863791 8299709557
9247473203 0323911641 0445906907 : 1008
9778623155 1834959303 5305923789 8175158914 5765040802 5109479123 4217584828
4188195013 8546165680 3017550355 : 1009
8005494489 4884871351 6053755934 0234574897 9516602442 3383214060 3009593710
5588457052 5157042662 8460035440 : 1010
2823678768 5509826781 6176552037 5795655481 6778960389 2749835560 8791541177
7494235734 0076416109 3294003899 : 1011
9821992672 5708695732 6068774974 2248020233 0752518765 0255968420 7606932299
8858757989 8896460744 3817881700 : 1012
8154889522 6516722834 0452772191 0699141576 4639485231 1267947308 6580319507
6455197675 6289574288 8179681209 : 1013
0026387145 2578583152 7761510908 8631740243 6956805678 7301523542 7804793414
2664952238 3370711751 1265375503 : 1014
9423720987 8466804913 9473446530 7140796225 9728713050 3077258714 8755705025
8257346686 6613802351 4260561161 : 1015
9740554343 6548698005 4448792959 7028759035 2258409782 6835986664 4658604569
4241390729 0952662499 3290297344 : 1016
0568160683 8057266260 5727708840 7073471496 0600645614 5407073443 2782514087
4742755067 2230484535 7006092214 : 1017
3900029929 8160821171 7047917614 5051910081 3267037521 4930740567 8533111060
5835291278 1007391749 9491978451 : 1018
1291591368 1107394055 1752080196 3053935074 0248509553 7725003670 5466516233
0430425087 4423242624 0463211507 : 1019
8997336929 9854070416 5626104197 6700202415 0948924118 5609240963 7604429612
0023645907 0644977062 7207919019 : 1020
2359648070 4892363697 9860198283 0872842285 6475235316 2882791324 2955248144
4750552190 9672046080 6895451817 : 1021
1220493032 1853740627 2474215197 4030576904 3602686360 7807920047 7623242955
1829473522 0272443763 3902772139 : 1022
2087767065 7162416397 5178585925 4426923428 5352743288 5633685078 9651962072
5194165560 6187037055 0218462845 : 1023
4342578503 8300009537 4518292958 4404649188 3868579348 3961151297 1605816657
4509670367 7495836666 6931218817 : 1024
6367964494 3617130416 0372430506 5848513174 9264055855 1940180051 8090847521
1868224616 9761492432 3831948643 : 1025
4415908558 0110730703 1120150224 3416073157 9295287529 3683582039 7003389112
1141706852 1936658978 9459503154 : 1026
3895890153 0382714300 1929589074 1499435928 9408309707 7078362875 9144840370
4503861896 6975811201 8523192318 : 1027
6865996803 8583812370 3291562075 7883594878 0941688205 5316051281 9015264759
2807574958 1545642213 4145937816 : 1028
7056992868 2998956119 8235383715 7880480478 7045841753 9466549769 0173220310
8900703033 6291176730 8448450372 : 1029
1456696444 0146954517 3857434157 8101586187 8383927855 2609399130 5702555755
5906094705 1498093487 7733200727 : 1030
9757303824 5989466809 6808222213 4848587382 2999281794 0908256652 0958165547
2475244566 7436975944 7468637633 : 1031
2428904269 7761067919 3391098330 0422310293 7282987989 0320939109 2682836306
1736101738 7812367989 8645149311 : 1032
7024371282 8588263048 6298884492 2074156406 0714705913 7405524665 7569718702
1735528724 5439427714 8091793644 : 1033
3765063786 1861324348 6357974112 5852086345 9927803688 7924983543 6329845768
7650165065 1153450086 9572123950 : 1034
7544785683 1736315571 5352704624 4235259737 5134088254 6160966144 0746675514
2268360319 5980107215 2463551069 : 1035

1718713357 3168548563 1280857834 4356236709 5965094994 6968820661 1851180860
3420282133 1801249410 9915026014 : 1036
3545001743 2730793625 1130702982 5049941799 4284451146 4793291545 9955590958
7807621636 6685917910 6543596606 : 1037
5253525320 2736507259 8912125568 6842802077 2464877220 1099663182 9559552903
3933122843 6486447597 3560859840 : 1038
7609472983 8954243393 2623153239 9189818522 6418083129 6333546356 8748288634
6561850481 0632288805 5967378445 : 1039
6200094146 5603499280 8794051153 1005758712 9552571964 1115068503 4077371060
4380371259 5755969859 4936205847 : 1040
7512026354 9473475347 4818926225 4190352671 6144292848 9985753674 0692165271
6300860606 5437373682 3556588626 : 1041
4863436891 5321809557 2204456777 1373683104 5807558452 9612832832 6063196297
2852796667 4362974800 8213186279 : 1042
2186904428 4342630735 7607039996 6943078950 8147269730 2538173756 9492275179
5354326156 9120405948 3286094999 : 1043
2366412287 8812264191 4850485632 8072066418 5570595203 7503032291 6894489427
5783060909 1085241060 1400683274 : 1044
2055839697 7382315073 4996108758 7637042555 6496408685 5071942256 3449667324
3065625925 0474581762 7332818160 : 1045
1701969816 6542426378 7636014530 3594653845 0325476674 9997373408 3566513818
6025156520 2836373891 7101654541 : 1046
4882674448 0091057041 8616262683 7971120886 1413572796 1109908829 2970229692
1281809787 9895139150 4270936786 : 1047
4449831964 2013456683 3908775943 0064424856 2301212461 4511697921 9396344095
0808322928 1294270436 5991464827 : 1048
4998437594 2113020418 2973084171 7881309037 9558545603 2471708191 9530277146
5794555475 5447542844 3440813938 : 1049
8908609776 0178573893 0751866190 6505018077 1650018407 4432585402 4184360501
1182429907 0232341724 3674525365 : 1050
3495947990 6333454075 4371812699 3998337192 1848541873 5979845348 9345922685
1506818266 2490078029 3350126588 : 1051
2497422624 1885352526 6367028276 6249934982 9488748331 0617642084 2901692305
2899608978 6041300651 0902817980 : 1052
5040587107 6711790411 3021748279 6682353001 9602202531 8557678984 3317586806
3783599687 9160153892 2220236575 : 1053
7655815866 1140919939 4861599209 1599175533 4178303334 7643131635 0127053906
9707932656 7812415906 4342847213 : 1054
6023521823 6741214733 1244999443 3415591527 4315931687 4778825331 5509277033
6202901222 5977948098 5539220006 : 1055
4527162280 8553982789 0658423344 7552821276 5176505726 6326769114 1075034845
8718969964 3487577513 8479148183 : 1056
6351006214 6681858509 6348887081 4569767220 2016799119 9462417776 6889079171
3686594596 0726468538 8107787830 : 1057
0216136827 6697026223 4594187374 7673353799 8884403427 0468030425 5169412715
8739320398 4443746045 4781611305 : 1058
6625176412 7598211819 3966110185 0562880555 9425660600 3231211618 0994622129
3010024709 1334715068 2268430458 : 1059
6803009042 4286168202 5562140946 0879000651 9109949557 0815816505 8289833407
3946608445 7565780636 6902728434 : 1060
6201858732 8252924796 5052866814 0850353851 9837523637 4519256227 9549029055
7907030283 9501048548 3592983454 : 1061
2814487304 3580470533 1508151050 3001521428 1171753936 4913316617 2621235405
5278633080 0208317705 5630294963 : 1062
5942016543 3309409417 7196326234 1193871051 6157010179 8053551679 3708602913
6675698609 7124120368 5838129576 : 1063

9530779814 1365700174 7613569669 8614606849 1439699573 8376316958 2460251334
2108072621 7136019430 1808720988 : 1064
8551415024 1638183259 7525959316 5531865833 1171268579 4152720661 2218422661
4118251546 5748487831 2610347834 : 1065
5467492583 0872998544 7421206445 0952332450 5087743149 6166555251 7971680209
9172002640 9374921907 5699368963 : 1066
3028139164 7208963581 7717355558 4859270652 4504862516 4195405508 0134351032
3389813378 3024977018 2275490638 : 1067
1499964723 3340796130 4146973947 6372650869 2733471084 1568560843 0921316240
4346298639 2084166005 5904598506 : 1068
4912435052 6476606760 0344441618 1864036700 8377411410 1094320588 9555986586
7007786367 1896944089 6223213740 : 1069
3411359719 9133135946 5536854466 9236765258 9012108413 7774324821 9181274784
7892287264 8929700323 7187345615 : 1070
7981599834 8391004126 0105074696 4599430331 9788106349 1392381249 0503061433
4079183280 0406390709 8672596197 : 1071
0983112659 6014747372 5330526853 7177421465 5400587392 4623727617 3649051987
1336806772 3952570781 3606866832 : 1072
6139501432 9509474851 5947246675 2720168431 6586608807 5127685847 5554118438
1169011622 0055521134 8448896066 : 1073
8259227431 3190079630 1158708467 0117654935 3930465633 5622531124 4727796669
0058311906 1610197266 3073970542 : 1074
5314398184 5737944948 6780134618 2178759390 7699960202 9083965677 2878469057
3640156401 5047696448 9939475414 : 1075
7460833991 8696889271 1569423454 9265124664 5507792554 0281050376 2203596753
0558601856 4920560628 7909076945 : 1076
3339208808 8494778288 9485112215 4743230191 3832455629 9388102061 4490266876
0102077532 1091568497 7830740859 : 1077
6498579671 5261701003 9475494539 9176987913 2354655010 6407355816 9994097562
4814996744 3278429202 7626441897 : 1078
9391815839 4562708173 3015821602 2551965989 8769376164 0198612074 6675504886
1110855726 7645070526 2244613022 : 1079
2335852072 2736204850 5728923881 5884938754 5352291863 9971438088 4061757286
2209501225 0651586310 4258884134 : 1080
3554319737 2985621775 3072022629 4755524830 4444534043 4888878581 1703413453
4252235431 9407877972 8467601815 : 1081
8322709774 5180929342 1931898158 1248283265 8950040704 8552060998 9378390034
1914163044 6391638805 4965878650 : 1082
1375046341 6956551566 1829887863 0705842306 9676602540 5302481147 1007899784
2118304890 1046405689 6539702885 : 1083
5955309255 5863605215 8957375114 0895649058 4415677493 7105859648 0143158746
1449125054 9253191164 6538215851 : 1084
9737009328 0194530320 5726284526 5804604633 7816631429 9330766466 4653076059
0548962888 7241897160 6022588261 : 1085
7577539922 0551315093 7720062486 3085562820 4935757527 2499556708 9221634233
9836025653 2873102919 4007041176 : 1086
9192208500 1511673567 0101958971 0017970195 7812089291 0969417754 3699043682
0256302405 4822625401 9056965077 : 1087
1058157424 0721496339 5603652702 8333440730 5750073674 5622605846 4988611510
1689612181 1190584717 1446106871 : 1088
9761017456 5873737967 4069713742 3238753839 0303172002 0020720592 8488785123
9117464716 7374373792 3283881966 : 1089
2016876221 9134623389 3762599527 0256721386 2211245898 0212130501 4072889043
0032253550 4095866818 7241393699 : 1090
3819306914 8744717186 6461831119 4260316166 4070377316 4870018647 9960024304
4003242241 8094022785 3330901150 : 1091

9880870678 2688353172 0076752255 3138008818 7804316901 9007280483 1799287414
1254761230 8960683309 5828377667 : 1092
6882875786 8868309297 6001011974 5338983319 5258861963 0132917094 3858166153
7417179449 6319177154 3125069598 : 1093
5348128568 4619377669 8942774591 7091880252 0012749905 5594072896 9659479333
1672243621 5678967769 6670803522 : 1094
9039018485 7308062756 7086765862 7104769409 2035655930 2535274341 8965927002
2270492331 8682999156 0936413757 : 1095
0049885373 0459639615 2734629396 9749517480 6269645179 3018719986 7885375814
1597579931 4806608557 2325683743 : 1096
0528276417 5670050288 0404894298 9958094810 3534833934 1449278859 2526219241
5547231997 1433850866 3732092663 : 1097
2728243514 9336407045 8968385234 5624744361 1752567669 8776759722 3439206357
5074715529 1810276261 4012992480 : 1098
4228839902 9787992541 8517499129 6302839907 2963558857 9890593317 7959087690
7390564602 5623533567 2215522594 : 1099
6883829845 2882922966 2751371624 2217295467 8670715840 9241840841 4755758253
9385240963 3020513497 0474069539 : 1100
9567897981 7278609204 6228683973 5779815111 8681526598 8460694975 8965481314
6511503926 2637774951 3761557248 : 1101
1951161198 7725034456 4710738513 4359273555 3871246237 5598193813 2142384415
8192907004 6389771683 8872079163 : 1102
6174143249 7079109658 1627464297 1707287172 5142745898 3568970955 3462682016
9085356108 9448984071 0058192030 : 1103
2176945120 7717745887 9551951047 3384184739 9807963067 6788584516 7575729904
3069715426 4238349800 9870869933 : 1104
6709121083 9445350624 5922432312 3482785496 6037465718 8014892937 9451478705
4060792457 5900601219 6221239287 : 1105
2001721558 8666345734 9714095337 2115165598 5757941724 4198890261 6701610161
1557834315 0254603287 8119842402 : 1106
7484608510 7224066767 7876085524 7617773833 0895026100 6438835055 0205456324
3461678594 5194179566 9874968515 : 1107
2448838475 1361818066 7108316165 5642093692 7052061189 8517292617 1417144346
5550870630 6063551012 9494003097 : 1108
5916779915 8426049197 1209543227 0267843265 4296572403 2720887143 2199964531
3202587109 6771651285 4966996255 : 1109
2698607311 7637182074 9882739977 0601991362 0930832307 3683820645 5732563765
9829125781 3149222422 0427971241 : 1110
4416299512 6594563979 2759380383 8047826231 6042432539 9132851123 0322470375
6194232173 3047854078 5762440132 : 1111
9171799297 9240783390 7157579814 2681686465 5382946847 3992058886 3165593491
9867896962 8404473449 6802407709 : 1112
2831376408 1033522552 4271740410 7673565424 4410044833 4744010172 6441052954
7872963458 9864050120 3608024451 : 1113
1903509949 7449397361 7181575277 0937802092 3666813584 1636268319 2634067141
8279742134 2546220705 4156000509 : 1114
5967404561 6840451771 7479527903 5325493258 9120483385 7465900967 8173041600
0521088934 6107467540 0424197780 : 1115
3082885181 2001733695 5912713771 4195011361 3044097532 7919050489 1583246399
1434835316 4868154857 9178632935 : 1116
1239255525 1021118278 8573696060 2769313014 6966143344 9642302114 3824837056
3353279385 8895267672 0766889712 : 1117
7443581563 2088106650 1495681435 5879657690 9857765902 7687074536 5927636497
5553449617 3080781609 8710324801 : 1118
3795136170 3677634575 9497568620 8013996374 5517624251 4778062872 2265971455
4829067692 9571364357 2152674468 : 1119

9878894188 2075129222 5756509143 5528288746 1419509786 2427527881 5715664007
6372103780 3194043095 8442725492 : 1120
6998716923 4331890022 1415031139 9876526068 8761566740 2101972017 1960239086
1082974927 6395695411 5303227546 : 1121
0173870795 6259935797 8530244347 6716399591 4623179312 3998998692 8437975702
4923695515 8729768385 4005227651 : 1122
4956144471 0597196288 9888157109 4151717015 1811474351 3643854005 1162462021
3117480079 1983749700 1004713634 : 1123
3252328157 8911355450 4533719052 7506822915 6185003328 4695679262 2620819044
2473340362 5038927920 7158596003 : 1124
9363153368 8427243753 6679969864 7934741133 1983286194 4146065392 2784099903
1438403545 6504705678 9552024827 : 1125
1760118743 3564369024 3503085631 3095590552 5039049273 1613311734 9225846446
0902453507 9190184411 2993216997 : 1126
7045183285 3586480428 5568222087 3721361649 0586303256 3689130841 0376021567
9927020005 3223554398 0465311933 : 1127
9775459044 0450785680 2139846500 9693429547 3102692499 4758646605 8091669984
1606846460 8729394380 8274308285 : 1128
8174796941 7287299031 1013192675 5738979840 9136425347 9694943480 3777033646
3495847686 2982590103 4707278612 : 1129
1862300198 6607987782 6842459338 3563891957 0206853521 6032116352 3006498874
4600200170 4130569853 6515466875 : 1130
2023859375 1832803728 5114327481 1699683692 8492204473 8057063349 6618711240
9478359158 6962685864 3589141359 : 1131
8542535776 8877493274 3634514754 4886408688 1803036965 2431755688 3002058607
7325695971 6086485415 8344684324 : 1132
8996307701 1371344675 1569302448 8548207712 4133557732 3069494580 6726784523
5943631507 8727281579 0157307003 : 1133
3178796854 4362795257 1902362327 4614262868 7327380094 9774112285 6237663214
9046532940 7202619753 9071740422 : 1134
2595392428 8816455979 6570030957 1413891069 3684503626 8231053986 7437532400
5270153474 5893325679 5149418545 : 1135
3780882706 3457295962 1690853835 3537038141 8115573816 3782090325 6151986974
5357646412 1254980760 0515614170 : 1136
7298046994 8135934831 5056811664 2793219335 2798227147 1576734018 6088721518
7996693502 5270075755 6099719882 : 1137
8630642854 4812827513 9280694702 7501481632 8972731434 7348528529 5046048832
7167397898 1563678804 7804436021 : 1138
0900732072 7369749344 6304997314 4257156043 3133690387 6181009488 7312071348
2710815889 8574832658 5420751007 : 1139
7953118326 8617080370 7093592761 4936782530 8583404823 5100363216 6378957426
2025503501 1686154340 7379504516 : 1140
4828967556 9835893552 2020173679 5480757819 0950269798 1271148703 4311903631
1224612829 5303820512 8704309294 : 1141
7197459469 0821025634 7889954317 7152437969 6211281224 5034260663 9926885213
3079196370 2777804488 5792057304 : 1142
6990800923 4401866381 1325209712 3096476059 9899479257 5985100817 3039606822
2199753273 0160658262 8527582576 : 1143
6950785472 6034938298 1335825281 7867060851 2656002268 8717811253 5978293373
4779141273 6284188656 1759208328 : 1144
7944741096 9703879854 7369840254 5806329483 5022359393 5435874802 2398976091
6296250110 4739311694 4910066690 : 1145
7230634693 1301697118 2063253526 9244043840 0937242844 2820970936 4856909468
9200873717 5325255703 0543539828 : 1146
7278123011 3980809386 7015474885 8034456318 7131960267 8548793893 3162050076
7526411204 4390237583 3427242986 : 1147

9965478636 8534102848 8573702547 2550236566 3418680919 0383886707 8790720840
3619402164 6701215348 3797815183 : 1148
2826472578 6288152071 0108149955 8980338118 9615694417 5676134071 7046538512
1709021237 7788433364 9651872119 : 1149
9054075818 7739439752 8364143953 0442459139 0317881300 4188791887 1145531482
6746998705 5587931040 2403888840 : 1150
8385068734 1625071657 2741851349 5208496367 0955542450 4394839480 4597915622
8282483787 9341527203 6226336956 : 1151
1805556371 0768148888 9361927574 2659935823 5594315308 8793305276 7558747512
3650658439 6947560429 7192002319 : 1152
8680243517 1993786810 0361102312 5683642560 7959741057 4153628297 1800464977
4857371837 8639037039 0153973749 : 1153
1165468549 9716453941 6112164176 1071714540 1765190565 0525206622 7788312904
5719693205 9902413753 9598386198 : 1154
2603205495 8395016755 5250964413 7118222561 4960140030 2303540789 9209698677
5078672000 3807426797 0530307167 : 1155
9322960156 4862280851 8403352350 1706085895 1291222324 6117830253 1636289439
4607365277 1336511631 6464461990 : 1156
9902122492 2412315168 9927678558 6373631552 6002503488 4878132330 0191018939
9616702731 4169996265 1194574263 : 1157
6761965002 4347371727 2902846220 9798394871 0659822700 0995491887 7696188505
4326532118 0221944428 2228425152 : 1158
5561411874 3401804194 6141394514 7128725275 9239125596 4437356833 9728963312
6767823491 0356332961 2947191015 : 1159
1571431157 9549093390 3261411918 6547523762 4721531102 0793691158 4874220582
2747343201 7355850771 2243796985 : 1160
7965491580 6279502740 9771688611 4807616315 1618553068 5669245717 1769220443
6684331273 9893379411 1629722451 : 1161
6999854685 6221570241 7594711769 9529165502 1168550010 8985761934 6394559088
2627077531 1465775223 8846343519 : 1162
3765397349 8480245497 6076024403 0808448901 0683878697 2612370978 3578245166
8011714859 8367940552 9046198262 : 1163
1656691720 2742628548 2393396001 8254599409 2543081696 9103297841 1234022885
6001905493 4275022318 5294712829 : 1164
6096939768 1373419770 4278121300 1473286776 0571940596 9979275512 4617184349
5698564171 2872481183 4654206423 : 1165
1871455182 4152867630 5675131162 6771773506 1751124546 3387994265 2912701057
8995671805 7214365579 1835069177 : 1166
7930704075 7329043974 9499582241 0623810514 9176502385 0418273009 6620171750
9405908054 0895728375 5406355152 : 1167
2199658207 5735131570 7592361539 8639459211 1558640009 8809755261 0538382568
9927215847 8504174606 5161511337 : 1168
8833609760 1211484870 0556016581 2492470682 5684427204 5472896309 4203066504
4529864622 3594226008 5549915891 : 1169
4995360649 8428034579 4927570094 9795945060 2378775019 4706246323 9495495782
3082283066 8408188025 2107663907 : 1170
4230973720 9162853371 7680621644 6935432317 9178553058 3317142084 7988630340
8465726426 9395570026 8576057539 : 1171
3478885870 9460058272 3230519108 1175142349 1268733658 5960799891 7329289158
9600181509 1816337400 8060354752 : 1172
0005151175 1029012299 2487096154 5928026206 0761698272 1810291673 1554892942
3740851967 4330791660 7849905578 : 1173
2101935713 6624359908 8361385980 8516156417 4769460547 8554008195 3530670803
0896976304 5294686823 3210532878 : 1174
2374389441 1568517627 1711636309 4014799096 4945635459 2950130739 0036268210
0732637008 2356150691 2696431833 : 1175

5171625439 0304698989 3142615442 6359511363 4660573786 5495124457 4752621678
9547036289 0483048499 6804037722 : 1176
5134319373 7344123661 8586944588 0640185840 7314763379 2940386340 4359194198
7235526301 5654608051 8686760680 : 1177
4316084512 8459160424 4132698791 2538560299 1599672787 6619519505 3176488313
4693257366 8946443825 5813910848 : 1178
6209663742 6745798313 0122234387 2583124422 0330945714 5754147047 9293875858
2389977385 1521352372 3895596643 : 1179
1223564326 2628601147 4890868171 5928106687 2708400820 3377186921 5352352692
6347226809 0825989889 8400262081 : 1180
5217828261 1229313118 2086600709 9686036540 9818326807 5582477670 6950410997
5861436243 5521619453 5302920025 : 1181
4667367996 4850433731 3349520821 0751199258 9266389956 4756985870 7901856123
7915788643 7446903787 1509500112 : 1182
5502100388 4531192365 2965599461 9004748466 2064234794 2329670060 5290037091
7557818870 8193522146 8714272352 : 1183
7763255989 8086948721 1138459800 1412384216 3827824412 7365424467 4883338167
9716201128 8619141540 1936712909 : 1184
4789902646 6644315609 8372961501 9686242282 5067230616 6720943546 5714251493
0864248877 8598682759 5887490650 : 1185
7726025095 1829536765 1811823686 1694472436 0783764294 7624692263 1949892196
4644068316 9287661615 0605081384 : 1186
6319415116 2025779078 6307180123 1159458603 8965625265 5422334623 4454507394
7886902681 5949751311 6885143694 : 1187
5210216883 1904461686 2976332522 9863851818 8500492869 3572764766 8238555646
3655449640 0631764828 5575785866 : 1188
6102285515 6485990882 0958689444 3625469867 9523822686 1159699100 5636608292
6791533753 8160661122 4786953132 : 1189
6158531871 7638859893 7792918890 2998793879 8100036973 0784895927 0625410484
8593158543 2339568310 4239029907 : 1190
0263443797 8756918554 3408976440 7601308444 8197862650 7947644083 0134942435
8342818859 1525929347 1436317533 : 1191
7495897010 7287350127 0788980481 6350456766 6769320755 3051840432 4461007403
2167647183 6083708475 0651269307 : 1192
0766084982 5299000317 8503058536 8213951273 5038638246 0564251033 7775580986
4643398017 1862081426 6307417259 : 1193
2226000511 0913426810 7467012901 4301654101 0649332122 8379082751 5001003530
0156545975 0832377296 5439697382 : 1194
0477416265 7106574082 1649960626 2274961879 5334790706 5988974871 7795643340
6484174564 5747906925 1701494998 : 1195
1009535341 3548908754 8363275795 2240720698 6291024671 7035792514 4176670388
6609906985 7262605812 4082533622 : 1196
5218992000 4189757457 6531512300 0064445715 9317017716 8863548333 3051921582
0559461173 5771632113 2233931965 : 1197
3203861990 0511617817 1334001070 5766526899 1970816920 2219464704 3237953564
1186606392 0558609034 4570641517 : 1198
9778214505 4722278852 9871210197 8 5884607004 7420028468 8737958442 2894997433
3656271877 9917211379 1616449254 : 1199
1329715652 8795295326 3975953853 5920950138 6333805075 6136953089 9547584883
0242619627 5898594151 3780515805 : 1200
0257675404 0178579585 2448831172 1050892770 8922727343 1973823884 6873071682
3024878868 8585510108 0735227814 : 1201
0537140652 0758107270 8481672639 7709873145 5162646911 4232861030 3693298433
0300323676 1627142640 6758780673 : 1202
1883971515 0027981633 7477907877 5038307986 7594045910 7392103458 7404219617
0349258081 8990720596 1291586420 : 1203

2028857340 0911495523 8865107911 3714953346 3976398818 3948804530 0750747403
7228093682 0535430494 9519483328 : 1204
3347007516 1979008687 2854399629 8157560589 1637624723 0691628711 1113767608
6480323752 4596649304 1175394613 : 1205
6464337804 6711650555 0467067183 6221285795 0480671656 3042762671 1429999113
4876984470 5037063790 0181096888 : 1206
6297217579 5173243380 2780617470 4963020424 9291661917 1886243355 5992820932
4391944571 1886321556 3201616542 : 1207
4705537593 8696624656 3341215410 1403228699 0930159132 8858088312 4124288287
6373872742 8380385907 1029274863 : 1208
3351503090 4453280525 9779565892 0554562434 2979827941 3489175638 2400771612
1733247364 2854016061 0044337641 : 1209
4572207859 2171559140 1037832020 1321338330 9638077890 4095723810 5588293927
9637438166 0686835195 0592770195 : 1210
1536160172 2158904287 8567848206 8291944169 8718192862 7308270444 1630396254
7130532843 8833791337 4768735826 : 1211
1221162583 6027289616 2455904189 6770247453 8275839665 2299371235 1630489833
0124214174 5578859159 4256059792 : 1212
4277218199 0855627984 8605617453 6844789237 9690797559 4555154646 8531630244
6232567403 4895845462 2567448582 : 1213
0204245739 1994253094 2642245042 0268903815 0152683602 4125598075 9752364816
2809304891 2746151196 2315461140 : 1214
0822056396 7806585354 0766868822 7542650381 2259991620 7601708955 6747446524
2344520176 6165032594 5665912966 : 1215
7863246213 7991922296 1458671422 4824928806 4768032108 6477994100 4100600339
0679275237 3625460277 4296007347 : 1216
8803835668 7522003482 4576949084 5686269605 7715701919 1748922606 3520812973
8797443835 4832861369 3956245039 : 1217
2976805783 2234021716 7655591776 6840375723 4844094617 6293128849 2689936871
3898388222 7106027903 7990019045 : 1218
5833600797 3927741092 6655739233 1470259092 3389065438 8422351324 1153880185
5923495613 9930223919 6450504503 : 1219
6935292701 1566305153 3519186418 6482344249 9919272027 2953459599 0630487236
0804159576 0029668121 1168317236 : 1220
6038110542 8035914457 2024825645 6105714055 4624208213 4352094810 8417158289
5724450720 6354681600 2305120140 : 1221
8480543587 4252617101 7681853883 5575587174 1542477544 9772221419 2613155252
6910917556 3331932322 2432185254 : 1222
2218272914 9159810583 6897025035 2281300214 1192486014 2480680797 5369964777
1939490680 4683552808 3473276103 : 1223
0604940973 3091690316 7830979346 3661183278 4531868716 4626807388 3365670456
6010423768 5058013950 7443647963 : 1224
9222841126 9794513477 3004924987 8649656367 9490992913 2712528977 6519181754
2796280608 4932375520 8153611132 : 1225
4033971316 5504391887 9601983821 3858500077 3242461778 8491875814 5964264233
7889793330 8194881600 4011312652 : 1226
5635693244 6593984006 3689031525 4722923991 4144743770 6963389357 6192603918
9247936317 8008310261 1419548543 : 1227
6051577871 6004955788 6565797066 5885510428 8246636305 7207778902 2667770425
1268157197 9533225107 6389036819 : 1228
7628440286 1025880539 2339329474 6720240885 4127649238 6447602161 1626208242
1299166036 2299184923 7822363009 : 1229
8347811952 2913821847 3263422857 5912097980 5478285250 5918379833 6801787411
2426447460 0225624149 8069140074 : 1230
0979721023 2785395756 1512834580 6165411117 9267104279 9057939449 7134946328
9504565128 6884784187 1758020504 : 1231

5832838748 5313736911 3510255062 0102775345 8094391050 0102183397 3245650472
8894768792 9892594501 9875076712 : 1232
2363791875 8647201214 9660611512 8048709648 8630562284 4083936944 3872169212
0849200851 5583812510 7074195518 : 1233
7208093746 9424597311 7281172105 1928903896 3703942357 7686212766 8210931827
6366498404 2124938144 0979598631 : 1234
1422543648 3965499983 4790843070 2176438555 4351257436 8282281530 3222238083
4767951113 5570148063 1820045322 : 1235
0723794891 8635721491 0624252699 3994671015 3668462341 0515333814 2684770627
5852035240 9920797208 6991453730 : 1236
1095516415 0331762820 0196916411 5460268207 2366925527 5141842996 9920539853
4330730680 5737238050 4167197221 : 1237
1273740507 8927266340 6388506867 3445856077 3266648384 5780277189 1147580132
3105519878 4133652185 1907146068 : 1238
1389868867 1031475982 6461129379 5439526672 8672759948 3359025974 4587868768
4964626834 8443441413 5917714587 : 1239
7660880778 4535718393 2937193739 3236408356 3375766884 6821111799 3505541020
8556188490 1020160050 5639541687 : 1240
4510822060 3555410817 6664605241 2496622442 2804545243 2160320360 1946413560
9792001959 0240497929 2367329892 : 1241
4553990101 9801121402 9086869992 0575891777 1880741461 2220502472 8585715367
5307478143 8973057178 7268366360 : 1242
1576136100 7722863196 3885264623 5125538077 3194595635 6796538236 2499926551
8043307963 5962110674 5528521429 : 1243
0262949826 5675533527 3100468788 6573104724 6649332656 7927331345 1229550591
8623293739 3326086077 4513507753 : 1244
0901574443 8294873397 7960532284 9358301361 8379586264 8032129736 8474817516
4769136621 1036036950 9106666505 : 1245
1717115082 7820093278 8358722598 3940463068 3763181180 8904423626 2199881236
8268078579 5262197216 6872017455 : 1246
1747262781 8032683058 5488039709 7704793483 1035439855 9078435527 7667603313
9884605271 5031388563 3246768892 : 1247
7104595851 9328951391 6782385773 5772658100 4798256393 5519352005 5204080028
7059678249 7393747886 0528356493 : 1248
5914978380 3779649600 0521244583 4779001756 0424658666 5199807702 8839438516
3809550430 4921960324 4360903400 : 1249
8517466042 9627430976 8387151945 9826447359 4023424821 1044757291 1177795877
3134155360 9527595708 9861258677 : 1250
1456252399 4500759380 2060935502 4892008476 7332293085 7422225502 0645569023
9126543663 5785242724 2905605320 : 1251
5754030821 0145123820 9021746697 5797653475 1725014658 3747884808 0537735150
4222240429 5760361375 4324861996 : 1252
5589193922 0504699982 1062931609 6756517907 5132296077 7857553310 2658584257
6086686764 5355209277 4827556754 : 1253
5177169950 8789411805 9363052499 4496701237 5980065534 9987396663 9539441701
7059698101 5127193331 1840767923 : 1254
2718539539 8097640485 2784674387 2316432910 0290654953 0861283330 2664007580
1296184992 0702200255 5972156957 : 1255
5883761687 8436434679 2755863573 9722535648 8413306011 9289574642 8093578580
8113233143 3115287482 1797660397 : 1256
1257952890 0364071989 2332813161 1640416937 7366280132 5973822223 7426818917
6489596422 7033803905 9295964969 : 1257
6482133114 4731667650 4197678110 8490966469 4257170694 5700787126 4014486522
4284694889 7617256746 5352205061 : 1258
6210730010 1926248314 6821203551 6995015220 0731638400 4132030333 2423121670
8268546893 1758436630 4307843507 : 1259

8592810447 8492663952 6523987186 4417338008 5681692321 3474297545 8326940216
1253332837 9009606486 2778549412 : 1260
6679513674 0458774169 4559614076 2656625029 9006922672 6787603658 7137932796
0418488393 9339346926 3543415480 : 1261
9518362332 3317522937 0352102914 6413312752 0371171667 5487206347 3892329378
5107290295 1446292741 5467619479 : 1262
4274716691 6030497829 2889614745 8702649979 7079206387 2408250230 0642554499
5904011974 1085351678 4440901880 : 1263
6462937483 5443961440 0353523310 3040411784 5722890295 8180581032 1237438258
9870274737 0401068377 7715925126 : 1264
4535706508 3009214792 5834989247 5127453622 0061058545 7599736931 3529707814
3742841340 5519544467 2148941505 : 1265
7452839171 6037154530 8252555834 3202512542 4166244575 2456296445 7910769717
1521470951 8505500355 0543906316 : 1266
8825810578 5074635656 2047914667 6805569843 8455202770 9969719889 8072337148
6956356703 1776877637 8974327349 : 1267
2829343905 1455670607 4460797047 6931646278 1214171381 8274378561 4621970880
8702106421 1057377851 4713588373 : 1268
7738824076 5280451914 2713748811 0559744718 3100939375 1976598021 0024101251
1230813682 6033847449 1087716132 : 1269
2857660263 9388492849 5989823656 5727204263 5720263748 2564949491 2629141917
1306462805 9566982549 3603261320 : 1270
1925280434 6170439028 9260279931 4043613702 6582012131 2851488158 5731117821
0413103357 2888718172 9526271120 : 1271
0081475064 0268304641 8988769747 8791731737 0381399918 8824241699 4212152776
0451859567 1190941807 3734793310 : 1272
9970928315 5468165639 5271010461 1376254066 4495861838 5463898220 8996778329
5501114314 9959368039 8222303713 : 1273
6329574232 1735744647 3421097414 9174364199 4731958840 0526387269 5923183642
3254918455 9550453437 7846709470 : 1274
4509594201 2021142208 6419127904 9359945213 7392487110 7432314951 1380429379
3655436372 1726348190 7571135312 : 1275
7093079527 2952211247 9531498969 9080894665 7476955651 2436056114 2008663990
5609900038 0302506124 2360775032 : 1276
9341347289 0501316772 8097131626 8349596340 9292243031 1950848788 6710353352
0023712730 2029165929 7525265703 : 1277
9210421496 3495238570 8560572343 4621576956 9851340683 0454833154 5907536471
1469968242 0910232143 1171769227 : 1278
7385347704 1779407644 1001301048 5960927072 1132052318 5382227444 8702433271
0398781147 9127546080 8361156877 : 1279
9215131131 0450083663 6310075175 1102590028 0864277150 2096271366 2397401075
2884454683 3161821150 2789264307 : 1280
2976355761 0551124620 3324800531 0599511150 5431484829 5534329598 3057427245
1737886527 1930007323 2173623758 : 1281
7327314890 9109455374 0270481185 5571990516 8393874535 2067970859 2118964078
5489504109 4056996598 8715988633 : 1282
6207795504 5219321563 3612468530 3174705443 9402941829 2635524015 5452316098
6825531389 7018801539 7045962501 : 1283
6917966481 2501555932 3114826730 0563383579 7260328601 7784741496 0045697257
8349562058 7328730124 5145557634 : 1284
5230298648 1495441009 0788352980 1207012654 1095251846 0666201767 4204525736
7994690771 9084537874 8206080290 : 1285
4825167017 6619820730 6183312392 1935356900 4070521549 8939034465 9388090475
0772416954 3651858075 0664904594 : 1286
4318886297 8723571603 0224813522 0460109063 5214508280 6397492755 1284769435
4996203399 1644887919 7437902095 : 1287

7188863200 2475020791 0237907307 2963746326 3366745942 7556378453 5691367345 5240148971 2590948036 8566282321 : 1288
0050039400 7310663207 5257283147 1151926332 8928520696 7239347175 0982952602 1254947643 3019535743 8350925828 : 1289
3111339115 3906337661 7373077236 3027988986 9985799450 1659237690 6754883798 8929400605 1628261400 4815046948 : 1290
2814033083 9164342486 5093963545 8909132805 9511163345 5036563482 4519150583 1794980831 8272813479 5050772717 : 1291
3359496633 7188214919 2837871164 6390356692 5779942457 3943554730 4493555939 6848032790 2086141968 1508260648 : 1292
1092468854 3383329866 3907454780 5263629161 5627988031 8782827074 5163032786 3907566533 6219750632 2424864576 : 1293
9459753596 6732006038 9826293000 0761251494 7980089567 1245256955 9827585485 7690124636 8659494224 2277271771 : 1294
5184964175 1071598416 3572072412 2437196806 7203927064 7894278942 1712842641 3342711831 8479441334 6064724314 : 1295
1150155098 5511712414 6682433123 5206284065 7226926069 0474791964 4729752832 2749569819 6327787281 6259540120 : 1296
2053807329 5825004974 4593080978 2409529912 9654233184 9879880077 1681631986 0865120883 1586725650 6594414061 : 1297
8446837496 3189291374 5993421603 4848228831 5828973094 2161473689 2558516992 7155311558 8887600721 7034102445 : 1298
8744020844 3428273004 6730979555 5666811501 3003388895 8380231464 3138290026 0076322850 3475830780 8788951803 : 1299
1398102076 2788985174 3534782251 2084675949 7430024437 8958428956 8075266320 3627696299 4601808349 4199491270 : 1300
6559130840 0058626563 9963911040 6851041282 0071532462 5642637145 6355757694 5284927112 6355771963 2506589654 : 1301
5536482125 4592633552 5729259528 1499341587 8776515692 2311915102 3373440716 9916564763 9820008969 8462984399 : 1302
7759385398 1121332181 0328198969 9457926176 4935829748 3733877523 5285946403 5138238230 6269453634 5810031936 : 1303
7250206982 8073843334 1175283157 3143426398 9641634712 7053034775 6991558003 1181591809 1137880268 8385475769 : 1304
7292339888 2860323029 9770430666 2886955301 2102727057 6339598976 8941024996 8479498168 4201199256 1348075644 : 1305
0406559462 3837087236 8881254894 9148794873 4808614168 1055211400 1845517008 4444842948 4755073273 6642827222 : 1306
0633658240 1745498808 2913018839 1401568090 5000084954 6573730003 2747797209 9175074617 8595157995 3202237285 : 1307
2359204007 4251522563 8616675620 3188398117 6186119602 2162847431 9079702503 6745928280 4678178536 6473935600 : 1308
3540382782 8184576694 7823374571 1382212193 2616729501 0427069409 5202650280 5228985909 3500239449 0874562620 : 1309
5345221731 1940957783 0195360518 5038549614 0621825306 1820365182 7337062111 9893902448 8975386358 1809944918 : 1310
1578487833 6528865436 5422483020 2789241704 9689651104 1727594750 1781226785 8143917486 4942435730 0909171264 : 1311
8771605959 2097445811 4629554223 1002200851 2052258976 4778114827 0394267766 6427827462 5939511743 8071986187 : 1312
2226558650 4030028469 1469278646 8003183603 4638172640 5702707422 6203429718 7555809938 6871240465 6223338914 : 1313
6465830554 3013155095 2851097263 0050805188 2652726853 3537293733 8569182693 7171677303 1611864749 4810424215 : 1314
1279159101 4606569795 3331337740 9593674932 6441463702 4275245393 3503013099 2833648540 7069840343 9912124524 : 1315

9275580299 7988240920 6644640425 8596620088 8741916498 7730275403 7292042158
1093781471 3136226288 6666945474 : 1316
2124495528 4909149219 3371936234 0294337125 5755699886 5296623645 0353519202
6777637942 4820828605 6893623152 : 1317
1523178850 1452131321 4914698685 4835944706 8658501098 1314205892 6764161151
6210940535 6780736810 0897342458 : 1318
7293270521 0853572676 3805642288 4092966588 4477795279 5467107351 9329547471
3015079220 8403282322 0442894467 : 1319
8218396547 1109021173 4072513972 4757357008 5553127432 1999675125 9582568063
2358808838 8436620326 2266191414 : 1320
9347404364 9800024739 8332092411 8386674296 0926946070 1418388178 1107142824
3965779638 8439864782 3137154249 : 1321
8947258304 1145149526 8724236189 9676305881 6820846327 4374412103 9055276521
8710735564 5257133601 1455804558 : 1322
5684558650 4328599176 7651961932 7114349866 5407774514 5004730727 1171479571
2227572018 1288644644 0777517460 : 1323
3282423173 3853376529 8981044232 2404677246 3204795179 8097157602 5800885768
9751340594 8054826877 2884776293 : 1324
8464549604 0270370508 5394190927 6993706680 4551719416 0403763511 8018551365
7545109524 7034602260 0207417428 : 1325
2384948178 2254906365 9920847490 3758320574 4677959106 7556606407 7500934712
9817005818 7694080279 9269046059 : 1326
4987211763 4151914882 2518670439 5573100179 3710004665 7292180372 8487979715
6922788883 9704198254 5657064289 : 1327
0898582795 8625659901 3759687500 7856985342 0944399597 1523667673 5599115570
9006141301 8853956006 9330508261 : 1328
1578831597 9018829128 7776539696 4067539208 0848582290 4755619051 8637549059
4176472080 9084852392 9966365377 : 1329
7468709856 8014236137 0763704674 2361802921 8679592476 9777652926 2929041798
3927505343 2943384476 5333985012 : 1330
2828362798 5150263745 4279667177 1484197573 3906572871 5430543215 7523544932
0534653754 2382048448 5088463459 : 1331
0853386677 2925385204 4498441313 6863751894 1176848626 1360368193 7363513393
2540806852 2692147430 7329134467 : 1332
6252932264 0845330844 9386471515 6181394136 3435036481 7794755097 6339255988
2786903696 3238633034 2579445292 : 1333
2923775203 2874489020 0405326681 3935475285 5017464531 7172145995 0814556136
4692526650 2271153373 8181759785 : 1334
5795041988 0754858113 3628915490 0903908060 7754157573 6137375598 8018757307
5362487370 0129122382 6113438103 : 1335
9234372313 5368988915 3374949378 6324984941 7642814170 4528408296 9399172432
3286772564 1504837657 7311449335 : 1336
2155385230 0178110827 6163630370 9020525950 3779092534 1104705700 4656525197
7925679331 4108866326 4059262317 : 1337
8893126031 5285758716 4242119033 3798725775 8742901290 3759362697 2723431489
3572572418 8379418627 6864566775 : 1338
8686920276 0143980501 6387143520 4776738809 0057892836 3381779738 8457344100
1499664332 3582225792 5351711059 : 1339
4856078918 2401521998 2852269465 0958763149 2471279520 1644676474 0270468954
5435103069 8261799914 0223407285 : 1340
4891546806 8420957432 0750662115 4487626644 6757986364 4388023258 6360886918
7594422715 2142965066 4161384963 : 1341
8150279721 7307126592 0578266002 7847181400 3420926569 3070309044 5702459646
7576490185 2781393148 1315092036 : 1342
4104984596 9060225314 4748229457 0702527043 6304061114 4551422276 6936650125
4252372074 3940182775 2508941432 : 1343

9152151705 9974545931 2594682121 4351062276 3303318504 3394889512 7672063729 1512493681 9357031910 4693572905 : 1344
2762887687 8250048505 4800597323 0753265227 7925524199 1315961791 1522069419 6854791873 4156699781 0967025629 : 1345
9399320816 4507174173 4905643398 6521998663 9055709352 1198524390 6798615021 4486239284 3873982018 7602285471 : 1346
2303949459 6615725875 0965032007 1247665759 3813721248 0113415355 0616754720 3695791055 9746106711 2541711745 : 1347
3695430147 1914199373 1972279716 9021161357 2625243116 4722893666 4414262124 3854981362 3694963571 2821160368 : 1348
5441607108 2317751078 0129830425 3814190892 2492085953 6461082139 5648113205 3160737077 7207605599 3498150342 : 1349
4064077512 3315121589 9924629749 7845474385 7855952270 8926710247 9199199645 0430401660 0562176296 2340149282 : 1350
1816115205 0464381405 1201017632 7979026932 7122270125 9270816304 5794086959 3885030885 8577776769 8805771202 : 1351
7746185837 2818585997 0177211160 3710982739 3241471979 3766386484 3160008415 7927253061 1640850151 5001652030 : 1352
0200142743 3763904187 8862263527 4702258984 8494690776 9474761327 6391052599 4056603823 8237163694 3555470658 : 1353
1748273071 8247418272 6362724046 2399440284 4447364245 8644475104 6902997652 6749734435 6985708539 0578191599 : 1354
5859960967 5061283091 0194748865 6507512613 9713632927 6415834913 0420830095 0851100414 0745574437 8492789857 : 1355
6072610576 9741819633 6967907551 8838322017 3443764398 0536829626 8732851893 9530815972 1384099875 3657746635 : 1356
4932531139 3625597895 4300091191 4267407538 5925496901 5797341918 3710401699 9179009456 7835962857 3224471479 : 1357
0732045696 4719786315 4908628412 3332517481 2784828809 8487610221 0097427834 7516462790 5539385196 6889569651 : 1358
0876062872 9574590889 2017023867 2074010602 4538941519 5473932814 2466223126 8923626502 7205640264 3021776903 : 1359
1895555206 1127114631 4671703891 5773390065 4528692327 2080811157 8757374991 0353244466 9361653517 5221246886 : 1360
6080593973 8054689486 7556025887 0687103081 1898922024 2174952934 5821953530 0991561355 3607315909 5673469906 : 1361
9924874268 0019538217 5246210534 9862701061 3215907572 6024080430 0827868356 2931983842 7105219835 4727511764 : 1362
2330279958 9268727305 3118355805 6875276124 0919742444 7633568095 6874844410 4554702835 2365141527 6562700804 : 1363
3630974774 5376780982 0873498038 4982599248 8106702977 5494953522 8299516546 5598506874 2831762852 0857196139 : 1364
3797828505 7790149962 3213922046 2341524168 2380388944 6624267373 0018965433 7647650363 4125182850 9512088864 : 1365
8562947143 9877956655 9280749164 8962562185 9267154146 9217676839 6054500821 6421626056 1064231444 3579823069 : 1366
1965780470 5747148460 0729681823 7228797756 0496089158 1786867293 6323790241 5792047283 6469702103 1397518009 : 1367
7841598550 0070553649 3875321257 4961674875 8725832599 2595761507 4339186228 4379883013 4604454088 0817809685 : 1368
4911945411 9347026896 5059919860 4109976532 1119658106 2966550051 1618365170 6202928808 7760914984 6167316442 : 1369
6864197089 2306484630 5675457388 7202476016 5257760852 9377210933 5844538710 7402729259 1915246267 6235381797 : 1370
8693064215 3401316337 0113573563 5111098141 8211296622 1073672626 9615672674 8307752488 7444841676 6573702400 : 1371

4850839370 2558385910 1226694835 8068391545 4791660164 5691486305 2393597793
2446725588 6717416048 5503871149 : 1372
0317607553 7321944728 3058221915 5807880752 4536969327 4460174736 0524205864
6968697577 0612186776 1972058749 : 1373
1045165142 7154954238 5392023252 6975123495 4654630906 1329460056 6507283098
7280338737 3515537522 3563183570 : 1374
2537006494 0926380803 1737463485 4036114660 0048468762 4231089472 3791650074
5179705248 6284672766 3375517303 : 1375
6873683856 4403704980 6617909200 8317107882 1049818331 5526148505 3735407503
5108223939 2474456301 0969204227 : 1376
8844737169 6889509111 8573692689 0336659718 5225377703 2962201670 8106551812
6758009408 5251506847 7579219138 : 1377
9321380928 6961195312 2090503801 8107658748 8368317882 7814252786 2618796676
0682197703 9093260067 2961512755 : 1378
7125278643 7069898354 4440961391 7379035454 8518040397 3331374805 2358791095
5583040481 5348045391 8785403824 : 1379
3236907304 3102740626 4177776265 7301034703 3840211296 6908481804 6162496487
3947345844 1215530258 1522214994 : 1380
5822249941 9419547256 4103175021 1442280865 2302802213 4240931939 3272767819
5990608112 5986239673 3945898961 : 1381
9071679777 7802595116 3147757626 4028588262 5148158216 4399441350 6196081175
8904619511 5853908261 3354960388 : 1382
0323713522 2451696811 8059751218 9590028591 7973908665 2449528040 7827130270
0453774372 6785553250 4850397463 : 1383
7573946460 9840856589 3018482234 1614986583 1503466082 1862236058 0194811455
4903515474 2662660612 9502687840 : 1384
9754779814 0726823956 9314724876 0982803450 8118938340 4096153431 4863011248
6764653154 7875845494 6522227531 : 1385
8773560890 8350438370 8112088244 1759938586 4663093970 4811725300 4020305813
4090447450 5115637705 4103501416 : 1386
6861912485 2526949334 8297851018 1114723298 7404539612 7540222219 0958440508
7230662326 8888497042 2345670001 : 1387
1949751859 7964940991 4897138536 2279458874 0760990432 8542281277 3058183040
2494510870 6336986946 8674008948 : 1388
1097539710 0908494768 3041071152 9550638887 6524905456 5999426077 3886347394
5525114489 7203610479 3757254472 : 1389
3966023547 7481274941 6069835101 3147640236 4194914610 5980556375 7044651556
6712365256 8282701574 4528476022 : 1390
0781753972 3371640969 8626492055 7668761564 4577446446 6492547734 6729725557
0538828590 7892317597 0676863982 : 1391
4966294555 6019387315 2710362720 1242931201 7642522464 4803181954 4683337639
9461313836 1445704160 8883422253 : 1392
7155878358 0701611560 2717754142 4723331527 8135669400 9898004445 8238998420
0640748958 9238923892 7522891473 : 1393
2945531240 4247755208 3805237951 0123938435 8587754549 9900127206 8286659998
5790984293 0384600732 9623842629 : 1394
0797218233 3727476694 6401526920 4881430422 7394388383 8698807236 5034008809
5245127260 0136152570 4157749789 : 1395
5464274592 8669621641 5427519072 0789657656 7620470876 2910259298 8877128340
5806131718 2068879509 6273552308 : 1396
0228036658 8530930270 4619400614 4644918627 8566424494 2081621020 3832761116
9622442138 6397311571 3011899185 : 1397
3169915158 1650258342 8128487414 9275360507 3550149275 1649655689 4986881445
7828072415 4009011617 6936589862 : 1398
8113745927 9032257848 9093397688 1608670857 0029953457 2157942098 0997220532
1457514271 5411220939 8869874562 : 1399

8011653320 7925455196 9851910384 2815726835 1201092367 9952429068 6799545683
0838859301 3667218521 1353641724 : 1400
4228370492 0603648154 4497177998 8618739061 9701265066 8437064042 5124459951
9090062260 8217984541 5139874086 : 1401
1561892465 9308440274 7014710167 2547160166 8601739769 1997662011 1199893015
5354062817 7813282386 7987398831 : 1402
8548093651 4175269040 5027399232 6953229393 1036045698 4252059471 0877602232
1016774679 2793562530 7683377220 : 1403
6929809952 1332754934 1076406829 3696256538 0979829922 1502007619 0656713323
3307191753 1109537696 7431445827 : 1404
0474521918 5656561730 5618532166 0425946455 3856168837 5993453276 7382788781
2223153728 1113417355 4517073553 : 1405
2082760440 7745254423 0785453748 1125966546 3557459604 3270368542 1573869622
4447960925 9367500830 9891400068 : 1406
5383635881 7787486427 1068825787 8740799283 4182519771 4084223048 9497915517
9876782746 8475408492 8993864763 : 1407
4983917539 2445932931 2913808073 8765005052 2006666627 2734384454 0498968011
8343255349 9976250119 2176787558 : 1408
0980672332 4167826178 2570891163 0179808819 5583791075 4011805096 2160109308
0422570180 5492976467 8411538769 : 1409
1430708824 7531217231 3794037236 5928771043 4554469626 6599992623 3933298641
1371001268 0408116027 6969402287 : 1410
1365072981 0644525201 6551733860 4686504062 1292457892 7147227426 7638614268
2367640851 6411947662 6514371013 : 1411
9385568064 2700778296 5968048607 7517949221 2156291738 6716354649 8898538357
5153249743 1583541399 1322136505 : 1412
1551384109 0309027554 3323644120 2253007704 2821114714 1918147570 9618331378
2294342072 5434103155 5828186693 : 1413
2838668366 0726916383 6967793201 0214202904 6813370491 5343805924 6547114970
8354012272 4100650394 9742164188 : 1414
6692274473 6899506252 8945027771 8989469132 9634675858 7926423521 1633546474
6864260548 5613157784 0361143149 : 1415
0269544275 0564803847 8887943295 6556048443 3918406020 2704514682 7824231514
0650702210 4851959207 2312004933 : 1416
7176738352 3709308856 5264344841 9467734538 2413296885 4306302477 8255435028
1959571754 3326873583 1728279337 : 1417
7410102634 7172525800 0551089980 8792042744 7783853642 7497206543 0922479605
7214003306 6159793981 5697061366 : 1418
0983964055 2028766999 1722547240 2063960609 6429945427 0591546000 7353673154
9880773908 3001581335 1603573011 : 1419
1114109280 1541228066 6670587855 5092703338 5009831156 7628516164 9242550929
2830390877 0988934946 0723490286 : 1420
5856020542 2067037156 8046350038 2605276371 0823986597 9318483093 6764165636
0790706605 2334341113 7793121612 : 1421
0205880951 4614377394 7683538839 5047212945 2834986548 0864837885 0194676769
4562326701 9987133184 5545348373 : 1422
6084512767 1800567875 4235887195 1058956527 9780453783 4484650468 1469516775
3813695184 5103083239 0374965716 : 1423
2143307963 8601544816 1449552393 5111212189 4430238269 5405786011 6467373664
7956520658 7250815927 5305713134 : 1424
3835699200 4899961804 3254950205 2195550206 1792779930 5642458366 5872167535
1928175033 4499239183 3256236162 : 1425
6502081490 3557861244 0518344040 3815991358 2717384337 3404529744 9996405991
8656664153 5612424308 0016261793 : 1426
3750921429 6580882832 2195705784 3171697946 2845513309 6838246000 3698996180
5929879506 6037607124 3272559753 : 1427

6508820386 3609588090 4003800176 0475078669 7443325877 2321543832 5998399864
3950114495 4150770097 2822653695 : 1428
8394380850 9128411041 6290966370 1274249881 7616344101 6674234005 0683616764
8232710388 9422394820 2530869672 : 1429
2292524340 7506026512 9885763587 8137500851 0056886874 3282747187 3232428984
7733542581 5041625895 5023854489 : 1430
0684967676 4892829707 2811584351 1676077617 2604891355 8510981478 9508429849
8360559365 9371053202 0599790443 : 1431
6973534016 6287645320 6371886938 2189780157 3219076299 8103612568 3876483872
6985360129 4481607317 6186580668 : 1432
0596837338 9411982650 0873262426 6960024090 8832076226 1178399915 7440210584
2789845063 0360141993 3928362455 : 1433
4027683509 9897204218 5962090201 6210156519 2235842119 4882020912 3783927557
1856055416 5620545534 7196978661 : 1434
2350583489 6282128608 2084034973 1199881077 2590454586 3376610850 5095823850
3075128425 9642859749 4715967542 : 1435
5924034955 8609796434 0196646672 1757237237 0707851846 4663837067 1702995416
9832988691 2472818768 0273812549 : 1436
6293898760 7223408465 7095098943 2016548760 4793394679 4685134373 2630392230
9331790687 3031699418 0074048000 : 1437
6872513659 7857958599 4780199496 5234272868 8988717813 5161715505 7783915871
3864040578 9565918232 1370814005 : 1438
8713808836 5230471671 2718220060 1860881125 7260339862 4035420675 2127690892
1081552260 3293004441 0189063723 : 1439
6591957119 5303028824 8586847825 6488300525 1812608103 5421351812 2471584004
6275105924 4487058370 9540835318 : 1440
9752152361 0342040845 0764137674 2347300588 2203432316 0474633043 5062814232
1082948724 0902594764 4118910322 : 1441
3374049794 7408578277 6220482618 2195142821 7981124372 6766258468 9519510699
8673740227 3230026026 1505970642 : 1442
1527460232 6999497006 1582359282 8222978328 6840199729 0365378168 1600288411
7306733244 9662838403 2435365041 : 1443
3975362055 0910521974 9095799860 5957269413 8402426755 5967486377 4293085831
4066480318 4453153290 8153215494 : 1444
3458288044 2937355680 0527667018 0009478873 3588609136 4949458385 2689279136
5594342881 7418645559 4102961792 : 1445
9958126080 9706454746 5090234261 8403450108 1240335390 0061073469 4120978386
7162772161 3708361451 5110500772 : 1446
0117042140 5751029551 1491370255 4533502068 1411652447 6917845869 4354034118
7913507194 7286833389 6624761011 : 1447
8301700497 2618956118 3989816053 9092008911 7277245282 7329958680 8380107378
1314001876 0672501269 2645464509 : 1448
7673374700 2367678201 3523567324 2624788804 8234362900 9996330109 7657305710
7450862132 1877968280 7434398964 : 1449
8355242714 4875730583 0321802494 5210923199 1204178629 8321106456 1898234504
9505439716 1803039568 5126538014 : 1450
9225169487 8479554724 1863827862 7582327821 2993978207 4286755471 0924982182
4468614795 8081408355 0046687559 : 1451
6261579061 7175902192 7186972378 4547241129 8557573179 3747953518 2955842991
3369281405 8848042157 1538074685 : 1452
3113023354 9462721418 4400563239 7445875377 2751807146 6016570650 3537500007
8000547610 0367863699 1113239858 : 1453
6213221822 4624643435 0103632239 8596701728 9928425234 1131543432 6293039073
5953429144 1393387428 2187214841 : 1454
8613127907 1626858266 8472059546 6403565113 3279272928 3670421533 3378156489
7878724347 2316577108 1189058811 : 1455

5922053413 4477675212 9774635506 5511098018 1145470892 1701244106 3492394924
2422673834 9439407865 4658363868 : 1456
5970026019 9154168385 5861557896 7012722002 3220031686 1954197028 9247574216
6676680152 4808240221 1115619098 : 1457
2909528829 3422784064 9039533967 2008649956 9654470752 1184613434 0977857777
3642631658 6916987627 4954188683 : 1458
1332475145 3159002335 4409517149 1408135927 3191146192 0067757921 5856331076
1254707093 3961164415 0880072729 : 1459
3945636849 2532718589 1551688147 2096011415 4056640038 9210281186 4854595041
1900558007 9283947161 9967600301 : 1460
8770007299 1661348781 0389918979 9277933082 6033333833 4057919338 6012599266
3543506471 0091260634 6252385743 : 1461
4635268474 9297906578 0017287665 9682562194 6854107798 7421844550 4710482511
3899365427 9944593202 4438989851 : 1462
3442567266 9327861329 5048517020 4267041681 0423988787 7662828350 1931254549
5101087037 6696381206 0312761799 : 1463
6218893187 7783052045 0194812047 4270520457 3212548733 9039302866 8085392898
5514539518 3070167737 2533915679 : 1464
2769039073 3624859034 3351476117 8705177976 6471010750 2450768161 6557253954
8200948091 1058631732 9891753118 : 1465
4160364021 9503463573 2195947558 6008320829 2675123788 4955166725 0649220720
6097412031 2931357435 3745218554 : 1466
5498302580 4156517986 2278016468 9374817239 7133811236 9536373581 1057393910
5369179739 2934319775 1880325243 : 1467
5258608082 7553740999 7210154008 0046979927 9434223454 4768970758 0313149065
4997645727 1996996280 3326920908 : 1468
9155838176 0321398926 4488023769 1008274209 0668080043 7399250454 1223684971
9409774670 4673167378 8785204941 : 1469
6564473707 1325437283 1395409623 1813376473 8488941218 2775687605 8275472115
3484064111 9286609198 0614228229 : 1470
5524907588 5258711407 2134140163 5238119989 1274778913 1397574682 8093424728
2311021898 4300702443 9996429064 : 1471
4450844788 0276686539 4635783597 8633014357 4307385522 4801180578 5516300305
9480351702 3052917619 3766804489 : 1472
7455190062 2981417402 2546879385 9809142285 8374494142 9466840567 8447862996
8730373668 6339751013 9100798455 : 1473
8831971893 9840420585 1783126255 6099075164 2566660914 4857660683 6793744806
5297240370 9933396292 8343483326 : 1474
6104136871 3447259629 4417153661 6832569298 7460751934 9004367548 7124501251
7388228959 4264322061 7183770595 : 1475
1665664903 8896234159 0342836592 4676238921 5431621094 7396500986 9257089507
5041141578 1971894579 9485168292 : 1476
3997676852 6059094084 7692555560 3209473017 9889261822 9473834688 6884787742
1474782112 4629005048 7616242097 : 1477
5722951786 0733959886 9641860539 9569127426 1105379964 8648272882 1472986544
7937270511 4310364153 9950430249 : 1478
2489038987 1904738048 1217370572 5663713465 1471541312 2205631956 9952971074
4845423257 8540931960 7037480624 : 1479
3288730574 0374143132 3821583556 2671427568 7557551361 8201917633 0108628379
7258551156 7417230504 7190608736 : 1480
1627708326 2964429580 4827975636 3082376436 1615455540 6169800458 1964467066
7810243347 8459880692 4847727489 : 1481
5298262045 1694370037 1120191295 3531129197 1380175955 7797453217 9706899810
7869799671 1614064725 8355731385 : 1482
2803781447 9461864582 1634745203 9855897512 3171364079 7468385145 5920414500
5217721229 1446699278 6476520100 : 1483

3653978899 7094195677 9542290004 1438454871 4348852855 6517630802 9925167644
4247682186 4906215121 9172342568 : 1484
6851600605 8597808966 2366883201 2839653122 7030746548 1821199948 2253881430
0401681144 5036211672 0244462048 : 1485
2829677761 6016563789 7576349795 5487255108 0910578133 9420347277 4484748769
8984192182 8085630416 4926029917 : 1486
6230362632 2504418296 2965215438 5628760703 7421868140 0473863094 5015910913
2542103032 5613511075 7558287347 : 1487
8656260809 3256450743 4633723342 2408558581 6338537153 0694587826 9202052395
0672724753 6900139801 1496431659 : 1488
4582971648 6863220484 1795219642 4498327948 8063134646 2010891393 2870531345
5615037887 6921145927 2685051467 : 1489
7135599589 0632238650 7647782826 9016803601 3061708569 8288633635 3398216641
1661335548 0403703821 0034458380 : 1490
8150558303 4017971208 2249390950 3856609585 5713953746 3476283240 4217519342
6566863925 5917743378 3255482070 : 1491
3861056330 1262376287 6981734728 2242509461 5318907021 5082050421 8103977489
4076572149 9083247852 8545951002 : 1492
4679597393 0841106272 5225415696 4938923682 7358143460 7727598033 4626431259
8278889441 8184917380 2687044960 : 1493
3886707186 4770831564 7875891178 0354308201 3186565820 3435407342 2928347455
7696514986 8391503976 1412613360 : 1494
7894809975 5916482490 6255168553 6794824740 5098464960 8568188917 2036998737
5796439800 1165295270 2772372260 : 1495
1935755572 0232631014 7686928476 2636285189 3048492690 9264098547 2493648181
4128316893 8328312579 5662135988 : 1496
3554452066 7408958409 2314862575 5911051962 2000503080 2042573700 2899660124
1363556488 0280339995 6946560958 : 1497
8576321992 6030004685 3975598028 7655583171 0706399750 6660476148 6777635632
2611612715 2242671096 7361840252 : 1498
9108255244 6153885776 6602779608 0898302837 0687781398 4923812545 1717898757
7906769165 1324603108 7551814796 : 1499
0012167620 1685543613 8875351111 4464644596 5948986286 8500384293 8167759796
1912729990 4591343960 4283622782 : 1500
1457438491 0806626737 2039815968 3311458313 2775573719 3964762139 4703694871
3448379653 3672088650 7609494431 : 1501
0674893862 8101668608 0935487620 4062953142 6836790162 2324344216 2500961919
8865282501 8478075009 3092989616 : 1502
8789351440 4852784485 2101949729 3149122933 6642838361 0958359117 9266973210
5032865863 7196191306 4985733208 : 1503
6615243198 9177517561 3307253369 0606289440 1403624673 5791686124 1907679730
7215389609 9260914778 0039218290 : 1504
9660567805 1574245394 8127051582 7865608617 6628088767 5485282643 5345792975
1091037432 4314804905 0997201340 : 1505
0938712099 6799226673 2745697219 9757397498 3529556634 4453243455 7032626027
8293136893 8896296769 1490051117 : 1506
9164157396 4151622345 9624143879 9849972397 2106259104 5242665562 8296014596
7901286176 4153524786 4330478558 : 1507
1496257111 3956032515 0363183745 0619425879 0732974799 0654033781 2932343549
6477095994 1597021691 8103681473 : 1508
3833330641 5138771322 1517339840 9381746568 3332375212 4521204263 5149480179
5737064857 4825588129 6241114146 : 1509
4692661774 7817386015 6155696776 8080635428 0813392622 2268057358 6043957391
6273877143 5084847701 8662653169 : 1510
7488864738 6824309419 6018928758 9120213872 7709615384 8809506565 3207344205
8984978568 2144810993 4432714379 : 1511

4129234072 9754793264 7618296204 0361443641 1274652404 3691754283 5856614059
5943326100 9132314486 4164204976 : 1512
4947955201 7171086517 0698122416 0848217072 1710164948 2479807749 1801666631
8076045716 3952518386 0958271832 : 1513
7208657052 9825589266 4923127405 0673123487 7203497799 8295609410 6360305165
8168190384 8011147030 4239018204 : 1514
5758372731 6520859225 3994751093 8900121122 1942666544 5908677926 9137115495
0789666576 6765460962 8827777519 : 1515
9570554507 2979236662 0852350781 6894340032 0475437404 0076217990 9188135109
4993966943 1342798599 2158062927 : 1516
0421382675 6214353405 9246720235 0206425854 1096859551 2829598880 1679474853
4882762322 6089882142 6027966949 : 1517
4883399735 3809115310 2615727526 0615166467 5747231126 7311304563 0210164427
5628278219 1487924669 8975320978 : 1518
3265292168 2584330479 0854783365 4269758433 0779557195 2000101207 8724019881
3494984438 4367638270 4117421003 : 1519
6951169011 1801683269 9946612010 0860532094 1579019288 9761397840 3516511159
9346420444 1482768205 4550634184 : 1520
8306161979 9460270489 6489524389 7025843417 1773190315 3309321479 8054202089
6195125075 9293649016 2781474077 : 1521
3224772573 2201913504 5680559997 8569277543 0546578798 4285946840 8586784134
1145382412 4072065675 5982648262 : 1522
5761903033 8341742518 4853854038 4703710069 0876508085 3508640217 6210101567
2829143567 3677110351 1643978363 : 1523
4404283023 4780735456 6914381770 4745089458 7211787839 1541665309 2472697951
9526863928 2330037168 5067876207 : 1524
8775481783 9108197321 8290478799 3291396078 8741768330 8186531819 9940659792
6782213227 1345963247 1409529463 : 1525
0761973967 4998463493 6360975806 7253661551 8078598145 3495358216 0148026023
3176252015 0636639939 1351428775 : 1526
1153532124 1122515057 0657231152 0853765028 4322101584 0618982570 0470439171
8649072412 0891714561 2024917300 : 1527
4379934999 4206586637 9857873460 6048061922 8119464331 5629256867 1087969712
3496236406 1937388112 1802073791 : 1528
5981801097 5908011327 2257843002 5011137880 3495792043 9189928830 0516242921
7600337641 0793371968 1331920675 : 1529
8299182607 8485247571 1775242016 8349348194 1400539164 6393521827 3710489150
0365804792 5976158343 6513553494 : 1530
3843191509 2146293081 9950183591 6709425302 6540329803 2496761584 3963471143
5324714370 3922148617 8438282611 : 1531
3866885521 5984613445 0580330263 6914394174 3559917537 8716668814 0045296893
4351987652 7230084584 6550156565 : 1532
9895211301 1048528816 9394156867 0635178319 2218559553 0500029864 8325444774
7771995501 6508265889 6713964088 : 1533
9880567958 0669160658 0609404851 3928010222 7697615613 8260831907 6033245484
6528661464 9429483966 7733008070 : 1534
7320067510 4262514142 9624471453 6875097068 7850660059 3940265187 7861032765
4702806325 7299061968 9759188738 : 1535
6672305110 1237949329 2597649574 8262551959 2739447176 4009255618 5211857724
4308889458 9313045709 7527258670 : 1536
7145565142 3603418198 9031519545 7218862114 9171034530 5965784508 2618680743
6497735831 7577008647 5879964322 : 1537
7445489500 7809667119 6162151367 6950853089 2336123866 6283481102 9398046074
3553427272 4428104903 2807567670 : 1538
0337727112 0949128434 4874508135 6882215603 3050438835 1754108148 3037534434
2084122081 6836058132 6234576775 : 1539

4279316198 6045430504 4485105558 0041167943 3767132055 8147058727 2088253604
7310649679 3184796373 5278844788 : 1540
5205873182 8660065633 4932560235 9088898353 7772507970 2005054144 0210559461
0720764924 4091363372 2789739946 : 1541
6397512341 1788366312 5090061416 2322765702 8541048506 7974498127 1814676430
8414103002 3752565373 0495276727 : 1542
5484545999 7871633253 3105061902 4021518146 8100146512 6285103975 9839412882
3698621131 8315247764 9679577744 : 1543
1913323947 9855287165 3023199869 8023983984 7319817881 7133310344 3398908379
5800005131 9653452338 3390109097 : 1544
0444714347 9426502628 5740315181 5203546507 2823118385 1986580293 6213522437
9754319380 1983432914 3125027577 : 1545
6675431686 9888602865 6770135003 7258969644 5868683417 6473878390 6654442181
9235857731 0787002319 1744542871 : 1546
4160030268 2837240494 6436034787 6903573326 1881143101 0813218855 2798589730
3450534403 3037227691 5140453182 : 1547
3618783217 1998898905 5082908966 2419765855 9805783414 2873730648 0985290782
1459411264 9499219651 1361256777 : 1548
3076994605 8020640723 9180866900 2017569564 1759552721 1359337589 7911604759
8231558725 3564456825 7143746585 : 1549
6688982037 3705497045 2907158469 7376355587 0609280120 1769780532 9357967838
0795022792 2001052016 7689887325 : 1550
4108389319 2509071742 8881081070 8623207551 0184800417 6969682629 0392399839
3811623663 8478713081 9320185559 : 1551
2678658980 7098502295 3739494217 5424696253 5470439547 3241339247 6485210376
1177731123 1385001618 7130471064 : 1552
7783932487 5850063619 9119677877 5326807139 2468984403 8826593605 1085465236
9222619272 4034991210 3831622629 : 1553
7241144083 5568680499 8074604837 1359252075 3901701446 9373916409 2864863919
0537573932 9455653677 5435632948 : 1554
9530854791 9735618116 8943469443 4436430308 7144425491 0609829482 8815811595
6356299337 7947392209 7851104067 : 1555
2166448032 0531067091 3037594884 3457873439 8473707653 7474047930 8090343824
4339705830 5326958562 9984793830 : 1556
4808177975 0890193239 7881964474 7281348548 6485639973 6790769039 3025212859
1950959453 3031379751 8529818662 : 1557
6201176126 0953213926 3391827182 5632758305 9118937210 6915776438 3887227842
2852900912 2612514080 5231508120 : 1558
2726247737 0667161537 2979623651 7171183091 8171522805 2653759337 3755812823
4864296932 2667847133 8695988769 : 1559
1580950811 5049936337 3569059008 4289200705 4825254617 6895416471 0778011758
6071432866 2404483055 2364259377 : 1560
5798552448 6960807267 3059076502 4885140814 7618917999 8962929079 5406069165
0986275070 3309100886 6119931836 : 1561
5347811068 9500553232 1232310409 9431566975 7128432110 5892729075 6266529830
6834612688 1743502763 4457348731 : 1562
3081278785 3966825948 0450244508 9945385062 6222815657 2066562590 8071060090
7194741580 6434289617 3131515706 : 1563
0558113998 9607656842 7723954812 0624654927 9224664410 8673930170 5267840652
2475041053 6043235086 8815254382 : 1564
1884057815 2295198789 5606499560 6982745328 9227327038 5375845209 2709242946
6734689593 3777896580 6769512859 : 1565
0449057399 1307948762 5397989989 4685344867 0842763284 7644098046 5348855120
9436064288 9373837105 3515595879 : 1566
5075103681 9995860092 4794052205 1548807777 4998306131 3790264128 2737157571
0612817362 4978364745 0207227756 : 1567

1952126743 2735816854 9611969888 2583112616 6950522240 2188114669 3062574953
8470869958 6574599887 8927868473 : 1568
8719864383 7904804637 4622281612 6871276345 1130947831 6617599707 5950853325
7460284937 4001043645 0345565804 : 1569
4944295034 5318338129 0785088833 3858378697 7108498206 6510206279 5707669833
4451779345 2718037691 1410207557 : 1570
4774315429 3290326295 3211497882 6203515987 4125464228 8439527779 5499289564
7543471058 9858515900 5508490056 : 1571
9690369399 4638054127 4407827207 9588120610 9501826667 5052829100 4286440115
9690915602 6024587211 7456045510 : 1572
9407684697 9736827481 4597904045 5219048418 0115456634 7833534380 8815341403
7239817881 9077576306 4723383684 : 1573
8076617187 8875254440 7318658305 0118647563 2030171398 3390078987 5424411026
2777492594 5578726315 1608748702 : 1574
5048062038 1626062841 5675429971 1008457236 0794368388 3177569711 6071774760
1977362998 6084709225 6124190334 : 1575
4303868060 7516077836 5027891666 2836093176 7596955301 4936812797 9354666523
9389865492 2082126132 7763789820 : 1576
2946799581 6243987059 3623917051 1750705049 3924429371 2287520721 0047900369
5203530541 7470268810 0313142753 : 1577
1174456246 4073545200 1303351544 1611612845 3063638220 6203182714 1203471057
3330570609 5610419997 7441294378 : 1578
9723336195 2936807116 2946497417 4674606161 9442841957 7150642124 4911540670
7312213842 0641412696 7154566438 : 1579
9159471777 9694935195 8346843367 8322141303 7431073429 1743734376 3544159073
7738077683 3554545220 4160755832 : 1580
4500141271 2899741014 9470548864 7243415899 1296092282 9862407455 1605891496
3102100059 5871347192 0979723983 : 1581
6831280110 1752643186 8611183517 0173586754 0649265791 5137405821 6297242018
8375102977 2027692280 7801732353 : 1582
6585248661 0373552463 6051974175 8718234908 7381977451 9960413516 0468808608
2725590610 4828222575 7674635819 : 1583
1662903439 0705475970 3480904400 4304333317 4134614234 5412745677 9872589232
4090915108 7305920242 7900149673 : 1584
7015134772 1514257148 0238781897 2789099319 2321188518 0400304976 2893873119
8868763397 7056903190 7414517629 : 1585
7505582950 7905515712 8977260343 5467222518 7519472277 7503478062 9888158027
6408830585 8873211408 9935625654 : 1586
4526325626 2930428543 9933282550 3295028936 9907770549 0294707962 2008390293
2214441126 5738208956 8543447852 : 1587
2535584373 1269337547 9376599430 6991005699 0821560314 5081988649 4389488679
5977365202 3776380526 9495558714 : 1588
5427065851 7474445964 6823526941 0568519337 3700491448 6237606597 9574374294
9313762849 5423747696 2984236204 : 1589
0406990322 3286254828 2233542016 5228291288 4434215751 2706020215 3831784521
8564841150 6693943643 6446339032 : 1590
9462869215 0012003317 3722315945 9937024404 6654644010 7095463778 6273667690
4564259977 5860341423 3762759258 : 1591
5363126437 0897307579 5526996850 3132069091 8306791326 5420306400 3148245598
6239265759 7573177591 2862530894 : 1592
6541251662 2840716337 0149790267 3843253016 1901013729 7886469540 3425694557
2630522038 7629423264 8064996238 : 1593
1630855003 1265168054 4788556819 9731089679 5755442683 9220485130 9190268824
0337712017 7863986046 3980025603 : 1594
7206069295 3460153673 5130093516 6490475996 9041534844 2284064946 4357839627
3959796970 1199959968 9705500713 : 1595

9802671431 5391239146 1161358183 4068087605 3466725530 5042239792 8096566221
0911118477 8965033519 0031281930 : 1596
8140470647 8740367155 5521140340 7030398907 2232339159 4235126529 7171121449
1591287469 6964544557 0922804347 : 1597
3384101385 8874280507 2514932018 3676549865 4426190687 6750303979 5693902421
3437475259 2028444937 0703219824 : 1598
0950852874 3929412781 5958647543 0366953365 4646504381 2295538569 6018708146
3036000681 0223193535 6775884221 : 1599
7066271778 7522895393 7497394498 4606881958 9926057904 2663242818 1883297682
5708783089 0164354054 6417536779 : 1600
7521401491 6981613499 3044910420 4274172990 7318379698 5131245595 8606399199
6596689961 0800504940 0729639709 : 1601
8959517574 6349501131 5239540543 6384247715 7673057968 9978093511 2310127000
6068315601 3470561688 4208186210 : 1602
5906584385 4685352265 3099408095 5068645181 9643104550 0569852864 0369722726
4496407220 9107280506 5651759005 : 1603
3633194257 1882619016 8520911094 4462304938 7276226013 0096650980 1815021611
6189314991 7554486648 4510193964 : 1604
0892424535 1858629668 5358807237 0252086290 3963751354 4240841676 7961062540
7745354397 1872022038 9829258815 : 1605
0488174626 3214401932 4591263846 7764538782 1490032187 3605288401 6158146769
3409724342 4966959679 7455129521 : 1606
5247541300 3838241759 6775542251 5486890349 8467584610 6631594198 8121179713
3452509275 3070140856 1426350301 : 1607
5271473708 7979022966 3568300178 7990288084 1938939224 9188448911 7670080380
3875888780 1697701113 4533491153 : 1608
4802106585 0875700255 1736325688 2009759552 7487122535 7182551697 8753150955
6908689854 6483794843 0351870614 : 1609
9231357340 2963136527 9127615262 3061043140 9239565352 2974932610 1802357414
4940020107 5752924889 5892932458 : 1610
0351889348 3362322662 1107047222 7951789643 1135332221 5133111281 3026996570
5654236666 0712427360 6758337678 : 1611
3519101251 0994437030 4629076346 6149644955 9967303212 5852284006 8128863206
0138439153 5239320911 5790604734 : 1612
1393629732 3492759180 8942336565 2609394831 3364810290 6435863118 3082596587
8597847150 2344907874 7678799566 : 1613
7424852051 0401039997 5710394022 0630691734 7420208969 9175600528 8898736759
3966293674 0172098219 5418337128 : 1614
2339328624 7731719643 8625665314 5512699222 3677667770 8198643496 7998415260
4519464045 9016395789 6049279139 : 1615
2910434902 7568381726 8404705229 8140890671 3149152625 0441754527 1011235786
8012989368 2849339139 6383366078 : 1616
1422917945 5434491680 9706649131 8845378120 2579621552 3212883988 8294830960
2541545158 3015564562 3132845031 : 1617
7418576979 9799107895 5656799608 2529165537 5861223383 8070069219 5796394198
3742611767 6769100507 3575014710 : 1618
4127391778 3596347944 1159241607 4096491892 3864162314 4315098433 7999995741
2386084556 8790650179 6604659040 : 1619
1190093106 4914597645 5087441691 7093615978 0546717465 8993017041 3753904682
5444198493 0697739630 3361433004 : 1620
0322637044 1388424564 8531960009 1024035914 8604341956 7188919858 5615655465
7750890144 3177712861 6456862190 : 1621
0128459460 7421607429 5710458314 0046201246 3901102101 9323688746 2318746390
6390518460 9082474661 2222586831 : 1622
7169890636 0640254048 9350875060 0193538323 5847779777 7718415669 2711224004
4551677054 1931073038 8369438797 : 1623

8890462421 7590046666 0910401636 2270064506 7167256329 8135619169 5775856133
3390398277 5996522509 9040387093 : 1624
2789862215 9549437997 0630648070 9649417708 0058122705 5933091621 1844004635
8937856324 3586419106 1540068204 : 1625
7879016214 0445787717 3981029526 0717300099 1217971137 5424333488 2266618671
8059345350 0359794062 6101694558 : 1626
9487985287 3823946192 5927383005 8657862369 0127192963 8265926393 7819596877
7634491927 8138391527 3468510317 : 1627
1283501167 7541289696 3401763368 8033476132 4250065479 4483551600 2423125664
6080107867 0258603760 9939080045 : 1628
1756260090 6555541309 8404273574 3005006687 7433135282 0617072990 3389370532
2254670042 0588964046 5239361428 : 1629
3079185404 1696667832 4070955958 7709423200 9415610955 5343344349 1385438840
8610824862 4289859619 7412565717 : 1630
0424006787 6712368593 5372271091 5670406062 1943472602 0409941395 4720131755
2449158345 9442749191 9291350235 : 1631
5834404187 2069743458 8605383370 1858976572 0622546686 3899147406 1713844091
1140542444 8924181252 8058737844 : 1632
4375990337 0271443232 0785204641 9314755947 5831429194 1697190629 7697904498
8213080192 5875904858 7570102804 : 1633
9890092764 6674318174 1931273879 8791908667 0564601741 4104518365 4736392112
0183127264 2135299075 0753167418 : 1634
6041139085 0791740441 7265800928 8966400350 8561829937 2472136843 4131495692
9570401981 3001606087 5412795746 : 1635
6419031797 3325939240 2107416767 0242353517 4221182857 1516183297 6814222607
3090296369 4871330880 7785556663 : 1636
2397283342 2527065650 7307251890 3090395022 9751455450 8141344442 8165414364
4104921750 6227064362 8610175717 : 1637
1120483665 8149705824 6357800755 0456264537 4462805259 3284156788 5798506901
0580452797 5626285722 0830478354 : 1638
3668131330 3172332381 3526470752 5779523301 5289166395 2865431899 9573174578
0167826728 1460222640 3818995669 : 1639
3799484242 1098248974 2008823311 1400134104 4095160930 8313090546 5503159555
1547397748 0221462406 7611052716 : 1640
1375799862 8354396966 5783552456 7007936049 7518767958 5004347859 4444834874
7345525999 6323925882 0104452878 : 1641
9576723339 1108520813 7994842347 1531895261 2818751089 0512135465 4969246066
5576734518 5717740511 3980900507 : 1642
4932280070 9405692065 5442879928 9768091385 3288239231 2742649637 9071979978
5249090304 6095850203 2813011881 : 1643
9189798753 8612770509 8311267968 6721178100 6094288603 3416074080 2044853244
1414457945 4721054698 9291664998 : 1644
1941599750 8117083997 5855253125 3479307072 3771948237 3833676065 5418502113
3373575357 1160498408 8630696264 : 1645
9890151155 6298277922 3043334984 4936783915 1985626850 4325200844 7985546296
9126299788 3130293630 6463370450 : 1646
3315526375 2040473922 4157072857 7799880296 3532890069 8400767818 9697563540
2176619429 4424753725 6498457226 : 1647
5506787359 0934057238 9793781914 6080319827 1139248049 7941100492 2981431759
4991993103 2808979574 7253376814 : 1648
6061745433 1326448924 8037013462 6426692631 7342443574 2705177475 6506755634
1333600591 7831376373 7592038902 : 1649
0426517169 1865422448 4136094659 8636267754 3332217290 9728067721 2181229450
1776642327 1673310919 2752337724 : 1650
2450590808 5892756556 4344115484 4388895321 3270285605 4060064352 4034011774
3942638314 9269420366 7677312492 : 1651

3344604615 2792228711 7473732068 2073795040 7753807024 8751397369 7487220807
9418362724 5792671586 0587568437 : 1652
3825696601 5288450159 3636057997 6487955466 6897963172 7641445826 7140983930
2605844437 3118321952 7357824299 : 1653
2380530460 9792532175 9529437646 6967178599 5655479752 6010481116 0900192559
8770316036 9289053546 4219693617 : 1654
9009854720 7855125725 9753257687 8034184282 3941930116 7647127802 0013244905
4170687194 0608748642 5099249004 : 1655
1243777990 2567236238 7521344875 4686801805 7002677165 9045717416 7509358537
4663278466 1471702229 1277538164 : 1656
8935814037 5420413163 1796620462 6016830058 3584027808 4250847061 0285632146
7634921644 1559656539 9512441115 : 1657
2722520988 5576318078 8086283744 4395373363 8662891394 3990459186 9356297993
1691320743 1084912387 8269670817 : 1658
3798380276 0073283523 7130570383 5388120192 1780745570 5392131250 4821976693
4063942802 3453350969 8054521919 : 1659
6416496966 4205192232 2293325224 9809906809 4382986082 3933868677 3954452673
6321944401 5986590406 6528670206 : 1660
5100494097 8671302450 8960506363 1147497897 5431863273 7637932022 7953001087
7176844026 1821800385 9090607702 : 1661
9345978640 9329695112 3385326149 4565585967 7117544246 1946208674 1789881434
7780270237 8918386547 5073672507 : 1662
0123164595 4210423036 8253015499 2729230497 0650068874 4552908893 6646551481
9384805637 4554470319 7171864227 : 1663
2096587063 0049834366 5718303094 5852528562 9892546132 6079017237 4623360138
2089476404 7217671021 0783635306 : 1664
3283918528 9426309708 3524184268 8066639724 5435194987 4594952723 6882351106
4759383705 3136149452 3326299006 : 1665
0613544202 0790080078 4435918514 2381095406 2463592880 0749174723 2601428291
5066473564 6949149075 3130404119 : 1666
4874612758 4251725422 9933765619 1322834413 6159688451 2305097922 9080934778
0610001202 4307933675 4606956718 : 1667
8678475890 2916716281 5104791098 4819968795 0537574761 0343839289 2981834755
9135337283 4288849228 5393965950 : 1668
1645942296 8490216357 6984603656 6677068849 7906149378 9586623897 8513950301
9552520711 5947916243 0380571339 : 1669
1044123512 7977174258 9499718132 0899397240 9457630504 3817654202 3774937292
9236666408 5826356304 7018894284 : 1670
7136621796 2807794758 1410647203 9686900573 3588378323 8393851564 3676929109
5321263095 3023723418 8776377595 : 1671
1325585719 8868415635 1143465444 9213461836 2582001773 9111963565 9736209174
8028951071 1911931216 1615049356 : 1672
6140019891 5406771914 7406045020 0848900785 2104489840 7155872491 3181424123
7453147390 9585928549 1926195512 : 1673
7528154045 5554894860 5304395183 0551638652 9635114358 5542678957 8843322470
3032239846 2969403700 3638667059 : 1674
7551896228 2166849472 1551679940 1023726052 7619206175 0456049663 7170762638
4595300533 4443878994 4328543663 : 1675
0146414207 5150267652 9987148414 8385924204 6843515052 8589264835 4129599964
1906383622 2550061620 2985179080 : 1676
7999551716 1428927433 2216280696 3512162902 9650503454 5598002429 2038066113
1224998757 4877781454 3334957813 : 1677
6558008300 4587905455 6552375964 3089947282 9415658468 9806294311 2725975546
9302188791 2731035300 2168642276 : 1678
3366103189 0511086335 9639860709 7473741955 2934178507 8013653378 6877679115
1473833252 5130023719 1023587588 : 1679

6798039385 5297049983 2183038998 5333735331 5103458044 3402573042 2758682609
7239834223 1501764080 3327317622 : 1680
6319675659 8977297183 9422971652 2776196734 0857344413 7475914779 3179333989
2435994058 1396032281 3465924785 : 1681
5875655057 5159428611 6131767395 5283481508 0851852071 5479514392 6672875105
7441386769 7189020877 7611959245 : 1682
9359290863 8396962005 7686509962 9303818145 4912807341 8097204032 0363366766
4994439199 3626414522 7145073503 : 1683
7059076093 8575244000 0947482297 1362377215 9263083602 2139158855 9094614074
0697630129 7086569690 6676244218 : 1684
6183635520 4727903533 1753093607 7977879847 0802390218 5958744878 9607452374
9056283747 4189310268 0628048174 : 1685
3381300198 2212777060 9218470081 5252327145 9867234378 5431049783 9043649305
8607645357 5602598382 8162540984 : 1686
9639983181 1297188143 8639542654 4082800861 9305729179 9568887988 1825724409
2308607770 8626351313 6094630976 : 1687
7740997028 4727668296 6685429084 5405202291 9003034322 4718982049 9382480607
5163872794 5668940849 7366651622 : 1688
8133694882 8583395313 5050217053 6111817502 1010169609 3737288217 4683892618
0687188721 1712203206 7336498312 : 1689
8125457269 2763105648 2587901068 5752080806 3392875136 4817583758 1096955969
9338484199 1062692241 1365611711 : 1690
2954796777 8123862393 4934140085 4705378458 3782851512 3279873088 4093735721
1004586036 5454494530 1868107329 : 1691
3761108679 7828280343 6613333978 0538614863 5943716350 0877119311 9559848021
8289926777 9769409139 3084926892 : 1692
3971610875 1625981364 6427951911 6303391796 1291900176 0953221655 7349110374
7120945790 0418602899 2215511759 : 1693
1568303624 9471608650 5186322797 4295613651 9830351897 1428760223 7507160599
9046852270 2848242025 7009629938 : 1694
2791478180 1540664169 1792596965 0230616749 6772478419 4741420092 4297729713
8883251166 5522273033 8964447305 : 1695
2704902414 7727564715 4092376806 6416528226 1122166055 1903510495 3216953829
9957017411 4651029984 8982516439 : 1696
0569909394 5986820498 5525831287 6445683898 4342109365 8943105114 0755583849
2789369974 0950138919 8821247991 : 1697
6209888392 4355873655 5452375362 3890641072 6631365733 8688715014 3766765743
9282983207 3461314039 4952617095 : 1698
3084489658 0056558463 0952329631 8712451836 6166304277 4985273620 2349067859
1557693622 0534724261 1125326389 : 1699
1430252893 7667373926 2860646099 1664258399 9874647434 1416395267 7732246933
3134089892 2821352367 1609562825 : 1700
1349234859 2680407355 1815360719 6567395027 0691353576 3434376744 3117249084
5537767032 6669884144 9114808884 : 1701
8013249302 5843817701 1878566635 7293539878 1140646588 3694172838 7365708433
7575104479 9123597365 9724344557 : 1702
4271838473 3620516409 8603931021 9592121122 5720343651 0013963890 6494529671
4205608908 6176982883 1638828382 : 1703
5229707658 1896118954 5729825810 7339454017 2774978345 4068776411 0778004407
4292966398 0795802668 9312946890 : 1704
8915006186 1841892185 3041643322 1694927821 3392118277 1902167520 2080596726
2749463002 8053887779 4596218568 : 1705
3074384298 9256463024 0890633667 6064738970 4968736267 7347143193 4643782695
2783760286 1465838927 8933623236 : 1706
1603686965 8885994407 1709014385 7650085623 7035707472 8812300427 7647474703
7794632000 5543727473 6584724026 : 1707

1839025081 8502039941 3109503970 8124812210 7761283245 9563995640 7303784228
2949417904 1799135365 3370609295 : 1708
3580412844 9019567717 4363265587 3343028440 1481499075 4651032818 1387821090
7214339837 4541509572 1808723321 : 1709
6393141184 8854048824 7613315649 9354030311 3131971438 8566683380 2176668360
8295032360 4059513677 5927155165 : 1710
6796802958 5973380361 3440693078 1375730116 1300265797 0242655917 8634319436
2646623018 6872587963 0575563660 : 1711
7828996953 6349814522 3886600730 1477187919 8641606614 3908057772 5519244870
7082910976 7355499112 0061231753 : 1712
6184781317 6543955729 5038545292 3653669413 3485621787 8926154740 4561504523
0885311843 8978335074 8069944802 : 1713
0815830478 0292913950 2742186788 6919801765 5546818144 5443074191 1022927219
3186444074 9790317259 9397713621 : 1714
8099711176 1468900013 7724070092 3484996330 8320304297 3219675827 8940008466
5235071155 1118348103 2014483529 : 1715
4744885188 6324133039 6067639585 7662392727 4353864765 5332592611 3916010589
7206949121 6041943843 6276922502 : 1716
0408360183 4811827158 5534392555 3745823628 2552372625 3314359699 6463666782
5593382100 9175987444 4027185225 : 1717
1290642515 7453594796 1305271895 9948884782 4353172256 2754131095 9985044277
4753526388 8711892649 7707170550 : 1718
2201056823 2530711543 8956475830 3122551162 8763968835 1432627286 2981524075
5887995962 0980439465 9688932951 : 1719
9044901057 4191419981 4985879000 0533489612 0691631117 5468253485 8290076839
5376626414 5205273936 0786513805 : 1720
7224150689 7185582776 5389521513 5856589764 0730148811 1133738692 4588890982
2760229733 9512441350 7510038725 : 1721
8948206647 8544539290 5511604925 4655683017 9223635263 7755462684 0904947800
3726847161 0264950882 6206935748 : 1722
4631643968 9789627008 3376303743 0175619457 3890788481 0424309285 5236310298
3551745174 5446606529 7670819947 : 1723
4320599516 5291596008 7156521154 6129673541 3957327751 6784434848 6459833913
7584856250 5546017507 9220988358 : 1724
7719338601 3948967514 8337638021 3903415290 4283626453 7637458087 8281537904
7085445759 6761421037 7236123296 : 1725
1964022920 2289514696 8811447058 2893098035 7001504103 6948417054 7286922751
9704469469 7292034951 9697662176 : 1726
6414362032 1098749717 2990814400 0371573928 1891528408 3197422943 7336774778
2587092187 1722529840 6930555693 : 1727
5225836786 2538776990 3710832987 7979505169 1696299576 0702662487 7256111532
1024522871 7970309373 8005633540 : 1728
5929310890 1874004957 5133056457 5546864588 1925344372 2717048411 0760458345
0525724291 7684423561 0013945668 : 1729
2456604288 0004740729 2619165754 4095336500 0544583331 3333486658 1972749276
0012423238 2251718468 9306940857 : 1730
2324615542 3928138874 2202766169 3797935663 4410450371 1559823675 7714312491
2796231411 5016452888 0440942390 : 1731
0517346456 3780362939 5327787801 6360441042 7591864202 5167711823 5910812494
7844895488 0725855125 7454960799 : 1732
5691801297 1317205403 8424909608 7363621351 3109752862 0986224262 4187424357
8376772912 6417918801 3767520032 : 1733
0563326189 2410186165 1303481024 4006856808 6061104811 2942740900 9375274857
8585868409 2297315779 3668860861 : 1734
9964429073 6157490533 0345107467 9213921393 5734146937 8267840533 1319923544
3303460719 5155205700 6610011693 : 1735

0106114645 5648916805 0091450055 6137849404 5436931348 5910618419 3301895454
8638521986 0080820040 2863322268 : 1736
5779286594 7499006936 0751057802 3474201503 5317737300 8364949891 6728935090
9354212097 5207472725 0436661880 : 1737
0613349590 9306114071 1046459924 4759717542 4019965407 3065408416 9735239250
4156355003 9026440029 2275155191 : 1738
0862941367 2360002694 1207127811 3087636531 6152244335 1622669190 1231017136
2950133169 8077009767 0312259233 : 1739
4099542352 7648984420 8910924390 2756481617 0040721739 2725660242 9583715571
0671685414 1871300357 1310230544 : 1740
7259377064 5420643730 2838147841 9102186882 1846656713 8326128263 7806975309
8860829066 3303966984 6839624674 : 1741
7688511450 6491315186 1552462947 9248111598 7311090797 7115298058 8092092858
1622770652 7567771953 9311203573 : 1742
1943345934 3473372951 8994157214 7227619036 7800430879 5904799926 6424281962
0163009888 3714884504 3980122446 : 1743
2455602660 4086161331 9972328497 8761593171 2681400404 5056189668 6909847082
3941370855 1813261996 3768897021 : 1744
2415231378 1873313001 5601121995 6570354141 0653535638 4524396556 4267272174
3450531708 9708620347 6547586741 : 1745
2814640719 7922805744 6954068492 7959945904 5820901873 1765866162 5178733129
6935726248 7630182748 0530656002 : 1746
9462418101 5143131869 0240874938 4112421582 1508373074 1283373100 3222668395
7690735069 8817682754 8481773049 : 1747
9539131031 8465327838 3865626717 4760018062 7887580492 5400887840 3927985646
4964515527 8927345020 0154100313 : 1748
1905096271 0809628895 2376898287 2651391408 1945116719 5916663181 8853373127
9822794780 9638418633 0812324934 : 1749
3682732708 8471684840 8230651068 0498401989 9661418486 8219292371 2432262932
8442483023 6101798391 0042699040 : 1750
7744799190 8376102111 1239606725 0719299793 1365177067 3164504798 6232255197
9702992566 5315101596 6045966901 : 1751
5088706888 2982527286 4059895142 5047655646 4386139713 9093020257 1945855825
2713927198 1132775888 6955446292 : 1752
0605202687 6752213679 6682746877 5874528760 7786344913 8269956348 0082544144
1318253472 0494801421 2654329829 : 1753
6784668605 4377906133 8910206076 5389746783 7990904198 2264282917 1356980043
4724666969 9301575114 9537152043 : 1754
7403191810 7954868943 2406229094 5862326245 2209667574 4952856601 6465787368
8424654026 5604576973 2900129583 : 1755
7874201711 5100574265 9749253328 6825866257 0245837811 5218220127 7576078722
7875362544 1767851681 8491979949 : 1756
4976491000 3693490955 0819450205 5638112249 6479676624 9650785802 3271236894
4622866979 6319715390 2499010991 : 1757
1767205329 6592010243 2741583664 6285103519 4005414571 9071486388 2469446903
8224568838 8500782441 0392501635 : 1758
9715374849 0569452456 0531254037 9176033101 6534775001 9986495805 8355366142
0716997411 7331055254 0420552110 : 1759
7731858104 5894610462 7355807094 7146683528 7835222442 4395195109 5964801933
9972822544 1237291197 5335233397 : 1760
8820050032 0948307780 6628336460 6324667100 1800870666 2889771576 1318039445
3085177859 9796791617 5623642457 : 1761
9913187479 9529518736 7560206724 3360786278 3164465504 7133342557 7456220329
7058370652 0846148146 1803279556 : 1762
5723112891 3791506107 8782367241 7063157427 9086027582 6804832820 4825305959
4486535530 5335573608 9436683787 : 1763

7887790883 5773316581 5665640463 3363117896 5577553867 4513596547 4379288244
3277617766 5299775378 8443212262 : 1764
6758789612 6638330684 3849005800 5776137309 4604324573 3141597876 1655537226
3016164233 5345100237 4635368298 : 1765
9424782425 5806480766 4336180523 7741563140 3789337126 9990081154 6084081424
0586928446 4087423891 2457751936 : 1766
6466994637 3591584411 9317795008 5848065280 5204513861 7897232991 0964611770
9762971698 8054741486 4040358883 : 1767
9279500405 6809668826 8252678332 5875358351 6005057945 8531484837 7702967618
3263606491 3660564711 8508049163 : 1768
5911181680 5735686256 7675748362 7962595423 1444084268 6944417808 4654590010
9830083247 0127327673 2518629652 : 1769
8101198756 6742512371 8547191741 9644610996 3814369225 2764876865 2429643328
4880267104 8804488801 5591064476 : 1770
9829183364 4325638379 8347892249 9242473473 4749255855 7293151861 1034534137
3356722746 2578276718 7551285229 : 1771
6157193501 8632517217 5999942277 9441251276 9491665964 1176453311 3076783943
5875570151 1268339788 0778230893 : 1772
2767292196 7390656501 6790988495 9899971836 2018377246 6979164681 5888400401
5083264133 9017024402 8639070088 : 1773
3106649068 3497676288 0088097131 5772643341 6470525153 6471773066 1392722405
6325710013 9729989909 5593747730 : 1774
5596363485 6006159849 6125351831 0745042828 0599101135 6152764613 7187323074
0548644387 0951037623 9129317441 : 1775
3926799644 7473236182 1363311858 5804069936 5837776065 5841495332 8326602877
8546968943 0022926853 1019343019 : 1776
8737058717 3582180980 0669389125 0766257084 7465950628 9918468346 9499119620
5056288100 6235243400 5024075121 : 1777
2565976218 3568345522 5766840491 6525157075 8414614413 2895209700 9306872902
2716370563 8590610592 1696945735 : 1778
1312296992 9258356753 1883445210 9537570173 5632618166 4424591830 7191732592
8053735184 8183098722 9456262172 : 1779
5404441898 6403975038 4513606111 0062107180 8868929053 8855653803 2123197766
4500797880 8922913907 1971832155 : 1780
3376607146 8815888614 6659370802 1811848640 9491244157 8015869647 3723909595
8580311735 4939639342 3239812188 : 1781
3858322269 0622730436 9154796477 3290362031 0231584622 8211866082 8589608169
0940900061 8964421346 1734468252 : 1782
1433863060 8641076491 3030963038 6061561226 9477567270 5661641983 2266128295
5940541852 6700993894 4181452669 : 1783
9815121965 3967190513 8431353655 0321313806 8242451734 8894759250 3124192484
1752557403 8182351139 0261635537 : 1784
0936864688 4710152598 6682006296 6604332671 5884702846 7252827367 5136369158
9349857215 1495769695 7393793129 : 1785
3333878685 8701558643 8472121908 8131194713 3708733823 2750005623 9923744771
7210347921 6899958700 1504698058 : 1786
9562365188 5426829398 5666712723 0583317473 9467989387 9179844757 2639669972
5651509335 0494496239 3298941183 : 1787
8095115220 2738593619 9162089315 5937352131 9380127029 8481882968 2456924664
0158910245 2240833407 3529472376 : 1788
7660187190 8356625734 3946835470 4836224454 6199371292 1994552160 7705226537
9834751066 6769463255 5115664949 : 1789
1168070523 0528173086 9088268238 0129412541 8146730584 5934358127 3433407463
4710981016 9733784511 3700036146 : 1790
6614777973 7566767622 1187825394 2360370654 9237225664 7519270025 0824888860
4062234098 1154511333 4223901773 : 1791

6841135991 5337237318 4676634050 7156896681 9381035854 8079907399 6134538888
2685756724 3504591899 7400691044 : 1792
7041116287 8652679201 0616132471 9998448237 1523349978 3637523014 3135132826
9553952901 0864942058 1860439615 : 1793
9053058259 7540015734 7529974982 7230953870 5772100053 9641886970 4874528973
5915687927 0799441658 1049442687 : 1794
9390222782 0026173884 2463895922 1139263874 9541141959 4330027084 6714237068
1281377822 9848743892 2580196067 : 1795
3229557664 6222560726 5008320437 3463689206 9742573101 4887778328 1459700550
6211252970 9439551348 2069706780 : 1796
9204578900 9005563599 3193030074 6710425701 7918474679 9520164509 8538151015
3969617354 5527780430 6267757948 : 1797
7710979913 6259366223 4937064837 0598168419 1440900869 2838417581 3696077026
2652637398 4217927518 6558553400 : 1798
1802494738 8424795076 3593696251 6581598005 4901107970 7269532448 8613743499
3884408366 1468592090 2013874629 : 1799
2972909384 5395689309 1552470325 4564948483 4255843539 2750026839 8089195124
3857147278 8922881800 4727979105 : 1800
9416493716 6417567650 9443374654 0972890144 0632813018 9143863392 8063344349
4240026022 8810471699 9725533939 : 1801
5707641070 6789505905 2416329022 1291761570 7202813379 6306498598 8232242671
0298264264 5482279327 1548045876 : 1802
4992124132 1268182767 2309034755 9579303115 8482489483 0141731719 3431046635
6199829342 2660845241 9772743000 : 1803
8947575190 6443642507 0401157381 3177948095 9953392691 5579884005 4782895365
2395636741 6572964880 4634630573 : 1804
6737117215 8099902098 9445373325 5406649244 5556570479 7720791458 1230450618
8806693477 3161154921 3528598081 : 1805
1109640356 4201032065 0313878329 8144430856 3872065793 8940705623 2795868744
6085284069 8062839012 8319940403 : 1806
1753698172 9101193027 4216487446 0186196321 5944684538 0755709872 2129647584
2610580437 1014414489 1074881337 : 1807
6672138354 5414247871 3666653871 8207128484 7617070028 0230771398 6200152328
5284674980 5171600941 7700848306 : 1808
0781630740 6741291585 7045857980 9143541609 2906134945 9709688925 7105679100
5567529007 4750437994 6338211192 : 1809
1199900912 2153965563 1726332987 3593583866 6500189702 1037681056 5539125811
2742565036 5892142910 1919356774 : 1810
0079666127 1382307140 8188284186 4932545670 0504789023 5799834629 6652053903
4526722973 6797112229 6475763842 : 1811
7953370703 0794156328 9311746634 8996286910 5186047227 2688877875 8797953654
8113309718 5257748836 2549950780 : 1812
8962383116 8239465051 1685470862 6136402178 2044527622 6218509468 7714584666
7658899947 9371028457 0278582886 : 1813
4945578192 1024708840 9805488404 9428920275 8632513512 0327683691 6550933375
7568774231 1036161066 8383215808 : 1814
0256433346 4542717220 2495621806 0593586057 7836839825 4618223644 9833541991
9081817549 2396216871 0528049514 : 1815
2246375891 2011361597 9984380345 5389874368 6379416430 0305130378 8958312792
4845498683 9906586006 4078993352 : 1816
8127851940 9840167197 2972706993 2213390718 4209551782 4752068026 8463616539
7716512345 7434030443 2466147817 : 1817
7119961085 5372824309 1711263519 5011915381 0332261700 9607819792 2946035526
0187876692 3621248636 2488512903 : 1818
5442839737 9232513895 5506401423 9130766546 7538114524 4024706837 6528064142
4872089134 5137963859 9944935160 : 1819

8677107460 1432747723 8510284749 4666363346 1941723016 0773629762 8897725128
3002580846 8772653015 1682029250 : 1820
8730013462 1992315653 8719904106 0550741930 3633901844 4239787442 3384998260
6967605702 0535368456 4642727270 : 1821
3489439236 6484590024 5979494739 4860416671 1335717028 1209226805 2781568833
5313264331 7590299465 3857485218 : 1822
4710972047 7182480567 2156192313 1996627637 8282067062 7977864382 2558087274
0355388755 7637225829 9905067359 : 1823
1541471494 7437264983 9787057663 3115053342 1161217453 4089654152 1554977462
4788862911 8303526040 3687328220 : 1824
2507089353 0843523458 0815071956 9588924126 0528757183 9649630550 7662860091
1167261753 0072817388 8458812373 : 1825
5985372692 9926264266 6002172976 9040932291 6645780080 2861573105 0138340599
6052151802 0233746749 3294109576 : 1826
9139999676 6385217537 4648850721 4642276836 4860918319 7332363921 5924903900
0696788812 1011297463 5837340525 : 1827
8687854457 0222146208 7368587279 6641453017 6263354155 8879405907 3212225394
6707378265 4675608107 4649604180 : 1828
4339579538 7211330646 4679928612 2948571393 3856329761 6178508911 5582766119
7902337999 8663577047 4963796822 : 1829
3993509579 5450820550 5111893034 7793570244 3035283044 2834702410 5904612246
8081137539 9707428743 4351207241 : 1830
7982710008 2933191371 4192887714 0989863705 4627113614 2170603160 3887715873
4107562660 3462603469 3205757463 : 1831
6326530612 0596147410 9678643663 2812848924 6217276990 6044003564 8313727017
1843261107 6286907062 9628767824 : 1832
8337252181 6784952087 0187388883 5266818806 8856155382 1029179384 6881259759
2238717575 6873776636 5217279182 : 1833
9359888912 4812904849 9965476445 9655545951 5319230198 6734214396 2690524533
7463449860 3717927205 4279681688 : 1834
9295558794 5755341314 6588128331 0245574792 8050200866 6957169395 7780153414
3906770746 8844437199 7229473142 : 1835
0962430984 6450531853 9652190602 6711006056 6217145056 5239616762 9158214100
3930733389 2918625670 3337144724 : 1836
1709240794 4822081957 9349698115 5249255732 5408808831 6481951994 8492518859
7971817916 5071886497 5353694319 : 1837
5763660262 4261722924 2548005605 9572174815 3559340925 3828324333 4477794234
5089465946 8295480156 1640088402 : 1838
3550373234 9654987866 2171076680 1062510274 4723405477 7387228233 7063244223
4657130998 3353563617 9045129664 : 1839
5359207727 9387939270 0954660146 1050918027 3269755513 5713654909 4051709869
1433383437 3538622395 6625316705 : 1840
0813221234 6736878144 2761854788 3058500581 0785155567 8807693973 2421220873
0661826200 9083050415 0607987867 : 1841
2078008638 7483147104 6796221804 3947575563 0990862442 4438280907 0751636039
2136097396 7193494081 9820051893 : 1842
0846341841 8513775869 4213859700 2519235721 0352359781 4756562837 0064989358
0619427747 8376736716 5686044012 : 1843
4253539425 4608374734 6222496082 9827247240 2187537341 5104438842 7140893032
9039663170 5985272743 5757224919 : 1844
8043964068 9369083307 0460690336 3403761135 6692720080 1720601652 5870166209
2465653183 2178359034 8318466849 : 1845
6336231773 5446303933 7934892379 5838233801 4835246620 7076888417 7564682572
7171361914 8355289440 3611579624 : 1846
6825347099 9577854148 1648466735 7356113380 3192065822 1354967829 6294583794
8992590906 5715085858 9924036877 : 1847

7247095602 2520603041 0594547223 5734307619 9202003870 3424402223 4909496718
0951194798 1181231766 2161328126 : 1848
5741888792 6717804023 8578005598 5292325615 6882467651 6359048340 5880044838
4582302419 9841762420 3975028214 : 1849
4203323781 3646956129 1816090880 7052269274 4785023579 4371561428 5496103309
9970139477 2146061745 0078824754 : 1850
1700679178 1881337307 3553878679 6010124219 2434173987 3289763228 0986762293
7453437289 9811725930 0822232462 : 1851
4375985400 1837266087 3832647120 7255449130 6433644995 1001947825 4452554256
1198544468 9633861923 3410886119 : 1852
0236636252 0061671773 4072684448 7670870786 3399288518 7857488689 0695595205
7560806553 5972362554 8665768065 : 1853
9973002696 1449979138 6394913764 3343951178 1865616972 4575011955 5271398766
6331024199 3649615936 7342333676 : 1854
5935189951 0821050854 5586590240 4524395014 9586570975 1688017729 8008199222
5972528916 1528326432 8713301912 : 1855
0720262250 5599302105 5200593642 7206720674 3658081959 1983894686 2415075380
2751656622 8260425584 4872369634 : 1856
2315273704 9647360124 7293747058 2351894637 7728760858 6271395235 9990692232
5870359910 7092753607 7178731275 : 1857
4815094035 1270138170 7948704002 7946364336 8842771692 4012826404 4475383002
1680605559 7399111532 7567430425 : 1858
0791689664 9365346106 6490303392 6454798262 4507527529 7035511702 9389549392
6050261167 3280508063 6191135041 : 1859
5038722553 5480524950 3072592208 3212991676 9939385789 6052191904 0226593296
8932015280 5385584883 2676736575 : 1860
6858379942 8685543148 8484598780 4319937107 8484089337 4197790800 3386369665
9632700048 0075341073 3130286958 : 1861
2860135928 7661350885 6941307268 9527062211 9444657090 1350002850 7817008173
2969360699 4478080116 5089977469 : 1862
8383275335 4462231178 9004142445 6125659236 1906713778 2218830990 1262050387
1386374611 0754713824 3333426066 : 1863
1119112499 6043119748 7300355784 6753855809 3194053656 4143872408 7159307028
0022336202 4034209266 9248410365 : 1864
4139246032 5281513910 6025806900 8679246947 8464151377 4253049081 1333748592
5659032521 0843787058 3690180305 : 1865
9338532970 0109696000 8742504481 4184589259 8569653455 6980827237 1276255400
4837927076 4101702087 0006758524 : 1866
4432577526 4579036182 6803605262 3878996687 5626868457 5871134948 2617027872
6420774053 2779178396 6059302468 : 1867
0537612528 7836224216 3181476420 4764333456 5869242915 6194617414 7929303267
2745331987 9627590510 5825564390 : 1868
6412796059 9605162941 0558357700 3536563242 8567139724 3309359986 1784854409
7181817255 4477791409 3209591840 : 1869
5016499843 8612807378 8718816754 7887565056 6319631976 7304705864 8946240459
4926976645 3285109198 7443373512 : 1870
1566448881 4513250978 2997998568 2830183029 2718126587 5799749152 5942142606
6384493476 1823669436 1301000778 : 1871
3474504544 3839405946 3825316417 4696216796 3754039491 5211600835 5340458730
0703416744 7688538635 3724175911 : 1872
9191257302 9628757699 8669830602 8450551254 2557781304 9196573537 0810975388
9805144982 8195851720 9632887924 : 1873
9759668785 8557626872 8363857714 2823352346 6567958926 9485489195 4487424195
2228540275 8101327257 2588484604 : 1874
6549518516 2225272148 5896972726 3289515266 1007419195 9717832883 6594559768
5705772628 4785595448 8372407579 : 1875

1628836314 8490647791 4553487265 7558501119 4226486996 2439109009 5948421950
5011828545 7021898741 0345718389 : 1876
8179048636 4649082967 7731508367 7699733551 5074170081 2202580538 8524519536
3983453187 6177812318 2923384516 : 1877
1492018946 7872021748 0245281591 9012422558 6516987472 5902155076 2497491226
7376594563 3076166021 1943484032 : 1878
3979914407 0248817343 4330292725 7109286573 8989424064 9561810909 7976549851
1842477113 3900728809 2990163018 : 1879
6941142126 1170343722 4966747668 2598881833 7774891530 0135800231 4602426047
2055275799 3198994096 4316114429 : 1880
8528316114 8698973174 8642308262 6493416316 8452780162 2286945217 6524878906
9955609915 1096015879 4169103884 : 1881
5956359668 5293612599 1245729283 7693574949 9600063745 4051029331 2523537194
2115650133 1547537362 8090714281 : 1882
5773185118 5927633100 1448478115 7730515727 7411636321 7629955597 1064374278
7164040829 8307104630 5419155993 : 1883
7148153916 1255478112 6437438903 9745212073 5757677775 0742115050 8298100857
3752383518 3835753993 3202975989 : 1884
1578248805 0412070590 4644073227 6884830874 3534512264 5470626940 9754445136
9757257089 1505730232 4257356727 : 1885
2108516847 0673901113 7210182805 8046032247 9166007383 2691454159 3198292432
5433746486 0496339653 4224817293 : 1886
8325475111 4037593843 7780558100 2679023593 3847989586 5486078741 9414168840
7304341734 9690424069 1742895829 : 1887
1133881573 9432277101 5619624776 3551902402 1712746862 7824721979 9676266290
0919176955 6438105385 6926401859 : 1888
1476166954 3194077693 4896555906 0313034157 9094455197 5602966248 7545487909
1117532699 3709371263 8067225675 : 1889
1463060740 2334459831 4820578077 8552538169 3436480535 6807945204 5386888722
1458052022 8137165269 8201162506 : 1890
1657297379 7480750029 7233921909 7501220494 7494170065 9392967296 0287387671
9522255063 0864385036 0228641843 : 1891
7662400917 4719032833 9083995367 4746861310 1093275450 8537010324 8816456357
5489558603 6799189361 1297876190 : 1892
8356737312 2748237818 2701853106 4263096134 7248714360 5493189037 7026133291
2020741185 7020396549 2336859086 : 1893
3272900347 8722237659 8941863189 5233975752 6448262322 8467814741 6783941758
7877928413 4112230198 8055483719 : 1894
1966208599 3112969783 0338865167 5854462279 1340538447 6083942355 5344903316
3300012475 7996527616 8526319453 : 1895
2930959627 3131467261 9266181983 2194566480 0489127240 4216573041 3638330422
2664897851 5562645655 3219711447 : 1896
2973260582 1321486152 8010097276 8015104029 8965202063 8606860045 9816142852
3749991208 5209347292 3079773503 : 1897
0153390597 7647834289 0747487781 5173481573 6268828728 8473096086 4188664530
3294760075 3309358538 0277049260 : 1898
0732884329 4119520864 8297113175 9375253443 8895881425 5548385173 2952836011
3779153290 1133575981 1747590808 : 1899
2955904750 6576584568 6049895198 6993050662 5060170970 7832987606 3046816009
5210972977 8751360186 3205458955 : 1900
7818005958 7571719117 2323509157 8713751539 9125515052 6069594760 5933157635
0909179773 3208336136 8071945515 : 1901
6407495330 3571884225 6369317118 3439732516 0573650377 4732145350 0563595638
2624749363 8224705836 8475214260 : 1902
7279199552 5107455130 4244338336 4054939700 3333713488 0029978594 6575449427
6518303412 1119702102 9987336199 : 1903

1177647930 0476493264 9152190991 7772362558 0527127257 7992841862 2127220259
4729578364 2415695183 8904261862 : 1904
9319498502 3980882790 1822770867 0870719353 9801835763 8280472176 2702614902
4918463402 5236129564 5126001797 : 1905
5449613122 0872848373 8833193857 4020189082 8176785029 8505262156 3352750833
9394101455 5472125637 6368546407 : 1906
9094647653 8165508050 0179673737 4314099089 4748416914 3006508118 2103990094
1917142905 5442874348 6917780828 : 1907
4127163283 3493337387 6018980519 2382363730 2197650070 2991998409 5395364081
9293935448 4433786725 2570777295 : 1908
9596138710 0715792147 2183707580 4150054413 4986004929 0749969893 7903488102
0827925069 3005742360 1746771263 : 1909
9825044479 8794775513 8368875388 8207757212 0635319586 5003008391 0654471490
7549279711 5572184609 0155394573 : 1910
3251863898 1285824687 7695980082 7417655549 9255652637 8718474988 7063291494
3908030741 7260367589 6802548718 : 1911
3739999619 6829326612 2412170677 1339713788 0925201702 6230147178 2080063916
2135982052 9738550555 8209594033 : 1912
3264708915 6195552235 6638062641 2574791346 3825374925 9912880143 1261443620
1171810061 0472258584 1850028634 : 1913
1562115688 4418566202 8272766006 5536243416 5318617270 5470460182 9523329536
4896057333 0745306473 0077394581 : 1914
7405562218 0102965869 5455429623 2136268085 1935845002 5873573058 6665195617
4463718111 3447756361 0293164229 : 1915
2999771284 8473992479 1499752469 7574616765 2401333988 7118993502 9199507259
4035417767 8878427586 3173386202 : 1916
1231433122 3245499952 1022646419 1705902063 7215636487 0441042698 3363333138
2169588483 1981209369 3683591904 : 1917
1149316234 7872763662 7592154568 4107024174 0529792569 4298191498 1740639527
1447869051 1842343371 9559261231 : 1918
9193757906 2117858090 9320588479 4836305795 6121560105 6518207521 6489529364
7504997836 4259687880 4760925999 : 1919
7018653611 1131360484 4810434313 7267327149 3251764067 7959128270 4180928409
9302148095 7457866349 3779227121 : 1920
3755371254 9498364613 2191054790 1195080548 1637782317 5531880540 4834474568
2348295528 2130638303 5954647975 : 1921
5313386037 1316576407 8334088593 5945737671 9674086252 5180978181 7880369860
1166138834 7129991537 7112383415 : 1922
4528740489 9564690282 6030068854 2763451896 2357355461 8158222214 0719678668
4310268265 5738115194 9371316182 : 1923
3492530436 5477918872 7730395772 9166760359 8906829849 7927532644 5793020562
5004198215 8178336797 5832458201 : 1924
6703344016 3751943613 0793606687 7060596155 0745818730 0740588554 1857077712
9376539546 1112355201 7737452675 : 1925
3650277123 6010262637 1140850249 3754519972 3881184972 0048529540 7605375755
0486334985 1760393434 0365895296 : 1926
0860734480 5531229553 3568821456 7118057604 7588419420 5834963384 5421653770
2022628873 2032814262 7192419113 : 1927
4698071535 0623640604 8801061017 6139643065 5066646671 4597747927 5127501313
3465967646 3960699440 5703118605 : 1928
6087812280 6328967816 5765372750 6296728357 6263974828 3846472950 1891798560
4824919850 0799160923 2399766719 : 1929
6476783301 2638465080 8804283111 0985025546 6129686185 5650350012 3610685297
4356644656 1984920921 1012663758 : 1930
3119546240 1126619489 3008382843 8659999992 8333379487 6598213558 8393330975
9653943516 8747702542 0380520337 : 1931

3382317893 8782825430 4773685927 3772357478 8865668587 3609256869 1056377446
8511315594 7867365164 8492178603 : 1932
8920469205 7392137396 5976293426 6179938759 8861055713 8473901586 9538001440
0337739425 9635248692 6368960908 : 1933
7053952625 1096127290 8873767982 2241074767 8488299026 2592141720 6513544327
1991645998 3330333820 5097023670 : 1934
3791891297 7711390002 1796464556 8170138089 4182584629 5987689363 5924393798
0370126043 7000195453 7532094758 : 1935
5656686261 8691377693 2365553855 3373664018 1142604011 7126345320 5372512446
8808392504 5538064254 7662809345 : 1936
0603910861 5119488314 2464773935 9453611346 2632539790 3053106155 1540475704
3183580698 8891168508 8278357540 : 1937
6260470081 3348942775 6419881104 6150339108 1966974360 3856073267 0871560877
6658858910 6089607208 7471582697 : 1938
0169056266 8719926815 8483351741 0241095060 4976133322 1030468320 0931629481
9666417866 4108925963 3540386292 : 1939
4130152476 4041519532 7618247072 3527897727 6957745431 4914572040 5417995318
5788374981 0850505715 7671110581 : 1940
5852167055 2201100240 3121471715 7984645854 3324890734 1098761099 2956496761
5654471880 6244284933 7194222747 : 1941
4044983798 5964755841 3348941074 2608333615 2120775019 2980151294 6567208421
1550763881 6459889661 7646436976 : 1942
0328924324 5105302982 5051224268 0703731212 8083935122 0225540841 5132029479
9980575137 6864933655 6761678499 : 1943
4713349269 5625773907 8483718248 2831556792 9819728778 6862903631 0560685800
9907222402 1538766147 1364480196 : 1944
5614811245 3886271654 2334288756 1979792048 5573019299 9750041758 8186220355
0843526937 4224183477 5423575605 : 1945
3467255495 6154189883 8178560292 6769085061 3136595288 0391735556 0245687971
7723106032 1745760497 5950232256 : 1946
2941963790 6309379558 1044909587 6213591677 8658295393 0306576530 9230704398
6757062576 0671427063 8526055475 : 1947
9595253213 0478006326 1071076808 3216210014 5794640977 6926800691 3907193727
2531922852 6274289573 8950413768 : 1948
5477459296 0335922726 2526666835 2170703189 4962827245 6528458241 4254606303
7280407774 7979885412 9463539799 : 1949
9246474691 3355243372 3183045353 8489080808 9315251813 5768485272 8589173285
9174645036 5612006882 9470503204 : 1950
7169041537 6867800192 9305063669 5778550885 5054236989 0122298087 7912670610
5235629735 8060222018 2943158073 : 1951
5552190937 5865774736 2647369928 8881279782 9333934998 6977352324 1375993155
4636311929 8207065372 7478607258 : 1952
9973120693 0627210401 5723943842 6087560393 2638706392 9022190308 5890987772
2019855938 5372688147 9322882922 : 1953
3698259046 4309339788 1652299859 7111438879 1916811255 6374983131 6110931906
1156325528 9261205865 1598514939 : 1954
7612705562 4087676714 0605906275 9367897286 3204655894 0753192715 9129511701
8443755758 5352369782 0603460308 : 1955
1114085616 2220429042 8905287093 4871938753 6681994211 9678716034 4751165632
1704404160 5351341390 1731366894 : 1956
6388738555 3138636824 3369975985 9706164570 6267041304 5912643728 4989148356
8904556090 9348101158 0923180730 : 1957
1845998408 7990904615 7493109861 4313315919 7840606356 8318841950 5707596210
3268508407 5395110460 7136774315 : 1958
0631865568 1175045684 2910985936 0948634686 9593672277 5807730607 2883798814
2468100342 6858744195 3320342225 : 1959

9222591131 5687185512 9884383997 7181848177 5752765286 8727478679 9756095598
1443326979 8023224692 5174800848 : 1960
0437354026 7386844464 8250945683 7198696619 8330889858 7835257932 3281004784
9800001659 2407290314 6602815056 : 1961
4724110345 2031576527 6577171450 5108046030 5129759639 0336904878 2270839013
3104005385 1493735374 9729516134 : 1962
8972263979 0211988963 4448662018 8190295769 2950434647 2305784526 5200580679
9064539004 9554274873 9603331115 : 1963
1334342323 9392815392 8575524189 2542753368 9936707673 6032707695 3407153977
8317693299 8580029024 7380912222 : 1964
7024700301 4973214830 9934933241 8808211182 5695862329 4651857563 6897541635
7468959866 0266517287 1063731782 : 1965
1154407328 3084095822 9371768628 0368564515 9152570329 0275690368 5712988312
7811874734 5960741731 0097884731 : 1966
5628386494 8619310435 0166181226 6303769593 7267645885 3838094304 9453023030
2680142109 7550250389 0721484246 : 1967
0093398754 3991538384 2137754597 2464098687 3792660279 4166204708 6632843876
6273660878 2721500359 8927765170 : 1968
7445477065 3839619602 8343102852 3840913387 2378563979 5368257883 7058304894
7266348134 8213171908 8833963367 : 1969
2412315363 9729520379 9561405420 2652355733 1822605360 3015161076 7270161366
7753472021 0899524060 1901907310 : 1970
7167115721 3153131399 1087346049 9485588793 0555732907 4866756924 9917791477
7762752572 1533153059 1915437576 : 1971
4020855624 3114944537 2545956809 7025647576 4244423090 4740701449 3872009314
8556612673 8641899425 4949313631 : 1972
0475961893 3034909499 3072843240 9009866042 9647764160 6362128947 6951726567
4169221041 2679197620 2629175585 : 1973
3059616058 8359815094 3813988815 5464739539 0022108597 8718592405 9647802767
8892392428 0477323241 6801150880 : 1974
9942907513 0067286149 7273785041 6001553809 7278691011 6538163760 2995600199
8756771052 8743417964 8634948759 : 1975
0228434508 1024845232 4285061945 6464928288 3380246745 3143600766 5393932531
6906934715 3411102590 9155950980 : 1976
9996077710 8192404340 0817401909 0499522416 9459367084 1551263350 4468374235
4082912646 5380354941 6953846871 : 1977
9159478644 8216907197 1882790453 7417589786 5653963543 6417496421 1383323912
7266085382 9567746264 2204374861 : 1978
3750869656 0381441154 4678174631 8241578012 5489762580 2405672218 1651902552
5646655104 1784031399 3155273497 : 1979
0128274640 7837967734 3103957500 1167643501 2323921872 1736939561 5725612096
2946586125 8179225997 1229360156 : 1980
0483252932 4660590007 4675382891 1358876966 0502304327 5464415772 7204135535
3431069230 2099040958 8280284249 : 1981
2545660922 5504736786 6335359776 7011475477 9378951221 6395039174 8837006069
2083214313 1056511403 2165914971 : 1982
6054503315 2608756244 3039751201 6270444756 6549744508 2910844914 2753286512
5788432014 3371916195 0742434585 : 1983
4267127681 1026007996 9773273109 1087404071 3888398593 0205685477 0568128370
0324106099 4880891203 7233751569 : 1984
1677129447 6770105736 2851752692 2673867332 4904110576 1883634334 3739931740
5736193536 9077705806 9918700110 : 1985
3875506825 8651233963 4192984733 0966787573 2032904837 0056903353 6216837286
9158682248 4931645864 1309955612 : 1986
8076135431 5839479796 5036457984 4225293998 0325213460 9728622695 3626724707
6289971779 6327633461 4112070415 : 1987

4148305304 4019675458 1623598606 3466572733 0574033424 6756825399 8855700384
2039565097 7199541002 6837628297 : 1988
5119707156 9287780588 2319026171 0147580089 7373783464 9921004305 7076158595
3225073361 0872957027 1507431229 : 1989
7920313721 1031512057 8694581824 2017418320 5651511753 3812845799 8173296130
0097228591 1308282090 9053314760 : 1990
1196781850 3836753034 7047000578 7483609975 9090912963 0344182765 5051198429
4261174212 5017453108 8376152772 : 1991
1032091623 3088335710 2085772162 5950992529 8643641820 6894396569 0856477512
4318290301 8353395109 1146751371 : 1992
8534246305 8517707443 4321613169 1304544562 0729557791 4988904854 7850294942
5186992301 5642048236 7299678208 : 1993
5432770817 1399372971 3647285516 2369102809 4394904980 9571114798 7353263361
0861544909 2136210719 5786271826 : 1994
5898464545 9587009069 2492488205 2343511286 8738626912 5293356955 6562440153
3344756716 2409478118 2657115595 : 1995
4756699368 4235624999 7922772333 2856784786 2452694981 3038295767 1588368253
9034846167 1496801413 8599194055 : 1996
5979179178 5828197578 4812372478 0229627342 7132738070 1712131593 4540225441
6861464162 0641854955 6220175802 : 1997
7171741932 9604030724 2855759140 3748752412 5583648684 7826530579 0211293015
0460093009 7911328939 1102092842 : 1998
2212628874 3972398792 9998722171 2680244269 5704364082 6917512394 7288580976
6317352190 3477402078 3010825008 : 1999
2306867481 6599291621 4204378559 6907008396 3431749157 0400704911 1330970230
4687661585 7483135080 1444759928 : 2000
5202072786 0406246909 8624581837 1056631825 4920666633 9286894164 2231681397
8537417455 8983550239 8141347627 : 2001
5686616221 1863675611 3454018506 1230145050 6414647662 0025479372 7370169115
0910570058 8058385528 7751553568 : 2002
3461355508 8814313744 9856363777 3694334730 7792236920 2328195126 0198833485
3193084139 1296921034 5115664615 : 2003
5817184516 0918653048 9711953801 1024852574 9893158647 2339992674 5372521914
8787799788 8075626737 5063872378 : 2004
0564697643 5268613067 7476116156 4030889810 7229900613 6202913855 3864683684
2458354434 2072490652 6943131926 : 2005
3630645579 1910328174 6224652305 0868114539 2237903469 9935761819 2283841178
3111273426 6093171716 0547230274 : 2006
8587000104 7866059835 3687620423 4909356314 6793544370 0708676044 4160809343
0388964169 1229384629 3502166110 : 2007
0210761640 5466145328 2613302509 8992955391 9275962994 6278263263 2116565874
3195517335 9427872479 9548287227 : 2008
8107931497 7711035342 5543816635 0502182004 7559845719 4707642967 8271587726
8483623611 1806592445 1595282915 : 2009
2301818089 7167227176 3496522837 5068073131 7414453350 9330105586 2157197336
7591051672 0488567454 1572816321 : 2010
7259397927 0182677659 2787907269 7595865244 4479862784 8766953949 1461017760
5776036071 1075086603 4557555712 : 2011
9623454066 3775844877 3140658050 2181444145 7012161388 9442942543 0127261439
9603975154 8809684175 3887787099 : 2012
7710531568 9605779553 6359670078 0699856501 1955361699 5819109185 3337403661
9990661867 7458653659 3782895158 : 2013
6192168358 3853720551 7181966990 0290622524 4297196477 6076579212 0834997981
4831084253 3800664605 6465462844 : 2014
1059597587 0105383783 7669513414 4117115765 8015291972 3932831823 7419072418
2705562114 2924812595 0086219348 : 2015

2545185655 3970125840 6477745909 4161077898 4486679878 7983603594 3067050826
4698506509 6507142428 7984166501 : 2016
3303364759 5971329458 3569058759 6970583659 8402375264 5595142841 5274309347
6002848059 7374451154 8230400857 : 2017
7453819441 4235491878 3809292297 8318441402 2384436112 3221688505 6243354185
8843251154 4720643284 9620845632 : 2018
8119410827 0588318935 4288454365 0548453563 3008842668 5693563642 8902027669
2308486633 6118299142 9872638798 : 2019
8068299808 6123949763 2951046359 1338269125 2518794669 4508941539 6493327345
4972994489 8362994739 9175474416 : 2020
4719717317 7987268394 3602401052 1661014981 5265541625 4038545177 9521584002
4958798797 4104952480 0475355816 : 2021
4544116079 6496743747 6718422118 3581573767 3704896816 5761864668 4473995745
7386389528 4956651895 7447866597 : 2022
7781950752 2588829870 2478900964 0653185204 7423769523 8933550121 8478599660
0740896503 8385951470 1804072345 : 2023
7768783856 0758095616 4533921688 4897542598 3059917537 6101323206 3543253442
4048860000 3090822619 0037306341 : 2024
8486886143 8763736494 1788740120 4826095051 2759863390 5097702424 7252980175
8826392293 8707936732 5221116705 : 2025
7926441409 0854374014 8530459025 0371696374 7745860719 1405425694 3815611701
4437888441 8883091592 2927192035 : 2026
8412987162 2866850532 4603894356 5002307341 6708375186 4595368025 2758240520
9237446765 7335127060 1601170349 : 2027
0806822232 7234121408 4695966733 2516156575 8066590243 1013032064 1153751168
7407756787 4060359258 7886171973 : 2028
6349367711 1426543048 4708113330 3231866339 8550949431 4397480484 0787647767
8327705348 8015967141 0169844356 : 2029
6978084548 7805182319 9575640739 7883177027 1135643924 2044520333 0076097643
6796999004 0958549556 2013135848 : 2030
0587537494 7256934033 0909172832 3941836921 9324915186 8723547739 3921275611
7946640185 1180013807 5010277721 : 2031
7130642042 5326555361 1432390788 2035094537 7075084348 8923010206 9364851728
4976129383 3257931632 8040240236 : 2032
6224770735 8488505586 1960214818 9507568896 1464986471 0858464453 7329496552
3337264188 3832621271 1782724069 : 2033
3226571570 7864175572 8961453382 9164489186 5204955272 9526330028 1049823109
8573394308 1602256698 1711150564 : 2034
2180307494 3611078136 1389682204 8773651856 6702091978 7109427227 6503470633
8508550084 2117094040 5082569924 : 2035
5756282826 2781375133 2708052945 5232216084 5405765437 8540071799 0812768836
6953749752 2864067146 1534564901 : 2036
1269387426 7114036215 1382047758 7549428565 7227853366 5848729086 9174951010
2375874976 6072301695 1857365090 : 2037
5794918186 9154204951 4818950633 1367232336 0017919244 3975940164 1677198359
4510693427 2172934837 1331527082 : 2038
5228587814 7644954066 1682660663 2817385906 4681708480 9801956309 5401910023
0303837721 0748322781 3901168208 : 2039
2582389277 9361395612 0621621339 1578640790 4096277774 3062394588 7116813593
2412443371 0944830874 2299489657 : 2040
2704969668 9190976787 2956785683 7491826622 8075947073 0876390942 9179184646
7289893503 8166571603 2383413004 : 2041
8221490735 5731011475 6043910764 2307049971 4171792722 4988936251 1853771844
5653611243 5366803341 5834710999 : 2042
9781275045 9310729492 0164004043 8736891084 8900002206 5896894950 9883554543
3034480634 6906836264 2692622526 : 2043

0480503822 2965665856 4454638172 5787202422 3930603167 4501605397 7551655424
6030743256 9145384140 6677000933 : 2044
4817262533 7857836954 9688018197 1420758304 7902504544 9329434408 0654706966
7092081966 8718095745 1822379033 : 2045
3116866601 0658854646 1622251368 0755807281 7839904993 8203254035 2222147912
7873573379 2405058170 4793436111 : 2046
6046575203 5096499203 0094306338 5151557010 3965436156 0042502091 7540836802
5107569627 2405400706 1307391483 : 2047
9978215497 5269620067 7717461253 7517747408 0770421469 4980724656 6921031380
3655901391 4463193378 5249560765 : 2048
1289588470 3956836005 2405603773 2266484889 7675986472 2223687045 7260025131
4653302789 4907366831 7542852793 : 2049
0436416844 9130901482 2977944414 5397767000 5047645453 9441997442 5340090220
6497079506 5778667625 6257904167 : 2050
8795171932 2821604842 7904222814 5745555525 8501105051 1185320512 8248170449
3408500651 1105859679 6611348054 : 2051
3157990100 2711637041 4625588451 4695315016 1376530986 3467935139 8306442172
1253914210 4848401806 9955555893 : 2052
3864698447 0972207292 0441600174 4645744857 8988521913 3254971330 2548209802
1992094686 7055130885 0411232159 : 2053
8940306060 7764070886 2153022528 3963061061 4984492974 7045128120 6439250952
6839331630 1653540689 2928056518 : 2054
7157265787 4119402174 7809172799 5418741181 1373735348 2320492402 8544437285
4241447866 7353172039 7284099921 : 2055
0753385213 7685218992 0275476375 1550880323 8203451410 4490336878 6105511397
4555644534 4133528058 9331495072 : 2056
4154536504 2536863587 6511464557 7638528618 4222500373 5443386084 1945720257
8083624670 5161354412 1936052124 : 2057
9265478557 9790112658 1591993322 5542147336 1025220356 4003582790 8575507305
2788354315 9467417937 4264974074 : 2058
0947948944 7795731660 9623021732 3972884026 0162155089 9074510246 2967183685
9160378905 9816357439 2667278295 : 2059
0299181795 7028068636 5101245445 1544131814 2965418452 4519788730 5202002880
2043389552 0952126242 5068207362 : 2060
5164648296 8883150509 5970100022 6437213534 8785826025 3357898428 4992642598
4938269865 5591574552 2772230447 : 2061
8367004512 9262032590 7284470070 7182646394 2993971057 9650492402 7215130909
0201632257 8929364662 0690791141 : 2062
8909170955 4858581709 9969398458 2418886230 4346386468 5370946920 1908664425
0014237049 0706054794 4016363622 : 2063
4484204946 1414540733 4077205613 6753779947 1743464186 9614416355 6429471591
9709591245 7298893923 3815001041 : 2064
2294395852 8812429031 6381893911 8293640475 6748013200 5483777642 2413083227
3379016805 5134561187 8652637873 : 2065
9084602983 2484496777 6765267144 6090984272 4092219442 0872905077 7247422712
8491998627 5288409545 3612244260 : 2066
8122367302 6362416664 6367695658 2340509347 8650114354 5223017211 0431829674
6118127124 7726747558 4183473918 : 2067
2964689242 4390835898 3041077861 2221646674 1392745808 4410934467 0914076889
0811548042 6990464476 6179037069 : 2068
1318643164 4872934811 6247531427 0947951218 3711895430 8016061368 6742330865
2068568392 6148047844 5664749457 : 2069
4832329837 1127834849 4575681848 2357381296 7298602509 4456310021 3870768049
0430110884 1043560659 5632913551 : 2070
3636595379 0577450863 4658418379 3785502138 5507306606 2032361892 0265343796
5542409138 8667805176 4866023556 : 2071

8680102444 3819982174 0818683080 6326579344 5013660695 8831163527 6590196371
0912216830 2179943178 1781159756 : 2072
2569334811 8175901637 0453954880 0254386919 5029394842 9633387880 2324540268
6831159207 7147266096 4081472974 : 2073
2564135237 7071326558 6567292609 3521313563 2697386334 5139232379 4912727416
0440716533 2837276663 6069920782 : 2074
8988515818 9007406817 8835600338 3955024910 5442191369 4943840259 2897576804
1647987388 7544190710 1007388250 : 2075
2600250529 3715712059 8821799751 9052515481 3512892650 7035031295 3887973951
9680714631 2979739398 8552240677 : 2076
1074781329 6611251424 4409425462 0586560563 8648411769 7376509322 2320058137
3898885989 3022336308 0952193426 : 2077
5228150675 3067731168 3499200307 4978449533 3173923562 8772498890 1104982913
5380994323 4673870647 9293918382 : 2078
9847365091 7415993442 2418013609 0702185376 8394823719 7255148813 8816352825
0823780875 6177303718 5933102376 : 2079
9015518148 9566802645 1066955667 6356270331 6375504282 1846935526 0793128677
1716300815 2297052501 3994404111 : 2080
0995237587 8216898707 2283241554 0437859493 6488165971 0601941701 1177530819
7796006102 0610758095 4184382263 : 2081
7717441589 3089344024 5480776358 9859838646 0044819130 6329182121 2522007280
6340890562 7313615628 2514259729 : 2082
1169096962 1167408247 1631451891 7473600695 9669914230 8087833837 8686590159
8670223214 2869157014 1424807045 : 2083
8972191054 2004790420 7261838945 6591675766 2433748165 2334310131 9777787506
2648144789 6237968544 9183339325 : 2084
4452263282 3898399552 1435086472 3998824618 2346783334 1203496969 6346523102
9709800703 1272981130 0298748758 : 2085
8451556284 4310131560 9908946158 7840584003 8361454306 2750283843 4516836793
9943115519 4067233688 0332618381 : 2086
3019065159 3168620191 8396364388 1182869704 1164945876 9422113657 6981495173
1860439447 6819223940 0670145512 : 2087
7928254056 5303246423 5241908378 9115209165 2075345011 4775133761 7613160303
4635001583 0432411983 0345045973 : 2088
1115480235 2914726755 6528539615 4982517322 1870281189 1475582192 5109751881
4749962701 8320123866 4665544709 : 2089
6270322119 6735206682 5688348737 5964507251 2079691451 6873963998 7295089292
8615057450 9391835248 9864171151 : 2090
5633710772 0704371942 9897852585 4106512202 0872198511 5201196820 0668515495
0907756992 1619316805 7612255084 : 2091
1079956447 3572362115 1384426059 1187852361 1115766746 2461676058 9490884732
1882511881 8916537294 1301847563 : 2092
6508362290 4096877270 7590630759 5173734465 3812358167 2056998615 4493374413
5511580828 5999797250 7000542569 : 2093
5844829042 1570329632 9695418372 0611253277 8185078243 5323918726 7379753901
0604218982 1333568001 4917629276 : 2094
3589739749 1510336102 9448548755 4126594588 3082627308 7297415813 5998785058
9708156429 3241595652 0572243886 : 2095
0158420781 0475042628 1129044255 2635054829 6613431983 4755788519 3222267186
9303645667 2710264959 9400511663 : 2096
0866373172 7404454569 4973748748 5211033177 5493646253 8061133447 4310806832
6308466220 3937077310 5244279995 : 2097
1374501935 2661423522 5514186805 5104005021 4387677859 2990110859 2518674991
3131450008 7258371166 9369824976 : 2098
9940841616 0624284063 0833289799 7161870505 7651962404 9243165999 5151896649
7547503900 1147398903 1896878326 : 2099

4557847453 7251804522 3597268776 6876242850 7538166167 9248800082 3409032034
8071465228 9022230806 1496574270 : 2100
4477221250 2661923714 2356260929 1226018250 5837318119 7103907517 5338577137
8077621317 7245287947 9158317148 : 2101
4322731473 5068371778 8157985202 3035280059 9998697766 6937008226 7088042043
3042717610 3604436021 1957405318 : 2102
3239775082 5376243533 5992587448 0669523131 4095082672 9742008271 9591871616
9601534065 4578147571 0124329470 : 2103
3404989011 7240314562 7070070858 9135551306 5947483050 1092675331 0504767668
5100687279 5324432368 9649387243 : 2104
4914018868 5802176697 0655158850 2561741520 7031509272 6514587358 8577166907
4118956676 2941681340 5784240677 : 2105
3388665298 4335828209 9209279600 0256053731 6119574865 1729717114 0435836830
2333102692 4475563496 3018267857 : 2106
3511105639 7494733570 8175806329 8707668034 2130966827 2612847950 6043615265
4421703635 5406583290 1954741126 : 2107
3216179414 3686238782 4468108851 0060879820 6571969473 1531688727 6558292548
4100600262 8870847072 6414636981 : 2108
4546760230 6906484800 0195089152 9208834752 0029483301 1835707147 4860460032
3180366466 3011378346 1481020801 : 2109
0408241624 6439862858 0275352540 5414811787 7257844982 4401215358 0883263111
5767938834 4399416742 5526718127 : 2110
0687048579 0500170018 8276611540 2598966456 3822695284 0861257000 0031201513
4146214627 4358818811 3752159623 : 2111
5509096186 9348253038 1968085084 9675713080 2652210017 5445215043 8824469635
3913545222 9483822752 1939781610 : 2112
0630815713 9473475716 4331002885 7201156174 7191922667 7195436928 3128266043
9606992546 3721960291 4253777973 : 2113
9831674438 1208097218 8188312362 2660338707 5326789425 3855916918 2977283327
3126155084 1748495123 5989157986 : 2114
0193104630 2040883658 1232828339 3282877527 4859787053 6473295156 1411429853
2461034302 5553130194 9643011670 : 2115
3792865637 6695698547 9637443740 4695144047 5248627476 7380255896 7408496302
7253885817 3832095777 7270442659 : 2116
6764502346 2419588725 7359338615 5268081204 7751364027 8605967148 9936812371
2011862123 4905481712 9245481543 : 2117
0238041036 5014875356 7454311180 0604500426 1307876822 1588514426 7302962084
0482261369 4974262081 7609999350 : 2118
0334461976 8841879030 4159595153 9264111965 4647748208 4960353618 8945761220
4857186264 6143232749 7191880858 : 2119
4172165024 9255612284 8670444079 4528091825 3914446987 6181366331 9439606463
7822450816 1381778729 2827839764 : 2120
8591104634 5562271722 2178176922 9741153867 8621460572 4201588982 1754945547
4948636317 6722743647 0898021546 : 2121
2007325013 0237057212 1626662522 0053039613 5167883101 3008568016 7987713860
0808744144 9608596103 0410411974 : 2122
8536983111 3671070824 7974741971 7080824301 6916661770 7713127633 3136381545
3158913375 2541683984 0847864317 : 2123
7506675039 4884663677 7214679211 2185361223 6316721888 0380661069 8593702379
0963186922 4025911914 6345846149 : 2124
7417121925 5019925474 7960048460 0633459818 6460801159 3744703731 6631953518
9087920564 8107281187 7724020397 : 2125
4402460212 9739110134 9926966489 8978223364 6553651294 9732934154 3406894694
3373818266 3778605034 7493433270 : 2126
2908375618 0110549346 9017933942 8739905663 7969763478 1069552896 1987646189
8507220863 4587475775 3558684468 : 2127

7233572491 7904765480 7751039237 3639618546 6753334959 7089174705 0103139694
3809023634 0457990307 0724852963 : 2128
2851430888 7866880742 4981635856 3633931419 4762523066 1525205658 9630703714
2091574467 8667376833 5155822444 : 2129
2263717555 2905493953 2882366689 6153326331 4935839281 2822458493 2540555941
0719507137 9970356374 2340097316 : 2130
1309864621 3937953087 0947165361 2565080331 5785044573 0000941413 9460014745
2544140381 6920993360 4115965838 : 2131
0050630368 2545663080 6282500948 8020034180 0214558417 5546348018 7653567764
4115164771 0438436690 0853706116 : 2132
9050325303 1468354371 3358180929 2400768050 9581888880 3131922996 6049866511
9235533344 2715995130 7690820852 : 2133
6629677403 1025947302 2591776820 1325910777 3158578447 7312075886 4509339877
5618726625 3938362357 5762515880 : 2134
5620309231 2138665780 7216261161 8127003756 0534462263 4949838625 2566652422
9234436513 9697208237 8259957626 : 2135
1080998493 7542273567 5122410923 2447930724 2828029176 2353753386 3708763873
5181552748 2111244800 2459124640 : 2136
5111511149 9664462619 8433900579 2546353949 6228889243 6232521864 0252481049
0595955408 3650286893 5748905420 : 2137
0091253386 7434313407 3422651959 9814488762 6448318552 7327749412 2878561306
2258218781 2001162857 3521338086 : 2138
0436525201 2350790830 1505963245 4682818922 4759891328 7169435985 1422675732
5815092498 2124899051 8465907278 : 2139
2376396492 3211904205 6438491725 5643187344 1622962006 0447190161 1612786080
6915970507 2338317990 2400106211 : 2140
6474775843 9023757467 8913169570 1182264621 7702894571 1913641268 5871868635
8249327174 6562706728 0751367431 : 2141
5975075657 7475837640 6338044944 8206683521 7833213332 7896776383 6574467462
0172883957 2367211098 1540162132 : 2142
7006816874 0231366194 8332501044 6485646460 3641253174 1333323796 0756729373
3052122974 5793335256 6168558920 : 2143
0437596251 3420306383 4294306097 1584740953 8019741154 9530010282 1650559592
5945919485 3348227327 1554448735 : 2144
2136534472 9423949559 6453047880 5317945586 2934189010 7779349027 6022180849
9185141257 1653165137 4508750314 : 2145
0146677425 1976476204 6166931133 2604538789 6451657290 8438615194 4311401615
1423070224 7163939901 0043790686 : 2146
4103416236 7907418506 4637682566 0389550334 7734896731 1334313629 4285431488
7603124731 3354196709 8000845264 : 2147
2740142097 6313695876 2258591009 3111299737 9360013553 3529207482 9853672042
7612698476 4006676698 6610534552 : 2148
0728721873 8180679105 8162907487 0107673696 5216687344 8787438277 1997327186
4925542480 6684238330 2741069609 : 2149
1855007115 3548924174 4407943370 4231825456 0683867024 2052339330 5803173064
7788593322 9299655466 2168705712 : 2150
8180663158 1075969880 3795419028 6710515896 8218399861 7226456523 7272159212
7269985616 6884308596 8396028717 : 2151
1538526694 1479317328 9354584495 3150218593 0086689117 9713664949 2410539530
1740136078 5889154713 4085003976 : 2152
8036453811 1157208612 9563947096 4557427082 3873126874 9887309705 9005337318
3461689693 4170930000 0861680278 : 2153
0058956741 5228443663 0022965265 0701385626 5684358886 2975858927 1228973122
5045019397 5398801959 9295859466 : 2154
7444885279 2346410372 4733413533 8390259480 7739551764 0674147646 5801453303
7551258783 9152060027 3054598058 : 2155

2800834158 6750878202 1829802912 4179773152 3538577064 0677116684 5213368665
0109064439 9184664729 1438415228 : 2156
4355957780 5241786922 1343902620 9703590303 5025270328 3979867654 8711129716
4150657689 1539350909 4042163002 : 2157
9212623423 4712852108 3954216649 1175188768 4890160163 5079499087 2514594428
4090769519 6996180377 1282792923 : 2158
3063139463 2150965793 6648852867 1853658985 4282324046 3873382817 8481530209
2030883156 9726734392 5583364321 : 2159
6320660898 8845807113 6277639996 6495706481 3332430080 4430706922 8179629683
2861316394 9834158178 8714262196 : 2160
6549905140 4499949051 3227583290 2039733890 2854257513 6640742837 7198389513
7584603568 5933196763 6542297879 : 2161
5979675682 8399831018 1525423666 5985727858 8886806485 1894597071 6203467370
3516804567 8974108321 0206877691 : 2162
5310505668 7668773293 3492002389 3505744369 5445160234 2979457806 0306718931
5767951908 9580811282 7048686785 : 2163
6517949494 2531798989 8545584635 1101662924 1506701611 7622197572 9255773222
2995795702 6951427313 4125870360 : 2164
2132593747 6429476772 3385539394 9608034943 2963081459 0799338159 4311461023
7436482609 0527489260 9114997817 : 2165
5992425233 9697286952 5241668731 5009238204 1212854261 3616353249 1366251378
6628744172 8736927773 2668533899 : 2166
9050914428 8059316961 7682577285 5927778554 8891224880 8866962902 2220090710
5319867273 3203501256 0832761865 : 2167
4686069004 6121765511 4103453283 1271204435 2295100167 9479031335 0534253556
7838691922 3431249052 1332794361 : 2168
2569046803 3045406425 9314334859 8935298788 2254953185 7424881037 6413754148
4499829522 7489027969 5089814986 : 2169
4690761644 3895752343 5665064979 8259415250 3242632552 9441165969 4055989586
6507612153 3992974864 1052808309 : 2170
8879197123 7287616972 9073029530 1586338095 4319401820 2669104693 1393035266
3628358321 9629341950 2205582156 : 2171
2811510082 7837021914 2231861577 5289443074 0125120698 2236257041 3511621279
3447479373 7507085853 4490402518 : 2172
9467769147 4206491390 2473152404 7392237570 3568331255 3974447363 6977591310
1672485564 2522704985 5871329918 : 2173
4758438211 8515249153 2108660870 9389477465 5589097681 5009091552 4531843711
0167970439 4227200606 5934727864 : 2174
9237655946 9584717164 2902578632 7183436043 8706061526 7993199251 7807196060
1819978896 1891441329 6815327355 : 2175
3656553178 2787898770 4548492565 6831540484 3368663589 3482791153 7849960146
2943301785 3591892226 8713560211 : 2176
5638066888 7360245242 8615177077 1110671285 1439717394 6256684077 7072585891
9518657200 2830268782 7488064624 : 2177
8625804514 3333445413 3086163786 8233257296 2579538006 7350910605 3396523255
7596824150 4827951961 9749459051 : 2178
0082179623 6567014770 5645902747 8980181006 3095188896 2137903769 3653372987
2681282088 4788701063 0825541585 : 2179
0421334101 4958285427 7180694946 3381388168 2451903444 8050492243 5510003314
1429208942 2576831348 0195104195 : 2180
3956483428 3831689946 9970689361 2395299336 4773605967 3795630161 7803184226
1826199208 1634867619 6602758664 : 2181
4711808760 3253007087 4535085357 5490894833 1667080132 5348249711 8067652281
5802360708 2333904142 8117022941 : 2182
3525360033 0633026112 4551686492 2753389765 3332750883 7308735465 9141118979
8341977081 2110908047 1374423563 : 2183

2419974361 9581423276 7405600444 6749156949 4557871493 5547922254 1764298223
0757366515 9603939567 8729520830 : 2184
7621299572 9056463332 7979056087 3601966838 0684152160 0534098228 7176820543
0304948296 4071437795 8967789178 : 2185
5265134420 9014796569 9695860332 1761028398 3223252420 9091874975 6952825023
6244494235 6873501034 7018741990 : 2186
5300293809 6986090876 1494567287 1126806871 9599242400 6465327711 5700461234
6955067259 6301566722 9090544556 : 2187
8896694903 6381979374 6846586653 4067955971 9446297756 3164582434 3862403793
4898047300 5757098395 1582161392 : 2188
1444041889 4226816655 3489541432 8206155392 6819933381 3234143139 8790872065
5644117610 0519791030 7921159446 : 2189
4124822986 9540395866 9789629636 0224807663 2631118560 9381709075 5322596581
7149254580 9500486428 1930723758 : 2190
6533109347 4102684608 8351017655 2329792792 5886429690 5772257139 0829119090
7196417085 3845945443 3599189629 : 2191
6182581379 5766195253 3777093959 3093755869 5979150585 4695906008 1600343557
0792205728 4184858559 9616477156 : 2192
1906337685 0432936554 5474742979 3082284034 0104214779 4004948180 6545729224
4834261048 0152048933 2597893682 : 2193
3575947758 4893907965 3986132009 7773887838 9002306649 6506731865 2650568283
9582196258 0338070209 7089887141 : 2194
4621585654 4262375254 3139384253 2127573407 4533191162 9551711879 1369927035
3917235081 4998662377 9442841884 : 2195
3345714929 2710333226 6309932715 9181177798 4273789750 1478943326 8497205154
3072375606 3998772961 6687253234 : 2196
7099071746 4054024073 9876530764 9992827255 5573339710 2244685228 1974406356
7415442339 8952240404 2548339769 : 2197
5537147315 9903911519 9581609495 9851210374 5365994424 3964558662 1895120731
4020177355 6781853195 7450015913 : 2198
8619106408 9978693283 1364839009 6137571062 7234780052 2824211842 6427552831
6128586976 0156604643 1833533610 : 2199
3972337460 1999153889 3157302858 8269160920 4948845413 0092262588 3777140487
9655160155 4359374511 0789847180 : 2200
8847009606 0778907622 0693684073 7849633609 6342509584 7082572563 3681267006
4291029822 2799915761 9394123050 : 2201
1066561932 4385291312 2708830715 6747196820 2186272019 4847446914 7750995873
7748660296 3126211239 3626268432 : 2202
3153391719 3569137898 9196606671 2770973432 2808251984 7506195406 2034493330
7037842679 8379941771 8823847785 : 2203
7304923986 2558566116 3352861527 9571343531 4524810391 6383517055 0778772229
7623979208 4070887115 8662399192 : 2204
3319336495 5741099493 7541006679 6880142650 2073106663 3219037296 8824698040
8070541863 1788519380 4782714122 : 2205
5654179999 4252084728 8328203476 8584897255 2574718194 1141110041 7415667999
9964197532 8403240933 1190631921 : 2206
0471346702 3378515181 6822986613 4384617955 9222892272 7247929512 6971190232
4963913804 4043995740 5009271208 : 2207
1861325429 4374946808 0349527402 8786638624 3934171088 5765745650 9859476694
8921845006 4054656300 7857601863 : 2208
3790396114 2713096570 4638609176 3460387568 1169616742 4770017570 1209622415
9952976060 3853488570 0148140313 : 2209
7001128029 6945431637 2351125088 0211913858 5426221056 8994899518 3018091417
1906159263 6934736495 3071541759 : 2210
0666788072 2820148829 1988205155 7077635832 9567219112 2035770424 9516850618
8295308898 8913377428 0092605574 : 2211

8231190883 1910313193 9299334559 2313428229 0824495258 0052392312 0354684095
9181180376 7004110412 4295206004 : 2212
1674976055 5822753840 2785572289 9442909707 9220373479 8808673500 1702235402
8870748724 1568779150 6214652489 : 2213
1733255247 7018448633 3604237917 4274985534 3362819513 7659386276 4032817426
3624814720 0965705761 7273393219 : 2214
7137016249 9437607223 2561327874 2493777785 8926933033 5964016213 3441364984
0271139133 8427470757 7695437786 : 2215
0117566491 0861942707 1829174412 4265444598 1363785943 4402043228 6589754638
6434827291 4836757909 0612462084 : 2216
3234390391 9234433434 9677277355 6111421320 0143944432 2732038136 9085729795
7363267447 7894386577 4890385918 : 2217
0992598862 9697792589 1374705285 7795461303 2054330367 7522033550 8550526418
5246835194 9293468352 4328602941 : 2218
6899457532 8382103070 0597142644 5390140901 8029918233 3664744077 8847072021
6230623856 0559758221 3448377296 : 2219
2995988321 1943413369 4583446147 8359693702 8326827141 0484814528 8290526166
4032814940 8184024376 8279808314 : 2220
9452046334 0131479318 7522373778 0641449565 7562106053 0337373631 4667499714
2819907423 9705585981 5350366620 : 2221
9046505844 8358290370 6278821795 1701095497 6396032910 4655406069 2645863021
2687402703 3337628709 0086360775 : 2222
7172312759 1619507653 9133776329 1958221560 2395743429 3446887129 8084612180
2689710424 3417090833 0991098588 : 2223
8888352540 8594227691 7768288120 7561794396 9011907566 3452417006 1632008101
4184753329 0811300309 3109758677 : 2224
0730363184 2545293345 3097666152 9175236632 3656474216 9042280616 9751560533
3059925079 1768250223 6464599957 : 2225
0337747610 8414750188 5998830265 5204068322 5323910587 2448941321 4920420150
7636619728 9000406059 2720424962 : 2226
7607199929 9976515689 8504788208 5190980357 3311574154 4655500524 1314901243
9899507673 7791147971 4212766615 : 2227
5536570002 9980643522 3585594633 4029151965 5744773725 7745255173 6846772411
4822876372 6800196358 4486242603 : 2228
7986498657 5821308051 2548677536 7180449618 7001591044 7387934243 0418785461
7870457854 3664944284 3850304116 : 2229
4819266671 8497525267 0736583993 0254006188 6594630044 2593498642 1887367466
7791401028 9921935190 3419847325 : 2230
7602258531 9484839338 2061146480 7036489978 6708653140 5317348151 4324651853
4005640853 0192899076 3601600914 : 2231
0767076874 8649878661 4472416438 4262549228 5981679108 1292052188 2291519447
4347041036 1926198224 9688650183 : 2232
2878812286 5526149448 7243355986 4067055348 8667642160 7699601535 5082328241
8270715618 1963143431 0929628040 : 2233
5256938017 2100643874 5609358563 6653375409 6152099368 4410900642 3455594967
8992586527 1737498298 0371763864 : 2234
4155408339 9332473281 3095490090 9116944267 6470996060 5136670340 1744118303
6622504899 1020282241 0449800530 : 2235
6393922465 1764328196 3200444786 4310710645 1818292490 1554707466 3013665850
2775050796 7666944709 2311169507 : 2236
4284579269 1986465479 6897698574 4247120250 2619936276 9049186893 8537969774
8241302056 0763043389 2247367574 : 2237
7538314713 4175467829 7496244770 6654093819 8182940533 9527866772 8983884828
2991142392 7736324571 6014373375 : 2238
2630480263 2494216545 5657671976 7519347205 4649944251 6009891508 5265375080
0251075605 4326553772 7234223071 : 2239

9696794527 2246615973 8660217416 8903912272 2547133825 9155322845 2515226694
6972817303 1755253671 0851911358 : 2240
8765425443 5790412982 4103543174 4232764343 2713706542 0996321570 6364060968
7138452462 4566335269 1301220789 : 2241
2078038541 2037602063 4119553945 3469466949 3091620795 8199116593 0757419826
9298778666 5036590825 8531021071 : 2242
7015018441 3675291384 8473908192 2356470866 5621950319 8651985556 9037476710
9471408761 3531548718 1593027818 : 2243
8382078139 4000869999 6704517400 5890292947 2049512466 8073909517 2243055169
3010480382 7814754464 1937702694 : 2244
2493272433 6812520246 0157153486 1044060759 0563320374 1788371475 3521439572
7778827463 8618416087 2134325498 : 2245
2369004837 3821826384 0092510215 5997628249 2414832391 1002469278 9253625384
8077699875 2416827751 5798144534 : 2246
5592180912 3520162309 2356187262 0356180637 1374370501 2462568124 8863511622
6947566896 8136190873 9138611682 : 2247
7810422466 4184881377 4916377582 3530717510 9336365159 2078320285 0748781773
2945679572 2880270292 5330393035 : 2248
6296096550 9081112345 9904500640 9831462600 1133976600 3729813388 1316144986
2460738400 4103873895 2334677015 : 2249
6047656476 7743753091 3530360277 3064948548 1818157985 5584587136 2783153768
0464822152 4841805002 4360485920 : 2250
4248195328 8367840363 8789956319 3321631831 7783975299 1937552421 9659608965
0655373940 4460898228 9650830886 : 2251
0509089024 9651264721 1910969229 4038660590 9137826663 5979448407 8326763625
4438297382 6316123851 2713588318 : 2252
9511072580 9941985723 9426389659 0594982782 4178092350 4759958072 8228798338
3670661002 0415953764 5690875936 : 2253
0820905304 6464560549 8351090377 8477908765 1974600057 4937826856 9622689236
5647368966 4002376142 1391404088 : 2254
5302285301 4229240229 3423918474 6072891824 4015840316 1965703700 5116503748
3283612130 5179276792 0294958449 : 2255
6107578731 2193123627 9249070877 4704940277 2076686395 1289959581 0379182555
2757370199 0356398551 2824029479 : 2256
3513430470 1498533163 4148827241 4708705113 7221073263 7816770795 7042443525
4240265878 4910323099 4421850476 : 2257
5710476292 6221526379 9117735029 4554041471 9797361893 9164136467 9582508105
2536221009 5673087070 5995351102 : 2258
3282255406 8808184242 4613290550 3411246368 2062595644 2929175740 1920097057
4675178378 7094973834 6200035152 : 2259
0235096582 2051323495 1881288097 4170138280 0727749270 6073842957 8676545651
2232806960 1873592793 8422502982 : 2260
3945265456 6153768909 5003761204 1625651830 1073537003 9070291502 0475371427
7894368805 9173203027 1185778991 : 2261
7666634257 1642695371 6695933183 1417686469 9203932928 7314780654 5499610556
3585878580 3559889382 5325627842 : 2262
7797527748 6959058290 1784353170 3864196779 1407650481 2980943838 7688116335
9953474978 3496325840 4256655648 : 2263
8352302309 7152638961 0852632841 3993551737 0055701579 2433145571 3392635064
9126910328 7457433680 1684708832 : 2264
1019831805 7258996356 4174994799 9141176464 0878309858 7388760126 2243929152
5135127431 6311424091 6595798544 : 2265
2311940742 6391419957 3700819368 6324395428 8891892159 0733557117 7725165886
9544944649 0515695732 4223604942 : 2266
9106113988 1878797814 5692300825 6816350893 3768853608 7849150976 1407672722
0176526327 0063040429 8829853236 : 2267

0410002404 0182990715 0582095348 6678658549 1475231039 1765304392 4444519665
1342514858 8659357253 0618789331 : 2268
7329084163 4035522164 7410415435 2613758218 1818791279 0652821080 5664458170
0818846212 0953276418 8236193739 : 2269
3715845416 5450046137 6347557269 1672524761 0255780211 1982212191 6167692479
9468148510 2211083546 8697760470 : 2270
6507970232 6979179446 6405825458 7841235137 8391598786 8576580174 7173575840
0554500216 9915662489 3432775705 : 2271
3162343985 7465121125 5669761595 7941655004 3092783958 0643678562 0176109369
5343227442 0323727782 9212641072 : 2272
7927331538 8054265718 7195231461 4760851241 2071162145 3707052346 0987352852
5352639855 5173759886 2152283252 : 2273
7062331717 7105764683 4420711218 4896971626 3292124906 1416662418 8760417868
3965335208 1340399319 9974584851 : 2274
6368676490 8868591045 2680787306 2160502149 5859193782 2714926533 3228609685
3650503986 1403799578 3932359268 : 2275
0910778905 4855586108 8592428224 2259277447 7365117818 2700198138 8531607305
6803365796 7627178457 7742916999 : 2276
7919369629 6290729972 6810304970 9697061750 3617848728 0491571455 3234024897
0086518250 5718413909 7089981443 : 2277
2108632743 0762953464 8301060291 7603173983 1629885580 7697144339 5677290152
9479249489 2573053103 6288092988 : 2278
5710977420 3433903894 2417749608 4967853115 8757524460 7210626352 2179995794
4832824964 9817968808 7770356049 : 2279
0697406097 5581511209 5162050132 7709107803 9134611475 1004969867 7195780467
2823682217 5885085551 2187378823 : 2280
8435502397 1353564767 5312848875 1114558439 4413075616 6908021940 4705402509
2561638873 0579959357 1007095421 : 2281
5242402389 7386614498 4302696436 1569759383 5035800086 5252066344 8232509342
8912815946 8246881311 0767064807 : 2282
2715392133 8085490889 3217446305 9788581127 4425344881 3196217550 7453904692
2922607786 8286365875 1566809447 : 2283
5047862672 2735707695 3714897264 8601362808 0150844226 3265972211 4711872171
5445818774 2615869707 9388695592 : 2284
3103553477 4484427102 7727918126 5419391255 4760484431 8093436796 6463340428
2833273374 1850629865 4994600120 : 2285
9056686091 0949503520 8441838991 6340306963 3435199713 7223404510 1839365628
3949057157 4119917388 1420686449 : 2286
1885648968 1633355195 0660009288 4333252480 6735584171 3374961715 0550934263
7189402325 3035425993 8439418771 : 2287
8742088145 5435435616 4303489103 1481520576 5886944478 2706449109 9533521284
3251910491 2469054321 7380510679 : 2288
4185988054 4012894251 2325899099 6231232405 3877398210 1446405849 6559741586
5952320581 4498852510 3769306549 : 2289
7489313506 0329360744 8181499898 2011182749 2778152011 3240464303 8340009302
2310805472 5959755121 6746706659 : 2290
2294443857 1075829356 8651598011 7901994804 5358247172 3450301763 9891490221
4494890216 0198684151 7587379191 : 2291
6826610983 8573845376 5280418900 9337550323 4876758875 7658350816 8084898048
8994613463 8467583582 7589450046 : 2292
6480260224 7079596073 1123470870 1901229396 3842199250 8876853711 1998543312
9372429484 7578836115 1740833584 : 2293
3753310906 6594270132 5803295439 8152692068 1054804215 5210247965 1145454331
9711530574 0995493783 8369320017 : 2294
0656410239 9396852034 1513173309 2513860829 8396103448 3756434854 7094563741
1060456166 6832802636 9760559410 : 2295

7860053014 8540321252 8253223272 5173232493 5578822659 3959508373 3400950598
4530084486 1549376083 0772932369 : 2296
7805390206 9489843652 2867928580 7815810808 5806495326 3317305646 8160917851
4712540008 8072257937 1359859196 : 2297
0203211769 8516618138 2057266644 8797145605 0564764174 2736841891 4506734245
6756416048 2903098189 7917595674 : 2298
4799704418 4815439560 4702337843 5681267617 7157987374 8731652445 8821001641
0619287671 5295197730 9612579504 : 2299
0132799512 5123044607 1737653304 4348897583 7750220067 4146778016 9732280054
5673449942 5372413845 8236775963 : 2300
9957228554 5930783851 9140395047 4413617589 1007414622 6819297696 9498861286
5298551788 0249933196 6356382483 : 2301
8294192474 3192355842 6763507319 8580303015 3430748618 2437832522 7933579938
3568537811 3275565386 4730024767 : 2302
4306723758 4455570664 3322396705 8378975019 4011098458 4530312039 7416081495
2863365122 4839511514 2651395213 : 2303
6199492804 7761456722 8484431285 6596154497 3138278595 3307673696 0149415863
7070362175 6586701043 0358696114 : 2304
5791714834 4582054822 9597116654 7021136277 2824935407 9462907060 1403720169
0357789239 9326303272 6072545060 : 2305
0403646050 2838092961 0076000676 2109358216 1548809682 7981804590 8769907558
2797111496 7485871036 5979817790 : 2306
0559920461 9921086218 8333938643 6767545357 8236336989 0881619356 4218210955
9511009398 3753747755 4658607786 : 2307
5594330622 4849127897 8754508135 5800095536 1863224778 9455782167 2858215655
8348574169 2055782234 3615032535 : 2308
5191306945 1960052894 9869404686 5586452883 9323919612 4043995947 7905755435
1905822581 2702468257 3216026993 : 2309
5312376217 3162563897 3247571162 8596069997 0829394959 8146454681 2429119289
4493216757 8936358775 2365870831 : 2310
2626129768 9522140371 2133337137 3636570097 4961114679 5473894021 6254866841
4635249814 8656843712 9932566103 : 2311
6903209843 2452443637 4578928327 4532540101 3787354608 7085784915 3391330184
8796502158 8810929903 7143501149 : 2312
6211919724 3727036331 8901179929 3100919897 2066058919 4991838526 9867800580
9392309173 7819542985 0851684668 : 2313
1299233425 9466707617 7767558862 0801261412 6146408861 5606370386 7564612888
1437886181 6884069210 5737310071 : 2314
4712755602 8255238461 0494287319 9498380141 9274943751 0069479060 9597627570
4074256052 7920403735 1322564372 : 2315
0532009690 2712617878 8419582439 2343316525 2466820942 5466272979 3482420950
2732777029 5359815649 8247338180 : 2316
6163938715 4774919753 5049321791 7432066843 4092062017 5808477830 5188754961
2442395201 1896490704 7660186063 : 2317
5733321398 7937346739 1490808812 3513455137 7407155868 2223545588 4575446863
4333775403 1387130262 6071462240 : 2318
1171706024 0106522549 1198646843 0964157219 4492446028 2817325253 6670353723
0042424986 6064805311 2750195435 : 2319
6523225687 3826351560 6179781774 9036314750 4957320325 8272282087 9015800370
3947220784 7114408535 3021626740 : 2320
5066505125 6166950057 3990832732 5068995212 6975261606 0272524728 3665244676
9954693569 4759472575 6685581189 : 2321
4258537772 5768098397 6858806496 4418575387 1729087162 3665842954 6000645728
3605313875 8138636294 4104313462 : 2322
9953741827 7627185306 1591934261 2177132010 3100211452 5657669028 7097455533
1090738581 1128955140 3927166875 : 2323

2247998498 7499178502 5892689021 4824595782 5905186445 4825308670 9605415246
3916486748 1996956919 6375971963 : 2324
0398096105 8033546132 7894359818 4350869745 2592000548 5900303729 6168307195
7622685435 3641731181 7450957993 : 2325
3164776907 7464027402 5905255388 0918629368 6045867395 2113310468 5547484403
8171072506 6369101455 7314738282 : 2326
5053570575 6613939175 5260695186 8964925034 4686647491 4652615608 5505037913
9420298199 4222199473 5438231783 : 2327
2230368471 3733024748 5594298263 8040651298 4891971277 3126979399 4244683681
3979719308 9451531301 0228207176 : 2328
0241132296 3912280618 1570953761 8452022878 6336178426 1035310734 1759203978
2916543702 3953430922 5910850108 : 2329
0565591877 2527775548 0027001929 4614144153 7672275682 5533214173 7201407471
3478844343 6316759161 4387225594 : 2330
3304949771 9612338422 2661604896 4396242053 2677970414 3080241164 0119680891
0109206342 9038027925 5156967953 : 2331
2441619283 4866410838 2860544153 6996436593 1969137777 8700593603 0482022913
0265145922 9613455802 9718272438 : 2332
4196876724 3708449263 6755070563 3405026837 1994493547 3265265632 0662743808
3699582633 5167607082 3549529856 : 2333
1583433119 5243922970 0398791067 5268314944 2248758705 9711975013 5716880807
7088016385 7843277818 1513027786 : 2334
8311685891 9461143010 8958921839 2897133594 1392888564 8854509163 7259398359
7420767157 0746071497 5246059863 : 2335
9896695746 2315777686 0487972014 8035767956 4845898219 7028876161 2319470132
0955924214 4883555057 6272323443 : 2336
4842642601 1753223061 8223035658 5080470110 1189919325 2571720549 9629266412
9773504285 0437026228 9723585281 : 2337
6267563789 6302038984 7435594806 1217387390 6685354384 5303929312 9881938833
0411834237 7836147805 9757505840 : 2338
6622541336 2933578093 1947819663 9297423503 9084805932 0069789917 6788339686
9131974825 8864747086 2799713132 : 2339
5613717273 0816533406 1394625685 5059072754 5864506864 6565277682 5553429721
4088338372 7882010289 0293240313 : 2340
2421020026 1063566424 4369661208 3041768693 2201048993 4515597321 1746630090
8671200835 5724205292 2510628503 : 2341
0294066927 0580504400 6818192273 5142563465 8435481109 5932073401 2749694900
0254472079 7360379164 6697031950 : 2342
3383284835 5167676058 3103654527 0857655498 0028239478 2231371887 0396521642
0784140386 3200501687 5592892442 : 2343
4891643210 7962003137 1107462606 9359189558 1823998836 5915310970 0423581742
9460073596 1247432905 7210929097 : 2344
6292410410 6566209235 0379244313 9268903030 6220340787 0584752136 8443498140
0664399682 8177728832 8306808296 : 2345
7474851072 6842285639 5031192396 7939970227 8280832904 0391879427 0125640317
3198670548 0903817290 1093826770 : 2346
3276181873 3382332992 8735425179 1214674169 6844438416 0995792173 4925475411
5169550363 2929460672 1879838177 : 2347
9848868362 7829099798 4302172041 7536252229 9672743257 1630803326 2679427008
8346679931 2372277892 8049072690 : 2348
6343593863 3448273734 9468718088 0694508882 4068997261 6587134375 1874071244
3535899935 7495057639 1055026023 : 2349
4884831930 1097762875 1845555614 2797284284 8760393872 1304909025 4184884269
7751401162 6937613955 0458568990 : 2350
4730039876 2225695695 2852270270 0707002236 3127827564 7209189072 3661453383
1506450866 0157166725 0304425313 : 2351

4573076142 4825299347 3550820094 8111074026 4270328796 1354558997 2387692438
8109759704 4445727972 2559558214 : 2352
8318579221 1683819202 2376660147 0535503329 9056638996 1139502003 5590039531
4314853199 9733956110 0645962955 : 2353
5821496162 1580455163 2496152498 4625491338 6661556613 0574710730 6606494761
2592513473 9867240429 4705271394 : 2354
5870057114 4617743592 4891999977 9853985891 5545801175 7075458419 8570746444
1715735287 0883181556 6490671161 : 2355
3720524842 1240675688 3333463263 0939467440 5915392812 4346865274 1507636710
8332946799 3079601213 2262362971 : 2356
9228890611 2943956865 8906746885 8225888839 8916501883 5533075233 1981579035
5358685515 5782065468 2183321590 : 2357
7429103474 6956756633 9248541522 3645371500 3886217890 2634313785 3026622744
8817999987 3853323415 2500505075 : 2358
9944529160 1038492429 6473792314 4851996764 0031204261 9311018390 0107455976
9324574399 6519682211 1570172250 : 2359
0007801852 0076909279 9527481957 2235224900 9245510210 0832943506 0470903821
7623401235 2784838737 7273143198 : 2360
1235331216 7350741624 7841954632 5344615208 2891223780 4692290850 9386280752
6773733648 9167527510 8867186907 : 2361
4857315117 9871911275 8973717212 2200697902 6862701539 7703337623 5391685730
2353277808 0515008525 9817532955 : 2362
5080787788 6672815650 9666916158 3911272169 8699388759 1126886484 8545345289
8384500172 0075317880 9612734774 : 2363
4030045241 6750323930 3836706170 7101305504 3805871730 6756683353 3745378303
6855999377 5908695130 6218465528 : 2364
5792359339 1741917120 5417969987 2561324532 6657739756 9709321705 6219380046
1482857499 8937523164 3513474707 : 2365
3658820981 0605778865 4165147324 8981787009 4630138507 9255922260 7297152262
0388194374 8439143105 9409599258 : 2366
4334465657 6817396893 2611045987 0100372754 3525116377 4416122729 9994101861
9566051421 5969412063 5513144859 : 2367
7195452860 8097486825 4874524459 0362604731 3806483937 9734468186 6249700721
5547106019 3500238648 3893437562 : 2368
2763501279 2584941732 6436623720 2327855359 4194930450 0111524937 0114763463
4157542640 9557473943 0694456354 : 2369
2362081212 2411763735 7697086777 6359301935 6383644402 8893630507 8333228036
6747439432 4865707989 5085250872 : 2370
7418326835 2719951577 9265271987 6374997907 6208438946 3472126203 6078308173
8142804787 8554978289 7862274724 : 2371
4177003016 3255013397 0537241768 2815323516 1769069219 9702556999 6205464243
7226357754 7251024031 2994355386 : 2372
4594831470 1949401560 2668494303 1837836936 5546618665 6625470825 8607489483
9728251558 9160385534 9506451384 : 2373
7442211882 7562986206 3313569213 4350535417 5325462294 2738570185 1422160479
7918123913 5818570233 6381354453 : 2374
5711277117 1943216604 6614310154 7419821554 9290475621 0901895720 8060623490
8802904067 8545667463 7241772486 : 2375
8119007420 6557848221 9295010659 6683535208 6790875855 3449009271 3251073537
8131123286 0041052918 8355048282 : 2376
5682124393 1809785796 6641441641 9743846503 5975431670 4183852145 9077943357
7314964845 7421486085 4886745291 : 2377
3145745893 1518483420 5058542721 1602752070 1053028812 1820442571 8504079717
7351938264 4415143034 0000389650 : 2378
8354760695 2112614351 5144940969 9151517833 2585172479 8947405242 0610045984
0736384351 1382982933 5370285516 : 2379

4153281846 8987804359 2175819760 1110371882 6011571521 2198992803 5754608388
7409473752 2040639123 3628982806 : 2380
6187319532 3552920401 4220009515 4808807061 0074538656 3972589708 0303279855
1240570967 5299487752 5034838119 : 2381
1484476396 0690239980 0858875101 1612900600 8076911943 8103026094 9480659847
6196904805 9321785213 9982865901 : 2382
6361397294 7333424529 7578429975 9023288921 2288761745 3643431583 7531437849
5746088747 3734258795 8758219901 : 2383
9353898142 2942391794 1515613153 9793025146 4137986089 5988767541 3694323040
4870285554 1978092295 8044698990 : 2384
1929045589 0684659783 8337994492 5127160494 1337790706 4865785894 9675759940
5061755763 2934756808 2892202911 : 2385
1549186488 2015921461 7765449921 1827254988 6765689622 5170636148 3219514060
3084486884 2947490817 1412276698 : 2386
9529766528 4671870107 2919337792 9282443532 4313828520 6357061580 7692592826
0322211940 2768779042 9240836532 : 2387
3232151023 5407534232 1094760532 1017167804 7889604168 5107197396 9399186187
9463461896 7971354678 6722440290 : 2388
5164443964 7829326694 6358491866 1504011655 0321379582 3884640354 5337067500
1468245089 3963508407 9633883393 : 2389
1644002155 7629487655 4514962298 4945735704 5563984582 8653901031 2031199558
6329789859 9642742416 5456402215 : 2390
5269311761 8219340405 2804977001 3958185699 5045062690 8321844220 9858065603
6039665052 0405092652 9449163112 : 2391
2474122439 8545523345 9397360215 8488959564 5760356011 2394722600 2909110232
3581832707 7603181928 9578931912 : 2392
0042282972 2719276801 0578564466 7334020318 6065797599 8976736300 4515534412
1222746492 1178419210 4299302330 : 2393
4754593408 1486957338 8558531187 8925724359 9624701958 1049408342 7130065971
6364375165 7463710249 8705362916 : 2394
9290060997 9797982081 4714713112 8995085184 9200398640 4614653025 0994914143
4035836955 6884216151 8200066725 : 2395
3985853203 5670787534 4741018213 4499703959 1787397534 9621147236 7771510750
6443419320 6709784810 1390611946 : 2396
8142996565 9469499803 0150150504 3949916581 9364340617 5471200602 3253305100
5685661995 3988521096 9917968103 : 2397
0651566276 1140012393 9441274050 4065600221 7098547779 6442468587 4863196946
1895510351 3339164119 5903971893 : 2398
8761054426 4230246441 2785966320 1484795527 3227434092 8634620098 4059824533
8576386151 9336443209 8339181957 : 2399
4962950525 2717041594 1103294160 5520070797 0044527426 5503291068 0168291821
5054886572 9790830657 3320056671 : 2400
1040393166 4289460739 7427613272 0699137735 8887764684 0767260164 5037969090
6737624912 3181569461 3258421465 : 2401
1124301808 8628385018 7273208293 0493205348 8349080225 7999620893 1538204345
8746206252 3629681224 0775571667 : 2402
6147083325 3074351802 8015646615 2412335772 6665459689 5015351209 6407409879
9335251124 2368393255 5080407795 : 2403
3890651359 8486931485 7268748994 9014085300 1062540369 8440243398 5738212677
6294549192 7728192707 2710075950 : 2404
5419454503 7909180516 3615808362 8887001538 6724394500 7027498984 3218556764
7403247143 9236648434 1116062093 : 2405
1019601825 0317806835 3985725839 1335713344 9303614491 7086659797 2333881453
0921740318 1174775203 2581674338 : 2406
9458264967 5275252036 1126273672 1097645431 3402380658 7201125134 5146117238
0163835947 2687522817 6383566558 : 2407

9618861321 6729989394 0149412510 3564658336 3688776090 6588376967 4182919192
0311945646 9780942498 3860904160 : 2408
3309637645 2927942341 9300230174 0054343225 8546250947 4354551709 6835436975
6035650199 2385147371 8492670597 : 2409
2332775797 9117381524 7435316334 2411729845 8941291075 0455504288 5877762737
3406633041 6039180826 8741726065 : 2410
9615989336 0778633070 1992223184 6664888930 4527152405 5117461202 2301603661
9219393661 5793786736 9615816259 : 2411
7300582128 1128255764 6795949402 8146674576 0455774706 7390220019 7698318259
7002938195 4149275908 1133733236 : 2412
0588587778 7161006725 8359623260 4960016058 9914889342 2047361613 2710075452
3004843943 1098999163 7221886232 : 2413
6257224723 0711979182 3049444351 4033574476 6397083610 6986071445 7006927639
6639734920 2921834629 7644381860 : 2414
1893766880 5345127770 3848156908 5406140328 0361502803 8609490335 3489323035
7925117393 5304158411 3326547129 : 2415
0567398884 4359308222 8303303222 1165929854 1919655979 7184885423 8871580891
4036930161 7172570015 5706148369 : 2416
0681274229 5502793463 5226450068 6934307748 2074666368 7476147620 0227501815
5179697826 6737414595 0438705887 : 2417
2387338963 2912139573 0399466305 4340289132 7746816875 4669502161 4124655037
0091265917 9830290388 7348417613 : 2418
9723433945 5693566008 3801609435 5613785537 4689207145 4423377646 7196313846
4652631570 1017132358 3974874665 : 2419
4423630277 9285419045 0156664578 8185997947 8971251481 1405023776 9026172897
9301308065 6571631212 1207914290 : 2420
7054215088 8983795453 6591643551 2334174598 7948092769 4175114903 1174605522
4557854581 3558670215 3090077031 : 2421
9556558995 9974680574 1613338361 6416911400 9923341556 4386836225 8664428079
4033626701 0522666936 1924674723 : 2422
7136409054 2898520518 8351003692 6818799746 5647052545 0682683936 2640699442
2311791299 7333641066 7817359159 : 2423
7162898327 4177288723 0205260980 4248757771 0069881962 4037291271 6284558358
4784034049 2436487818 3372432037 : 2424
1618788149 3183663213 2424242420 1471879866 0129082954 4902098739 9595428721
3906677698 2756308916 7942174016 : 2425
8823587653 9750420302 4489864188 9636909631 6270120557 6819699291 5499277514
2543788129 4676650832 5035126716 : 2426
8466448444 5472404101 2452806421 7832732227 7604369161 0288078358 8718437100
5180840179 5801410835 2816351636 : 2427
0380534630 7638919476 1501869867 3670605014 7556545191 2556348547 4406162027
3938350356 2785615295 8894681701 : 2428
6999401433 2311095287 2124482704 7206054602 5850066704 0757911141 3682790697
8686587117 7920435611 4892996871 : 2429
9888032590 3495462586 8507864515 6073721715 3995339107 0545742084 4700489981
1282899421 6021222092 6244947274 : 2430
5405610358 2092642512 6780398190 5452659443 7375194281 3217137033 6125910575
5169899284 7294695342 4298072325 : 2431
6290258883 6268426784 4702983631 3294966054 4162538614 7488283479 8167322881
0978487694 1323436718 8334829751 : 2432
3277555209 8111835661 2998485687 0217344971 5945581420 5167601363 1681044749
8709163649 4315666700 1634124731 : 2433
5263146646 9447022286 0280711813 9928158887 5163721426 6832121415 0923172057
3189111732 8825980525 2015609004 : 2434
1554777595 2404089135 0094036519 7084870074 9678332743 2335886946 3126879009
8502313172 0661411321 0860760486 : 2435

1957063566 2452304872 0492970167 9457814582 0090361928 0678213945 8937433777
6931269876 8681171248 1640849105 : 2436
2538842393 3369089463 5409235802 3108172557 6349979969 4364597544 4894856647
7328044988 6762357891 7302150269 : 2437
8796454984 2771233602 5239601367 8890263912 7631673348 6900994658 8810286310
2237495535 9950171618 7794059542 : 2438
7203256807 5099172304 0604059244 7593475587 8192311507 0860386436 4001669769
3588444107 7368702845 7703794092 : 2439
8349414028 2212958640 7527063935 3993044472 3850843968 8527577798 3552083175
8107094826 8654551492 3467711645 : 2440
1188567223 8076006299 8781844878 2700527203 1293884799 8209719432 0227571363
5203988800 7560979354 9685072221 : 2441
7381909642 7575684664 4078438497 6235954164 3789860716 6734860499 5364292157
6909269615 1709528254 2108602668 : 2442
8812876213 2282887012 3941112135 6084998485 6022616743 5034883052 1151995222
1309472231 1882454739 2608085441 : 2443
2153442104 3454311042 8353361072 3224461095 0475490308 2384976233 7877239798
5764670714 8507270115 5033507917 : 2444
6889428553 2565557856 8914111053 9376812300 7641727332 3355555569 5819795161
6787651611 2715880238 1737058125 : 2445
8433764454 9639320903 3630888423 8313464741 3254157583 4085328701 6214784675
2736603532 9814219899 9001039965 : 2446
1663985781 6270835896 2481375811 2852050274 6831434621 8654210028 7357984530
6419721733 1119032520 0734619298 : 2447
1287229517 8982451117 7032832347 5986403956 2706190854 5507358079 1658971007
7764022903 5197705516 5146315695 : 2448
4288414243 7557975709 6894223288 7312501544 6591323568 2234856323 0881862148
6917525444 2042503115 5171125209 : 2449
3266720935 2445385328 5759307857 2051963112 6771596563 3535956460 6638121569
9176134271 0503579689 3469256097 : 2450
7592291135 5755004495 4680939594 1988076919 3752888650 2489711246 8591619511
9118057366 2336507492 1836732839 : 2451
5749066939 6689948638 8125465885 5888383033 0864279223 5459716140 8639132801
6968670609 6747793497 0251369670 : 2452
9492118518 2680683710 3932976818 2793490904 8809926852 9794978557 3337154568
1229119082 8899996496 1736727582 : 2453
9677225427 1826422328 6640013272 4327309242 9509230566 2213469775 6027497131
1377496402 1604518693 3589599433 : 2454
4517013147 4316716699 2535535262 5191822960 6855110255 2106617693 9130589930
4704401305 5539478586 6316843769 : 2455
1828647234 3532485938 8779733700 0237434440 5223057833 8504233697 4867005016
0028663716 3548072142 5727423634 : 2456
7165982592 0005995273 5028634294 1390667926 6972379873 0437353937 9577587467
0438709507 3567124554 4966030978 : 2457
9611819455 4170245592 1930096405 9380552294 2769217350 9881950338 5424390196
2235565665 0959811895 0849558347 : 2458
5832679441 3719433477 7064417430 6876072873 2386031909 3764674529 1892183927
3405652449 1250586956 5176115620 : 2459
6981250039 3153884581 8440649081 9305513822 0680810239 3363085653 5953832850
8151852860 2490738088 8193971974 : 2460
1926645634 1614484265 1154131695 6283525195 9112442838 2628810103 0847554548
9736902540 3588236483 1424405950 : 2461
6043336372 1723113697 9737662537 7898329148 1467685475 4118971023 6465877932
9452455366 0846298717 0976671452 : 2462
1539359362 9565084168 6793887474 5177684647 0197056012 0629111965 9392716942
8782001047 3842269120 8420374736 : 2463

3388386274 7926634381 7074600861 8165177012 4738002689 1028324861 4546728946
4377033943 4204648424 1967025618 : 2464
7916489725 1838674622 2304165160 0018564312 9965411759 8208500563 3239524163
2046675535 0013353729 6849174676 : 2465
4631941349 9223744247 2246330220 2185954740 6463788211 8823459394 0899689958
6677663701 1429528531 2707935566 : 2466
3237832561 9667821366 5709220602 8310256539 1354011906 6214292393 8164112080
6961721604 3809938799 3003279135 : 2467
1939616054 5906725965 7242443886 6730988394 9480405001 9958699540 8776106691
3890684279 9356469502 4599087865 : 2468
6104815262 6194880291 6220377285 4404310761 9152330967 6134565789 8664927602
3103467080 7839099262 7547645000 : 2469
2311159881 5152501667 5637395740 1905770341 2613420416 3590448083 9076537485
8277752596 6628543162 9883314207 : 2470
4778261209 5040776043 3388366358 0243089244 8403488354 1028706147 3396328346
4657835796 6974592587 4011346345 : 2471
7623216081 0423976222 5168959734 7681742851 2737721348 8843126429 8689167069
6316238734 2001469489 8521423020 : 2472
8355107107 1050255718 8462778564 4037605341 5487371340 5703530471 6047710677
5293200799 0830085796 3350589136 : 2473
4869774715 0937612256 2844835142 7933875263 6745707222 0025676912 7483422379
4366061319 8626760944 0621051523 : 2474
7198485974 7379297406 1772433307 7735380254 3022189439 5766769509 5667279812
4850084862 6425884845 7967719356 : 2475
1466464626 0149649514 6347149006 1886726013 0216748107 4660541119 2668918406
3788353110 5630441708 0835780199 : 2476
9223295643 4743032959 7913588943 7938009727 0442582150 6927979988 8314467253
2976896720 9243310967 8778704372 : 2477
5470404969 3782685353 3277996781 7629151867 1277541365 7266836934 9108729256
6065622815 9152633504 4669497567 : 2478
9229497645 8396040312 4782609680 8076324572 9179631357 0638055301 8179506155
8934619200 5525020421 2768920472 : 2479
6523519590 8441637059 7622758052 7335399057 2737729245 8984311346 6208946935
6846280770 8795934236 1434261835 : 2480
7397284121 6652601954 3848177450 2442968737 8704478184 5808456698 5918167574
5936300712 5099299455 9021579797 : 2481
1267979286 8141836179 4529381147 4383459113 0494949062 5457775739 6574482504
1893661050 1567223911 4063379044 : 2482
2693276717 8357282347 8402429229 0403767470 0397134683 4385546406 4270710261
1753009130 8476127357 5638893444 : 2483
9578014367 1978013898 2653424377 6067204873 0565920693 3281697377 0772050673
2140005736 7553449808 9554053868 : 2484
8878671591 1240760240 2876493610 9146485632 4351392282 8961692053 8474220896
0466080590 3823099659 1893342589 : 2485
0790062237 0408006699 7920297981 9440927717 3502701273 3684682086 7383102707
9479355302 2082277521 5446092735 : 2486
6207151719 5538748966 8190846802 8606626805 2662617307 3955928932 4327665608
2055892649 2281145720 7893258778 : 2487
2368082793 0505003074 1774353514 2587643209 1818543266 9406906760 0791908213
4203963689 5309452563 3402213073 : 2488
0209864586 2976896554 7248652624 2846110473 6657509041 7717320523 2374140756
5848993239 2708682167 9426432687 : 2489
5694735191 2174769111 1577540799 9719992668 2888507939 0393406103 1042132964
6825040770 6477052176 9095572432 : 2490
6859664717 6986382914 1153779769 7600025819 2723944692 0104966004 2850854700
1480918081 0817266504 5679671866 : 2491

8806462058 4788093007 1167141907 8497133939 1499399525 5245452094 9465078434
9719810361 4287781840 3322057069 : 2492
4639515476 9469727677 4770642486 4607939235 1956543663 5083070252 0798246537
4274256996 9457756462 6119873862 : 2493
9434532805 4150827620 9906622774 3584448627 0376709248 8431396731 2656356805
9785342858 1984460900 8250228150 : 2494
5106367269 1418876039 7883197731 8626572931 4218073290 5509353856 2444488808
7051585120 5561944137 3703285405 : 2495
4757221463 4071373693 2655210869 3222709423 9075429940 8994425444 5906867574
1143225242 6167234352 1912782585 : 2496
4388455951 6797829932 8323642737 4574525445 4605293989 6806263513 7335872148
5080882020 5518659958 0340814883 : 2497
2970125378 1235056793 0508188185 6850573123 3257555542 4196054273 5831944797
6432499228 8226604355 5852334960 : 2498
6680905502 9052163377 8474635193 4749713022 3294939655 1041598783 9740167516
6185936051 7933895039 2466205245 : 2499
5112688373 1112078525 7244245799 6232944501 6834171359 5140252095 1792646811
5682982031 3618827396 4266233216 : 2500
7644152469 5487558164 0843582125 8504424767 0699693803 7585730039 0579011051
5414779557 1791693127 2909599822 : 2501
1364115981 5952014586 1367892066 6663532183 9445791129 4294937274 6424648239
2154779756 1573367089 5761840575 : 2502
3220985047 4858357089 1766352727 8094953542 7448252511 3739382912 3783351841
4718278488 1809377625 9467255433 : 2503
4206902383 7559765846 7444988572 9710015336 5702593838 6098378883 7055966165
6612261881 2454638780 7403643775 : 2504
5829259340 1364517385 8446245540 7649029616 2202292447 9177890142 4327249245
6246105728 3299442767 9678314481 : 2505
9346705517 5708350294 2567326335 2649065141 4121023786 1093296718 8631037171
7046176289 3116167259 0290677122 : 2506
3985883659 6414924553 0812072857 0841006607 6168543516 6635303413 8280113381
9677912289 9741266552 4495134833 : 2507
8934636181 2822564990 5341150317 9141167093 8307677687 7423256980 3429140799
8029191076 1139653077 6180407622 : 2508
1944515194 0260406347 0356799353 8832743785 8815201108 0406490885 1752700820
5623802051 2864218424 8230026324 : 2509
3205599799 8346926232 6656447019 5635730067 9539057244 1503981642 3908213623
5132717714 5861912103 2811235726 : 2510
9933087662 5534408941 5120517990 2731473868 1826266444 7528040672 7464085723
8015503894 1891259589 3739926501 : 2511
6877527437 6974153374 8172422037 7071286644 9077116260 3154417119 4141083486
0689952950 7444772203 3627442668 : 2512
1847119656 3615713772 4246154560 7047965087 8312900133 4349111362 9297558360
9060175949 4537968615 0681790850 : 2513
7607566212 7381001179 1829307611 8629911635 5745026020 2127565436 0951138569
0948154244 7672260734 0061037334 : 2514
2612736080 4485531214 7578890237 5590577113 1745500941 1859748652 9627058856
3917389715 9515988987 0141758696 : 2515
4865418532 4863779433 7805069893 4555388050 5233124949 8418875730 4644473314
4459850552 4739865399 7073462338 : 2516
1939800857 7304356954 7616982826 5893810030 6024112186 6568598020 7253371656
1353350992 1885956010 7881521955 : 2517
9929848307 3711416176 4839950330 0378988247 9034541053 2502054955 6935880161
5459891893 6886572124 7489636136 : 2518
7186281885 4644786179 2435817101 1255185131 7871774504 3073536450 2976150729
2301108330 8025515349 5186929484 : 2519

9716900991 7303947697 3337895650 2295614877 8780483665 8283482754 0230192303
6903858197 8853430382 8558273006 : 2520
7215613042 4767965099 7367389863 9630845953 3099446736 6005278953 5100775106
2354051809 5062072959 1214778792 : 2521
6626338542 8792589775 9586305806 4650448452 6239133538 3426270504 3086700946
7036220406 3397672529 9136518784 : 2522
2306583966 7022625805 6212221073 3541161850 2936356416 1665577923 7766395860
4946932445 5080590361 7986442755 : 2523
7412949830 2104696986 1644931370 1037027750 8486015396 1665864512 8535453048
1559829638 2985981545 5625924865 : 2524
9186328817 6301101499 7372069201 5386987741 8621655782 0878850289 7085678297
0192695827 6952394082 5795893466 : 2525
6666883918 3588155490 6943683070 3532763207 9349451093 6539945097 2042836730
6703514419 6315528875 3214822189 : 2526
3259671737 0781271405 1334747386 0809636945 6351201901 8439160557 3384080516
6382914886 2479351379 4037131979 : 2527
6687585625 9482942074 6324161481 9626828884 9800968875 6413177902 6576910555
0802543228 0312585899 8458287208 : 2528
3257358894 7631349260 6249627183 2200731813 5424395364 3770564819 2953995700
1445543839 1087844914 4193680471 : 2529
0651634740 3117037448 2458505185 7881806866 2884417079 3566042698 0031632363
4912030291 9753700996 0106661938 : 2530
9621731876 2267071826 3148522844 1727943340 6818103101 8384175349 9734969790
1352604608 3898649384 1708529346 : 2531
9279158345 5942477874 1475818626 0672246624 8117722498 5686229897 4404384392
1840245603 6091912369 8959782488 : 2532
0644631955 5555930832 8167346023 1204066700 7248774759 9806326845 2732025570
1562168766 2840583268 8949305051 : 2533
9390050495 0495870154 0034854277 6024624858 8466667342 3859744545 6716114198
4303835706 3974266667 0385556096 : 2534
4523903570 2010736523 2835276920 6772213666 3585746080 7615994825 7589026155
6442866496 7372569208 0468511746 : 2535
2670246787 6686032287 9651197857 6164426500 2553662207 9972039998 6561469155
1199659189 2609987569 1957219827 : 2536
5509506475 9786156264 7423557864 5011389704 1993509976 4066765571 2085029584
2115591494 7290752355 3499274100 : 2537
8512949193 8559625940 3263820252 4988224921 4444755882 7002900367 9518705235
7627644235 5841833307 1204601246 : 2538
2993991548 4195813551 2551467709 3447144330 9247637321 5011861279 8381856025
5716314174 4264421039 2318412486 : 2539
1561304709 8148024733 8812569605 1967726943 8321490104 6524099815 0118339414
5060084222 9131941609 9500996449 : 2540
6196330766 1716802799 6614596490 8485717408 2378057131 2943966103 6877272697
9043490318 9674932321 6657233190 : 2541
3721541461 0364718842 4635680197 1257097712 4204559927 7189401630 8075557915
3180388638 5226329349 1228689445 : 2542
8712440718 7398513109 8072996000 0540296913 9086326671 4179236497 5629719250
2128839909 7084846804 3907176319 : 2543
8298386258 9760312738 1810275493 4261012824 4583510397 2461726002 7124726441
0283930603 6777543984 0384623746 : 2544
5571177660 4274794044 7110253227 5260708819 1525962388 1035944912 1002592156
7550999035 9849028736 6394653336 : 2545
2227856019 8785244807 8120000922 6725563043 1187021878 3254738688 0440918833
1048255150 3395062370 3534591157 : 2546
5694871584 4081222535 4661461213 3683291417 7138712079 1132563299 6961058630
6388145503 8293070650 7642500409 : 2547

5978377200 9135428432 8731106694 0704199932 5305683169 5331854406 2180960834
6131977993 3817165917 0654879552 : 2548
1144399346 3691039132 5853497773 8053801424 9409345036 2761658136 8950030951
2610570841 2344562960 1328070394 : 2549
8714675890 1016641151 7039393214 6989030266 7266058467 3505964752 7480561780
7867953935 5103268491 2986766565 : 2550
4263127329 8852919270 0824708774 0022137431 1565869690 7606589908 5477980877
5648655941 3089027045 6897729741 : 2551
9559655010 9221935693 2384978162 2587517646 5524209255 7409257176 9546886051
9010003160 8012897289 8705286108 : 2552
5422973909 3968150775 0096597173 7146008611 5220926226 0852708298 8364373624
3877981277 4511708223 6808061077 : 2553
0774136633 4795574353 3547250663 4409792898 9918408218 1502006262 9005813678
1545284857 7595273359 5359748408 : 2554
7245005388 2741039998 7019521262 3316986282 8034388497 2691416958 6295036202
7229748868 9849003974 1471616745 : 2555
7511413346 0273449742 3550587807 2186655258 7350641253 0832457388 0356085157
6626591008 4790720477 0453688975 : 2556
0719974356 6506306631 6758761134 7516441890 5099495304 4117199851 4991673976
6229426944 5166214080 8774913553 : 2557
6734530651 8299977582 0146575308 1579408167 5035725631 3082689752 7686949131
7516603141 9627412271 6209578299 : 2558
7451259507 3689499764 7865130983 0445539167 6187931636 6404096977 8731171580
0412265552 8863709140 6258178846 : 2559
9239036439 8767943899 4419596332 2773315106 2417111111 7589582042 1382268247
1585586231 5936615312 8943219165 : 2560
4892821195 9762276658 1435967431 9046931897 0709546254 9848023495 5018692311
2936640292 9099667008 6387840042 : 2561
8904420862 4836617790 6430206330 5933920322 4343651607 9432570246 5868466897
7153432807 7217098798 0118148551 : 2562
5792816444 9213543001 5252996137 7236010772 9210859513 1459952461 6594227164
1574763236 5702571880 6117063487 : 2563
6292627323 6008312525 6996543432 1893745077 9674452915 4278947127 2289470446
4813147441 2422116659 0081005721 : 2564
7233044387 0087373605 3316468302 9287005557 2001906994 3199870645 4465506242
8217271171 2459206812 4294810550 : 2565
5040470592 4105288357 4006564845 4724560748 7562476347 2596201955 4163080869
9130865678 6967875539 7008127911 : 2566
7686691949 6838135150 9880852095 8276792948 7854818158 4339038957 6480289850
9257246086 2530061488 8628650306 : 2567
5719865793 6561579559 8257299189 4328947716 1896205693 5467280544 1856350184
6263442674 8571556088 8443376776 : 2568
7751811195 8796316841 8536391233 7497661237 7125870557 5367714255 3545280102
3619128824 6608468567 3608493413 : 2569
3311957993 3540423335 7735889637 8053183909 3444280492 2703521622 3087149443
6067300423 1179796828 6390517195 : 2570
1575052097 6559027309 9670998902 0051300226 3326473818 4520239976 9112952460
6155729336 6996541826 7875614644 : 2571
7436938872 9088789425 9922714756 3262066667 3290809469 8629295343 1110762432
8164327360 8630864133 8648646683 : 2572
6833403411 7417243361 3790860478 8056800459 7543289332 7214060803 4447503284
3441146117 1909670176 2539842822 : 2573
6686468388 1706100253 6499007431 7384700086 1481761643 1964214609 1993738188
7765482706 9979398415 3938974909 : 2574
4610308060 8952105623 7233733955 2990648545 6547771113 2351150583 5187239748
6970763522 9334354972 5610030112 : 2575

1589126783 2849264645 2926571161 1514653003 4496144130 4070786937 1417923311
6662476964 0876354873 9901747753 : 2576
7102018211 4281421448 2462132048 9013665523 1442441340 4287752981 1835667348
5565936917 9625585315 3675107980 : 2577
6714527966 3745899421 0311881547 4548075246 5185317021 8249967058 2009281703
4714330564 9061103029 6600778862 : 2578
1864395862 0309126219 5374593191 5501116913 3155954733 9411720861 3535884052
0458592736 0463219827 0224071542 : 2579
0614033109 6148629907 5908103313 3065914757 4953943653 8701843065 3038342790
4014305982 9881096866 2879396068 : 2580
4213401058 6681368770 0862550410 6696553622 4307609748 6920666744 0684275559
4708259403 7595454393 2812646151 : 2581
9786010940 9221003946 6239381000 2488578082 1530539641 2362630356 8044502330
2479433432 1344188084 3146928161 : 2582
8392368201 8691893983 9333078257 9391518765 9886158526 5883031305 4820647419
9238611662 1691904597 5656253336 : 2583
3184467689 5075299258 6777308978 1132205545 2689323411 9637741580 7042979172
9618493376 5166937562 1514648813 : 2584
8417266217 1532362271 0802784181 3774596097 8655724521 6534926787 6088009188
0707514451 7955918932 0746484076 : 2585
1990517355 8584889133 0328063507 8797052313 1676769315 7737318795 9490721237
2637992597 1534942241 6504918609 : 2586
5916392980 5153754153 0560108354 1412442663 5088411708 8954264409 7702742282
3212873781 8584813773 9355097493 : 2587
3555114062 4466436894 2045355237 9302295569 9025688892 4724764828 5698792777
1770439584 3692472400 6222094132 : 2588
5554943292 3268062651 0065606711 2487799788 0399882214 5863452959 1716662481
6532287411 5532752641 1248966562 : 2589
3653627179 0517082015 3100267353 9588247022 3528163997 2401534641 2203202579
7780827313 5512050193 6842815520 : 2590
8185549975 1491511016 9914127116 0844540090 7620830051 4164618825 5296346203
6087371679 0190582518 9468389540 : 2591
4682662971 7486680083 9302951260 7766929924 0694352257 7743863181 2759679506
9437001560 6250562785 5914341512 : 2592
4133940303 2771295325 3107118617 4802577223 4948929252 1980974308 9521223146
1957666207 5923556359 7266907679 : 2593
8666123329 1059527961 0183431069 0770203222 8716252083 6119564948 1175299713
2749730597 8355285212 8578547844 : 2594
2861816852 5710739979 1644850737 9463019479 4860109379 3836405400 3035089249
9489138010 8931322703 0643660409 : 2595
2136152251 7503364759 1255299336 2345087462 0625221161 5213453346 4059073152
7324079559 3956003274 8790973869 : 2596
4260663614 3145093479 5793642528 2076057673 6682245561 2779788579 8509050746
5575999523 3257680197 8516473222 : 2597
3573444661 2494779990 6429335103 2029241706 1814769571 0507728019 1727166542
2720280245 4806556829 2656244457 : 2598
1074844343 8092473558 3240595727 9281370093 1794958428 0200678166 7030234830
1074054742 1926860540 1978802767 : 2599
0617733116 9854901005 3252658070 0391933221 8325517622 1950049560 2329543188
0724870989 2249330737 5904553488 : 2600
7851895773 4282512509 6765197185 6799652910 1719951017 4647814302 7813335716
9564223193 4075713767 8346086967 : 2601
1224381217 3079896938 3121704204 9112414515 8622120573 8198926028 1325336165
0633270961 2681127354 4576450343 : 2602
8627183739 1993894379 6958561167 1266838339 3759855826 4615427978 1331791205
7829123789 9892276277 2561595125 : 2603

8427540001 4463204457 9106546866 7324140533 6586191842 8042262582 1688273710
3215382122 9001605389 5557458048 : 2604
1497079514 2882874275 6657075814 8260548242 0221061203 7688341073 4370446169
5313565847 3158464995 2332889740 : 2605
8613892603 7436545571 0313573097 8703051579 7674186488 3330833346 8306177619
9649533323 4340591686 7838864524 : 2606
7041275531 4395794027 8842161375 9684918282 3286006692 8911605076 1181598098
0572296761 1642356090 5478275530 : 2607
9990228360 1182556875 7238788125 8582934211 2120643535 1362342333 5454800037
6373539228 4413374664 4754648997 : 2608
2715324870 6234324739 3949407436 7849041727 2542657426 7589518279 6020334362
2606184340 6548292910 9694732775 : 2609
8106315005 8025056894 9213398370 5710619553 8103699251 0061604500 6231958956
8527763384 1453387092 1568787580 : 2610
3212746031 1284924887 1469759238 9661216541 0078451665 1875999266 0729902045
8345627963 4204397156 5244565003 : 2611
9332697584 1615274186 8905281033 9622772868 0145702700 3189627877 0775137289
5137492853 8916013451 1814790912 : 2612
1245542835 1145074766 2061450207 8740552719 8310649131 9508431939 3794051393
5608624487 1206328233 0972563106 : 2613
5680671593 5871233992 1409666332 2511910445 0832165354 3621993777 5858432812
2723097176 4972700282 5335202360 : 2614
3346945160 8228728472 7522818468 7773750722 9883391318 7683690263 8244934888
5643460614 7021410159 3355370833 : 2615
7592611935 4381437532 5836805068 6926516021 3196385900 4249450260 7779328982
9297493102 5747485191 5475821236 : 2616
8427556373 9778101515 0277718846 7396734397 4182564271 5865300921 3673388007
9123112661 6041891722 8468906382 : 2617
6787172246 9773414700 3033770948 6294246783 6229172179 1257397858 8957230493
8003585912 3639968963 1216138583 : 2618
1046483707 9637662679 9297615668 2119846593 4155933916 7444688620 0355689651
8406189650 2099578794 9475034213 : 2619
4510646829 4189135762 4099495577 1883376474 8449361489 0337338736 4084487665
1285799060 0569180355 8021757432 : 2620
8223720982 9640541398 4917686142 5411357801 9232843223 5662201253 3756997103
8210371450 5361135215 8087544325 : 2621
8875177314 9812341597 9007748415 4852468747 1869828437 1642775679 6612188225
8983635864 6123372708 7316163958 : 2622
7829938155 2734158028 8062228960 3227447919 7315134195 8948838419 5292905675
2913582847 0288997290 4682421781 : 2623
1588125445 0027577348 9756610693 6993830600 2844248830 4085568975 6491161569
3828782862 0459017159 2066183555 : 2624
9705573502 1830926911 9605068711 3637921989 1638826470 0303239855 9982585297
3720675968 5021223259 4796092137 : 2625
0133154369 0047347580 0526697316 3662808767 5468684315 4412005445 1810963963
3177996327 0733270078 4242615943 : 2626
2871983671 0018530522 1100049935 8589809347 2727826132 4522255474 4663365234
6902607995 2018829848 6579293564 : 2627
3341058619 2063576580 2134949712 3815423332 6330818249 6330203863 6180607430
0789362848 0494572747 6555968976 : 2628
9047963077 2584358960 9723556268 8527717695 0957548546 7415631893 6544434526
8252226873 3165858336 7174104535 : 2629
1860168973 9003705114 0387216074 9256572866 9414463434 2819342201 0787994447
9315289080 7044167837 2085914038 : 2630
0787192020 4687148954 0429657782 7423277263 0067548268 3925720474 2895691916
7600520523 2153821140 8873240679 : 2631

7255883699 7229770397 8174778655 4451339369 5280467309 7919875464 4054050153
5598421490 1764939708 9933682368 : 2632
1797861826 3713774776 1899242139 6475468152 1802356570 0846506246 2125800933
8239375893 9853525532 4737030726 : 2633
8761318693 2612577333 3729027491 9695015084 0481799877 2837366525 5064027193
6147773259 8808908149 4639422730 : 2634
7546211379 7425254478 5730656206 1623275845 6633687160 4105456555 8219632284
4425800161 3092292561 1695217058 : 2635
5617429297 1169937298 7985526865 7367981622 3076859491 7332186376 1507735171
5337805336 3994725317 3790467038 : 2636
5755272237 3827813588 5645323766 0838981202 2949751795 8499014168 9663452187
8608358384 1189313847 2832576864 : 2637
8734746219 5353899780 0875424150 5867497801 5601593113 6540552070 9508035255
0048121231 2377181521 0729800323 : 2638
1017591837 8625405659 6253994854 4710762023 8523408341 5014218901 8389630276
6908646062 8899731583 0500060541 : 2639
6610521126 1833245630 8874942376 1321117383 2359910267 1544333398 0903010767
5192156068 6091509929 7579489847 : 2640
0913404847 7603725331 6486633273 9977457417 0787058858 4989036478 2505006075
6527667766 6730181427 9834629978 : 2641
6311547247 1904638130 8270269502 7155243458 3777132888 8401133228 5612327642
4758054914 1453340043 0735136820 : 2642
0167103048 9674079132 2041732936 5588638081 9902402504 2475897990 6199739449
4240613938 5900204374 5081712616 : 2643
0362783912 4114726820 9085690526 8374225068 9109919376 7722077768 7371267701
5290712968 2261584375 7149665346 : 2644
2961540352 8980698498 1990238158 8132490072 8284203166 4545864518 6778718177
1772779283 2125269683 2297664124 : 2645
5496739715 2787968043 4765895761 2653385245 7391513438 1378450051 8738591532
9634140536 8948443972 2550801196 : 2646
0792690281 1622936704 3437115837 1953865778 6003419467 1309665344 2535523561
3503926374 3335590248 7780093167 : 2647
5855665020 2614245175 5202310518 0379792416 0186816532 7213490744 7418792630
4637935701 9572546568 7076964902 : 2648
5628311394 9083065981 3925877165 7534329051 8298830744 2209315394 5626718913
6509327785 2758561418 8691505843 : 2649
1128180621 1645338346 1456499861 0279908878 3159992332 0834970349 9009644828
9736199726 0841303050 1613834375 : 2650
7350335026 9679199100 3945764850 3139889980 4053476207 9799510355 6280094271
1980771413 8625374689 4200671129 : 2651
2903794021 1050993128 8176786355 7121288220 5845259023 2988278448 8972855767
6433765513 2098372084 5365197273 : 2652
5662945407 5207868377 4825993769 5085474585 3778544015 1866870321 2703251083
7885755352 5327422467 4561655301 : 2653
7529469704 9286034935 2376631937 7581531269 1121571250 5456493662 8404613157
5493234361 6114386894 1551917955 : 2654
2116040327 9413870405 9736596828 7723555493 6953672492 6033527449 8928882204
4886844365 2751589546 8955885890 : 2655
7183172928 9129234577 4428441927 2505276847 5503870270 6328297997 5853882599
3879067899 6367663472 6367997091 : 2656
1370005004 5191515070 5020857447 0532031134 2837530396 4506837349 4746515254
3161640695 8839656960 4776248100 : 2657
7698125762 3240276563 2471455867 8116653563 3573841332 0375632857 7111457947
7361177589 1097844959 7487134549 : 2658
9454050089 7494312370 2669160022 7796215160 1644314463 2155674657 9869691343
0209173753 7932953736 3102934825 : 2659

9418485153 1345770064 9437390976 4208589573 1774231457 6728829679 0675029922 3152573286 9833026341 2335263163 : 2660
4902064904 2970821006 3264881567 6763242544 4687039213 3376789489 6001251362 6523547256 5170222559 5569986284 : 2661
2510886689 6847107872 6001673324 2215625124 2927213080 5593262213 0721409368 6435499689 8787430352 6768849221 : 2662
2318344924 1907637471 5747446252 1597457646 6357242752 7952228915 0406427767 8656611519 1933191817 8305671646 : 2663
5304813810 1066673424 5915686417 4457688390 6241920186 5410225266 9706153890 9990725499 8428548419 5668192454 : 2664
5197470930 6142275315 1298445309 1827577151 3611816730 3580932146 0322584723 5281182550 4706062154 2622432455 : 2665
1446896457 2693823166 5552509598 9504109342 5374308599 9797137004 2588583403 0449726710 9629969763 2336077767 : 2666
4373479878 8356730102 8647138454 5928791637 4901454066 4751939489 9352212362 4743661317 4783048688 4631516036 : 2667
5922435767 6276234466 6539589796 4687905529 2390270201 0757218919 1382148316 2685249049 5848675432 9312418264 : 2668
1346627228 2093653277 2839766755 7267289731 9381293419 4305723962 0723292007 1863867466 7030636460 1331111164 : 2669
2546802512 2894305331 1250985386 0120123607 0449699785 2109585987 5329303277 1622679823 0551076769 2680002207 : 2670
4188490301 6500503853 4475971018 3016737826 8194361241 6569639252 2947410357 4318517658 3656034123 2764339009 : 2671
5651186326 0791733899 1262772072 1351617522 2255241829 6124339628 2518232869 6862544411 8623812330 6403453315 : 2672
5601640695 7472320383 6514566355 7498734411 6859941616 5518249604 2597983926 7816131483 1809025345 0716466644 : 2673
2670262761 1859764913 2476829527 2780570322 3834351506 3672177066 3763740249 0304659096 2859602719 7972553780 : 2674
0141820199 8101813981 2595042348 6624834404 3921136487 2366629202 0639396288 4531448374 8901026084 0361484073 : 2675
1200674156 2291596696 3669408360 3264334149 6371209854 5475250177 3669601971 4617846451 5555994167 2637397085 : 2676
8649598779 5324215832 8218410939 1640528356 7907068642 1078034660 7571979891 4815540054 2005107300 9627962347 : 2677
2724997011 2217781656 7984491943 3222633415 0338567530 8244677341 0455032742 8561157455 3874214000 7192843017 : 2678
7447314230 0983657607 5155127779 6281014722 0530668174 2035059679 4105098046 6563136377 8251724709 1409925552 : 2679
4710368126 7051382467 5211720052 8494295219 7488628489 8527787835 6210600487 8127114406 3499088164 5924451898 : 2680
0104429357 0832904722 0160726966 0461984260 7722478310 7171439093 4928973795 0750564710 5380291618 7491886994 : 2681
6353013572 9350187320 6687311501 7315310912 9679486254 9795815121 6822075712 3189190913 8338353447 1023659794 : 2682
4808047123 3882744403 5053467999 5291335460 9413927284 4651391190 8360762265 7983981564 2463829159 9904416284 : 2683
5276818935 3279135674 7403227351 5068909877 5472181557 4998488346 6946222711 9434351395 7275609331 8776721574 : 2684
2857830330 1830422251 7049632971 6122968367 5274898329 4723150697 4978874140 0219267006 7206567729 2131084935 : 2685
7294076359 2895618281 2900110978 4724328303 8465197437 5857368591 2309817836 0303140528 2300813026 6313304139 : 2686
9253399217 9415764798 5347081786 3611377200 1408570838 6394377035 2918349740 3741835116 2237004017 3188269393 : 2687

2630875054 8756645290 9326566503 0243944536 2272791708 0038157851 3252902365
1056809625 8917942400 1638714961 : 2688
2121694699 2544239867 4726206005 7131153878 3883388307 8016537883 8775211593
1194493595 6949179394 0578848860 : 2689
6239594441 8497289287 2308495579 2607211329 7712137238 8696986360 2368291622
2546471088 0621094479 1323990154 : 2690
0667816028 9346942821 5506272126 0541798291 7817382489 1997338295 3016826679
0617780135 3365047188 6339784273 : 2691
5335856232 7918535797 7922662703 8024456968 2968625491 1874868530 5497857965
9891848621 8623748556 3935321563 : 2692
0489928348 6556154154 0649512210 4661037654 8180602506 7654913403 3273862941
6911776263 8138347851 1836410569 : 2693
9966094920 2044895026 2944616846 6855106066 2420883145 3740126877 9478139859
5777699039 8770739941 7032565319 : 2694
3556130005 2814359945 6061261608 3223589903 2489565956 7527576985 3245744035
6076128860 7968185761 9771788765 : 2695
5619852375 2744722359 9272020602 3891671879 0814708867 0683027939 8976837802
3337579683 7484716792 0420561188 : 2696
4614435084 2383697378 5948258849 7825952141 4316768984 9592131938 9412875069
5961491932 7114703587 4533660814 : 2697
9437546971 4291952903 1019389435 6371835937 4914830742 3044934029 5962811166
5238995811 0600093862 2216226552 : 2698
5297661065 0745271358 9497334474 0728167392 6348622262 9713471555 3293624445
7994650823 1979087690 7445852537 : 2699
0345709007 7440815341 6786387014 8099674124 0038378085 2342739778 8174691080
5353305091 1944331387 3040838043 : 2700
0675060306 8626953212 4522901667 5038563185 8585931743 7769494157 4108105727
4944438400 1399152295 2924016806 : 2701
6745842469 6605569107 6975969873 1095078451 8251857689 8000939428 6371219101
6698078851 7105711446 6950703127 : 2702
3706962004 7300356753 6823520581 5249186823 9083974088 0926485270 4581680063
9153401349 3735254715 0923527044 : 2703
1619265721 1004234584 8085322398 0930819701 2158641732 9130532589 8717158855
1684206065 0340556996 8593715915 : 2704
6219395459 5558557009 3477116811 7983599584 2798195564 3563653093 8905094196
4641889243 4176612177 1175457371 : 2705
4429402729 3771776591 8310744305 8151531596 0948263506 3365572386 1413920813
0754146107 4051274134 8138890687 : 2706
5208965175 4728644348 9020150187 2018366138 4172807988 2729582018 9774861263
3836037110 9414086804 4146381899 : 2707
7551441905 1152014024 1876289786 8823386652 8874956474 0110724599 0553799217
5155647819 8091849558 7675277828 : 2708
0803822629 8180439415 6397956172 5694090929 5185774478 8365159447 8720682678
5963694547 6370623820 6696202396 : 2709
6206659210 8127818321 9127468081 4530314217 7986735336 8489380826 6818969129
9983519942 2321272638 7715975764 : 2710
2852132151 5883717146 4854288124 2312246840 2839056157 7968199897 8556251027
1070628379 3994319073 5797975362 : 2711
2873719947 8452103831 6866852142 0822019266 7231558101 1737244237 5609151489
3863436665 2657942603 7168289281 : 2712
5806931590 5715237948 0256619126 8708876475 0695085011 1370257880 2333818019
0302100297 5975592681 8216359535 : 2713
0706418857 1900594974 4467974174 2025213094 7246191950 2772132372 4702570296
1631681464 7621846436 4465179953 : 2714
5877759048 0917246955 6739679455 3734971032 2193694559 6277893779 1938340697
2537884155 0206295838 7483096195 : 2715

4204615469 9022268434 7461767711 3197374866 0008793544 3607302433 6632808653
6847335066 8707408900 1847030676 : 2716
9821475313 3731542862 2151551318 1409541497 9724670676 3436976964 5830928679
5212019941 4066540432 6668344081 : 2717
9686918622 9176544103 6492080785 7292423388 7550618098 3659122265 3797288411
1201306910 1857603049 8329532694 : 2718
2141884259 4286621469 5276880632 0825719648 6713422469 8526419419 0222362411
8633913028 4171844724 8227557233 : 2719
7996970748 2002437580 3717921807 3420208053 6935740618 7656641696 0773912090
9813494702 1207251972 1369964234 : 2720
4209305478 4650692374 4649042088 8732630226 1563579196 0630923699 1602782364
9300034497 4712377945 5951240858 : 2721
2397099465 7027536675 9813304777 5050505366 3457471551 6558372773 1007857817
8715303161 3276848925 3576078462 : 2722
1147886035 1804029765 6960584867 1756763665 9308748016 0999279507 8717891310
4203849478 9432860847 9705150428 : 2723
3326524571 8864231983 9993285634 2268607883 4437453092 7289314609 2544299060
7871117367 6695984963 3062177514 : 2724
8848993377 8786785978 5265280570 5486612173 7921355212 4702395325 6081906788
5280383242 2968075544 7174377489 : 2725
5014302315 0146961225 4948953383 6275694486 9304674198 0229225550 6508742977
2758076095 1068798271 0919383714 : 2726
2290968268 7285963219 4283672724 2477443909 0600368048 5278454385 4819955828
7433441890 9552309926 5929588482 : 2727
8977719675 0543920577 1668938552 3977360925 8209069343 0578986742 3572953120
5148509038 4652493140 0689961737 : 2728
3173581622 2294455416 1493578714 7750627037 6192498036 3844001609 1361171372
9557661808 9263864679 4027936567 : 2729
0385305779 9129885739 4478375763 9092679443 3365054967 7074228596 3808721870
3995827147 5800044022 2404214003 : 2730
3035903609 6054800471 8847304678 2868077409 8983222526 2453168032 0340844351
0937431949 9380299081 2417921108 : 2731
9542392709 6542582195 8485866799 2411578844 7815219557 4983222583 3567422679
8960098032 0093548651 0854946767 : 2732
6713405310 3434998643 4975800021 5286835835 7213659782 0843573260 4661260570
4644092005 2064374880 8684041999 : 2733
5854086974 7731601750 5390253064 9036204494 5847644088 2040053860 5715251822
1779351801 9414711660 0865329482 : 2734
8106002191 5944692783 4603798292 6881867778 4883782713 1481606681 2848087479
0430234200 3771308964 6478178559 : 2735
9461837510 6206884413 5862845063 0346441913 9428937623 5474277758 6769014678
2289070060 9268325225 0324639953 : 2736
3375667289 9766025424 6597951963 2609027426 1515748186 5278192977 9836811013
3133965162 5793318419 4070269649 : 2737
8895138692 3961261275 3695920602 2969008742 0834720840 8331884158 2683801938
3358973224 3351364122 4432117497 : 2738
9404766824 1678096352 0356641543 3254150964 5019779105 4146094374 9815990445
7928380288 8013356248 1814072611 : 2739
4232759728 9482414188 7025957454 9342547227 4698997687 7162316099 3228850420
2807083810 0814091887 3526333183 : 2740
5842074074 8446573397 8384298053 4710600237 4219987211 7626833349 0920907386
5337959074 9280892830 3010720755 : 2741
0472450851 1833346763 0475982066 1789998004 4627448033 7019655502 1320441396
4236745069 5370878169 7379969379 : 2742
0616378482 0116979627 0721270358 4804794885 8065830896 3231288673 4029638482
4112876595 2185362411 2569697491 : 2743

9905747828 0326299861 2317247930 5032363770 5845698785 7745316103 8667067555
8440682408 9105118184 2902580329 : 2744
8514096157 3315387563 1143854779 2152983663 8382158713 5882408201 2778384097
3623264758 4435263028 1664756079 : 2745
9932214839 2715632124 9990837098 9309463295 5985992872 8433521252 4274334943
7902382494 9445785164 9361270326 : 2746
4233909454 4808620028 3535262617 5298183552 5297880465 0281353991 1284716128
1153414460 3897031654 6773952587 : 2747
6538384445 7461103515 6164180927 3346254142 2179033107 1472031059 9294953895
9584368857 7348949522 5982103831 : 2748
5964206232 7307148371 6691798967 4445418418 9037251127 2835300592 9827393747
3757109927 7652356370 3606473487 : 2749
2478483968 4203742309 7589988743 8787654284 1593565973 5883450609 3612992449
2587467691 5428045981 3281582587 : 2750
2999110300 7806315924 8172205221 3206010771 4923366010 0318271006 6727266488
9495509423 3689793548 1055796423 : 2751
7715449541 3717740799 5177501466 6954655741 0080155793 4179598301 3187154617
1383822033 3287263136 9978080937 : 2752
5628169857 5352925390 2365681143 5865539828 4283241700 0516419900 5176438351
2005756933 4304218029 3152368541 : 2753
1424059868 0573897377 1720903281 6486238395 4980500843 6023535858 2554618855
9424426129 2892143414 7898267094 : 2754
1767604522 2513492987 2997435333 8206276224 0633100488 3774527381 1887225818
1998221942 8293676666 0000403799 : 2755
1487001856 8674554412 3195735187 1237930599 5148215954 8677010540 4782025853
9083335640 6182622252 0802864866 : 2756
6809763156 1071376418 9090236060 3954124553 5703806675 3567652472 6680367517
6738455646 9669359602 2634258001 : 2757
5557208962 3840364771 3214296692 1934724207 9872962986 1675796746 0829597127
4857067903 4670391578 6585811127 : 2758
3825743251 9039782995 4457674305 7529928302 8634186242 5450224962 4197916398
2734904359 4113958984 0434895757 : 2759
8332464822 1652497253 3181193055 5856140503 1050765485 8991552554 2652886287
5288955457 7367874202 9770378468 : 2760
4756366424 9476708485 4307354132 8403616349 1347107446 8329858980 9511102012
4254488464 3071227174 8865869643 : 2761
6722375125 7407663807 5779685938 5182321580 3790138851 3245670422 5278538766
1013519568 2865233946 0402003567 : 2762
3386025205 5134753079 0074689452 6143616381 2466020943 3968818299 8572534654
3552854046 3610413121 4993769216 : 2763
3026148315 1823469420 9627911549 4171946607 2066552844 0044356575 3266414389
3427722090 5575184236 9120803473 : 2764
7988670796 9228398693 7508881614 6073838246 4200081539 3674001886 2573073695
3499730836 7252810149 4304364563 : 2765
4975213545 3195195003 5076482370 3618453849 7563616339 7442943098 8638719898
8081880867 4749583176 0222984672 : 2766
5019591837 1787001546 4719437744 0245879644 1934330527 3778617450 2452497071
4990700051 8726929283 4587178630 : 2767
9174843878 5506397547 7813979761 4710297480 5258930692 2166622252 3537344901
1353986266 0281419264 7629370976 : 2768
8013187204 1406668762 5545954222 9249384946 2711775501 7586202137 8876760029
8051574112 3780955192 7818159082 : 2769
0663636540 3568683324 4556620095 1604633752 2568825585 4582920019 3063815338
7365651794 5743702588 7562647321 : 2770
0773227646 6231522699 3795825381 6250741193 5992575434 7032075189 6392792921
6230912990 2590445512 1720931896 : 2771

6179934694 9541502186 8337701522 0759113008 8868902385 7991528263 9867824654
6088746278 5262268142 4733188588 : 2772
5724166512 6195900032 9224404728 4089619602 6492377307 2793032869 8350719950
9173362220 6904266211 3793573787 : 2773
8963398219 2711117924 3751868381 7576213472 9227304841 1090528931 2739756654
6440159910 8920563595 3955062268 : 2774
4903481778 3401638880 7477585910 6047358664 5607299400 9420006312 0435623081
1649814551 4965551300 5854611735 : 2775
5240521671 5566604133 3475875987 9204455921 7567756328 3767227690 1154164921
1224642236 0395403368 4550113426 : 2776
5424744898 9545996792 0364424296 6524827350 6879964950 1574046214 8251111674
0163812882 3705492676 6797280057 : 2777
4632906194 5617997309 4448723746 7063062834 6193769263 7284371025 0629430239
8387471804 1127794451 5182108640 : 2778
0015584757 9128464012 8739950977 6297708262 2634588250 5207818345 7605053081
5712768164 6160126745 6151310391 : 2779
0717697384 5578732241 3300300055 3471951166 9012581135 2080156303 7304690809
3097927353 6585649135 7471135090 : 2780
4412759076 4902991938 8200826217 3939592861 2333657297 0664641027 0587838551
3189346579 6268593304 7956026011 : 2781
5450359677 1014005799 3336889004 0220753848 2513993086 3716343366 0079237124
0645761765 0036410612 2054356886 : 2782
8817740625 3057006023 0189829110 9153407117 7517124423 7036436371 5890220116
2317102635 6501302439 9121540427 : 2783
0127303916 6043485289 2171767800 5443537960 2681447698 7479405571 5993778356
3996621006 6927419271 4681089620 : 2784
4073611163 4720258986 2464744081 9612040336 8752089701 0880633542 8443692521
8017425121 1967856991 1058334944 : 2785
9991683094 4946984707 8063675466 6776782538 3723040528 4892911730 5480298931
0613282285 2430139744 2127840108 : 2786
2297992256 3749918616 1909539509 2292352403 8726563349 6244744690 3480575135
6594650462 5030962501 1185996363 : 2787
0240365418 7824457074 0245894880 6050741683 9071505803 2424183755 8626796044
8940311842 0715618426 6389930059 : 2788
6835196088 0991550054 0819116094 2615617799 6494555738 9362335095 6021693845
3029407415 3542201700 8850593410 : 2789
8021537744 1689697655 2390007001 1310946928 0003444356 0636076613 1030272873
8927422665 2498990981 5901237651 : 2790
5704327731 9218502844 8811193320 1103571057 1944438712 1835232255 4867726440
8667340454 4135367403 9901046417 : 2791
9288114132 7732957052 3323399878 0091602670 0289290467 0034550632 1135518225
9645456365 5802704621 5314706032 : 2792
1476780387 3454420398 8775731536 4197294374 6586782763 3623111986 4674608317
1624959380 5163179101 6021743160 : 2793
0363721351 3550655568 1162767164 8322879623 9003714331 6348095868 9243847116
9048307896 5100591104 9650159928 : 2794
3143831201 8932525166 7689558973 1051802070 9156128212 7947857682 3150309965
4870137801 4203423508 6218894451 : 2795
1309174155 2012125037 7976572630 5117588445 5791816612 4319147934 9987937188
9746676777 8272433292 2702482645 : 2796
4802849998 5675549452 6946870327 5037839400 3665144268 5682081309 0209490578
9962210081 4077366965 5662797895 : 2797
8759938160 3739294081 8983260231 1979060514 5978038449 4121855073 4723444046
4136333171 4829781976 6986696551 : 2798
4005181845 4197633105 5635044884 9713422360 3391300589 7971734678 2373472329
2305173885 0500463602 5681998062 : 2799

7282581124 5559158601 5018439090 4098641809 7171007546 1884773934 9112735711
2710753309 5079036197 9461708733 : 2800
4466480524 1788806067 7311064558 8414287431 2055368645 0754131237 8920501641
8245598529 1702855298 2349175681 : 2801
5198174953 5650404537 3588004097 3693100210 1619740994 0885723368 1398906852
3058021522 5783079858 4444988490 : 2802
0267221549 2888861292 5028852813 5271737803 1820762808 6658198702 1339186121
1336024618 7362649128 5983857042 : 2803
4605478859 9442082401 8091973627 1175154047 4656341180 4862886439 8751105260
1860076320 8664032080 0588098124 : 2804
6682872769 1582888514 5355599297 2145134318 8177166455 6450266633 6275157142
2612127028 2902358703 1467862427 : 2805
3023359989 5133833106 9080367912 2897592232 0900535339 8361052808 4879743470
5051051242 9799469695 8773290081 : 2806
2070797287 9653583923 2426576733 9214438047 0361706529 5956729932 3441686930
9201866257 1582035045 9222746011 : 2807
3349178476 8678310636 3023672435 5370932562 6949823072 6186313109 1050164320
6126742460 8679167037 7930940669 : 2808
6071354477 7204124017 1387152541 4787133745 6602291427 4536828100 9292055889
0079508483 7232678718 6595562128 : 2809
3765493043 1227464459 7738111563 9667409274 9919903096 7831570443 7927396416
6675109789 2640931174 6824187884 : 2810
6539287943 9142807191 3722819450 6211199604 9420141675 6751415522 6569328596
9399005410 1116477675 2925649440 : 2811
4287958357 1003684509 0703458019 0874999930 9273423323 7906647410 7462898117
1010402778 8338214509 8316061371 : 2812
8505842790 3895394961 3459869455 3433217338 8380442292 2186848247 1011714851
5834710609 9757869761 9681601243 : 2813
7330230684 4692710557 8932616600 1295993498 5974917184 5033446105 6240840010
9524903112 9151310207 3536606699 : 2814
1425097441 6710891804 4279263850 2557662206 2566434705 6888812091 3431296547
8161984539 6751548210 8102441606 : 2815
2444931858 7351214286 0108581558 7151941939 7655261062 4780925408 1424759646
6270191943 7855071869 8349687692 : 2816
6575171350 1764020035 9938353017 8302781767 1022024492 8865565462 0105595674
1577115904 7285830165 4225614200 : 2817
5482685137 1916276898 2527266000 7703368359 2676892711 7466145886 4432562954
4170512168 6083735716 5976102782 : 2818
3884860670 1446329636 8213637303 3174648717 6320142788 0067424934 8568445726
8867825525 5509250061 5469758288 : 2819
5492108122 2247668229 0277511682 2369502543 9873245618 6120999673 8050145752
1453467701 0802591529 8160421223 : 2820
1163287602 6457848920 8814442541 7823517877 2946368491 6863787103 3559880293
5287975131 6600965034 5021350087 : 2821
8614816527 5693425491 5758254478 5878977900 4210159280 1135480971 5815493253
8649021151 3898577566 3927058200 : 2822
4783308103 1935861720 9592850309 8371977956 3846649873 3455490133 6566062958
9933126670 3542551795 8589534255 : 2823
6852221670 5720637316 6820932241 5546565287 0620820268 5332600866 5800583966
0906950497 0302254534 9369418434 : 2824
7991814854 0317521615 3188936016 9898297123 8272732961 8815135404 1870492734
8526265666 4081364863 7887168029 : 2825
9743419921 8404526700 3615580203 8750040963 7218865537 6610564625 2585967623
1120914555 8061492374 4622486559 : 2826
0525941467 8341230133 6488120864 5131781450 5464179416 4567238577 5090452177
0549975833 2360916182 4686637311 : 2827

9959742563 7392431936 8360663346 8788836648 9399770870 9923975176 9429327043
1571634050 5835198994 7721259861 : 2828
2465956758 0313640200 7793328797 8651130119 4767901228 4933455937 2745446777
3069942456 2602023887 5493090223 : 2829
3573983039 6642856599 2346239434 3075435576 6148585186 1284466173 1439799759
7768447092 9792773827 6470935627 : 2830
9494509375 7497580940 2297195543 7014385922 1216058081 0042397438 5330454346
7119143871 2266270914 0126153844 : 2831
6277366108 8651827155 6640204899 7387185384 2797408717 8039858785 7487216892
6362934079 3705516018 3714050877 : 2832
1496281607 8738336233 5559788371 3608096663 1521893228 7510522740 3710184125
4829712856 8954164194 9279438506 : 2833
3945483861 7154528632 9870074344 7464614650 3414460256 1936493892 5571934232
0962385728 4093622072 0551764698 : 2834
2530400643 2287560380 6977314699 9660101861 0184090834 7452808928 0983391290
9149258303 6511730299 6765473925 : 2835
1518450277 2448449537 6804763886 4019063487 2967747990 2124856127 3166399844
2736186230 8855173182 3996788171 : 2836
5818320630 9699648514 7295737236 9464794425 4825014483 7278643035 4266996443
1539815277 1686798446 8577777317 : 2837
6724214993 0635976518 1359539276 8068710323 0458025191 5603646418 4552722886
1482514597 4092997199 4529105998 : 2838
3347241041 8542027208 5136054307 3574876227 3840792001 6763466151 0906147191
0813300876 9243989050 5428382858 : 2839
7174596002 0088457644 8251903137 5548086017 9403410944 1898837265 2319407183
1370537998 3523443759 5489813215 : 2840
3424084287 4824428098 9888047197 1054529233 9984765517 1775144109 6350331443
8415742836 0807901341 3016396157 : 2841
9445590873 6627890914 4275984522 9763054393 4086667826 4314016375 7170561881
3450653637 2888736845 7730018975 : 2842
4353864153 6393817376 2901822963 3304944189 1940659730 5753851213 3986275646
2498470327 9184151114 9121135250 : 2843
1046851190 0896117079 0218889188 0624882538 4228364119 0655874808 8381207312
3231413442 3335314443 3609656271 : 2844
9210824764 0392720608 8886262852 5885199283 0133305890 5765272829 5714261949
7916499589 4363177324 7495809598 : 2845
4149163996 0872405594 0589740951 8518453701 0842391107 8235447953 8977220797
5226175997 3799318017 6602584167 : 2846
8345852154 5313578584 2096991306 9952099187 8609886124 4401060741 1986374471
5309935103 3428616375 6809485035 : 2847
9275704744 2658967956 6193382876 8847466738 7627035779 8755596549 4014662899
8920998697 1648540723 0339888393 : 2848
6761101330 3784045113 0783799704 3311605332 6219954425 7703071039 6843975279
6919730812 8025112622 3600777540 : 2849
0051308597 4983046454 0495130970 4803426138 3540913445 4056413410 1462193716
0565528044 4840088045 3039649492 : 2850
9738268650 2274528229 9484577467 3433786755 0280099756 0510091528 8666465879
0262577689 5712418793 1583948729 : 2851
7388771483 5384248129 3119168316 0601354302 9978483686 3527731202 9030297107
7830277473 8958134651 9427561606 : 2852
6742843607 0204002387 6861045920 7769656762 6787819706 5606120339 7304722965
4813734461 9132198858 9232186743 : 2853
9123224152 5774192907 8225709141 4018156957 2845738336 2291885079 4868329493
3053359319 3572091676 3645955813 : 2854
6799238696 3556749298 6511324827 1394607316 2855012413 2311737264 8773982965
1492342674 1322472886 3284602104 : 2855

1366966644 2677281041 4959430276 7238763428 6066448079 0484267719 1598564512
6086187040 2572744277 4514307901 : 2856
7361515617 7315157500 5988399640 1418804973 0699755066 9101292407 5303749581
5578462768 3114837351 6100826421 : 2857
0568687863 5684085892 0119268252 4370390352 5176669009 2384082646 7526170926
0269710407 0471481531 0205739799 : 2858
7681579182 9812892353 0414649198 7593615632 2124516827 4617227796 8157330253
2557352230 2296833982 7799416034 : 2859
8264985693 8263973605 9056232139 2948550742 7648532942 6710589699 4589264214
4119600084 5353331145 0406865373 : 2860
1319571484 3485415041 5172347068 7159665889 3468794776 1605065252 0532551887
7942762000 6779291742 8629514803 : 2861
6393715562 4921492192 8994506784 0972054346 0019562984 7440967486 2465363711
3020873814 1754833381 6616561518 : 2862
5111911346 8473236553 8248531987 8581818145 0105386941 3158042894 1053108502
6258281571 2311114555 1238854904 : 2863
4534798670 0257077621 7413802918 9276234523 8939140280 5293096864 5560208707
4750296305 6856668723 9774998591 : 2864
1356208348 5942647022 3854033139 6655512294 0520677229 8210771698 8749068312
3218656678 8925348437 3928939824 : 2865
3039270631 0460167855 9287530601 7870222133 0681129914 2564872664 9716803285
4943908954 0115982149 3770170327 : 2866
6762092987 6361529476 1022963864 0039099466 5174286052 7160651172 1401325095
9270559294 8397361299 8179810256 : 2867
7138533177 1066750331 3178278732 5121501327 8377486502 0703313550 6227558130
4810800294 6051717988 6418646938 : 2868
3014272229 1579435039 7991778016 4905282771 3029562457 0910278494 4459005025
0012647562 3251401612 0398203255 : 2869
0275269695 1967074235 1684211910 9820901473 4553452473 8516054502 0344886526
1194845794 6739103249 4601754606 : 2870
5949133473 5648784812 6818801870 7352591838 0839036790 7272198713 6512691128
7379537715 9952741342 6674052986 : 2871
0582672777 6084199746 9706419659 0299595629 5605560002 2176378828 1809646584
2429443116 0434101540 3241618371 : 2872
1241183341 3369086040 7381821867 8592966006 0152609792 0430269051 4322256814
3657469655 4200716104 9260710551 : 2873
6213629879 3005559121 3266525433 4472375154 8317961256 4078677742 4307070087
6622029181 4065502136 0191663843 : 2874
8599988612 3275152903 5298703495 3210752896 9061140401 6598021882 8037680534
8714902083 0871917804 7753136085 : 2875
8414106596 7519604324 0179858915 3532443423 3629910033 9036772618 9914046681
0276148578 7215430327 5752435932 : 2876
0503117165 1703430242 7376082327 1009896214 9506384969 1002902577 4167136585
0448982075 5135344694 4194185199 : 2877
1214566815 0684357307 5876541271 3166542356 6812227372 2233387587 7673936228
7460341063 6815651866 4932281342 : 2878
3042420400 1730539139 5804503405 9680448257 5134055549 0464163357 8438160588
6860227991 7915156776 9918483857 : 2879
4581978981 3620129705 3878672660 6889518850 7220074426 1029707371 2835694266
9779338208 7809127052 6740228190 : 2880
3448878106 8280495959 1079430888 1004695618 3587209343 2323304409 6976123771
9689629821 1991687870 9833961134 : 2881
5262017695 9400345867 3833978234 1473219138 2499749914 0658967734 4743402835
8031537479 8489967619 2496998518 : 2882
2240176819 3052100224 5510857860 3846905687 6636412890 8975155436 6506561650
6192218185 5860639525 6352043478 : 2883

9459161679 8123236057 4968374823 4389055596 4335042927 5477260719 3219830825
3807155388 5217730924 2994131441 : 2884
9026355802 1105357988 6653626415 1014646071 1985598292 8954907551 4813694409
3607176494 2629103403 8171821614 : 2885
3960417698 5281732820 8821036906 1259770468 3140598765 8981426853 1700310662
7425740828 0910233116 8159759586 : 2886
5485617816 1460643447 6519302871 7343009218 0010058739 5483454403 7627646013
8242763405 3294655808 8847737466 : 2887
8362561690 8343097270 0699748178 2424574194 2626077709 7989142290 0350084332
1192397735 9095594574 6815664694 : 2888
7811010103 6963569468 6789030933 7571102762 0866070878 2106555537 7926088164
1337529391 5596153910 6238153813 : 2889
1481317662 7603231988 8049792167 7961104910 2388322750 6396471007 6965247441
4609822625 9447672975 8448101388 : 2890
0841014521 5932988735 3851807306 9810094601 2616786893 0860243784 9860720828
0266924451 0981539169 5973601821 : 2891
3287944079 2230675841 2984849036 5630343690 8701425831 4198991054 1398522693
0754669501 9990277201 1993438099 : 2892
8096571948 2859198767 2414559171 5959557500 0602439147 3464999909 4962280732
0890185317 7416621570 7333892838 : 2893
8639164413 7593974177 9799619064 5277409657 9769282536 5348782886 4697225365
5754525228 0316884710 3926942994 : 2894
4017744150 5654108392 4818538097 2246265514 6890300020 7821275494 5052791543
6981754966 6187513478 3191865741 : 2895
2558355397 4077373341 6015611445 2815017161 7511799963 9940611911 0086304704
7957134095 3158279114 9697550614 : 2896
2525966187 9019404745 2875733438 8929908002 9769877893 0869873399 0273247229
3618765193 2972809463 9215880105 : 2897
8120917330 3560696088 2552321797 0057600041 5904488179 3922984453 5803797460
7129470760 8200651516 8336564512 : 2898
4128112940 0207911460 8274243109 5203360028 5057851031 7812969011 1967320860
9949003674 2606578833 2367599890 : 2899
3174678418 8273762112 8226150437 3578261282 3923583260 2350621502 5388120503
8087610757 7923411020 0663883596 : 2900
4616931815 7504286066 2121240253 0812757970 0257872538 4490578740 2407677517
6118282802 2057007680 3317314373 : 2901
3222497208 9532223176 9914830922 9185252467 4905187716 5871928339 1451272224
3910768803 8146764776 8256051991 : 2902
6242899466 6415868570 0335897100 5817274611 7001313172 7201526453 9575067017
2388733144 3852719496 9975372458 : 2903
5045185112 1245323760 0924304723 9954398933 2763258474 1996602126 1252980566
1849768230 5041205726 8302788959 : 2904
0129847903 7010123634 7773267203 9081071163 9303289268 9698589802 7604285309
8125791957 3240805314 5359995068 : 2905
0281647637 6786162049 9087220505 7179263264 4701380210 3274475785 0959815376
9279437353 9955990692 0110868457 : 2906
2761587374 7414978132 1992210097 9463616836 8837698006 8019326724 6356331393
6198022844 6602908257 4970887611 : 2907
6126193917 9889891414 7203605599 3690883893 0580535936 9339114503 1666583767
9068253381 0154946336 8505270216 : 2908
0528658989 6942257096 3534549240 8795324498 3450152302 3103683349 3083408235
1682915189 6416671575 0476290195 : 2909
3467655050 4543318915 7265705149 8776384149 0791267283 8031790537 9403906551
3432425793 1330413249 4807608810 : 2910
4697312495 4534545785 6264329245 7539754436 3110660436 5289403443 8429341310
2992185638 6196903953 6229361901 : 2911

0163993528 5350105729 9327718394 4687864902 7719241196 9477667967 4321691661
7401837190 6560463900 0765211961 : 2912
1483507207 5559291017 8537877056 9542074600 7253475463 2987591800 8302027150
2977497891 5283989453 3255407195 : 2913
1666575323 0926495139 4211425540 4511537786 4569662346 8005001055 7665686222
5655975320 0069485364 3862230379 : 2914
8485693682 2387490319 5490049166 5783397436 6986091833 9199872371 9472588452
8872540128 4645056630 5472362710 : 2915
9926427857 0245829223 7304220010 3989251437 6074181197 6799800496 1158488903
1365744048 1472776934 9793351969 : 2916
0791241286 8049505017 7445358305 6740426732 8578975726 4025168112 9144017289
3889390600 7862203398 0666196507 : 2917
8580853482 4907943715 1059318692 3206404967 3865635312 8130407910 7222213576
6548218780 5198585300 1988320719 : 2918
4602635121 4279937006 9407085655 9587246813 6554341671 2160070267 7482923620
4014529850 5602122441 8548337825 : 2919
9554164191 0011069844 1606111936 1341572843 8557376822 4370273680 2105490498
5965165829 7294455519 1824151604 : 2920
0655118397 0720272020 8464020439 3072986300 1390554348 6080572720 8711812587
7938449849 0437052921 0374970100 : 2921
1663998151 9494762949 9986428493 7367525363 1752188331 3087108880 7978839241
7704627889 3607737691 4701380205 : 2922
7889504947 8115887563 9904502685 7550561741 6055899462 5034600921 0210935213
0947675934 3508224228 7365273888 : 2923
3742321134 7106010920 4939561731 7488537802 2731466288 4160388678 8153423753
9156003774 0786683286 9398483480 : 2924
8067071923 6001585719 2029231113 4173510221 7455941199 5983544456 1375619179
6311070401 8047448038 0943839754 : 2925
8267445519 7750593665 9329500786 9513983479 2987338878 1017794560 8175447381
3559180829 9812498231 5003735066 : 2926
2654337764 5218316617 2959235665 5050362988 7119356012 0416793837 2520077159
3141903519 2724580169 4493938939 : 2927
6886129000 1191170558 8515157980 7832197586 3643962234 1155912478 4518708290
4022020705 5268885676 7767572084 : 2928
3301962157 9008529471 2798233970 7670466783 4310190431 3793909567 4184931794
8755991990 5514096196 8939225573 : 2929
3193871822 4016540438 9424297616 5912825960 6455657678 9626950067 5457661057
4970349472 0985496417 2219226415 : 2930
1810279891 1059033065 3915466596 7022021495 4542925225 6801199732 2331862993
0128897726 0054888305 1801907365 : 2931
6178487244 8961557321 6482574547 5381614346 7084018257 1136375339 4201684005
1144296003 0308232427 6272440249 : 2932
3439561055 9393307378 2790939544 0108058510 8538114412 6655161542 8095286811
7050960782 8910789971 9529989342 : 2933
1677946200 2016998984 9651440553 3694909314 4156637478 9827892780 7417117097
7983171522 5227691017 6290637536 : 2934
7829868692 7280589881 5004823069 7907349216 1899553670 7907033479 3754336049
4530720796 4877016335 0386612477 : 2935
1679889404 6172089012 4337095817 1600709412 2493621549 6495754923 9133890521
3792848132 5600654070 8721929520 : 2936
4117451146 5783621081 1062428541 8078166194 9418458010 9484822606 3578640095
3180380553 1752590882 4674419440 : 2937
8371697512 6383772221 8999035166 0818503784 0103696419 1489110716 2792024978
8407850357 7014056163 4787422640 : 2938
0006355817 4689957745 9816171765 4247358213 3634390557 4633670041 8263131436
1141816033 2966762967 6001679942 : 2939

0553340364 0351816660 5499089721 6378910193 3149352978 8162093954 1968206581
9064283642 6624132370 5903925680 : 2940
4645463665 8820270576 3264918291 5871363604 6873638450 5449748883 9362556344
6902584799 7339724378 2686679200 : 2941
4894254402 2238694891 7200466558 2573288023 3249943539 8108946466 2938621067
8261585278 9957371136 4825491949 : 2942
6669990046 5148330478 6736213896 1073979915 7349933726 5679173820 5067948903
3578517034 8015684871 1905733150 : 2943
3073236481 5754371927 7078267888 4988301848 6465398211 8322884774 5990682339
7462156125 8538266237 2283196986 : 2944
0252043378 6277355751 5550720101 6059787121 7465069736 9997748850 5823686182
3974059628 2389906171 8458168239 : 2945
6699394042 3403802715 2149341645 8066506094 2551663306 0493107167 1973308360
3311809122 2672826165 3649177815 : 2946
4547133361 3951369357 8970790381 2910081720 8674970566 9639627505 2837274397
9162876486 4557681076 9790438777 : 2947
8853689843 7300360143 1294866492 6231077274 9037059562 1425875142 9326687828
4788078762 8478186245 9567816866 : 2948
2583026820 3644597885 0982950892 5259441721 1353558594 1142399515 3512414886
1005811481 3371447620 5748937224 : 2949
1692219063 9131378281 1620178787 6086438886 0506587083 2408698639 4651945116
8883795277 4635359778 0059964225 : 2950
1188127560 1601791590 2247521397 6910679863 2092833840 6008610218 4671398118
6612051037 7179678586 4715889119 : 2951
1978082065 1104987209 2732936744 4664555227 8333155612 7984651988 3483619760
8015831317 9067455124 1002086775 : 2952
2382206554 6145593974 8978692163 2321555311 9290602588 5276633670 0281085004
0367760202 4625577852 6526697703 : 2953
4006959215 7761564764 2596347433 5647160218 5512767526 4892216759 9801547259
1183530177 9011021484 7022673250 : 2954
7958525275 4842316261 5892967491 2801498005 7541492894 3724074644 3811161096
8912792553 3486650439 4777647016 : 2955
6897046624 9702173347 3368079477 3210861934 4342264216 0858013035 0246371110
8941668102 6503673752 1403282434 : 2956
2933691937 3726378916 8983387513 7555791026 5452952313 7287857195 6727250352
3272501149 0885240111 2212315223 : 2957
9535668141 7563608279 6889420199 3232452749 1150005680 6619067100 7305621131
2564018295 0499334281 7836112004 : 2958
4138708304 5411777990 8282985239 1031652175 5322173908 3863377240 7026085160
1865097547 2284511975 3912039762 : 2959
7596019905 3838294949 8226841606 9423707468 5285118596 6687687972 2986460275
7184502991 2469206489 9469489774 : 2960
8305050135 1974592227 7894280494 5888936266 1755755850 1668491138 0345406553
3840450925 4779564836 8531413220 : 2961
6728052953 2217757679 9732975000 0772090302 5851044324 6399340674 5109433124
3587323598 6285879361 3228624008 : 2962
2781507656 0815539481 5807462635 9271910622 8645291219 5489912738 8993476898
4063069350 1530580839 5651394402 : 2963
4323250666 2242987659 2396826941 0311308341 5119675553 6137839854 2211792194
1144956191 6538849185 9795707643 : 2964
2673697459 3593810668 7751104593 9056158759 6349661795 6775816312 9073143952
0021171624 1602363878 0963299013 : 2965
3247037859 3865529140 5189485402 0217477434 9411807469 3780365326 1313994620
8945557490 6039769709 4955008026 : 2966
4968790833 9222077306 3033152719 9447948997 8198182895 6639787262 3599650538
4508402261 6071288706 9191795534 : 2967

8341272915 5634507839 2909554962 1763780941 4476039267 6202991209 7978526101
2616158712 7075355867 8860701332 : 2968
2937414800 9805349019 7876511723 5013926032 0282831938 1375304617 4386185487
0731522878 9604380162 6021519321 : 2969
7278125010 1536609553 9593948266 2978500964 7214769630 4142092856 0836755369
7145767446 5825137792 6893003498 : 2970
1876730953 6656154094 0181812138 1451571893 4852114137 6323974759 1249359704
5511811135 1262638521 8226841422 : 2971
2905761672 3362217416 3859695942 8555841526 6032581735 6968734787 1528253854
6866170831 7156789798 6920796685 : 2972
7938520386 7327936313 1109701750 9091536397 1577478546 6923898418 8000859849
8493115913 7283608134 1437393598 : 2973
4087726813 8815763164 5299040033 1700614355 1587132104 8490860408 6309955458
2932498512 6941508965 8896620196 : 2974
4410303356 9857578797 1913718747 4794198902 3985576445 4275317591 2782182067
9194009597 9448898021 6551994749 : 2975
1076702194 9659610071 3430683605 8843980625 0293333929 1805043787 4958457839
5597521918 4993991566 5684207644 : 2976
4015770263 7349736079 8043894373 2337103577 9462949595 6975286424 1690766716
2597431170 2801304335 2721392098 : 2977
9591016293 1503144061 6760330448 7131088893 0329252095 3246042438 7158071374
1133334967 5218946784 2043520120 : 2978
0510171558 8810140727 9033709013 9060829623 2573654349 0225158571 5973438396
4325496828 2425392777 1242235747 : 2979
6365147462 6863042160 0367373905 7280974151 8026332536 4778806297 7836503169
6247887663 0494658904 1398145368 : 2980
3496483767 6379724030 1315465398 8084173969 3614258671 7938191404 8143126221
1849010477 1070020651 3634019865 : 2981
5824595491 5189360886 0207754337 4425879392 3503783385 0250116772 7052720609
9193802611 7950159381 8710043945 : 2982
4786426117 8425146172 5602723804 7632869979 6611311614 7174598788 5542037896
0304851193 6947192108 7749508553 : 2983
8628995850 3777799044 7987976819 1744941429 0009803597 5024431445 7216387987
3259333474 8433045739 4081255124 : 2984
7937720838 2988604655 2487144520 6812031443 2872021824 9171333241 2389413032
2035955727 8051440495 6058136514 : 2985
0986160079 0510074644 5574326539 3945391647 3024455409 0449359189 9500930070
1806172333 7167982022 3173261954 : 2986
8655260653 7659624069 3939127964 0892608416 1148803306 4649940540 7464597314
8240396568 6343215931 2994393707 : 2987
7821600270 9558712592 6393806265 3090637227 1590302144 5408278903 5367016709
8152389832 7042384001 6348101274 : 2988
9869965938 3318771950 7936973083 5694367288 5650611025 0945352047 0419059542
4682019898 2796106997 3226213662 : 2989
8167363122 9890562473 7776292375 5761011654 0065593852 3727391506 6906457679
4639205897 1652286497 7434502284 : 2990
1403508880 8309344618 1683666737 1572514163 8260562684 0092010884 1378388920
7601230008 6402723310 5874037792 : 2991
3680777236 3376099963 3454914229 5140263407 4069500035 7192016355 4103905240
3529072732 6982389146 4585498060 : 2992
7121971683 3951774684 5727327057 8300165043 4385226732 8906604188 8411382508
3413613799 6198101697 0527367724 : 2993
7197959326 1177622344 5398195320 4724407339 4552129375 4849964621 2828779894
7592635564 7112480984 4788635485 : 2994
7684404007 9645408653 4311784469 8666996315 5071535534 6753318278 9421749052
9540847002 3651559371 9130070765 : 2995

3206101646 2428460575 4459136274 0802494426 4302474203 8723113681 4035011649
1673880339 7962812882 3708626314 : 2996
7773710950 9252116696 6772840005 6696523553 3123572734 4712250584 9334210006
5438046715 3351524918 2317846165 : 2997
1025808818 0164460499 9405884890 8523540861 5183894043 6071934067 1292367260
6944806459 9978077724 9220902903 : 2998
8624407445 8697014205 9022198880 6846096515 1709482306 0009646570 1436440766
8066372796 8756940713 5156782799 : 2999
0649079063 2772090445 2450324857 5792807082 2520326239 6833548515 8606931459
7838528361 6945633618 6314956895 : 3000
7512520542 7583408433 7654387735 7558113323 3224458741 6913044623 3328853040
3416818532 7650796295 3325671968 : 3001
0304662622 2693929242 3381761474 0621769438 1301816446 8362506028 8786882638
8486228440 7686498705 0543822925 : 3002
8065848704 0403556257 3256749520 1114319375 4775407709 1962079537 1814882506
1663069254 8318888716 2368525155 : 3003
4848108075 7453534756 6506910899 8902579006 7471165361 0446354848 6258453296
0111660552 4812366581 3348443402 : 3004
3033838766 9473085311 4682295290 0928520339 7072972478 4595032639 7304709692
8772755667 4107879212 7067696914 : 3005
7191629646 7784842643 7554185756 9851081658 7182715147 4955036156 5003713207
3805452127 3860737134 9328329950 : 3006
6794838104 6672169613 1667455649 8386410665 5385195711 4779798978 0405313153
1304695355 9560125012 2307301351 : 3007
2282359030 8974131531 8724606576 7650692731 2075405356 2280539695 6670940455
4331001699 2009340160 3015109670 : 3008
0708703308 9628658695 4811171164 4722242956 4592624702 2843837363 1682761826
3814545273 8190115715 0895220166 : 3009
9555895255 4622067920 7427677769 2308115226 4751106824 4339134165 0025240406
9330258598 9456339270 3641944075 : 3010
3978120082 1821358504 5547352157 4804387050 9653774461 4781134687 1656555888
3519727921 3185045506 8355394307 : 3011
6050369009 3629623878 3640866701 2944398938 5444418785 0881588375 0876290001
1444690128 8812955855 6775237521 : 3012
6598684934 3242206433 3126591574 8872953995 3386225175 3806821534 5051124474
4094691104 5116323995 8515057551 : 3013
2312582940 1236747715 1579026785 4663437832 9976843751 8167132466 3954643020
7887683845 1413313112 0043836272 : 3014
0947289667 7974394888 9034039333 9347714916 5560987844 0626123581 2235129549
2562595858 3191636244 8449112395 : 3015
3657658710 5078807396 7088109766 9121051336 7202108849 0924983481 4108085147
1979311983 2560149134 0711816926 : 3016
5377176694 8863256773 8508113099 2939242797 3112696774 5834906089 5901467971
1219754532 8687949100 3849694060 : 3017
7005645672 3771053491 8906445065 2802955744 7570185785 9353330253 4373141442
0530358189 4725025659 7419922390 : 3018
0855033384 7929662370 7672334636 2903759950 0875430750 2579639741 5020398849
5903758576 8291001734 4100301639 : 3019
0174848563 3064220117 5201778857 9842745795 9125026072 7814703573 1188031082
5407223370 5618403981 4643086670 : 3020
1326413991 0735002441 8877255635 1318020125 1858081489 7540973697 3081487305
4088717373 4744284091 9291643424 : 3021
4021024496 4263928735 7398240381 0584200373 4586953107 0327979850 2293094662
6806901792 7241787098 0625238297 : 3022
5492674271 7401093133 9084400603 1771503988 3971756517 1566425066 1408635197
5457345974 7328546035 2577081637 : 3023

9080538731 5806922805 3306986610 7176170472 3189417223 8541326756 7686410850
6936197728 5088089991 2059322947 : 3024
8701723659 9579112547 4049023042 1735611643 9548935384 4038661966 7832273836
3099111005 2858370782 4962506145 : 3025
5188256938 5163576399 3030755907 0740917791 7689600909 4216626863 9945930987
1665187527 6280612167 7655991791 : 3026
0290609779 8876029113 1293858955 3501801828 2822842512 7176741423 2794377249
0834468421 5709467901 0491342939 : 3027
7381565935 1333607061 9125183966 3489878904 9470872644 1344580810 2413965253
8971443908 9177752252 4117802201 : 3028
6887498243 0162373290 1654238958 8802987575 0062831044 5539487277 0124930831
5249497974 7126764811 0417923690 : 3029
3256497908 6214759140 7233856998 5684899320 7228126831 5037098991 9313076022
2768091775 9601941966 3136065337 : 3030
5426514541 7078972656 4216994912 7767201935 6518712974 2389742007 7422706008
1833146868 9266029409 8085839553 : 3031
4529813264 3374294839 7137157826 6348938881 7585285996 4321524684 9202217050
4236438629 7153170378 6120257828 : 3032
5472396855 0109472648 6865273393 6132705317 0918496084 2867973063 0043616542
1346267661 0101700359 8757979069 : 3033
9862232054 8802641853 2486292510 9616879659 8076953897 6545361454 5744554001
6522391424 8148929729 3814279062 : 3034
5588597012 2387283489 0240573855 2464234439 1199345027 2065771715 2104991279
0899211699 2426409704 0941620723 : 3035
1803949694 1688985426 5615303280 7224682554 2458111142 7009573232 7190155988
5378957557 1161924596 3123390013 : 3036
8923872721 5278612420 3816814896 4678214166 6758766918 2854585244 3941373067
7146403734 3309404136 4476929357 : 3037
8325756754 7224604923 7725453066 3122614055 0175638115 9994319702 7883656146
9974535618 6625199217 7475878966 : 3038
8022046667 7625977438 3389956603 9040362829 8614827021 3861905360 6636684579
1514514912 9662414918 9690080815 : 3039
3987865583 8537811570 3426603443 0482255013 1978660476 7627111519 1413296063
9612967956 7514855605 3596642717 : 3040
6487338775 4842166807 3267934468 2737453566 1080150860 5743399198 6215295787
8761118559 2447235271 3169009007 : 3041
2760229277 8572040739 4928408102 8003889856 6540215556 3337562229 1458982640
5817184880 9035219592 3229845591 : 3042
9169463929 5796753009 1549871090 1410398873 8347924936 2893105797 1150462061
7690105468 9301366912 5649607645 : 3043
5191053362 7317915600 6459648274 7654805723 1889471398 4109860130 2864866561
6266629576 2500981783 9445743520 : 3044
3937949163 1861623245 0841043616 4555398170 2333968280 7540816067 6789235051
0276470520 4099569714 8193078321 : 3045
5993225562 2579133690 1779370937 5425041782 5765707059 6223970542 4120671641
8742464155 7566178175 1832110091 : 3046
8462648717 7650911903 3457230778 7317880484 9377654425 3945247149 4240914793
3707351348 7876314576 9851002496 : 3047
7498296725 7183895783 7846494786 3985440231 2145464070 2316093210 3605559461
9547608318 4107815497 5855244947 : 3048
3221438932 0523373497 5829477293 6978542447 3319216585 3383175552 2494658487
5939746120 3136817692 4912879005 : 3049
5178403707 5161061508 6328433445 6738495665 8915034942 4052007897 4138322121
4924671798 0852846342 8682044702 : 3050
7578369882 8657304473 7017987549 8833918216 4436320438 3602752611 3090044610
0374779902 7949076124 5993811240 : 3051

5161919960 5965139007 7909634293 5831190343 0562435671 5734095056 1636287482
7820587615 4899881362 2840063195 : 3052
1119520178 0809674906 7049765894 2820319324 5913242559 6171141643 1669441618
0155240661 8863311739 9506879643 : 3053
7781553809 7222929745 9868674364 3437724460 2250082170 1314936991 8140245420
9157676839 5501316818 1074034283 : 3054
0411268652 5498680326 4579323184 5029509774 5201389805 5358194141 0191319848
3386798555 4819940171 6615848836 : 3055
1181485049 1864675639 1772863305 8374665125 1909951762 7862182217 7837392160
4284381236 5855035987 7683167988 : 3056
7691766786 4037696003 3967280640 4573452597 5201932854 0390384327 8475185561
4753102263 6373359384 6397999451 : 3057
1977345685 6654467428 3895819110 3508750244 7542075011 7547265579 3430406416
4481640001 8548961723 5736987650 : 3058
0209146312 4400680506 6561821132 0293368595 6754722666 4660246868 5420080392
6074056629 8299662282 7886673306 : 3059
4506550328 8416293882 9562588755 4096946807 0706581219 8050857892 4056678207
3019132006 7060364216 8364916525 : 3060
6313537762 5483089594 8420536098 7229555827 5245931500 9433419009 0781546711
2257271079 8521222737 5561583261 : 3061
0302392053 1679278861 4118329822 6459755765 3404542346 1049029572 2395875331
1957961950 2289460503 0835697403 : 3062
1984824793 7503897398 2795389920 6897189129 6270970268 1601799948 8236851544
1024906345 0379330463 8505980500 : 3063
6893688509 2159609249 3454617018 6966470225 3261938819 3016821226 4843687779
5239618136 8777350178 3987601428 : 3064
7972084836 6410773287 6428869253 5871469783 9726188810 3384503337 1181517114
8471575357 2829111377 1363131977 : 3065
8212024568 4609697774 9219637968 4737496910 9456442146 2352746752 7261285301
8007895735 4303275355 0589002760 : 3066
7241710824 2779772273 2735900626 6638666960 1352230256 0850972996 3153757824
3379250716 2172074406 3331379638 : 3067
7517144392 6623811455 3939004388 6785184247 1758753790 3066366609 3268883193
1210323207 2351409070 3360541657 : 3068
5082209060 3372016681 1388503146 8464451916 9504365588 6615212595 0693828434
4581528708 7122829314 0727555933 : 3069
6996812109 9035159101 6421256071 1075765634 4306351170 5673167435 2895819475
4952161142 5193110025 8904134528 : 3070
9007507758 3181812267 0748616937 0511374740 5144794546 1979530174 7600610679
3453483773 9405521359 2998818346 : 3071
5458679825 8758604317 3404055460 1182336493 5533063590 8322439166 6806121029
2859293099 6275724503 1274894390 : 3072
9642963320 8730774671 5007773300 8339343158 8596370124 4337695776 9454826077
1609767086 1548166679 4238910350 : 3073
6090461460 4413039687 1368948867 9983508780 4068064381 6217724063 4791781191
6200629577 7701399370 9343944321 : 3074
7249722182 3195212537 9413260275 3367456855 8608844105 9085112302 7060655379
6894861190 3334311829 3391087619 : 3075
6185654145 7096893874 3695706123 4280197733 5573896240 7681631584 4335848770
7336072070 6401263672 4168412550 : 3076
9830095138 1957688615 1246486601 9104419000 4053873335 6712015287 8262611453
1444001949 0501156417 1801462355 : 3077
3003346080 2176758915 6147995037 1467332745 8150272811 2721182646 6922555444
1318839859 5093341962 3985945561 : 3078
1849476747 8653220714 9201414043 5873483891 2071052581 6364490692 0398813879
2728992853 9884606794 6999733862 : 3079

8784322251 0037432806 6659264199 3060846936 1675677484 7179955538 2249785746
5567250556 7489493096 0003881116 : 3080
5025990005 9937401738 6660470626 2123885284 8170109467 1013876835 2200253700
4909446671 0475579000 8627486998 : 3081
6075801005 5989739775 2748532074 1834661939 9378999761 0753993025 1144261568
9204855197 2307840758 2278483831 : 3082
2358647816 8286347239 7050703377 0155108037 2168639415 0717589120 2523520030
9364453816 1000890881 3050203916 : 3083
9341159108 2327549299 6997841354 4833229671 8754241822 4655200379 6227904310
6977024167 6548293949 7616404950 : 3084
0283098389 3942602243 0461690484 3558047477 2240387178 6693491539 2885786302
2989243143 6841730470 3015701090 : 3085
2306067503 7024472003 3264134872 8560100321 9723656520 1590949293 4148262122
9998231733 2073064879 6012037972 : 3086
7647315563 6303760929 3837342346 8209183320 3420880375 8319968924 0992749363
5290735649 8472392751 7964835646 : 3087
0381131807 4452718462 2345859794 4972284318 0527406250 0057844200 4830052382
3875108485 5426648661 8405878804 : 3088
6412068103 5919898396 0987271311 5064108184 5490455579 9276094354 2184006717
6453548615 1052824756 8296268598 : 3089
1806029377 2829879244 2529438708 5412073102 5294049832 7891791277 4900315217
5521488252 6034714160 1819538454 : 3090
1767118062 5218368758 1941540703 6768156157 6618172047 7998692331 4640336138
0334652040 1842615803 9026418253 : 3091
6185722468 4486061288 7368699927 2027416268 0637666211 2069290346 1969545811
3643447415 9871401892 1166046622 : 3092
6582661590 5420697639 4359312366 2045528756 0342165003 4736011943 4225614914
0320157941 7118517154 2756396517 : 3093
2568645384 6709545248 3713059489 8582559745 2775643783 7209390373 7606448757
8053808966 6661399183 9630554346 : 3094
3515315481 8588677926 2912725363 4262889852 5685446469 8144974618 9241495863
6636719814 0065068588 8608602242 : 3095
6733798812 7687969406 4970299154 5245272132 5754281953 2491731150 6620858665
2077490952 9651007534 0404922735 : 3096
6548282957 0256906293 5888169041 4651069717 7724209554 4613025854 3817863048
5080605899 0637380905 4306950261 : 3097
3842422270 5375354598 5909932669 6733216515 1948172534 5327473336 0274472585
2724539478 5787049054 8475863311 : 3098
5718366332 3591323475 8825934064 1521039072 8719363267 9637592847 3313161233
9781549856 5077459574 2630192501 : 3099
3613442181 7786573268 4594980392 5741969699 9987645982 4956709409 5595490645
1431929975 3296990292 9018113346 : 3100
8491893973 1673374047 3761021534 9790280131 7223379127 9986391471 0105736458
0882496403 7793669144 2602252243 : 3101
2918220359 6947965229 6324150462 5930376366 4328408656 1602312161 0990271779
7940482442 3743772421 7545327436 : 3102
9030749262 6172588806 5223326141 0603381653 2093232026 6991087084 7586819856
3990498575 0117619963 6905969925 : 3103
4104368753 2918190720 0415759826 2345466727 0157369711 3335704140 3209379345
1266060707 9906558687 9616157998 : 3104
4934109540 9032124365 4431087316 1586375727 3748174501 7866557379 3984866922
9117599204 3422476006 4859760054 : 3105
9782806294 1873914749 6645660197 6892636165 9828965655 7445804099 1426890947
2497067352 2047011619 1536009452 : 3106
7363253066 6440201002 3201873227 8197614868 6634898913 2734701444 8203242931
1784100915 2833330376 9111970512 : 3107

5251189029 7082942974 8839813714 9977805278 1649343705 6043260053 5269810691
8986588689 8161088992 6992043454 : 3108
7815535740 4652938225 5475792396 5125781698 4986734218 2185312407 3431152960
8214112091 9999406160 1015821912 : 3109
7373016501 7695811861 9036689779 2756904678 5710181059 3937314388 1192914749
4435652218 9626028632 5866365190 : 3110
1745365921 2186387681 0777420915 8364690909 1651827398 0753103066 4998062448
4927747561 8845297329 4727139148 : 3111
9972684077 8589778686 5604872330 5752422857 1724734436 6674181812 3270841591
7914786168 1978003287 5242194648 : 3112
0119315937 9351523941 7409041909 8412578099 0903889407 7942046727 0491534500
0248604274 5530730683 6472220758 : 3113
9308219944 5234721452 1842825620 9291437681 4387802139 6936399786 0221263221
0982207735 7144129412 6406537652 : 3114
6428543482 9706936551 1680672830 6148155350 0677992734 2874671740 8366660205
3729220484 8408702523 0125857179 : 3115
1456966579 5239635968 6270645903 7207268058 7943981340 0667670141 1765812522
3348104838 6867717405 8797368961 : 3116
5599621784 6549730739 3409546604 3145860154 0495615776 4561673447 2126427468
7461408302 0638793598 0428496622 : 3117
4722325504 6609523176 8172458636 2611284833 8740765075 8288956774 5895887361
0951974207 2321262333 5561519338 : 3118
2471760218 1868389530 0679750212 0769043883 8128356258 1450125007 1190271565
1443462706 4814950590 1969961390 : 3119
4056079067 3726072411 2924471994 7702838483 5318633298 1761999473 1283314491
5968777504 2903279770 3476193806 : 3120
2955123840 1384229100 3576769296 9859590544 3982892666 6087801034 4059670559
0520766837 0210159521 9513945447 : 3121
7311874107 2352794560 8443549466 7607928826 8193567646 6658916136 2140424785
6427926815 6254485063 1668526832 : 3122
7524656002 7477652412 7942705341 9348280546 2670281599 2453912487 3937558094
0125919778 3465336557 3562593766 : 3123
8087699752 5746027021 6696492529 9837753819 6938462547 0851886151 0475226475
1364891883 3573818916 9712195832 : 3124
6810041957 6537702119 2118272566 5088966824 6750486669 9659885042 0411916153
3208988567 2380926018 1428117623 : 3125
4479558294 2880858136 9886073727 5497491213 0687437421 9430148667 6625969169
5195368856 9117493823 1969755955 : 3126
7940217938 8737225151 5997140544 7289708395 5103654866 6281965036 8486846610
3927455684 1436359017 7744038756 : 3127
5120807714 5636487772 8443833156 3572496836 7563148103 9438968092 1192823374
1450761863 0565777737 7941453330 : 3128
2578207887 9337135523 2819306677 8103474487 8814533022 7671159824 6301467739
1318064699 0171453231 8195729640 : 3129
1713829934 4586652810 4239029204 4042050301 0850372288 9707226673 2041640758
3835211625 5472935420 9770671865 : 3130
2620762946 7204363364 3273505663 2041125212 8722589414 9976280468 9148555075
9736146505 1176412579 3028369745 : 3131
3203604650 6156923811 9721325105 6180613452 1343817681 7327628414 1810216441
3417024917 5584365681 1454797775 : 3132
1958280662 8442497503 7917477236 2042042450 5736306091 1154840242 6995643360
0353014366 6597511828 0328039534 : 3133
2105404919 1030567314 2107541302 3579348772 3423907395 9389300436 8913281485
4688219705 3482680640 6133447603 : 3134
5746590906 5646098009 7171769957 5204375563 5452168504 4224390827 8083480721
5680031213 8051374475 5738663358 : 3135

0661289825 6952535572 3266307395 1954063697 9469139273 9195092970 9929134880
1486760727 1497851682 7026905071 : 3136
0767896576 5304404633 6262688872 2627429312 0287517384 9755421431 5049989074
5646121569 5542535701 5201474626 : 3137
5873588291 5448693671 6821097263 8770366929 8762703332 6926764051 1235659178
7736324770 4611611875 2837864088 : 3138
0382801363 4904145085 1318295587 3803365715 9237554043 7403660094 3126624937
4473501836 9408835012 3180570255 : 3139
1089418469 3666883038 0623604361 6899868181 5046281454 3993708439 4672379528
1953032886 0609963154 7805411053 : 3140
9798317391 0481917987 9337630991 8240071636 9535925678 3586699908 5256828346
1799920448 4921582868 2554266066 : 3141
6948065905 5883754767 4778900630 3076397732 0119162644 1931231373 3282236418
1193438305 0586585544 9829986899 : 3142
1146681311 7119421891 6017937360 2575963218 5321048687 4992047347 0299426787
1276133342 3766834282 2565756501 : 3143
5748972028 0343180320 6244849572 3097390057 1509314538 9184344943 8283973734
5157998051 5625916432 2627014186 : 3144
2062369447 5930214105 0850281203 6049109939 0536847805 6021662704 6368552572
7413290604 2289956351 0252284751 : 3145
9070308253 2738499555 5330449578 0330902592 7531635221 8098898826 2911598033
7112570172 1767669045 4568064922 : 3146
3051574746 8915571710 1756724035 4189350611 2888730240 4314431986 9586521866
7607330385 4903602774 6096354501 : 3147
9525296534 0703015970 3240985115 0252930588 6567190112 5083284714 9680648943
8130078371 8623924468 1790216217 : 3148
3569122887 2394802164 8464737751 7681894212 2207103565 5596507979 4844990082
7193552814 7914340403 7138717208 : 3149
1760903198 8564587461 0810991059 3691774379 4712879368 9503247741 8648580648
1967995614 6436708248 7089936839 : 3150
5131507203 0565303007 8868598203 6072066991 6737616414 7566542877 1935359104
4521169167 6928163963 6456297834 : 3151
0737064788 2740618340 5435707741 2161383202 8737879037 8585893274 5536149564
4612550505 5478266874 5445509088 : 3152
6889469271 8598894942 3449507483 8218501843 4130323620 0466807001 9175045926
2840838505 3643126768 6980340268 : 3153
1158070981 0345898604 1308417355 0995694415 1799175432 3348063307 3261339139
9797880003 8210913276 6014549611 : 3154
1570428580 2816067516 2338135386 3052429563 8330950320 1928781641 3249223004
7611795358 2495605914 5300042464 : 3155
4788062130 2468978655 9692629257 6357402879 9401356921 1675539040 0269966455
6025682972 3695049590 9960271709 : 3156
7381666168 6780048327 2929905942 8102968161 5746963060 6061010622 6717021253
9667479513 8938137485 3895351757 : 3157
8329451364 2600693613 7777440005 6699301740 2011366773 1787704469 4506012922
6074969110 6157620763 7893345041 : 3158
3733651195 6003290783 5235964653 2657474974 3528672111 6298075675 8510838501
6069989693 5867152259 6463057009 : 3159
1390876287 4164925417 0816909684 7309548860 2332982888 0010165296 2808976987
7231969074 2102700934 8880934455 : 3160
5121881883 3651978431 8535596067 4763507221 6117872873 5639465655 4342761406
8559412425 9121417078 1163030801 : 3161
0192587621 9938098958 9430509396 8251827713 0330334926 6488532956 1879526644
6319063493 8964927974 7796305818 : 3162
2377606580 3540227966 9108180932 2672514261 4287685085 5023403639 5970562141
2515137620 4672244428 9799785545 : 3163

4131901372 0560029456 7063403824 2991780751 2572448446 3577152589 3472233684
5268000504 9057940765 9261183404 : 3164
6276469983 5321989331 2711732710 5112938774 6272008131 7263208671 2472478310
3695325057 1166846693 3919993838 : 3165
3416530133 0924772942 9358470743 8826342400 7013217132 0972827494 7961163566
7826150715 6628250212 5206276389 : 3166
7515665851 3406045529 0109261126 3816622423 6689927106 4904188627 0014421128
7359239399 8250056580 0643250607 : 3167
5094589358 5007218072 9322308246 1082581858 7467133406 7334432680 5154756527
6112290094 2154655618 3103412687 : 3168
0172286541 6226907574 4666374025 7850581983 9033726685 3912834251 1388977610
3701554705 9697842843 8121104166 : 3169
7496423619 3689840635 6138762279 5954521514 6892881328 2394396366 1294501387
8167659092 9250897532 4716691236 : 3170
8348275154 6078272804 4518243453 7596550492 5680648532 3099928145 1475345509
2755527796 2794142455 7440525521 : 3171
8825323955 6250857211 7996353019 5675811774 4367625509 7319751155 6514845137
2854240749 0483808199 5588091516 : 3172
1179392191 0426190928 4857816139 0514823147 4455310151 6159178414 5999471223
9051659694 4103972729 5741158339 : 3173
7459390620 5007758070 9596765898 9249614373 4348781563 7744208374 9991461834
5134117857 2135657651 0028821653 : 3174
4378838906 9722998862 9462268685 1956288323 2057053789 4217895936 7844899972
8380822512 9585737429 5653148704 : 3175
3040321954 3301647220 4581397905 7452880207 4728588103 1140858798 7472118632
8648984473 4053114604 4004800111 : 3176
8964742165 7199931158 8903720590 0314664181 2617920911 9408878500 6677801099
9185461909 3489266918 5091189985 : 3177
3281758015 6149634425 2727420213 2308326262 5378297537 4780849648 6205747377
2980595049 0214555010 8969348064 : 3178
5109743330 0599798555 3203313182 6917519159 1950065584 8843655165 6335235079
4974414869 9554593468 2666761749 : 3179
8419319232 3919858493 1429070339 7249074104 3353155071 6185771284 4527045561
5148720821 7879010758 0994599078 : 3180
1903407684 8455908034 8525612112 4463883238 8776007725 6405950585 7614506172
9291017634 2106201632 2763133570 : 3181
8698416113 3384335191 5955219934 2391045655 6597310082 3169947997 8456319598
3411430563 7058884135 8572324394 : 3182
5999371608 4515443224 7333257474 3269029321 1134055360 5260907241 6883753354
3614128702 3423782527 0895382357 : 3183
6585165964 1732811004 2367022479 4265894895 4200246277 9779309893 4507602136
2417137031 0651327597 5961348992 : 3184
4270047047 1725333264 2277426234 7566320225 1798424952 4982127655 9511394568
1412700766 2315485158 2573596335 : 3185
3826717922 6699363339 1806685509 4802896639 7016335803 0635777920 6151292995
8681816023 3278268287 5613869534 : 3186
8983385807 8404867756 5029162328 1022321262 8623703647 1167259091 1162207449
9444703371 5467171935 5796034064 : 3187
9572183991 3243157175 8689056357 8201193562 2777763421 6236724756 6768761722
0610152133 8942884427 5546380391 : 3188
4298293409 7063768466 5116996845 4977492484 2162345498 7901900590 1223071102
4880588049 9208202673 4152133452 : 3189
6194822718 0404524659 2288442955 4190815449 2145808904 5704939272 9832168834
1342995368 2586746410 8367284683 : 3190
3132874321 9812300745 1303144400 7683574024 4627809770 1300214218 7123511884
4390781428 6457148671 3031627438 : 3191

1405338857 6038852946 9050776911 2439640284 4275392116 0112468485 5885908711
1204331369 8335512983 1770447543 : 3192
9527351180 9456262736 5072134238 6754878162 3509663987 5898907810 5843337220
3975442049 8618992145 3439457001 : 3193
0766993023 3340450706 2355822832 1979391210 7163390447 8457406495 9683736835
9996102705 5981093432 7545718226 : 3194
5819162737 6408326491 9446339254 1412570472 7893001218 1409950288 1776632184
9854530856 6679007037 6265770583 : 3195
8273174956 1209175232 4760127284 8854422298 9867998826 6283950867 8945771438
8158096755 0580114816 3295101341 : 3196
7845494953 2484743502 4616682604 9086054349 8626070769 5220684576 9308881019
9363886020 5690810166 4505187218 : 3197
1500574347 2746924004 5662801352 4424290432 3796847683 2649959785 9954187679
6098882406 4974266522 9584406972 : 3198
9776191462 6489783151 0287400669 7923620332 0604044535 4794419819 9894752531
9571705208 5531617779 1641077276 : 3199
4613921571 1774029913 5555015167 0979661965 2406222291 4160997227 0298654087
1469079193 2911101746 0401206520 : 3200
8119033479 3350740933 9673354381 0667764425 1621091064 0978277403 9347240922
6971399864 1932401833 6762578795 : 3201
1661669211 4857084403 4731109857 7186041438 0657030581 1408965227 9903370706
7927777173 2401960636 6905990239 : 3202
6600617475 9660274364 7723341713 2112564069 5730430073 8718969760 8262537784
0761689045 6503696412 1017260551 : 3203
1050631805 8783071771 0457167891 7209315551 0051312624 8850749712 5708827760
8184629735 1566606413 8131857756 : 3204
9232194221 6169819818 3386159109 1112129640 6834764541 4987489259 6769391552
0889997434 8348251097 7171733477 : 3205
4849042415 7044765736 5745775288 7031370873 9190721825 0577299317 7204212596
6177890946 2781073748 9396727533 : 3206
3669497759 7757614013 3909640059 9495991247 7424058226 0227674347 9140436597
7500711749 0687276269 3456375287 : 3207
6279153861 0334809869 2957428984 9008704137 5521603759 4678764398 4362695913
7289972377 2007314533 6673352699 : 3208
6836629260 8565705839 0388235938 3118961476 3613317043 6652976443 6074940164
9697173956 6002324847 5312782613 : 3209
5116274516 9349859149 7284257801 1566354011 2574883834 4957820547 5651348675
2535339309 4929877785 6682473832 : 3210
2471363412 7815663824 5905023073 3983253608 3003873024 8396399541 8402866298
0876689960 0543606746 3747817597 : 3211
3859320010 9703840094 3290848252 1486078558 0072030283 9262481484 2107356768
9436508478 2917239431 3537730783 : 3212
3828620452 5607937143 4798902477 5448000157 3891165778 9491143660 0367935436
3932067762 6302110521 5210135921 : 3213
5469454499 7058887628 3657933406 0613239110 2338123478 9116513396 0648232734
3027611585 3432570782 2596745669 : 3214
2639944506 5492819990 5402788150 7062990362 1350504523 1732501367 0624394106
1807667613 7866141456 3785766044 : 3215
2074924229 2697703498 4501265149 2167176963 3105167672 6784883729 5490056786
5723978442 7631134732 4977198906 : 3216
0076187591 4089607306 6582151384 4913456155 5571151084 5132159738 2911001114
2873895994 6162393850 5836032323 : 3217
4420828814 5043391507 3780818631 9372078380 3113641758 3734875197 6507933520
5354261064 8396468002 2831803234 : 3218
7662678243 8970343828 2856767409 9388012245 5841828250 6165887184 1917739713
4842495575 3553615428 3510624483 : 3219

2821141075 6609679699 5104825227 9421587069 3151868682 2890659905 4454289767
6834078428 4686635169 5073592900 : 3220
5944652517 1865184120 4544327116 3745245912 4940510317 7497432716 4330478610
4252035794 0327121038 4808446436 : 3221
3330137152 9014988427 5237794983 4574799875 2815119651 5943440298 8721491706
5555918493 9637362023 5346368882 : 3222
1813143720 8134936589 8041885261 8146166856 4638974283 9150595583 7039471238
3217345273 4821361569 6128326308 : 3223
5165038215 2956508420 8247108348 5563684152 8927797777 7563665982 8621565022
1250746116 7590321165 4209654147 : 3224
0122942838 5457567214 1738992979 9800524641 8168473348 1802173252 1988211950
3518467058 2808429271 1592599970 : 3225
1537509747 9007950930 3318805655 8010139808 1229845654 6816187153 8579928128
6103766006 8440830849 8397279763 : 3226
7004166030 6524614826 6042831177 9328935587 4205593451 8813264760 5029917983
3827163739 5986023588 7851346092 : 3227
6732319515 8872972924 6677097635 0989467701 4028229224 5909364931 9251931742
8596941523 6656465995 1119254685 : 3228
8864800103 7779316307 3879444378 7115163435 8086838550 9803616734 1177467955
2505415696 3255517565 9300110987 : 3229
1034383335 9200752031 7718713211 6942973736 6650481616 2281397436 9194360219
7933189113 7147757928 4604851057 : 3230
4542832874 8179328722 7871096158 9682604151 9103216035 8867794747 9259298850
9994990492 1648497138 5441255339 : 3231
1673798326 9837424487 4127454709 6206337139 0474048360 7941575293 6336390448
1795414111 7349362026 0485120123 : 3232
0536055436 8887938629 1448078791 0052555989 3282527733 5435924706 7399972692
2739927756 5339725669 6715240967 : 3233
7307351761 2992214202 5550653142 9549087426 1705335850 3331777460 2589152893
7886543076 6613971869 4743502046 : 3234
9696687505 6766378250 0133665443 4120149780 3953340948 4305880408 0871386953
1589891754 4323055761 4844599297 : 3235
1287256207 6470023808 8330825578 6375448067 3545385987 6380858476 3650695267
3628098611 9900402679 8113020461 : 3236
0101944598 0359274353 0574492962 4227377339 5520169158 0053320669 7551301973
1133703912 5611913305 4329794169 : 3237
9192948293 9055726795 7952817726 9687932911 4257773205 0021476019 9698152202
0615245179 4238984581 8526827978 : 3238
9797440541 6564805621 0083302860 5473010316 0503201457 4051888761 9845350296
9999264657 9565001904 4827824173 : 3239
8609540179 1037025463 8143624452 5088702922 6639233664 0435196780 0356654544
3642503625 0795296489 2139065648 : 3240
4140182736 9714502145 2643164662 1003738099 5041855887 4896308246 5740733426
3002530994 9781559142 1287519705 : 3241
2710011571 0171637749 8833787956 5686539344 2661195440 7741443992 4874232060
4226615977 1478006548 9643349623 : 3242
5839622113 5312572898 2013559848 1565702045 7482109173 7378662802 5393070446
5543688974 7490658477 9494095981 : 3243
2244787432 2186601530 0973454802 4696172671 2947787951 9441513440 1730118832
3674487204 4780623663 7003524258 : 3244
6176568380 8485736885 6902370922 9088212227 2083417008 9791292565 4354194078
2689930557 3151916990 1180705018 : 3245
9588596474 0481390462 6097073419 5449422706 7754033537 1029648360 6203275561
7310215914 3468441553 9095659097 : 3246
4654992791 5376332947 3599602008 9802640794 6829253612 7789832058 5360585618
6118945140 6113222427 1286111668 : 3247

6725025703 3634886315 8571739711 6032882123 8663749187 4566260429 5318985208
7917692798 5320596870 8273206077 : 3248
2049087562 4976239850 6391873788 0636107372 1159075733 6298970566 5746883773
5159852306 2078348566 7267399450 : 3249
1972457061 6041702865 6143126685 0797557168 9508067386 1961335761 3077933856
6529426117 8995576128 2203962786 : 3250
1630366976 5729274631 7600522624 8813002620 1593446959 9332301304 4439085141
4406961294 9095143511 3240437281 : 3251
8037803052 7016147459 6669731823 3961655755 0910094477 2011231301 6992708211
0437284835 3646580227 9843531764 : 3252
8760439520 8073622595 8640787345 8191531059 4558252403 1075377215 7432145713
4477818430 9844749991 8919885723 : 3253
4881539219 0452015661 7192555429 9875334540 0238110974 8520270053 7114712389
7974701015 8547538626 8028132317 : 3254
0286201290 3865851442 3644864928 6552358171 3720293880 1752941010 2252028569
1187186079 6450108765 2984253297 : 3255
1631174418 6507520187 6929274642 1061313176 2030850679 0665882560 6161624512
3298333856 0212251996 1320728581 : 3256
6407020906 2312718344 8482230078 9404092439 1632745240 8745667813 6735570039
7551427807 3920034353 3132128228 : 3257
7328689542 0618818832 9141890423 9295829349 2930594813 9194192101 5430324356
7447699243 0684895495 2023954564 : 3258
5503065047 0971258953 0864100115 6968732728 0575346288 8209289499 8037694276
7152372469 0745276874 9411737611 : 3259
2489070191 9879296423 2474948371 8639132922 0403340476 2728529048 0570200588
7722635976 7881747157 9821421764 : 3260
1766434900 5485153222 3351305020 0474266084 2602758111 4308115877 3578536814
1365683667 1339174031 6070565037 : 3261
9085284322 6782164643 5930151920 7099123433 8444761974 8970136375 1895168498
6306334712 4393571993 4533251192 : 3262
4340248672 2685096712 2444228095 4862096706 4207146038 9235934010 6844004753
6963864975 1359735833 1093783260 : 3263
0190925157 4431920376 1229377089 0557454362 8477384533 5137564981 1641233689
2295228886 6851999159 1629961786 : 3264
7208191837 1734730700 6822812701 8606502753 0298339014 6699957479 4461070267
1611160278 7170650023 4454852655 : 3265
3180461527 9803013588 9431096643 8222475439 7716230467 6463531802 8199644937
5662371151 1605197875 8708342921 : 3266
4674980001 5297141091 6725705796 9361548756 1571782358 3111771013 5901253539
5568712745 7997201759 2606546190 : 3267
0593796089 7846190272 2181450723 5879548427 1499133150 3962030512 4110916504
4196697558 2513218186 1783569427 : 3268
5540614559 7270535238 2673571102 3180819720 8539486018 2205482689 8663602869
5668183486 6852454461 4408406952 : 3269
1826332804 8760444691 9008266967 6296454907 5457223692 3327441664 9131955648
6439945889 9338775870 0988054331 : 3270
8639985540 4932306776 1591828592 8743896057 8046409842 0897405506 8329611397
2392222269 7903646977 6787551730 : 3271
3966644741 5747265846 5478065963 9564895358 1955700357 9716689122 6946992715
1284486477 2273981174 1814886633 : 3272
1928994659 4706008921 1318989429 6771965704 8618527686 1342368815 0000417238
0028297670 0527792276 5540848554 : 3273
3334861688 9849738718 6788618987 3232380042 4009638640 6798435171 6251126972
5924658678 7211070538 0153194957 : 3274
7164948506 2981579894 6941714282 0421641655 8665990728 6198493849 1754802695
8461964229 4779314981 2238364153 : 3275

8557038089 7890076139 0103234971 7969632547 1965649122 7455826354 1323414243
6435745947 4929792785 6960776359 : 3276
1484728012 1218205712 3722912544 3324556605 3407484951 8144676905 8959806952
0034923001 2498661937 6210850051 : 3277
2364425478 2643573382 1329660966 9731653535 4256247308 0902881776 1133739720
6298364305 0540861940 6221838502 : 3278
4498547566 8721260067 6339743731 5325783835 4874824409 7809739336 1487310202
3904533809 4741597766 4560313768 : 3279
1106298921 4090166123 2700390050 5022947613 5188591241 0647065603 1298014608
8899492786 2354781233 7075637352 : 3280
4321271800 6105308551 7170340536 0330737630 1871136693 5321769842 8260176112
1860063584 8965341436 0670914199 : 3281
7792464097 2114274495 4698914635 5548286434 0110401472 2300847400 5897193942
5556775578 4403993657 0126377009 : 3282
2330401772 0157097140 2261897254 9024996396 2566890848 5897750415 7130429271
5928933801 4627628178 0424512433 : 3283
4611567291 7087218116 9866958713 1261066558 1097155155 5696334819 8442249377
2789984900 1691334065 0139225583 : 3284
7452536445 8711315377 4964284515 4005364042 1859769093 2979007200 8362962024
6732239658 8374131752 9058662626 : 3285
9446735210 4426037919 3152110356 0615613271 7795833242 3894101137 8308624546
2951409581 7189416538 1882609858 : 3286
1362550702 8147207441 0103228385 6977012126 6793214647 2801159682 4377110164
0588292038 1222982282 5505648850 : 3287
2000903159 9084096698 0142501599 5974256102 2026318361 7255711149 1392144361
1018533845 5682860703 1576644860 : 3288
5768250919 4621585069 5082849408 5301806644 9912071428 6759898668 8412790646
5394835715 3197916296 8219164287 : 3289
6269125672 1694722877 4367226907 4631085341 9544061133 6084871630 0832207815
3715481543 5464583702 5948503616 : 3290
7470707302 7584996723 1328096016 2936123357 4085051675 8670470322 8598724739
8086473267 9403949139 3791308497 : 3291
3632414139 0294392845 7663835198 4218676346 6430180868 9606921433 8319604221
0094720984 3976665228 2254436830 : 3292
4222054701 4325651194 2687039719 9293424806 3911831147 1596792882 7659016065
8481161372 2944199433 2637690168 : 3293
2224343559 2607990882 0340023435 0990585913 9297715760 4947270770 2830957584
2707091369 7704371357 5902026722 : 3294
7121355344 9733043050 6741037054 4077345958 4309227693 7484039563 8204885926
4704386362 1119935422 5500002561 : 3295
1449708505 2705009162 8929604984 7398305977 0893141204 1983770187 0006385807
8442617736 1278758099 5515950384 : 3296
6662074848 1572501821 2354082543 3798202752 5680745741 7795530258 9468194577
6460653749 9326993288 8205395151 : 3297
8474085147 4436356821 0924250171 0525863495 3457915902 8715221299 5365959158
0769737340 6864938135 6148346862 : 3298
5939742529 3493723922 9911260953 5278901474 1520946191 6937523396 0791800588
8183785066 8857882217 4089302237 : 3299
6707892649 0559017835 7330290498 6104756624 0884046301 7442091646 0792765924
0335005213 6975053666 5803340501 : 3300
3128071242 7989700062 7372915259 0701666746 4372456678 7432560019 5554242094
4533633439 3919567680 4590871330 : 3301
9834479621 2966325811 6376546648 0890071124 7402815156 4343463108 1445327339
7154323345 4492686193 0744837353 : 3302
2903424290 2270632344 5090815537 9512377571 6104927121 7981853535 8509941553
8718219449 4270861149 6926208222 : 3303

8311054505 2601764694 4214984796 9162493833 5086438997 9794372572 5850349473
3211236420 1564544595 8323210258 : 3304
6733821208 2720110236 1718132916 2813476936 1336231630 0055518541 6994512370
3087471769 3475306380 9491488282 : 3305
3181146014 8473591765 0196830571 4704713971 6805417120 7608541527 7906148408
1543527705 4711138661 9355191825 : 3306
5549673768 7537565590 1891589207 6743526148 8529376321 0787312387 5720648408
2373753032 8846402348 8292875867 : 3307
4575174119 6425955473 2547129988 7844133769 7174478289 5148060629 7501598104
1096517924 8737254241 5860607092 : 3308
6343451122 1805151576 8050395232 0798390703 8445590802 4897675124 2811186831
1223535944 9536233280 5156428450 : 3309
9116382532 8468276903 7798940560 9730496598 2167747942 9060522906 4271154090
7596304907 5004578669 4806442407 : 3310
9141604249 8973639622 3835534265 6882489249 0309401353 2275982922 6761802139
9446801899 6032067580 3429397947 : 3311
9500581123 5633986972 7902367019 7762898407 3319861429 7089945534 0903726057
6822344748 8692010297 6488719461 : 3312
8758629342 1317093272 7776916518 7100249899 6665855083 8113356789 6174811392
4008069204 4146256653 8294522339 : 3313
1551604594 8248702775 2402656030 8024160840 6335831049 9159313085 3947426907
3234720884 1191820248 4707573291 : 3314
1507212454 4689852685 5311599851 1179393810 8020977347 3817493399 9888973856
5369940387 5952533621 7482394715 : 3315
4783480058 9460393665 9188928996 2175210471 6304660384 4441237345 0910382932
4838945600 1298364934 1732042243 : 3316
2165642758 2862696462 9870549437 0864746342 7711852938 2480439358 2160198607
0062171196 5918325918 0757449365 : 3317
5660319057 4210330697 5370732230 1404429391 3607299742 3148218620 5712797240
8432297422 2530647847 0222877289 : 3318
8891720445 4136416830 1862059266 9478106500 1577273034 6839982955 1105996427
5198344209 0994298239 4136155832 : 3319
6538840685 2983750196 1002674327 9608322676 6089824373 9923045838 1935994455
2036471095 9082518991 6629159319 : 3320
6398638609 9844252455 9194442374 0980765055 6117505789 6146435919 9255642675
9631196277 8375128746 5379652386 : 3321
6525681695 7003269642 8785072018 4716605737 9227252095 2321099076 1127024194
9139628074 8169651494 3204319306 : 3322
4899696752 3523778013 3601715355 7941527226 7440438354 7747834615 4171361080
7197104730 8860237450 3121389008 : 3323
1325624317 2049768868 0228292550 2433493959 9178274676 9594115544 3098503616
4313163944 2573259674 6141758066 : 3324
3424492506 0402322028 0294698737 5182931270 6554137782 9886195848 1093502663
6452075093 1961525082 0150239512 : 3325
8106961883 0208378779 1753142473 7800666136 6107125561 3809568754 9097630224
8182547336 9577744250 1363334650 : 3326
5272962419 5150307001 6298234339 9091000615 5502494673 7128843219 7235339945
2938067223 4196217016 1634329247 : 3327
9375744591 4665771562 6682851107 9209801382 3602779085 2490861406 1323820470
2960336142 1239678916 9499234232 : 3328
1715288832 9799391129 4665253847 2686405318 7319793226 7770276017 8715157112
7131902216 8364169453 2904519520 : 3329
4615033553 4734469787 1415224470 8770708456 1412083149 8011066671 6520746555
3671878728 7374690498 7162762468 : 3330
0530857583 5228041919 9560732766 5917569577 3002879207 0628730635 6228290710
9317225412 4410289965 6219439303 : 3331

3935979312 7298249018 8505998207 5302805818 2687345362 6207698842 8838928963
5517726997 7552708928 0711983832 : 3332
7126416498 1358505660 9297466819 4143322037 6360160317 0238645401 0380419337
5688745593 5123839827 6056529937 : 3333
9694631115 7362367658 0351575643 6802079790 9806973589 2895033363 1027509478
4781357000 6470316765 3179843748 : 3334
9385931992 8446797050 5014658422 2778269606 7591977743 1422848983 4622963809
5752245329 2343592235 1304412034 : 3335
5909610127 4485891405 0582747677 5911036197 1874811262 5960272458 6676557147
2754939412 0555133622 8916904263 : 3336
8208357399 5206157239 4645174449 8991297021 1065709594 4985564756 7105391013
6404655906 1659117790 6243645957 : 3337
3934607185 7771170611 1845170015 4545080998 8500405931 5587509512 0616554563
1462007387 3944332743 9156540822 : 3338
2265516671 1498136135 0737399548 9339174808 6374196648 0932781710 0263955025
9140477751 9347205269 3611721552 : 3339
9192894891 1285123125 6310527709 7234938677 0893098856 2479735893 2759808463
4232185162 4985305036 2731645550 : 3340
8600244801 1287948708 9021875287 3453929413 1614662088 1482680861 4162015915
5491220419 8602598488 6099100100 : 3341
8935521986 0042743405 7310121427 3402947594 3567269776 2854277727 6759795406
7832154999 8708026058 3813286902 : 3342
8183862100 0610033764 2379198001 9442370442 0633319989 3214651697 4334576991
2822318261 1409070898 6144154081 : 3343
9914737474 3368164498 2432526608 1666966973 3619695332 1277769129 7726357984
3015097108 1585627795 2410140312 : 3344
1972539950 0985483700 6991572638 1749334231 9841708678 4859633091 2936697738
3562887078 4080023922 3578231036 : 3345
1229231334 3138708713 7560726479 5530687856 7678761408 6797853884 1839750850
4683509284 1447196834 3566924539 : 3346
1453969726 7038657031 7296241838 9792354536 8707062910 5358409586 2528817292
8169247104 6395913765 4339775103 : 3347
3038618690 5416278540 6796718856 4523146253 4683464300 2030663624 3007728041
8391505048 3997746390 6523527004 : 3348
7668220337 0156915852 3770139905 4126383476 4846638411 7910763416 3945096265
7613452483 4091389875 3793488871 : 3349
0844082251 5024794471 9876888399 2003573792 6073657685 4930155343 0268438483
8893140272 1966820387 2768490406 : 3350
5078641498 3954838623 9914432710 0354841428 5714663675 8141086857 5684974964
2492058856 9843745946 8657448018 : 3351
4934228027 9582356377 5643882682 3262418776 2216226070 9045198459 3267734773
5018285436 0693935241 6589601174 : 3352
5073761140 6406895988 2929944645 3818686066 4747288899 1909822960 1789297804
7357912479 6321869153 6870365595 : 3353
2944433499 5425160580 9083049273 5940881012 5128045910 6505047664 9626758222
4133933158 0270942092 3435482413 : 3354
7545730590 8715656756 7670109209 5054671117 8377321047 5976697943 6357024999
1724776409 9099618422 3422593896 : 3355
6846991544 1887720946 5300703443 7183115728 7057320673 9875957821 4079407363
3423608496 3832333591 7902277126 : 3356
8263032769 4877532004 6801847577 0539434300 7951196667 7524396159 1663082780
8395905383 2295112723 3820755074 : 3357
4153078897 7762867716 6251881091 1875351973 3009863771 7478618815 4764116022
2390303195 5967898153 7333332582 : 3358
9836004518 8973413138 5796009297 7745880614 2401045915 0147820679 7394363629
1993558227 6023675103 4782755648 : 3359

6627121882 5552853178 2535860103 5108225814 5026112047 4709240171 8602564690
0684617317 6799057349 0110072728 : 3360
7689261945 2758833587 5822194738 4320834578 6330528075 5745493828 9523900598
4568272914 1346432348 8171485846 : 3361
0678305948 2601453596 8762159670 8124955323 0576378156 4934456525 7825522938
2496265751 1707494988 6687654410 : 3362
3827053374 0898921040 3686773656 0454285845 1695031402 2323633757 0264179657
7852081754 4892465570 9924081323 : 3363
6655786985 2645353888 0111889193 2862519259 2221355667 6815576092 7630615575
9306626473 9260898327 8347680214 : 3364
6055571315 9391575134 1973623814 3794497888 8581196334 3728292321 9663203357
8012611307 7010857215 9898202801 : 3365
2452701419 4405510821 1091262826 1670708062 7427106208 0737562374 7391179018
8063039151 9189825171 8666137575 : 3366
7275971038 9811462813 2094118243 2120115782 8817875559 1908312841 6589810129
5993972458 2828850347 0905302829 : 3367
2225178972 4313294878 9737432150 7341389531 9923645094 0330089444 1779949785
5061146956 1529485655 3112229462 : 3368
3526306051 5601499246 4164694165 3581717906 5975346477 4747518944 9033886376
9454910133 8475379570 1243435328 : 3369
3214492982 7573206494 6005703587 9741372847 3085568500 2414057806 9949444209
6564545420 6704067126 9177034205 : 3370
5893545921 4751399465 8656379532 4498939480 7296345359 5989773250 3522425366
9755202625 9661902239 1137446470 : 3371
1515637749 4681734403 7945328859 5394926777 6387559086 4797247808 8680068172
3558531314 3563297543 1766043925 : 3372
3838540575 6899742985 0951712778 2488540660 9203265598 2812701957 1356428139
2540661309 8387251913 8828743230 : 3373
3850700660 1992157068 8871565313 6986458667 1792836573 5525658627 1881440144
3171436793 7481059691 3709321216 : 3374
8062424716 2372966171 5393439953 3680541456 9867938971 7848891015 9860309541
2935298746 1691091925 2380974294 : 3375
4551488558 3184994985 9479590242 6366425548 6555633149 3468935615 0341488374
8844312946 5300565980 7467697736 : 3376
0194559815 2863062761 2626766355 7735976758 1138421612 3733145970 8729860471
3841740124 8888918797 1332736362 : 3377
6251131133 4676537629 2998408903 7789520000 8539399763 5847442819 7985767807
2104896345 9907780154 2669717456 : 3378
7542617238 4022032773 3647639755 1754916653 3844727336 8535249669 1562976924
8343626504 7461988233 5945593854 : 3379
2388739013 6417704694 5395798118 7527121597 7684425117 9958071694 5466861749
8200389141 3675742529 5199235363 : 3380
0286409958 4773808066 7594169715 5808335426 2399667913 7191811974 5650950142
1574144502 4655942823 0866854834 : 3381
5475504175 3280904924 3969757733 8340692003 6596269823 8063210584 2116831153
6080639600 0302983486 9862501418 : 3382
9519615977 0985309893 4159069182 6447462124 2983607543 4644624346 4183758912
6993823538 0143849573 6433235890 : 3383
8803400365 0356294509 8957173182 1193875360 6040337502 5733118252 5704669344
7027575665 7246020243 4358051761 : 3384
7980430500 1576612206 8591071191 6447836787 7552875557 6514955386 4629003147
7843352420 1822322818 6288983604 : 3385
9955671009 4713073625 1175046568 2887493345 1108004370 5700857207 3227508319
6736841092 1087264603 0862698701 : 3386
2176044090 4028069794 9111839643 6605218431 4923073516 3158890444 5480567978
3225559628 5773250306 3907386237 : 3387

0333133937 9486490735 5954685179 6504296661 8255310526 4598245173 5375632502
8373943551 0180592720 5309010514 : 3388
2909243352 7312786900 1515130558 7405395535 9526490302 8131275892 6687850268
3802775398 0878419695 4244470759 : 3389
8265277147 6368013445 1227046469 6586160140 5000598135 5426600455 5304125645
3856489931 6031280559 6469709727 : 3390
7176477513 6237931869 5356428514 3044563012 7610428264 9403679748 0293301901
3443998609 2840586881 0026888328 : 3391
7786069070 0575222916 3788459542 9511271236 4162904172 4925928703 3518121777
2457206347 4441407104 5798701168 : 3392
3486538940 8608793464 2885814053 2372205220 3338827824 9160655075 1428498895
0267304733 8689414665 7715191706 : 3393
7007154628 4933413054 5953261782 5762474557 7860528907 0556812093 9213636020
7948400190 4534822717 5019919985 : 3394
0351721819 8291555373 9405244748 8160840114 2868298542 1981541739 9461519446
7566539911 0862570266 1728915721 : 3395
1616708612 8078644225 9687806540 5524084076 9809262297 9890897438 8648718812
4121528586 2107144171 3146827394 : 3396
1512454023 1527184641 6021045875 0969483759 4318073620 2192937400 1190274750
2717656734 3594247367 0661803986 : 3397
7767305606 4018589905 3575123002 3066083708 7116869147 5791336384 0862325384
3492671860 6613904668 4337879651 : 3398
6790070950 9607700445 4535563136 2405135816 1617384399 7008720780 0572797374
7669733000 1951815716 6514482156 : 3399
3406218186 5695556952 3059973209 6135279604 3608491646 4544009853 4029608616
7453343809 6899823943 6938249495 : 3400
5094716954 1670641283 9424517141 2601916247 3827086697 6931180696 0971015572
5814785319 4625746145 3503976260 : 3401
5546512543 5021452414 3432982492 4138121771 3203403237 1867152062 7359281633
7441031547 1046396311 1770834535 : 3402
0154482594 5760866597 7571739145 7660958775 3667249840 5172457008 2595821245
4852196164 3789313109 8824817003 : 3403
8422332063 2885004325 4208492159 4423734706 9301246933 3391888763 9562414254
0683244296 1396568624 0316412840 : 3404
3058782564 6523428183 3656651895 8550369909 4325953094 2506080824 6841425531
4370373906 6510989676 5398087358 : 3405
7499377033 4589530194 9519517021 7522645207 3203602754 6394131137 1161162228
9900457080 2847540361 4063814770 : 3406
8904136189 6398146793 6090582948 2054185806 7653745684 5215201141 4513584740
0424917141 8349791085 2675578692 : 3407
3646084051 0296405685 4716291147 9605166051 2986011642 1892358470 6774478369
7916343692 2030219888 9638490152 : 3408
4172648351 0873673134 5839534421 5022462615 5336633614 8347864323 3561380530
0749667620 8428757530 7448476761 : 3409
2212952046 4026842561 5740219898 4963409036 1092184760 4312888296 9266380818
2477416243 2149180517 4576051652 : 3410
7671485956 7236058544 8148544021 3290753231 1933789705 4213766925 9894146244
3758062516 1532179973 4696371796 : 3411
4554733535 4924900140 5607436631 0647417667 9985725698 6302528399 4443463799
3051824259 8412579835 8649590627 : 3412
8906953651 8549591816 0282523112 9648862454 6624741498 7802348961 9904934728
1966658307 1475772532 0158683554 : 3413
7077836728 2575623992 4355463937 7454477973 3829602922 3953910136 3924842457
9908952046 5639804512 1789111868 : 3414
3684611173 6749565952 3732158831 9222476966 4295949694 0073685050 2840377689
0626580850 0246717295 8976399152 : 3415

8855287127 9469207921 2206720132 0528093964 5226322708 6822123147 5800660317
8586184106 9345045525 7909777395 : 3416
6618012334 1874179413 6221578605 0078339034 5912525245 4048575436 3277089363
7348137625 0658388020 4845701532 : 3417
9172877536 5851028430 4303738094 6458279446 3170347696 1344868288 0549387474
2078360720 9181966956 0779780786 : 3418
5955740709 6044302978 6093090797 2073074960 7010815586 8501594809 7534353052
7293411617 1483174733 9661005175 : 3419
2170023014 7990108690 3452297704 8775984057 2862989418 1021529690 0745556597
2433483618 5037771505 5086033011 : 3420
1505317616 4169618772 3661381272 2787109865 1734520385 7378730216 6127222580
6239456342 3895118271 0638999382 : 3421
8193946808 9089171426 8678748870 3236974236 9825435588 8832460845 8202840234
8362623588 3643493266 4801755904 : 3422
6034282151 7219563963 4953043008 4841662178 2715144945 9540901179 4488525950
4947265569 1194579203 6793654037 : 3423
6113867493 7619135288 3897654886 1213181153 0304716061 9686704364 4288434988
2255233129 6875029163 5458654268 : 3424
6608894604 6929370596 4951284489 7404856780 0358527040 3993562604 8982822152
5557199699 3525794545 2617407432 : 3425
7085989113 0014290671 4002259432 7469021898 2995179553 4427487181 1164211742
9343466185 8125957750 1754135341 : 3426
1801559140 0125239948 3939017661 7521611923 0063203926 9350307440 8005564853
2173648116 8158330203 1705897646 : 3427
7823292023 8182476764 4909239971 5748466900 6626848707 9269797454 3610502665
9791718725 6545818722 5995671847 : 3428
1895396896 3563982739 1169454008 2867722083 9735648520 1960596067 2645551934
2925230681 8637594677 1657471639 : 3429
8510375801 0266451314 9058946532 0117025939 0298092672 1553261881 1216870595
8416472932 2719691537 3233155149 : 3430
8788130347 3988948962 5427170463 1085200203 0893497607 4748096952 3143814485
9523362875 9639315703 4764290005 : 3431
2519090875 2531657407 9244929463 1761411281 6050060433 2367814921 7024471810
4090002523 5729074509 4008754190 : 3432
9448501323 4277377902 8203768777 5988388910 2242894658 6307188578 3632584011
4004029582 0151572777 5055220497 : 3433
6717418065 2296812814 4535963077 4683991974 3615077556 0849014830 4881526622
6168875496 8034628330 4068496724 : 3434
8845831448 9610898411 1641853472 4679049542 9842336879 2950285053 5622738086
9303629064 3496106389 0273423981 : 3435
6444371298 9675246221 3499807179 0983253853 7518245143 1822981701 4980471474
4052082067 7173482930 7574244607 : 3436
1778472524 5946185748 9049305096 5097953905 4225306921 2360701703 8379108571
4698302257 7248525173 8459145607 : 3437
9105588537 0470662930 5268611514 9625701677 7566050987 1522189311 9931908608
0409589327 2714652003 1599118043 : 3438
6374064955 4498522211 6311079240 2530841208 5874327508 3025736046 7355905024
2876200960 5821782540 7072535881 : 3439
9424274822 9061261156 5067099069 0000964622 8666919350 2613356980 4484990306
0697708791 7964203449 4706647343 : 3440
5831304985 9323970595 8907652120 5938976976 1799954609 9025575012 9252950517
5646332819 3778481798 2728921626 : 3441
8839791503 9028415489 2848405010 1832934301 6939308597 6918820760 9832728889
2113551698 2344564447 3332530729 : 3442
6239857923 5645767684 4465574078 8184753280 0320620409 1248503790 7903369696
7998575698 5481175481 1838668849 : 3443

2826248933 7313463656 2096236436 0176047562 8848255746 8798352316 6892032758
1208311926 7273870776 2830879194 : 3444
4164060207 4628031822 1576402945 6583397476 0879869175 2555031704 9629196191
7121507212 4527733136 3754728630 : 3445
4990037502 4593485960 0321151449 9284066215 8257436742 2744755010 6391222421
8890391206 8857149902 8125033222 : 3446
9301019625 9879383127 4820795145 7466369086 9011102131 0530573875 0610287625
8248047297 8297597037 8866527021 : 3447
7441124608 3737007276 4091503713 3336149717 4009051602 1354287018 6599060553
7125909298 9698875726 8780006791 : 3448
5869091084 5740780273 9901018725 8340250270 6752349279 0845564584 7233838793
6948393212 1937056631 0273581109 : 3449
6309442346 2935733587 4395461017 1509748417 6032594835 3621751671 2490048287
8786934431 7863407778 9561343147 : 3450
6530447271 0158730835 9186544227 5335060945 0045427094 3829595234 5006179508
1549912226 0677053695 4034708723 : 3451
1670377358 0038588820 1853606077 4078592030 9100130736 6861533205 1430948329
7128610836 6025245559 2697326600 : 3452
1032976141 1191743742 7678278974 7510302954 6503081040 6048421266 2927492258
7131958043 7832558381 4279728206 : 3453
1046716445 4539667827 5066337611 9561547180 8114106637 2890445086 0711216506
6033989238 5555376753 2053879934 : 3454
6850503492 1858653621 5611625607 7378507833 6813948450 9250569703 4605431168
9014345656 2307724317 8045128441 : 3455
4990211879 9309048288 9896661964 4774295261 4868975745 7320468170 1313930905
1170561296 8133624656 5767032752 : 3456
9978842563 6726104684 5139557876 1745542614 0794992788 5159419323 4557306458
5536376766 6277904565 1946875235 : 3457
9050707029 0264235937 6892174112 5833571439 8554727179 6933471669 9045245738
5765734636 3234020958 0112235447 : 3458
6444172330 1996887594 8411158859 1938802652 0824126254 1577592395 3557139009
9406192578 8576243834 3967082535 : 3459
9850867717 4520306477 1259716871 6292719811 0872264071 6731620311 9950574953
5335078557 9058055228 0567687094 : 3460
0035886214 5084193945 1102129664 1803010250 7190041435 1802625839 1841696334
2871083924 4701121728 4273032774 : 3461
7013437984 1173301244 6913775974 8817280837 8086328358 4806041092 4220865767
7287522099 6324008042 9944929304 : 3462
9868849898 4582499837 1385891669 1314115948 0537977042 0015970689 3471118315
7338901047 4647987808 1565219264 : 3463
4112417536 6266821681 7707694323 8146633641 9486790863 8258471341 4390786785
2662542025 5079875005 9834420864 : 3464
3353203403 3854071697 0048589542 3819416463 2023644992 1869693519 7625148758
9536447516 3444940641 6198941671 : 3465
1341044350 1482484379 8746391600 0978580071 4886541351 3572346046 6234792972
7283142415 5920800251 0346789545 : 3466
4275219424 1325704026 3069769465 4016135485 4687985714 4294868030 3910184410
8638904144 8112375443 7128533082 : 3467
3993732836 6819623130 2956918569 5856627411 3770338858 5366462274 7193167250
3361104073 3156570076 5207124240 : 3468
7975699950 1517168219 0064511788 7028746352 2929808818 7710072903 3972992256
6421130560 1375775977 1901399412 : 3469
3632672808 4538940031 9096154214 9931926133 6411225553 6011836627 3278385267
4019875478 1876353935 7339492847 : 3470
1029582528 7103809975 6543973256 7129487558 2247836268 0745273903 4903745390
6581151941 9572645585 8788269618 : 3471

8599474918 3952654963 5447571365 0412286059 3117832774 7043171702 1755542733
8113164461 2205777914 6073657791 : 3472
4630762301 5698777942 7994708000 6669333908 0866312852 0372580428 7139455275
6934418643 8283216307 5424935767 : 3473
4340668984 2924817540 7624563484 3585997894 7995073584 0897211272 6010980185
9131872698 5820436021 5449353734 : 3474
2282099832 1512799675 4771510867 2556888219 8976906794 3231991850 0345646597
5468942090 8586518688 5415650053 : 3475
0177043477 7447943867 2703930952 5080717481 1438806676 9440408803 3700276228
9229403949 5464568694 6736562765 : 3476
1215744427 2755618547 2729709231 6077100833 0327612046 4400195010 8825543666
1183840175 0643307987 8960184957 : 3477
2564092270 2645363838 4878282644 3778467646 7889452186 1373535543 6563776064
7667817089 8450543551 1469123142 : 3478
7414164836 7976459749 6100751751 5958073916 4799319511 1269366016 4858482939
0733187973 9169087881 9561867435 : 3479
8831373515 5390613140 9862755152 1295444877 1045808977 2191058276 3398947484
9102783391 6995225437 7681439407 : 3480
4186682375 6712442332 3514823465 8676499645 1945247633 0870536440 6871414568
3260663976 9454801930 9437100867 : 3481
9575123989 1190608598 0795618977 0286170467 1402026390 0407552116 9679039679
7200717133 5597714677 9184571361 : 3482
1149407966 7124629229 9933147763 4216541227 7835748258 6275349900 6792111978
0603078857 4954697328 4196464487 : 3483
2481549238 8450416874 8844032656 2774709529 6067748119 5278592514 8107028490
7091501865 2287534294 1836314061 : 3484
1237085873 2629402433 9098198386 8089791860 1274620818 9395020988 7488319592
0209202041 9914311024 3288618404 : 3485
3867214698 4472180588 2747761188 5531433454 7759499157 0811215472 4304881109
8267530850 1879271223 6067265424 : 3486
7225549511 6778349951 3760704930 3136759892 2164576741 7615608894 8829950931
1427656818 7950445739 0726038660 : 3487
1558121381 6955577135 8420435493 4789536420 2338974649 5492766853 5362013175
2865705505 8809440977 1668261877 : 3488
8503565693 2488370061 9168688132 5769898892 1537701642 9415477035 2800561194
8422472985 1874877749 5354659986 : 3489
4731283766 8102316847 8386420803 5715066104 3760802759 0099241762 6841020731
9104116864 7523492506 0364563867 : 3490
7729618049 5366105618 1453874745 3647356295 5760076828 3850026539 0338220423
5925539829 3284193944 9420500089 : 3491
9887248918 0621128704 9039130029 4855651474 9543445752 4482287158 3406545423
0746784942 7493096347 0015143263 : 3492
1241612982 1109765779 7808646208 9636434720 8805915536 4232264831 2645150216
0519650265 8472066706 1301204933 : 3493
8696992206 0721205507 4846983132 5044567903 6197937511 4510561099 4059722706
1024553164 3418423159 3135390530 : 3494
7277364315 6367264580 1322676377 2668618633 4792960994 1243270017 7449539823
0432040025 4498544641 2582181492 : 3495
0560149218 8788848500 4281841829 5383774717 0367919128 9376308701 0427207210
5279340761 5909556056 9028788415 : 3496
4359370412 9448706773 7642712638 2152837911 4630861459 9688138515 4958969349
4777530091 0908956450 6288798749 : 3497
4998791897 7330055395 5499696722 3113032996 2237357438 5677800288 4728964213
3583266974 7258346061 5280371262 : 3498
7432217243 2529339257 5924924474 1154505859 7603142539 5401902732 7179534824
5344711818 3267533772 5688313193 : 3499

5700208316 1783185793 4695550625 0987414808 5073837262 0144835846 4003533691 2705562119 5890629893 5556277917 : 3500

8399088576 4956240379 7143088390 9711102642 2589746293 1766896722 3404012789 2499944003 0146467922 0203830421 : 3501

6198807717 3466473516 4681098205 6584456771 8498749974 2916295695 2777441709 5631284596 1026901454 2233395364 : 3502

4324790898 8275294511 6320992753 0749937938 1898314756 7944412459 5963726257 8848779824 5921701280 0557580271 : 3503

6147579216 8773028767 7241427839 0459073912 7392942678 8438577192 3968052299 4034053347 0735733345 1836725155 : 3504

4265826396 1999930983 6730795037 2448686461 1493730495 7612941570 7070662032 8918110723 8915527538 3366638170 : 3505

8043033055 6707275261 6774194630 6060870155 5657445230 8746558396 1240503014 8057910614 5816309013 1489886188 : 3506

2107938274 7513047641 2486827801 6019084967 8442189011 1018392559 6780158884 4050853939 7683873601 4191230960 : 3507

6008688848 4095875909 3979887545 7125098902 7721154014 4019262217 2796546564 9895992143 7569114290 2020411115 : 3508

7924873126 5080755959 7284727869 9682789162 6827869491 1524767458 5192281108 6526770098 1927943533 7913553500 : 3509

5146898579 3823713882 7353572717 1789383014 2216485717 1410290069 9728235323 2884846219 2811289408 1707974024 : 3510

4219054436 9303801749 2997032084 3401108733 2111453684 2359999320 9089515669 0859649152 2776672296 3969422424 : 3511

2341883180 3521099116 4028064348 4735443798 3200003628 0819097685 5849256117 5239297741 6582041844 8109089512 : 3512

6836470846 2474117396 1271488101 9329455867 3819432508 6126535883 7368559920 2590781539 6082128790 7306065333 : 3513

5449059883 4747010168 0869637317 7373258291 3940201499 2329209692 7067831675 9681476696 6752080774 8268071111 : 3514

6784929358 4619841862 6172110925 3221606852 5388159185 5988062776 8721643818 3516049553 8527936421 0903131036 : 3515

0781624389 1471254331 6537004929 5435731311 9180428419 6503216157 8090425201 3312485605 3203608591 4481767170 : 3516

5497669073 9276489514 0552152060 9254132916 5702491817 1664437203 6661368578 7673325110 8285903819 2154476652 : 3517

8622524635 9219175013 0342843933 1954912297 2792697639 5317486223 2078789202 6844508352 0412496126 0947859764 : 3518

7640505134 4616457799 5797419209 3941387311 2767240579 4054104620 7559701574 1827181915 6561183209 6658777408 : 3519

1467691697 7483766418 3860817525 4785968515 2469480775 2817906293 9927065252 3129308948 6645014790 4547107615 : 3520

1553997996 3895457354 9956164209 6460474745 9853772405 1945204244 6951551105 7601494221 4459850785 6289800520 : 3521

0150144217 2360303944 2834244487 0888738111 7569548458 6881015884 7305082029 3605143013 7010450699 0268158198 : 3522

9459515045 3748572348 7980716132 3689988928 6204993413 4778445964 8262388689 3449564130 6886939680 1624022569 : 3523

6097259058 3690359144 7830464096 9184582654 7581534945 0077651607 5819566412 4007433611 7286666057 8064097906 : 3524

8506843839 0865501366 0341715661 2177413653 5246662924 0385273163 1584292475 1071820737 6199742570 0526636287 : 3525

3424852769 7091241463 2604391164 6825719384 4749765274 1279128833 2764753400 4587978756 7219728508 0251587765 : 3526

4905655892 2047149620 5252104012 0166987644 5381749387 6153962176 2099818669 5643493775 2040786704 5502757495 : 3527

4208283438 2652727990 6640404630 8555601579 3541580183 8870887714 0523005777
7479101336 0758346370 8160374140 : 3528
3358166217 1052377954 7780310828 7893113898 9447958611 6939228137 7248209766
2231537416 5551870724 3003784468 : 3529
3873444610 1858764966 2483023535 5290685620 8893670501 2914503374 7129770829
8592055240 5187831056 2165213224 : 3530
6122880034 7547325593 2571175091 6176594166 4823949729 1335237054 3886272408
4837055281 7002962980 9058650804 : 3531
7949546245 8224013571 4158490067 3308445738 8181196744 5142911521 4342793926
9965562257 1986439196 0677530383 : 3532
7468667222 1703556890 8425034832 9476678415 0367836819 8974756263 6076056173
9594959045 3787288708 8011965418 : 3533
7892476530 7820255412 3421527555 4653752784 8511642193 0021040069 6015444285
5007174370 3540070335 9356505398 : 3534
6517857070 0658209980 4195744951 2358764478 1248610414 7556590450 0553778745
4257327663 1373882502 0172648969 : 3535
6161291010 4812793263 7456731347 3871166387 7099845420 6702700296 0111722637
6272726745 2101811865 5910525861 : 3536
4533881970 1289911963 8473502901 7400070680 6314623013 0805397545 7760287247
4099097506 9178826192 8319716472 : 3537
1284958737 9180354955 9085450061 7988132119 8211429825 8375305635 0978991235
9405448601 7902486260 7671948351 : 3538
8442349058 7190753372 9443395924 0918535280 2957201038 3942096233 4277562087
4772317011 7527568724 1012245302 : 3539
8153879584 5096348579 8391045192 1318634956 9030391256 4113905964 1254019436
2929494919 2135307171 5344840968 : 3540
4207106422 8228981186 4026763507 1607969350 4702100231 8781098561 3567910659
3066007897 7625531978 9825006631 : 3541
5156298335 4439164449 2580570078 0106162803 2102201957 0475798656 5811680933
2423000560 6648967369 4876242617 : 3542
2410119907 5962069434 6859404061 6410905711 5534545376 8092327128 7235399907
4435919539 8397372110 1334949165 : 3543
6927142993 0280203147 7456050544 6618457532 3059017078 6156443038 1970179149
5676539404 5943560606 6050701739 : 3544
3893132134 6258503810 7319503867 4483556019 4918228051 6773057685 0268843254
9589089560 8141995075 2565189355 : 3545
4022636610 0848161462 0386544647 7107121879 5163069833 6202140746 1243273856
0733007417 2908534137 1467646620 : 3546
8233265300 7577845780 1275507872 6278301618 2508700842 8187745332 0787977894
2964858597 5819284204 9964829620 : 3547
4273540733 8531005469 9395412546 1947347217 0399527022 7035779385 8512968366
4406788983 7465656361 3544478066 : 3548
6838466976 5176614999 6185439896 2350237176 7110274089 0919399583 1451468723
6128713849 7242069080 9504422367 : 3549
8551028762 8680425188 1635840261 6298518072 1152833223 7580970905 7453569178
4019381600 2439654325 7825045445 : 3550
1969403488 4092434085 7899982771 9151230309 4738055403 0823155608 1860716158
3433955617 2810637331 1060601526 : 3551
4964148156 8404319462 3560436301 7431750920 7713049088 6856047273 8551730953
8084750144 8651760497 5677360783 : 3552
3154472292 5343359638 4563021521 7104987385 9253102039 7895814333 6638415592
9925508944 6027807017 9622745210 : 3553
5550671191 3163262652 7993669635 9892383006 0969816001 6708813443 0020391177
1916307016 3778003830 0371126418 : 3554
2414601498 7041714805 6626033849 7751756038 4954919157 9141792953 6849597062
8632971745 2214945264 3640004011 : 3555

6851275873 8794348666 8837572288 9961342993 8664023640 5894482452 9124282016
5800478166 9418478063 6604784838 : 3556
1761856515 5840174603 7289782158 5659083148 9306511779 1985723171 6476472418
9304315319 9908815499 7137742072 : 3557
1011833196 8619684944 0518471348 0510376244 8875818172 7723344272 1570874000
8524939194 9339810308 3199952288 : 3558
5426263085 1814551410 4967489642 5768172031 4577419760 5540116651 4371933763
7218818650 6524854450 3999376609 : 3559
2267777074 7939980142 2580866214 9871912470 1387469895 6765809816 3424079513
5703733486 8399460740 0148381880 : 3560
9109122785 0498742256 3947102858 9036178924 8694551999 0491711623 1709297408
1951721636 2506237142 0825261190 : 3561
7131791876 3800373820 9989415516 7362903105 4940290537 2537989577 3700088178
6588704904 3920910261 1278111129 : 3562
3551942277 8192147007 4363737351 9008329473 3687322396 5494763299 2827531835
1812624007 1189203456 5885546809 : 3563
5030263042 5192198797 7737443263 7260928911 1090052198 6875537228 1053055764
1476148246 3985863718 9772977761 : 3564
9373087670 4264970441 1511412890 0621261138 8530603736 7229595816 1170213950
3007414176 6112848332 8860167367 : 3565
1673203588 0470158024 7848646398 0799957676 4707923310 6445662603 3730736158
9922617875 2674260135 3600729527 : 3566
8511447312 9927914524 5062334029 0963971759 2321979580 1119146699 2839606609
0546200798 2374521224 5036016910 : 3567
4115632219 5329726954 4555151828 4094539550 7867404093 7185323660 3877962805
1431267662 7403643982 3983188176 : 3568
0597852813 4663946956 0555811845 8939807050 1135751682 0680694894 3855291350
8828544734 8281517609 7154238595 : 3569
2213273086 1322041292 3307765655 8692790957 0047843039 5568199401 5964223359
5945502281 0565437799 5470862448 : 3570
5590800370 6950156080 1930721464 8972673163 9389255381 5995861702 2059139730
2709006385 8841495346 1627642604 : 3571
4283738912 5191847819 8500254982 6303585100 6341034437 6334073346 9903127035
5830801124 3548158661 1646799047 : 3572
0409954792 3812878971 8678010981 5204978806 0504766682 0366296725 0936939607
3136693374 7203672430 0318966620 : 3573
4997506525 2017547135 7865734643 8364676379 2685019584 6151069676 5363902447
4899543219 9723190611 6932962287 : 3574
7894612265 6683064351 8956163899 5745077221 0945127991 8853244493 7648778904
0544224233 4709795872 6521769377 : 3575
7637079604 0732789702 4558295386 9616412632 4657549210 4481223808 9336380764
9499671963 4861554060 9091415949 : 3576
7678087746 1912277082 8684460532 5727180192 3246319994 6676789260 3035979578
1182790004 7139409217 1939987407 : 3577
6000999706 9581634479 2336051061 6646007787 5593954578 5813561911 7973484724
2756439544 4690018537 0303889947 : 3578
2199830805 8834058367 2047301059 3371645853 7773337382 6188199786 0740674509
0629225548 8990834435 8447071868 : 3579
3462839079 2947080116 9686394851 1850811643 8134602812 3477126650 0937086434
9808106119 2511699839 0691124112 : 3580
7884922501 0740467001 0024987554 9803524056 3673715644 0483483522 4610219675
0403342268 0901960919 1836769753 : 3581
9184488898 1030769313 6036846490 7518519208 9980796748 4952064252 8150398821
9929450163 1224441929 0790582175 : 3582
5002124641 5643878011 4736353486 0323181769 7699256886 2342243651 1614137835
8486034572 9184564597 3101458014 : 3583

4233910343 6619091211 7514143224 0043381471 4649570280 3681747815 7339925527
8767581363 3597536055 9539384017 : 3584
6867674304 7462011227 9164213879 0162717829 3633202573 5406498294 3159500554
7141144066 3728022590 6499077297 : 3585
2153184835 3962105920 7509481870 2230994825 4672970184 8378246112 1232263919
4350073177 7620066954 0599603740 : 3586
3765441062 8719241786 6624208821 4648278279 4381136197 4467588435 9564005433
8403532921 2785957488 1400073749 : 3587
7083250483 2785755627 8773866528 3977888843 0937242195 9455553543 6816532937
1988636343 6817907518 0196127350 : 3588
9345804109 4465517258 8496802612 1015346697 2738116149 8560713593 3184212517
8208697644 1366280313 1959266299 : 3589
8008004999 3801799153 4491839141 8600149176 5200955505 7429439030 6323540512
9132722852 8953318386 1280090024 : 3590
2018592068 3012455768 8515998603 6670882144 0361799497 5769301015 8168404896
3695057071 9416181922 9869985461 : 3591
8110664316 2421543438 8574483433 8880719947 6674230925 5414252204 6461326333
0219254123 3265542251 7900396237 : 3592
0011877224 8801218997 3986745402 3278310111 5581388876 2532752346 6564979284
5609117583 9397247695 6922958164 : 3593
7917057753 4606301700 7527431401 7131409347 0826207580 5563651383 6577977785
6686672314 3389971585 4051283817 : 3594
5395172886 7267924335 1932427944 6404274845 5722042713 7038338428 2433585631
4255103756 9101814745 8356414678 : 3595
9534508685 6290184867 6091677061 9988006592 3053443639 5717282508 8842599663
9289712477 5740130935 5439247332 : 3596
4079182045 5538176637 2353322209 3512540132 4815902989 0264297425 9278789454
7500654994 6921212466 6206626082 : 3597
1434932002 0805522405 7223984383 0449363282 8192284548 8932895874 0225793208
2077749921 2636085911 8902964660 : 3598
9838958881 0029238820 4924622971 3322929703 7751993136 1009803526 7052448685
3226374478 0463843256 4630100081 : 3599
4486291219 9457156447 9009944608 4293682847 9652744612 7653332448 8110332420
2517473862 2234490697 2367553788 : 3600
0339959666 4030721437 5360233103 6965733231 5237722987 4426905668 9617796282
3218987189 0962510009 6673816079 : 3601
9485616991 1256164664 5110175597 1637994994 8745173586 5867708404 7168447477
0849909769 9794470710 7221646280 : 3602
1394627545 9233625336 1858445760 2386869037 3404023429 9795866218 8971415880
4575375650 7850426102 1206974369 : 3603
7096921440 9411346899 4263087437 8569318126 9940438714 5948959983 5002482339
0435296021 4104347987 1030683534 : 3604
2977082983 3207751807 1528480882 8337014599 5157366923 4089522074 6753110902
0004087766 7734520830 4107255415 : 3605
9390052576 0346091208 1402841774 2037713070 0842437158 6729360593 4806649256
0890600621 4007352462 2034669434 : 3606
0437762973 7141729565 7738443594 0054753577 8336478717 3858831995 7253806398
0996026454 0532720562 8731511229 : 3607
7815716086 2957374684 5665423600 8290043057 6533154127 1689897462 2851338510
7670664275 1110612635 1131616092 : 3608
4243466569 7007093299 5217809475 7112910514 8114698468 1636209949 1211915737
5909346546 6617608592 5249964296 : 3609
4904995300 8495876078 1963600471 0267394007 4073208436 2442003461 4494703242
3951767853 2424534809 9377528028 : 3610
0525924048 6576284671 4121867299 7407703083 9307426174 0145003221 0497262071
5170167184 9128134389 8195596976 : 3611

6521920190 8137000367 2782082548 0844407705 4889497735 2900740939 6411521592
8130985243 2760682475 9522533080 : 3612
2824831409 1145503496 4805588336 3143637880 8692685098 4022754356 1095303877
7070121200 0289803251 7901049232 : 3613
5077885025 9083263414 3548516877 2200998294 7719960230 5243885580 4487383316
0358617226 9203006710 1878742606 : 3614
9417860407 0699902507 5004932336 2692993214 0946580996 4653142858 9402950507
5993837424 6713892699 0433088969 : 3615
6893171221 5225093296 1528322314 0401522472 0069622519 3121876948 3286742002
9173664435 6806635821 0533280417 : 3616
6726151562 3021242888 5543250193 4774716280 4383419392 8479334861 2220573524
0128775008 9411533192 3097650828 : 3617
1811189635 5024958467 1068452926 4484555742 7712134742 8588612865 5316929049
7678456625 9641824726 6927783440 : 3618
0266072794 7414440837 0664575851 2376709200 0483121940 3483729208 5600229027
7977308648 5414746185 8811157024 : 3619
0287314904 2337471022 0512254210 5996607035 3839506482 3345926097 1244250184
7780809771 7267392473 1940165503 : 3620
9607867012 3047395765 3328065850 8348412526 0088946584 6731191572 9174574224
1779493354 9798919007 8206213086 : 3621
1080386551 6223758124 6483554065 0724329292 9115450926 6713185929 6769309032
4380500504 7057980295 7471634654 : 3622
6186859666 3348164737 5637582530 2952862923 7343678949 4660108835 2913613146
6638328071 1983717491 4550006429 : 3623
8208172745 0617511533 1643178506 8605599580 4144705051 4513443466 1354349132
3777336900 0477575850 4046993549 : 3624
5286508212 3106885520 9618997124 3934341116 8098795396 2481708529 3385576090
0270227949 7412417974 0475717835 : 3625
8915130101 9483114208 8683249433 1481780233 7650953324 2995528594 4368343551
9727318517 8049465583 5099194309 : 3626
3702256819 3180246844 8831867708 7309347278 0380591552 5023120406 4223609470
9885301607 5837054367 7837414644 : 3627
1160534217 6004593264 8901251010 9436888394 9375010807 8235517849 5090823589
9540726474 6204374288 8105083964 : 3628
5588426654 8692765048 4032831218 2091922525 4911472664 0636220307 9345506603
9883857248 6005108528 0789875092 : 3629
2954152053 6859292619 8779168922 1592205220 5519853201 4792614710 8591884318
6865741159 6378993112 1609989761 : 3630
1097963356 5444902734 3590983289 4786788397 4966109031 0579656789 8848146337
7937809720 6339058401 0251514800 : 3631
1880401230 8861312975 5420721536 9649270580 1455275021 6146780869 1366074981
2251621634 3548510063 3443491035 : 3632
3420533934 6945249747 6955086265 6936276212 6109157268 9185127730 3763839795
7159366930 6794929864 8386633844 : 3633
5448631276 0855100518 1894581104 1464708454 0153629145 8227457290 1534105798
8169354055 2831867360 3825088352 : 3634
3047392021 4607034933 1910672726 5127717991 4138372267 0503004288 9515491348
0292363239 2882422079 5773086577 : 3635
3544300787 1499407960 6303257695 8992623920 5966013855 5418034196 6989324367
0285104942 8362179024 2444777381 : 3636
9256545997 2644214164 5880183324 2657385618 7867895985 2363478423 5368090599
2711999559 7854014483 5966062156 : 3637
3295163890 7300427568 2540019235 8883203001 9113497523 2861300651 0978769294
5531139668 9555834764 3708070331 : 3638
5692464770 5668317087 3581839480 4538875307 5171478337 1195076879 7640772884
6486649791 8231845023 1538105517 : 3639

5873407189 6253879853 7621437172 4368207104 6881420524 9239365508 7044891355
4165562666 6713154192 1763666664 : 3640
4546134197 0658844048 0395862840 1161360336 8548134550 5093590314 1657252576
0552586987 8703051193 1907553636 : 3641
3454402515 4523395823 3658716038 1608528218 5007141035 4993608466 2784075061
2368387644 6214589192 1779142593 : 3642
0064973124 1854636559 9567512958 5020308227 8410326137 1511351983 9061241596
4633398239 0087897387 9933716728 : 3643
8720356789 4869612784 6918029884 0007702404 6141684357 0964611623 7948458002
4369153446 2319297027 6370458105 : 3644
4668618648 9200648185 7884470461 3762942820 6142107034 8036308441 1167800614
8150239136 9013966575 8030939659 : 3645
0441591251 2096952973 5812730979 0325832927 4793996928 2438314920 6259029330
9286010332 3917403838 3427457113 : 3646
5139845774 7832024488 3860219680 7671633165 3498673034 7454013999 2159822111
6711151254 4258537609 1673432769 : 3647
9904005844 9743475905 5642063076 9469347283 5650649910 7776795892 6507983234
8108401822 4489117074 9665104061 : 3648
8759184748 5769721036 7432681514 8420195697 2207473448 2208792869 9608362935
8763165389 0478390048 7427479351 : 3649
5152841972 7078614321 9658284170 6939049681 5772568280 5012202075 1092363465
5176931515 7232381050 1802258551 : 3650
7518247822 3413643178 8165206850 6613942262 6534469728 9359780575 5792607832
1667388980 6612519238 8000169534 : 3651
3252262005 2889956108 6132879950 7595735547 4955247424 3083287019 1803056477
0827367014 3445393759 3763155017 : 3652
5093518515 8798505069 4639052319 5829252309 7514074052 5459563140 0743663136
5318598857 5773788784 1902614118 : 3653
6895445650 4216033706 4786252627 2470854341 7597795006 9264902695 1120275026
3095436740 5801647213 1592679453 : 3654
7943694975 2610184466 0858000783 1271913160 2123509138 2970425870 2336416046
4484612847 0918634315 1986536546 : 3655
6090267618 2247471801 2253559968 1323522386 6795597369 2580689597 5303712648
0244820458 1370182034 3286908637 : 3656
1659833775 7108583301 0368743465 7741106218 1431765563 3962100638 5831674674
8091887170 7737653558 9866220294 : 3657
9987382744 4255479241 1634654679 4060365024 7460245618 9525909349 9667359512
2712732052 4011111391 8523027067 : 3658
6142772309 2229472881 0133192963 3399781855 0318293443 2632206469 8010129517
0455704300 6325015625 0431508365 : 3659
7628326741 2300204950 6397173678 4188910051 4252530524 9688625673 8738477796
6286621649 7362402883 6930961735 : 3660
4373373136 6775835835 7017944714 5747081249 0698022977 6815688257 2257089498
9089912066 0554025206 0801322450 : 3661
4825120813 7567437661 9867964264 1298605943 1128300054 0888188123 3133472974
7312126744 4431186759 4320910668 : 3662
1841781026 5573353262 1387737473 7553457827 9041511593 1321851828 5887774417
7229662046 9138660209 3496942560 : 3663
5364506621 3331732596 1987682588 5942987942 6709380114 1054153993 9189608363
6192487418 7916930646 3578448072 : 3664
1565839931 2037940118 3244802015 5671414285 9863727859 8469707404 7554361823
4815879601 1413662352 4548067203 : 3665
4561475693 9780122895 6585216225 6169671298 3690085390 0438345103 6299591188
7655548585 0103824200 0728257383 : 3666
1915399075 2707127241 2720139581 9955356637 3551289090 6976251070 3048920279
4525915565 0048553046 6782494374 : 3667

1234171084 5111272716 0221094193 1455401579 9975257597 0623571196 5092077965
3514558284 9694641247 1272627520 : 3668
2638568898 8076835163 4588214744 3822121862 9646612627 0826742263 5057195047
3923863507 5412116648 3026200971 : 3669
2767100489 8267161324 5191035278 3620701285 4444670111 3052325226 9301508703
0633845197 4189270634 0692454588 : 3670
0913768037 2957533468 0677935046 8647874587 7073762192 5305287481 3619851138
5757845429 2910766542 9283407134 : 3671
3502250882 8188929152 4452778398 7472190081 3143807204 3020724546 7931758778
4666266568 4652886150 7688403122 : 3672
3212187332 9607244268 4943391394 8234400487 9473181001 5672345261 7702576708
8679077948 8435759761 3518177223 : 3673
3032469991 1665759663 5615488804 0497320129 7560095265 9882349602 2332949604
2355117872 7855292408 0285877667 : 3674
8063370228 8000221810 3354707338 1937382622 6757192925 9335637095 4061572778
0572441483 1116404280 2001173781 : 3675
0453773569 7122670282 2063493126 9054549816 7184995028 1156890107 9688029001
1256589179 0555923454 7229213785 : 3676
2872328218 1212503451 8205954809 6772277660 7131017786 0343882957 3318206423
5872369934 1009810617 2349638246 : 3677
8683009133 2565344923 4684068069 3901747353 2015044406 1138347551 9042887754
0607758191 6178115413 0039925740 : 3678
3786443591 3019078461 1315598122 8743338330 9853783972 8020727286 8923202989
4397339481 9817757139 2474916946 : 3679
4666678961 2342910937 0338793251 2337739713 8802549835 0706646550 1643565968
5314582650 7561070572 4702998374 : 3680
0904860172 4376419814 2270431480 3738452936 8743600536 3912672650 7615680781
3142004122 4119594940 3929770163 : 3681
4281240787 2080852166 6306459230 5670320709 8161722529 0269602306 3081292602
2797817743 5716138102 9192980622 : 3682
5097290234 2140272119 1693803271 3298320083 7285066796 1628159235 8286686576
0999044159 9158720394 1836822473 : 3683
0903669496 9071774570 1217771446 7268351119 4579923576 4378946935 6535420860
6297301046 7307198227 6607743930 : 3684
2309468861 5482810951 5202159705 2550236156 4783557941 9688175556 0913857785
2192225964 7769941023 0570038366 : 3685
7476235506 9788231896 5569824146 5028678631 8921131224 0606180938 6048832645
1730840169 5014208215 7326064292 : 3686
2288691115 2502549936 0218005104 1932609690 7384748832 3914024155 8533260115
2850704306 6093224442 4825241441 : 3687
7460744488 4453418552 8241403762 2464301160 8492948664 5829985520 5426715174
0545442020 6280703433 2941069337 : 3688
2726429706 6891081535 8484014909 4569382462 1684799297 0706873787 7406289283
2545134226 0408837187 2199038355 : 3689
5267472209 4123126765 9633815572 5024484611 2695927469 7523814493 2164044468
9895162240 0692837095 8627380688 : 3690
8336291637 2400418759 5864552046 3746275695 8305079845 3815347152 3066411007
2569236293 4927639367 1091313512 : 3691
7040305457 6969966845 5358471455 9181928377 1246258352 1412445812 0274966078
4771810569 9457043360 5081685095 : 3692
8945374630 7951839500 3471725194 8443599494 8544685268 1297359019 0690653598
9033569560 3100644796 8263120457 : 3693
3191011544 4965551122 6841732515 2277386303 0813597582 1722905976 6137627641
7661634769 0912507016 1999553759 : 3694
6432115575 3439662168 8271084036 7511929996 0008862463 2199987541 8094360053
0874133962 6929982076 2966801548 : 3695

8090523558 7186807616 9855606043 4175392524 0496799964 6517600026 9269165516
9875265250 6773950851 3040805388 : 3696
4450609260 1096436881 0894202685 6159073899 8509908250 1549463918 2209700648
4553663668 9968589228 6382159711 : 3697
0767179767 8975011542 8087230391 2887886996 1867893071 9828675499 1974326270
9341000268 5103259296 3268818524 : 3698
1489579124 4327648721 8014529821 8574977142 9294366689 3783643822 2965859111
4179405184 9420735706 4528325988 : 3699
4591225939 4927444557 1466776515 1145692903 1395253758 0463375857 9641581163
6218722766 1518982412 2053513224 : 3700
2248825616 7742459676 5412540331 7844988384 1113760735 0077200856 9727762648
7365278338 9163614249 6421063088 : 3701
3493089033 2084877942 8644049722 2682861859 9466893437 9485133485 5905828206
9646123946 4975741223 6283139773 : 3702
3819087455 8033167517 2722386647 3603363229 4221653127 7635847730 6315342973
7277492314 9739427339 1681398751 : 3703
7165017810 3261317455 0637765770 9788695976 7574221493 7163528845 6874197560
6953600393 3733202065 1649369708 : 3704
1492295834 3311631037 2740909866 2574705051 4702272309 6296228767 6893693773
8712390204 6516729559 2539637570 : 3705
4229682701 3721597849 6180362348 0433790639 7035855824 2210882352 9872394968
8535770504 5182895099 1937634747 : 3706
3203519382 8111442092 5467393367 8229258496 5322008001 5263180291 4086468882
4845131277 3062679808 1559547648 : 3707
2850417271 1392640315 5385040583 5488218720 6324982256 8308378114 4749672788
8333200287 1767003436 9690112633 : 3708
3771406814 9248560992 2237095518 5473844571 7905417687 3927917178 5165931986
8288756418 1867768768 4101914918 : 3709
0793996112 8907075158 8545524699 4765082176 7567721164 6636427681 9985224100
9652130504 6508340727 5744849661 : 3710
9761614608 4059060741 0221444976 2236679944 8483777754 6120077234 9048346804
7995536374 9200222871 1271851356 : 3711
0661682291 2323921800 5582781418 5815990440 6240526254 6367710924 4383173398
4763285147 0653500079 7831849158 : 3712
2401175956 5900535027 0640384087 0712334125 4904300589 2552683369 1094710188
5680630974 8638940766 0730095765 : 3713
9146215650 5019760734 4889921706 7459599010 0871084361 8344332778 7653682587
7230822334 4345432927 5264503508 : 3714
2751036885 2787696779 2475290543 5163484717 7830600434 2306802510 4169270984
0429423261 5492831376 1218792561 : 3715
5980554961 7993488223 4043222837 6565318861 2575026407 8045152954 1971941954
7797740615 5462426767 1433647051 : 3716
0868155403 2166445251 6231071226 0123089371 0475506825 0860324046 3908671443
6907732441 3439849284 1467882847 : 3717
1479332996 2152726567 6834010400 3538027202 5275418435 2253722834 4897065811
5608571059 9336667745 1985660472 : 3718
2008416019 0860000788 0595268182 6691313841 7341819905 5901058235 9290201354
0514610372 9259840892 4440579962 : 3719
9261209829 8196317512 1453533210 3905565185 4357550155 0633023039 3983217424
1638876581 4182837549 5738248135 : 3720
7953537641 5648600199 1020976353 3920754849 3483886928 1611240904 2860516889
7241562593 7579583189 8950072924 : 3721
5960467988 5544190931 1233298527 1844623356 7773599333 4804074703 7205328034
7069763700 2715728202 3073057696 : 3722
1414871966 4204255727 5049195036 6323004651 3954181407 2510893070 9443978312
1107332766 8759280377 6507172513 : 3723

6087557193 7648563674 4855888825 7887661339 1844930211 9188229270 9019500146
8423436369 9275319954 6115671386 : 3724
8570450744 1432163390 4078029992 3490581446 5652044962 3351543572 0105541820
6044026298 9131818118 3565214801 : 3725
3865548465 0393917442 2768998320 9599473453 2614851533 8090881844 0673726935
7842931940 8508180054 1112501203 : 3726
4574453033 3139738044 6948852770 9899805696 0005191411 4142279477 6296903922
4152446159 8136812557 5139849403 : 3727
0158908912 8390396906 8237159676 2282654375 5905616030 0204572745 6152794596
5191825007 1334871565 4510175277 : 3728
2344850870 0166716220 8813471455 9577331251 8768056485 2910540739 2802221120
0328627810 5640243980 0380701905 : 3729
6977396529 7877909880 0316200668 9840372155 1962433429 7992810095 6086160201
4910881911 5097937929 0036944991 : 3730
2062176537 7482371955 3245806361 5013009808 6504424252 9347835698 5149428846
3384019650 8628261926 8181813314 : 3731
5352271022 1617868856 3979618042 5572709401 9690419527 8221897237 1157604600
1639909088 8725718251 0468842834 : 3732
6189131812 0296964133 4893176005 4804610495 0927989751 7631114740 2264219675
7645198488 7245324003 2533025182 : 3733
4830774912 3252566884 7561695423 9348897784 6191575794 4227708558 2057854100
6163487982 6139855509 1924933763 : 3734
4419693864 7024154343 2823282768 4568414269 9438372354 0254093937 3073804634
3438282071 9655314032 6715847016 : 3735
9414833139 1499674109 0812530998 4316312073 1085050169 0632931910 1821105395
5855670398 2419629169 5207657199 : 3736
9272307176 3730813642 3894014823 2626547055 0903841961 6639574583 6874762673
5252558119 9071918362 3735843687 : 3737
1834305909 2217995449 9692223300 3428708226 9330565446 7683677675 5877960837
5718328106 8255595685 4316804574 : 3738
7689684479 2012444397 4874700573 7572457408 7492178275 6424733258 5933827183
6011855063 7271168238 1424624589 : 3739
0459076029 6921428081 8057785601 8865526909 9927922154 7127089592 4017947650
7844514414 6455171727 2454847694 : 3740
1607662478 3260657354 1389446198 8583756749 8476780536 9839295763 2660226472
3992041365 2162373636 1466640323 : 3741
1551854151 5481118581 0844339855 1504367348 0709801630 6252475101 9715046695
4597491414 1011308661 6816368604 : 3742
2103388074 5651994932 4586116048 7111648627 9983802481 2539418990 1363772731
1133834266 7785325923 2453334485 : 3743
3759664716 2083385374 1226355530 8937443901 9515415756 7994245323 8033408307
2168419947 8996812426 8884616597 : 3744
0608333973 3717714436 4986798767 1879725126 1506319728 8554063191 5126103864
9620313740 0515441345 7210449044 : 3745
6705194515 6922739366 7324976857 3916031054 3114816892 2251592576 6878095346
2048818191 3512012841 6258147210 : 3746
2780965979 9919441603 4438418232 6231448081 6732189123 7994659746 9447371994
7344754614 4729069858 8542010162 : 3747
5230641968 0597237228 4233732451 1406540283 7552630397 0784269804 5973210194
9345700252 5055959469 8145422107 : 3748
0283690676 5111338008 2719704304 5192478092 5117856272 9103526279 1697425806
8402627307 5841902146 2844091789 : 3749
8442561629 8673909100 5370029788 5506985823 1909288554 4099345771 7507878492
5479171378 5432263146 5566615358 : 3750
7021169160 4317209842 6523231396 6065488930 8538301968 3401924136 8671420697
2763367197 5814778723 6143301726 : 3751

1255520558 3002475666 5571711557 6955972073 1113661647 6492124100 0743253672
1117181902 6986473496 5813010136 : 3752
7113732229 3076821469 8233095862 6017552721 6725842477 5944321583 4483251667
7179538137 4496444065 6738138326 : 3753
6457517449 0574800465 0568402111 8589827064 6002549842 2293207274 9822069747
6980622608 2660110961 8485569352 : 3754
1180662292 9940380157 2610103842 8296088987 4606563046 7985022929 9020929193
2461771606 1603848014 2250886912 : 3755
3734248586 4417312671 8474663451 2149080855 3212758118 9474825178 5940175844
7229142593 8199664790 3390875103 : 3756
6666615416 4686547007 2022022545 9753480984 2350136484 8574947885 5474592133
7509637678 2165427914 1354746700 : 3757
6281276744 2222991188 2799813572 6907378546 9091101556 8104297173 0577884247
6385374026 7974152370 9462463943 : 3758
4605121639 2451482105 0797603268 5986609400 8732341564 2353566766 6375312276
3480101077 0454890510 3240579565 : 3759
9667413083 1157927974 8096695434 7260724741 0692015339 2090933458 2777447339
6165009357 5247112417 0750635032 : 3760
0886771741 2193495146 6676387126 3737284611 6040877063 0158376001 1153364921
2190203318 0685003246 6637922173 : 3761
7979268046 6363761955 8362047195 7455881725 5112000804 1991146136 3395552011
3003942391 5975356674 1136702232 : 3762
5519019341 7645573146 9482202225 2295483332 9774556051 3730774251 7709774446
5980760759 4546606593 1031273487 : 3763
1555326438 9083383702 7738220743 1456894458 6437175412 1066049337 7454249047
0014903959 4657459437 7123789728 : 3764
0058429396 2105526750 3956632149 2376729814 0663500612 0236079359 5072500261
1882298817 0244093230 6066540097 : 3765
2298243353 7652437799 4159185149 3390417225 0146997425 3353780504 3621410935
7723549008 8186907390 2695140166 : 3766
9721483626 1672332385 1117638972 7375572281 1689919980 3995729374 6032217281
9284534093 3258441169 6304062383 : 3767
0035426887 4290211824 2985645124 5795207347 9846273461 9510455242 5613736299
4330149839 3729111087 0529969519 : 3768
1835336505 3484424284 8046310476 5220834208 5936517309 6584496130 2858336548
1954359818 6756220294 1613843211 : 3769
6325768598 2562967201 8289067811 2418686878 4597313387 1404187297 0071191986
7701293106 2930969498 9935481392 : 3770
1875928804 8398451387 4075864302 7656157147 3351835440 7350625531 2343664960
0531091488 1303238446 2652043513 : 3771
6290262659 3330681205 1158885104 3585699456 4993886957 0706111923 5727571102
9419692899 4187655436 8842569280 : 3772
6386143513 0310638444 3343133961 9697973356 1523242666 4170663902 4036236559
3574944251 0383568249 3854306636 : 3773
9565628381 3497578319 0902427704 9909445141 1112412302 0604342232 8892597491
4382315759 0798442351 7784541128 : 3774
7242594935 3333098571 0978755386 8398175482 5212175181 0308218176 5051555634
5242994845 3904577562 6744657855 : 3775
1759148916 2251178070 5328811704 5974059759 8645830080 0561032233 2538430075
0981134310 1204508331 9042286546 : 3776
8587799440 1229656165 6389081595 9223940343 9222600101 9722176573 6217116359
9025757529 0003386602 2749259421 : 3777
9355961294 7551851369 9393249692 7866350652 3882676894 1167638908 9823146190
9596833861 2311858671 4957572705 : 3778
7568639974 5427113036 5918183847 1780818084 1752010065 5662130476 3267524946
6820599746 3936880649 9907641342 : 3779

5808930820 4090236323 8351783063 2217617206 4336320110 4034409940 9158405146
8783262604 7756804071 2615484260 : 3780
3468338802 6809404479 3089193734 2390363864 4025492448 1157390763 1680266467
9967901755 6718706413 6332402887 : 3781
0508745716 5871395916 4292536144 0259784029 0871343774 4417589566 5581130737
6296889347 5271737113 1377900031 : 3782
0082729387 1224879869 1412428008 4502725122 5463527219 9201876242 3508027844
7967843370 2368073614 3992859048 : 3783
1111125419 3110135095 9312727661 1248889546 8919453556 3621879329 9748845867
8348144862 6980470399 0091628952 : 3784
8836891137 4616731561 7024100451 5775700160 3778370339 5725392613 5405325011
4657419224 8012182431 6985303536 : 3785
3945988575 0411326222 4005410970 5194916292 5743792819 1687620492 6756847774
8603451383 9694679909 4353055861 : 3786
5004279444 8311206309 0593443247 0009375367 1005604492 6556034727 4778079078
3639504518 7624483926 9798731353 : 3787
5284348463 8881332304 3979277480 2201529619 2355486000 4734273095 0786780625
3076736433 3228241961 3242705655 : 3788
7794522660 6569501618 9262562694 8238005695 0336491504 7116242857 9689682908
9690583637 5827828389 2895520363 : 3789
2239309604 2046228781 0637658111 7827427625 3234050556 4585343287 3936828810
5558230201 5544340256 6605611991 : 3790
6083312452 7637674938 3219523349 5631786258 4221341665 7482628877 4471793387
0329083625 0399594515 9111325505 : 3791
0506004055 1933106785 4022140139 2893077471 0317833410 5594829183 7130769245
6979522064 1402279584 6705868857 : 3792
7578468725 2894700750 8518500453 0687570783 9746663845 0736019535 7370737853
5670036258 5582066737 2433386238 : 3793
3506635732 3527255029 8747371571 0495782761 9393563501 6759283674 2308538385
7351276128 8261732009 4878087312 : 3794
9781099401 3720875321 8796217675 0723431042 0409830054 3075454308 9347560999
9670530062 6008939587 5398030047 : 3795
8168171271 9509393419 1068331792 4294515954 3851121090 9172267321 8321181622
7957059544 7822536126 8295095486 : 3796
2217231236 3166507257 2186345317 4107239765 0382647261 2065862723 3900281950
9088574096 0057687874 5913537314 : 3797
6151589768 1028944673 4608601099 7367803156 1291557189 2379694137 8351231627
4075169456 9490868054 4178759792 : 3798
0036554692 6344461817 7373227167 0034442667 7604609492 4812058797 0646584973
5288079241 5315639324 3441257891 : 3799
7793572590 1786375825 6938063425 0443054588 2795813741 0756022875 8444781096
5427651767 0622237069 7241979180 : 3800
2152291454 8339562562 2846078386 6292668106 0526376677 3584382487 3378258642
8850121233 2924720709 6075960682 : 3801
7560580289 3569806800 9042477941 4480522461 4080192982 7445359142 6130067315
3997422039 8080445375 0119539726 : 3802
1944848449 5199114028 1906600326 2802697095 4487165779 2318799155 2650311232
3553639686 6845302430 1596197349 : 3803
2262352274 3230536042 2337560641 7777316238 5607180504 0559182350 8941421612
6043893732 0034975326 3396931768 : 3804
3963974083 4087614896 6053467650 0201801809 5017901524 7573815905 1446684042
3320462519 5859647960 2895315590 : 3805
7754720418 7516844221 0488273979 5248273749 6742588221 2290841842 6732735405
0629747395 6251860978 4580701322 : 3806
7661151733 2515928012 5006610307 8456552279 3390661195 5467698467 0693114454
3416538587 2999911772 5534075836 : 3807

2673072059 8192318641 5819683344 0775617358 7601158541 0628859025 1485970663
0256627278 1881934320 4435440594 : 3808
2641747287 4446384970 8875389243 6548269542 5677051955 0050304857 3967192593
6918313224 9952382903 9519078750 : 3809
0747741928 8341283393 1514360545 6554992740 0340005147 5286011764 9136881975
4058351813 0106428191 8542772397 : 3810
9888911556 5442061329 5214834039 7465595374 6931767971 2605932622 7357889369
4395281403 7155498125 6229183602 : 3811
8970988448 3519995909 2724761429 2738476113 4273942707 6465965860 9967512305
0324376092 5283745354 4759855871 : 3812
9279745135 4560409284 4631238388 9192976387 2245094869 4664331645 8586117087
8840725956 6344648877 2838981448 : 3813
0697593346 3489209208 4747933664 1147691694 8304375959 9830298448 5104669711
6761180315 7567043074 8683191350 : 3814
1510866268 1481108066 2676444881 8763734119 1046301486 4339513331 6943864867
1912530511 9771222260 5129272066 : 3815
2888283651 4602219447 0819695463 1978248180 6230642959 1934380319 1070756728
9113034166 6939068376 6514373340 : 3816
0982917691 7783350235 1137723780 7008013449 3751248050 5007321973 8123851625
0632081049 6669536517 3928688581 : 3817
3977490771 9078245297 9419561688 6972231675 2104506671 3791920865 3893674998
6314866410 1700444576 2995894532 : 3818
1955986639 5809179418 5105600868 6137283520 5544642033 2754284596 0428433290
4254398613 2359098896 1610491104 : 3819
0370538129 5507746816 3875576813 1629919025 0815702719 0764907757 8436021697
7227904172 8321524831 1154673779 : 3820
8675348633 0097098407 1885291772 9363837896 8670454493 8022017157 4243059909
2277663302 4297441704 5007458994 : 3821
4751500765 7427829025 9389637763 4935315947 8320022542 4326725184 2300506882
0208254972 1292140563 3725581581 : 3822
1271047415 7018538931 7829398787 7127982748 8656715046 3224664385 5553484887
8687664135 5612761048 5824205891 : 3823
8651972343 5336395723 6099480816 8173974031 1903093045 1000439618 0691234477
0855514924 7296678508 9521986109 : 3824
0165524898 3577004961 9203738616 5583736313 2652345982 4531995105 9160583790
0039799695 1777069795 2469522983 : 3825
6760743149 9008548564 1433272200 3498903384 0428531242 1123402100 4981624291
8589242224 3495753608 0826030676 : 3826
2401546957 8806472961 9266845275 1116009590 5378548316 3671245484 8677882017
2520546398 5586500358 2350020215 : 3827
0851642366 3500787760 7151985614 5256264951 5910555074 7727853798 1540632716
8195111754 1022889298 3782225453 : 3828
3675665190 2512168230 1691819839 4259899975 9411769587 5215092794 6376619633
6087192509 4446859022 1362185710 : 3829
4072204996 6386358712 9905305552 7028410467 7262021274 4667454521 5457723708
2364445335 7064522514 9756787051 : 3830
7194800500 8025678246 0781818548 7583892258 8816417620 4903611290 4312816748
3207669348 5228577606 7934846458 : 3831
6576690952 0661008787 9064301681 6753391621 1520568108 4467894159 1233741463
6821138833 7577326140 2746246664 : 3832
5150989210 4406381808 3056080401 3701008907 4171937662 9584442169 1227908932
8319826772 3332127920 0768091661 : 3833
0268387238 4324544206 8679756039 4826736582 0673451550 4267630881 8158558039
3191617327 5442955764 3651620151 : 3834
6644668557 5937259409 7115187836 9217329993 8415788276 6022670925 2914746182
3418663672 1931122698 4236086854 : 3835

5204188312 8019978033 4878756588 2710587023 7096130274 7991323482 2581376961
2170462274 2249610167 3375478801 : 3836
0634085514 8662286995 6915271626 3350178980 3398273644 8774282315 3814829001
3432567888 3228760716 5110115958 : 3837
7187145010 5783476297 5288743392 7458182768 6000452775 7177247880 2098965643
8529423881 7387999187 4712975874 : 3838
1165413323 9155651089 0877525768 6286616807 1921090343 2650301463 5427492044
7493124652 0147323197 5770361204 : 3839
5011747939 8628215253 5497202671 7908950519 5085909795 6215389121 8056487899
6785730688 4996390327 7813229401 : 3840
5207305052 0209607654 6883252517 5264101355 0651450298 0213435336 5875559161
6836749428 9441214804 8659982256 : 3841
1119879720 8669302128 4995016647 9523817064 9564273515 5258468462 8920014188
1385743510 6708068039 3400649398 : 3842
0278204084 5015597431 1129777968 0235822043 9971891827 4081952024 5230161759
9479708280 5860032889 2886741062 : 3843
4713286909 4694668439 0420120577 4106699462 1572385110 9156923759 2785586844
9397736068 8001922934 9147175801 : 3844
6316525474 7272854041 1178250031 5404941649 5321147450 7549665480 2310185518
4474204933 6821498678 6738789249 : 3845
5751470690 2466239047 3966302006 6157810935 7920463627 7135039469 7843435917
4142643778 0311902195 4752269208 : 3846
9547968878 2548564570 7135566698 5023275461 2398823027 8689013175 7321917167
8493016365 0138955291 3159011742 : 3847
2489607374 9460202948 5127985592 4507737935 6908133386 9747037415 7396558258
8573741349 0358278143 4074625542 : 3848
3551643968 9046077958 3866499664 9445074756 5796993514 9351359732 0013303372
0811662891 6765098694 8072944032 : 3849
9917612907 2301114046 5891138242 7272632968 5176094284 1879398026 0523956540
6645933007 8965741576 6970867192 : 3850
0929452539 3155274373 5282800482 3457509144 8533542404 9182058302 1173669376
0497384227 2159134835 5204042576 : 3851
1699280642 8547140193 6855614110 2765828085 7040342325 8758656946 5009202308
4983082678 4327277486 9277955040 : 3852
6611100435 9764768678 6538495502 7225588542 2498613657 3631739614 8099893608
4740971764 9547154334 0623707326 : 3853
5869752764 5921337429 7141242070 5112256449 4990791441 9244483283 9379138042
0636238002 9700282062 3931807561 : 3854
9226485397 3949330832 5197087313 8458119857 0171002491 3652522719 8684083184
0478469364 2902511129 2751766993 : 3855
9886203438 7891751726 7572217442 1453857611 0028348743 2190524333 2939928094
8839061529 4064110187 9375394113 : 3856
4303171916 8526355316 7352985363 9867761635 0661196822 4723486403 4096810385
8504606729 6014761310 7402400742 : 3857
6534024368 0379210684 4615341970 8080881220 7680711064 3890588634 2157852825
8228774367 5470140831 0031138429 : 3858
8534935282 4935850403 6234592392 6688311448 0775467105 9106940654 7237346587
7894192487 3236437024 0202401751 : 3859
7597046829 5402550796 7730984443 1798635413 4645953902 2765743203 5188431115
5975463523 3045235505 5292850636 : 3860
3115240811 7678464547 1413279813 4065927467 3208354528 2276634305 9530547343
0938175037 7872317646 4890964910 : 3861
4425587969 4041170527 7662970232 5695246367 1775845878 8220284605 3053054661
8172209655 0981839675 4440119356 : 3862
7027827210 8335560644 6657522809 8058628703 3307578738 2108211292 1021038961
6523122848 7920328037 6653872019 : 3863

5820518561 8250144240 9628366098 4633225284 1101574173 6148823414 6028737749
8491498459 0462988903 2849256083 : 3864
9573991405 9102412068 4557624803 4590472359 5368839274 6190063530 0602528684
4345466005 7857456872 5002887974 : 3865
0781269950 1435969649 4764149318 5049625552 1694445285 5745294807 2324330277
3655574561 0429660883 5697425307 : 3866
3611130310 8238000315 8538736286 8849210917 4918033905 4828505462 7864100013
1694719989 4889420953 0864465026 : 3867
7961007150 4843926206 3311711860 8612081871 5732734551 3700141867 6869219670
4885901036 2423315167 7601727174 : 3868
2233050495 3061585247 9196288967 5944421346 2356519312 2194284407 1430978955
0559858862 4420173628 2607350511 : 3869
6384635711 1878468347 1588349062 3342588343 6142747938 2954834126 3417416882
6028887355 4781665323 8321540768 : 3870
3590968120 5488454817 1892692062 0367096536 0564089875 9284508157 4274786219
4136716994 3229458693 3049535372 : 3871
2967997800 4871910203 4782473949 1797171976 7262538849 5281730513 6008922705
2430855648 7450412160 3355151493 : 3872
1039641863 8089265951 9034350197 9503662299 3710706109 8032834728 6934541513
5967181506 2976637443 0994932369 : 3873
5704953495 9141935272 7151418737 6586945122 3636710793 8011524216 6244941198
0044655721 2597535804 6399910239 : 3874
3279770838 7184753479 9854360306 6177468548 1674302535 0911414225 3635633838
8354354468 6001108638 4802147927 : 3875
6315847935 1071062272 2123289000 3040085551 0346076806 6519552168 6567189181
3348506806 9377003906 8701376254 : 3876
0552466849 7671014459 9349618422 4689804170 1860235101 9649901337 7865083234
3686944378 2163920218 8508599658 : 3877
1311126588 9482917734 8973611017 4739245996 3978823770 8464359325 6464055446
6354564506 2781650789 1000060357 : 3878
8958471411 7881140014 7045565792 4684793389 8590443560 9519395598 3048411907
0859683908 2696964630 1033129195 : 3879
4054835663 3470754825 5971622409 5724560789 9522751374 3172920432 9442145717
5702111966 4927329504 5892435115 : 3880
4224989284 1829309680 1627909990 1439100065 1606646547 6572330384 6462439511
3830677701 2416199103 0939928940 : 3881
3981160616 7420760338 2917593065 6604400871 6224229486 4120927680 0548402635
4604602128 6431296861 7604867685 : 3882
4587332459 0490445815 4468976001 2917675937 1782877815 7463447497 4431747224
5714178249 5933265895 9901232631 : 3883
6543180978 5578257753 5504957177 9819902461 2577728494 4635352521 7033433931
3436451383 0425898151 4773623679 : 3884
1994181688 3285991871 5721722876 5199470314 2482487875 9641163100 8191276660
2885664731 0961164666 7671189367 : 3885
6347212909 6300869239 6234207088 6293386312 5561287340 6534679097 4199382881
6638506027 8732800485 0544918201 : 3886
9933787113 0829493390 2544408454 4528311079 0671755781 5800938675 3600386035
5768063981 0642357510 9973318402 : 3887
8068507709 9666201399 3830215705 7490439491 1543830905 6158300113 0141099139
5832903258 6958376761 9707448919 : 3888
1132631216 4098184347 2560087662 1115969513 3389046258 9050402103 9716887529
7634115653 0163761401 7241741275 : 3889
4156775733 8515633409 6892443014 0617974975 3717442566 4734934545 7924696799
3011026194 3376364486 6753756108 : 3890
7691613043 6137483304 7944122595 2612415148 9816998076 8101848189 5780038325
7633878043 5311856971 9951740840 : 3891

4042921573 2071789387 7089593155 7022742728 2658161497 4080133206 3433840086
8902388939 1833314200 4515961400 : 3892
8986112156 2924364169 4725431356 1701811792 0055959285 4391833343 6945191945
1431715413 5007420063 9050607167 : 3893
6581905824 2162028976 9684409055 2454470012 8188685814 3549545420 5566898387
8624483207 5212174109 4873647410 : 3894
9157860409 9102585558 6394013091 5517560027 9765865384 0756823639 7358347233
6482553352 0524772296 0881930540 : 3895
8210532813 1130488092 6255044062 3955759039 1944317141 5265275408 0895298657
2630716340 7322439503 7998714881 : 3896
5126244606 3128138524 1074143597 9408576657 8525169438 9927465562 1805886920
6623250449 3702921288 5127459270 : 3897
2137674834 2927777412 7189221554 0626833008 2860090179 2187538294 8629150162
4540724777 8866998890 0709805056 : 3898
9250698581 8373901350 7639436541 8723293031 1836581686 3336447793 2900707857
4510611399 1480911099 7127929438 : 3899
4939136701 5842418969 1093862380 2725764521 6678140223 6132290468 0977248778
2686418812 2877993432 9767071638 : 3900
7056951199 0189632309 2458999902 1797571313 4517425936 0325136902 0883958608
5467504777 3229290628 4648172031 : 3901
9507281491 1800959505 1258134475 7157055467 8484436237 7691481917 9984052906
6771547929 1174266933 4885856708 : 3902
8139473448 4862143811 1159547544 9852804057 4225069514 6472008424 6244722993
2157922646 7217881452 0648208302 : 3903
7693611921 7385236785 6740962791 0852342295 6090632666 7104802881 9226371390
9073768444 8093184449 8969929310 : 3904
6036870095 3087272410 2621367887 0697269856 9734002280 3635047552 3688004955
8697359039 0206919312 8307071061 : 3905
7913556767 4215877257 5982219129 4318645745 4772051328 9458782737 5775210307
1194255451 7528858363 3445935800 : 3906
5524008931 2099188257 0655486639 2314480862 3744145953 0559003065 8095292440
4261767525 4206771033 5536849552 : 3907
4654643577 0101789796 4163130386 7495239610 8289032577 2448991862 2996519842
3278274995 9589340831 0312932676 : 3908
6102724108 0737954628 1880763426 9861761703 3100934680 0830651834 7713287054
2549962326 0325011619 2827302129 : 3909
9349341397 9963426151 2648506347 3483275913 6317868247 2894449321 4660618479
6505730406 7187482841 5191474168 : 3910
4169089722 9561606700 7281977661 0831141061 8692743573 3553564627 2541437940
9813463832 2862433641 3347897889 : 3911
0912641531 8730406053 1724773810 3537082669 4106544406 2266129376 6233632590
6138381976 3194013039 8985582094 : 3912
8209539186 0248199098 9664862713 5664987570 1497953400 6764478586 8204910762
7096977570 5491508796 1976919441 : 3913
3901594125 6832679426 3957901827 8254556896 9968032231 7815897782 2565761387
1677013214 8021155272 7673111734 : 3914
1653118922 0384514736 9902316477 6475938806 0160755374 8338232854 5825950629
1023936001 6544735394 2982876682 : 3915
0737121927 5299907369 7440811342 3720201314 2340450525 9316972775 5905856570
0313202246 5479086657 5834512665 : 3916
4568263055 4619314754 7187391279 2397211846 9217783762 0063262466 8145632249
6877656028 0104550069 6828152865 : 3917
7542598234 8340092647 1012832843 0643025697 6526446941 3502706214 5831911049
1793194143 4930177034 8702633346 : 3918
0168710291 8360837412 5327587763 1540223039 5628877684 4723280302 1429872949
4647493950 3146491804 1335142023 : 3919

9388152487 5937371592 8383266003 5406428953 5416985061 9922353038 2921861048
6918255269 0255749889 3197784720 : 3920
6199198050 1797226224 7637181445 9796413713 8462079820 9083958821 1870480155
6318520813 4305168523 7415842174 : 3921
7814580115 6305831176 7877089770 9425691536 0787809113 6284354824 0228283945
6580419522 0130311977 2279659598 : 3922
3884093635 8355901185 7427044826 6505634384 2162536049 8340854389 0402854304
9268993066 1353029562 4400202826 : 3923
7343672871 9262079402 9735061796 9254101291 7537626608 2539682218 0621641812
4621731189 6733474309 4220887670 : 3924
6063054671 6996314182 1559029243 5137843643 5063226209 3120680143 1691638497
8101227107 1502167924 7205626346 : 3925
8706739088 7567539442 4482708382 5087882653 5655819744 1663577849 2417631881
4836216441 2222323635 4993894299 : 3926
0784092165 0996112353 2512011031 4233835506 5493889092 7596110536 9398083023
8610371304 7566762605 5583092784 : 3927
9435785197 8563905347 4326745056 0490940770 8591886510 6554530081 9826445252
8174087746 5478991615 1163487539 : 3928
3067419302 3342758365 0496415686 3538834492 9702698830 2128385075 0117307225
3194741218 8603060660 8665654771 : 3929
4244902618 1091509558 3074276082 9485811035 2011070330 3058709257 9512353389
2215724742 4785671676 7883589737 : 3930
6761772117 5130586934 3355367649 0367437388 1704440543 6987385939 2664944715
3503815259 6325055300 2049067646 : 3931
2445852260 5213931219 6299705782 5503270457 7980448589 5607020990 2153446695
3706842845 2178214452 3866710469 : 3932
4081075061 6731747856 9718979427 8788716052 8345042902 2122439834 3390694767
6464131458 1807048184 2165818603 : 3933
6693764098 9433649381 4267999197 1798152335 1084157064 7616502760 4538632999
5224430338 9384486986 9487964925 : 3934
4150996947 6646546897 6921648614 2392356827 5318509654 0881335332 3603150184
5430918115 8979530096 0882380182 : 3935
2954836466 5073083461 5875373909 4675311255 4261080409 6592752714 5627225527
3501217657 8243220128 2217368454 : 3936
0188340911 2563696058 6741735441 8830302910 4229540931 2139163251 6652716281
2140200393 4852462926 7702533556 : 3937
7801321211 0007468751 0492729215 0677332824 6340543305 1464110169 4580152392
8106487165 1193748496 2847145205 : 3938
9228602216 4309227073 2911314850 5514626087 6549103696 8970857811 2513994774
8938328842 6598056121 2183161153 : 3939
0304191551 8697722069 6407051706 5583074542 9556802055 8606479198 5611372083
2021397337 1589718010 0934195299 : 3940
2408631804 1862299694 4159001484 4534898620 6796450530 7736118399 1361144007
3059304205 7496786395 3731729127 : 3941
7835848215 6291480030 8429891763 6403425821 9775859099 8861542291 5064881854
8939194649 2442442707 3460653662 : 3942
8241104380 6874634642 4452489217 2700800269 8896840097 7049906879 0618994349
0599879123 2993931686 5429778734 : 3943
0536374108 6041168212 2328032144 3018450796 6819142023 7115246394 8220177093
1427298845 5433627298 6473983606 : 3944
7619735520 6465441486 0620658273 1163150571 7724208876 5562487222 6829266602
0371485112 1462441496 4999515202 : 3945
9735696346 8957934911 8486138732 3567372723 6286722956 9307022272 8261893624
2091446834 3426468002 6607719950 : 3946
2216509365 1912429912 9890949673 4886916099 7557084153 8563266979 9928723106
6968120038 7350435700 1390292512 : 3947

4556125259 0033221973 0742628557 7090519268 2914207816 0159206846 8518949602
6803241645 0232179784 5493500368 : 3948
6853112007 4399878675 3531880447 3478513954 3177396060 0553611144 0635364216
1812602126 3865730946 9059325590 : 3949
6397219492 0380562876 7882874309 1909615406 8691521231 8936278924 9870124538
8290743083 8160748627 3896787745 : 3950
3782363709 4678891160 3417540455 0891697540 5922739404 9266623912 0406468558
1911120898 7613022144 4569725686 : 3951
2545295013 0411217199 8091066911 5415746874 8584959986 6199083983 7817126331
0681533641 3320408361 6858950592 : 3952
5150168276 8475240163 4734415535 7829189499 2138591091 1224831818 7172194868
8386571304 0482947774 2444060109 : 3953
9690156383 5237827612 9095967481 5107225829 1813299828 7427054987 3596774842
2085620675 2067159942 1809118851 : 3954
0214643881 5360796234 5415092134 0343006129 9786868257 9602938497 6591871270
6805152707 1418619605 7390318885 : 3955
1282433872 6837664102 0767175712 6610927458 2233522905 1299485282 8161359254
6990669160 1850275940 1681624498 : 3956
2303860880 1382424988 3069963917 6230939613 4854599451 7800671078 2255193242
0508739012 6795497844 0156989887 : 3957
5079123165 2774721289 4791531731 2845796906 1288370019 7518951091 6690850679
8567605165 8730747998 4144568492 : 3958
4921279790 2881242410 0884088477 8373159193 6132045482 6303567474 9277565413
8559987303 2159054076 2951436378 : 3959
8522298669 8185149670 8348605274 4652644981 7637532110 2923062590 3686358567
9948914924 8808960412 6511073389 : 3960
0966344359 3384412495 8326982682 4276010209 8965945460 4773652398 7118017518
6513673393 3944944267 6775423211 : 3961
9108230942 3728968680 2203474395 9324862376 9437410866 6028201747 6556319495
1087225889 2022152498 0324583927 : 3962
1724279586 6773783806 5801661372 9297771184 4665025011 5258124073 0709630180
4069407496 5568499308 7263042183 : 3963
0015704120 1379224567 8731940258 2131094546 1093699749 2226185837 4651973232
2036812782 1679785482 5403853579 : 3964
9586852096 7119563235 8035567447 0587232686 2792820603 6487179366 6055944117
5390180898 3779028084 3442247968 : 3965
7088565952 7161278836 3404327608 0005804509 4816137624 1757342033 9477986303
2236738284 2571940605 8354743783 : 3966
8904464154 0657909999 3185608124 3084635339 6799172288 1263788799 1600338337
8858845547 1982316793 8893362837 : 3967
7732064114 6617025954 2085088518 0512900843 1630715043 3107539545 5419907964
6509023164 4288344740 6871971893 : 3968
5466735649 4568123543 0496129888 4537609297 7089319942 1463048220 7660235398
2432157616 2548761284 2541210616 : 3969
1133133648 8224541324 4897545356 6265349142 4082249134 0206617500 7211269430
6124966341 3321878554 6802945912 : 3970
4917217464 1038186713 6215145716 4950732198 4896957276 8726883643 1105360442
7157154289 1366815712 7442833213 : 3971
3397633430 0050812739 5671874826 3504621039 3143243199 7549195809 0961365625
7002432016 6580709518 8730589919 : 3972
7870679368 2952440485 3420238652 7588450776 2927871463 4221930150 0563804572
5131759401 2285611355 4368542010 : 3973
8182371586 9013394106 1317832929 7920410991 3274541054 7130717453 3004723383
9565461871 6344570340 1855604718 : 3974
1381578089 8159470368 4942578682 1926363634 4774282218 6059642474 5794721701
5002581733 3302273204 7386573215 : 3975

3469489387 7232700555 6946006475 0620414873 3161156862 2852690741 6880934990
1746912760 6927198938 5055648400 : 3976
2433703265 5975293686 0238742696 0159164509 5650888549 0208984849 8876250095
0696676359 3812316977 9034511780 : 3977
5540667349 2287980647 6973399138 9207356808 6405247056 3221867382 4785071644
6114309809 5494880924 7373180071 : 3978
9660589269 6476743801 9702682241 0328865611 1365849274 8669631268 0737684133
6743444843 5491569475 8172853359 : 3979
4401006332 2752305783 9938753032 5507216780 0795954106 7981610512 8701235325
1890481593 5617217032 3986252992 : 3980
3914117857 1556013166 7449541431 3623225134 3090938117 1389724401 9512308210
3278926583 6285024029 5904940492 : 3981
9415327768 6423597978 3872458059 3951390567 9556048902 2429600734 3177928261
8475193395 5564593715 0468756199 : 3982
7579743169 0153572144 0668758086 8655452485 0048273571 3085317312 4413579232
0993581940 0847803553 0266617899 : 3983
9034001645 0047094910 1383340177 2294629926 5642334547 8810460564 7714030170
7861814478 1116162479 6255323924 : 3984
4068009690 1179092285 1352731479 5503645016 1301300382 8067884533 5633911351
7314102426 7843977071 2448559939 : 3985
2643077445 6327190903 2088393518 5236603305 2369976131 3173348345 2682577284
9605743916 7155395158 5295633300 : 3986
8194184226 5634997617 3206683909 9307221655 8028540545 8697118901 2219660240
6694608294 2278143215 4324365761 : 3987
0175020591 2960184367 1261833291 4534588474 9623927097 6581693940 4028638746
7768485985 9221382070 3486498305 : 3988
2250484391 6857753454 8195511374 9471895212 4618141523 9902153365 8511335253
5414111656 4403339910 1481716030 : 3989
5596064373 4286803239 1003140709 4111326232 6323998396 9953761922 7136973500
1483985838 8497144816 8151714974 : 3990
5907959017 7449277445 1113062728 4201316358 4376417920 9332924316 7440055461
2311429161 6263976070 6635603860 : 3991
0656083714 0438303004 1343474639 8954845945 1126970333 7758329055 3364122253
5927524973 4553483337 2325862570 : 3992
9633843342 0359664089 7285644317 8958761351 8100942754 5203740088 8010727884
8188959516 7231262118 9125001920 : 3993
8737338613 3650137343 4886404252 2520017704 6207266882 0070446561 8473896823
5564714710 7826826300 4521490432 : 3994
4105536355 4525591842 1354196865 1137397886 6630975008 1425643198 4740377591
4587895906 0984574260 7885786320 : 3995
2314402576 4605246372 4839202431 5470427111 9003189042 0403338170 0098642286
4341807477 7777983442 5559893089 : 3996
2290697457 0187202046 8182941675 2491348559 9606198009 8948447489 1662876041
9800602597 0012736569 3936297540 : 3997
9320859454 6675623408 0461501354 5821550863 2072266038 9340137673 0576253406
5551698152 7778559929 9882419464 : 3998
2665167687 7611917362 2270209227 8336052507 7048070759 0718034363 3570756382
8365968139 9539076072 7068181365 : 3999
6575919866 8375105461 1521808378 1191964755 4096709582 4956017828 2456727368
5631218502 0980470362 4641761986 : 4000
8271774847 8222463490 3278108854 6314151737 1814329792 8832562499 3711562971
5737390115 8363108704 4860251030 : 4001
0496946914 2583869370 6512037704 6630824216 4894433580 0059686873 0214852492
8795382422 8610007364 2036496791 : 4002
4869424254 7730644728 1042550872 9193419606 6705256450 6409608790 0244040642
4731141356 6099006514 6788809327 : 4003

9138493846 4806546101 7890562764 5635564452 6787973176 6008564598 5904575945
0452936327 3229140340 6240934385 : 4004
1631402526 0021020853 2500280314 1809837523 3896395830 7623736733 4254811893
4277189269 3033982841 2036495177 : 4005
1760100346 7519208158 3382936321 2820663131 0891456020 1482252304 5528829442
9174005143 8913118279 8098198484 : 4006
3229029838 6962825148 7394458203 9109406532 8018875407 7209490747 8611791577
0017190387 9128063762 3661744014 : 4007
4045207022 9245232045 4057628069 6579308502 0398121837 8402067202 5012026675
2955313083 4943534719 3634177273 : 4008
4063602625 7960313651 1978554856 6937284640 4204684892 7715778043 4586776100
8528960736 9314413346 4873773525 : 4009
0159245211 9765975459 0876950206 0561757819 3591077403 6258357653 6008089376
5328137084 3694390227 2298653222 : 4010
1828843740 0138825811 1629715534 5756740321 4986097554 2868865798 7436900949
7050979860 9377027835 7223388331 : 4011
4539804939 8921017143 3582618967 4003122527 9973033645 7106160728 4968264026
6823477045 5830154585 5748271713 : 4012
7243584709 9486137265 8713025494 0244957385 5889966053 5370903389 2511454055
5812456929 4137888271 6519900043 : 4013
7610796725 7280599874 8204798956 7855938858 4994834696 5194930897 8149972776
3473305857 0717902709 3568227576 : 4014
3063930497 0229663395 5287633799 1307858593 1420781133 5111432012 1026019873
0421670626 0143575841 1797707904 : 4015
5808380884 9808816662 6185358835 5924200630 5302464346 2899230820 3070806494
1073041567 5977100775 2398558686 : 4016
7594573174 4767094556 8426890385 3112849498 8018144774 5665050961 4898991517
6299241642 8780004741 3850804520 : 4017
3295305391 8409768994 6319969559 1278676949 3195927336 6205430918 1205566924
6215274078 6651432352 6592070708 : 4018
6787955864 1686045277 5357502074 8767143337 7060119129 4031585743 1076777779
5213590261 3080828983 2488394832 : 4019
0949988456 8307672417 5929943034 0209439932 2708275483 5738850741 9917136940
0498798586 1942344627 9608414447 : 4020
3566520379 2829531701 6335118153 0293127230 2543562910 5545863957 7778022116
5886661126 9335740729 4436145574 : 4021
9056372007 1282544811 3557834029 0160485176 0524329698 1355027471 4705263542
9352648136 6238869584 8981951679 : 4022
0476124747 4468008477 2588713945 5273671088 7847508425 6882598396 3683066764
7664513308 2342995384 0637149396 : 4023
5512602596 4126916639 5532942221 6277976078 7495529174 8568842182 4863746324
7477832449 2983235440 2571567607 : 4024
9286742595 2849433898 9676434365 7548230757 5478403350 3696537687 3654980223
9878011920 3544049128 8268359419 : 4025
5397184364 7255409053 1421055666 3207320463 8848382768 3792610550 0380573953
7940215136 4136624967 4935373241 : 4026
0440434862 3823362492 0495354428 5790530654 5277265072 2034659290 4432022017
1632423583 1378351252 1095764152 : 4027
7412446577 6261675436 0947097433 5640076904 1436221806 8299355151 0913855657
3734119489 0321845622 0443877152 : 4028
7004821101 2761208140 7824526498 8636103832 6508480852 5295149522 6355426460
6718445430 4265338266 6861006655 : 4029
7716951714 4295655590 5423681933 9387175320 3864115522 4288474087 9638726559
9650354531 6017872842 9959062489 : 4030
7569431465 7253297995 6564427538 1025956667 2558761130 3086354595 0868484208
1702309037 7601073137 1062342933 : 4031

7807454750 8237856054 9479876902 1390566558 5892860091 9904560260 3206378272
9076155397 0383110180 0844901121 : 4032
4811927779 6748391027 2882057559 7820535088 3461500219 0348376576 4631105684
0142504210 6378331650 9790934725 : 4033
9499426617 0452072326 9101718680 6893159895 0080623997 5869483897 0524161223
0171728940 3904669984 9427213392 : 4034
9568126161 0046509028 4562126757 3941439279 5031958650 2350481104 7168563578
3540426485 7212754026 3881287194 : 4035
6209203813 2546481161 7031358676 7106436587 6605516551 3311331702 2718232156
8773621958 4821685646 5284606970 : 4036
6619054395 4014065106 3097333651 3811963331 6594903039 2164270853 5422804979
8026714911 8956364251 7489134412 : 4037
1426361554 7808921452 8367082216 9402598711 2632114388 5299391696 3048048178
9296298820 1123807490 1305294249 : 4038
2948016114 3533023900 8067065721 3781679719 8568613029 0301299399 4451249846
9010019891 9360598279 1697305147 : 4039
5943464960 2883328969 6608150563 4505660937 8129236133 4905857805 5094564210
3530907360 1958446371 2165073198 : 4040
2015642422 0132684566 8774183233 1024731921 8685156434 1203271703 0573066078
5175385097 0691717079 1725285511 : 4041
7436278713 0160095220 8920242405 0305756402 1537273695 9266799747 8107072793
7239123557 7709346828 4756010763 : 4042
0127913119 9539176281 8615943038 2077839824 3261731966 3133362063 7934967687
5089524023 6424692319 0454167386 : 4043
2358360482 8374392788 6654775948 5902892040 2019395937 7065673211 9490991043
3528551798 7140350203 0760557820 : 4044
1914838828 8094649648 2084241766 9924567583 1226247807 0390557653 1412632602
4292243620 3719532918 5547180915 : 4045
9644318568 5205788235 0103091076 1280604457 0442514799 7589608880 2812599786
2387743549 6599049296 7322084497 : 4046
2443458243 5036897803 6518490995 1214229401 5669174534 1683830903 5284779643
0676086115 9976367872 0495505795 : 4047
6365166938 3452102120 5712467189 0236358379 0833911908 0206899596 8969901881
2232185525 2869348573 6518886301 : 4048
6045294102 8179736080 6895495240 3606648894 4683485357 3711706079 9430547192
1648759431 3141269759 5251661025 : 4049
2290957537 5509509337 1854490007 2907676126 3467652916 6464558037 1533060205
5347416205 5566838087 2331011456 : 4050
7060821971 3601991166 9601177265 3512414405 1093620360 1001758405 3344689875
6534900244 7580184990 2851129056 : 4051
0362815437 2796762883 1238165774 3751766245 6404578370 4964856909 0428184674
1434107660 7549841146 5742153343 : 4052
7962825237 7393517758 7703994255 2131816901 7399018616 4214135439 2779733470
8765973694 8171010331 8186376892 : 4053
7283763660 2301920591 9792959179 1482244163 9403180414 7790028285 7125177644
8410593156 4467536330 9241579702 : 4054
1262648130 4280838933 7706723982 2865434173 1736481424 5629661807 9313695325
0911287546 9498015503 1799451669 : 4055
1228413844 6463087410 2798782095 5877346176 6677933200 6361614129 9836112387
8526984496 7622494946 0162224198 : 4056
4818828441 7597250896 5043238838 8267762115 3869449072 2314080038 6409667479
5565960336 5865500834 5015746681 : 4057
0037154981 2154559177 0828552690 5878274626 8018954840 9854806477 6732259308
3364643266 6789519813 2303438478 : 4058
0554257118 9332448803 3710276608 0664261976 8000401457 6819261412 3421421090
8378826034 8803987158 9674691868 : 4059

1275950354 1904068967 2781395132 1988421183 2561094874 7352764866 4367133593
6837371907 1671361534 4289207252 : 4060
7305707780 5616065916 1544235891 0784646554 7369563439 7073722178 1859123010
9443692313 9522030101 1367407345 : 4061
7059526133 0293674379 3212040615 9970890681 2035078623 5412780541 6826582353
7425938569 6643576271 0973540865 : 4062
2303333957 4924977199 5346662569 4281212119 2667488866 5256315169 7066072400
2193962668 4282515447 5614963579 : 4063
3336584523 7724099687 3579532275 9190097974 1551721334 8453335786 8142287399
3851902093 6782740215 5999142045 : 4064
6446438381 6000999065 0537188148 4938160865 5035722706 4177438662 9751678966
6554999878 8957217902 6230908454 : 4065
4806465185 6930925569 6453172241 0894516454 2679676181 9728832958 4139351338
4459604167 2854573991 4150804959 : 4066
4466135343 9845014276 1805422096 5984867109 9440825081 5132392521 3606951062
6733736792 2332214259 9523022293 : 4067
6409047664 5961545055 9484204881 3114413172 0464692670 4975974905 9935116920
4390276051 5744667739 6870803247 : 4068
8040634377 7841672502 1988849435 4098282116 0007277291 5050759869 3656847220
1694104618 9444582618 5511600415 : 4069
4945106281 5887248514 0345190055 5634666152 4473749607 6611357787 4837400388
6293884886 1019502812 8078179274 : 4070
5034958405 7529284529 8389091576 4913247310 1056333147 8134640265 0462629156
7537790921 3724782897 0031963259 : 4071
6891251330 2152445612 0543583762 2686092820 3077741687 0045904352 6358174946
3672455178 9784931750 6753904640 : 4072
4160336384 7240546498 0750039300 2457661071 4660605719 4951091402 4823273526
6912214960 1607089722 0722054628 : 4073
8100387307 6229689062 1526297111 4289273463 3921437857 5838167995 7096512975
1212882470 7622937565 7213489062 : 4074
3618601418 9959500029 3934330117 4633003329 7290783402 6382527837 9605300004
7355927546 8487189299 7206561365 : 4075
3375153747 7921962495 5179692200 8557314794 4574288225 9242287677 7321288598
0653704654 0246199387 2964993594 : 4076
3563230213 1108482424 9501800675 7189398611 8972621824 3077831783 3445857036
1181609413 9763446516 2725658288 : 4077
6168782130 1342558907 3818405734 2227527909 4401507963 3506963068 3158584259
5975834413 3931666799 7304805147 : 4078
1042051621 3562175409 0487773302 2739698065 6495900945 6956985365 8432083562
0615934529 2542418929 1617305222 : 4079
0979352465 7122706640 0541353921 2620953741 6070259881 3126795666 7461709323
7174052362 9631960893 6529844425 : 4080
0743022804 9766416403 8282925713 7163603061 7625967249 9571761536 9585248664
4931720109 6085345723 4236254503 : 4081
8544414412 7163847672 6283333081 8958559364 7600616352 4985906328 8744503255
1137768181 3053346646 6995015477 : 4082
4932420985 6865935049 0106211412 9914177309 9804599788 6539985559 9720886527
2973882165 0877480019 8668603163 : 4083
0561230114 4493319357 8407633418 3313859772 7323452702 1265265772 9626488462
0440503237 7509270264 4091599212 : 4084
6524862677 1659965913 2457154139 2540015381 1699661401 4497922059 8528654631
1988145874 1918733755 1855095811 : 4085
8710196924 1766429242 3893754945 1631594772 4531101984 1450800876 1556264407
8821720935 1125934261 8446830352 : 4086
1073794000 4183828936 0585440706 5172644916 8857872854 5265072810 4911722412
9415223468 4844898973 4965331556 : 4087

9393268554 0211665594 4907515310 3970832462 3445957019 6856432675 6803854451
9358687335 1496819597 6960082012 : 4088
5379900840 0105463352 3364189127 9605446876 3570371065 1413568371 5512448361
8491925094 9941414462 4632178459 : 4089
6766719116 4877674448 9599464431 5839584871 8188466274 2027844189 9928803275
1244966696 4867934589 4132986023 : 4090
3034829288 7626063713 6445807371 3401017269 9240031409 9962898759 3282399732
4878713822 6525474190 3488221774 : 4091
9819545570 7963780042 7801458791 9441189077 0714358011 0302662454 2936251505
4346165151 9860793423 8562390664 : 4092
5515459086 8997009872 7578338564 7691033468 6388994289 6361916953 3138310635
1444319469 2997895215 0427343027 : 4093
4505489128 2240465675 1683738409 1737414843 7318197118 8226411967 0295140010
4844973686 8836048926 2885407453 : 4094
7124601578 4688794778 1317083920 2770185008 3959940135 0787510645 3561461548
4503534678 7490153402 7514090183 : 4095
4645675419 7604548330 8692169390 2489806750 9229922940 7155069237 7787826669
9123015899 0938081337 2850555299 : 4096
0599347167 8423507867 3905803655 3895201811 1477155275 1613837266 5668705503
2514568315 8295906535 7006080657 : 4097
2699022721 4337914923 7524221958 2555155273 9047664151 5242308413 0932793556
1940500532 4441453950 6109491632 : 4098
7038715303 7015281008 8754080933 2947909865 9178396540 8974119198 7143734113
6512716438 2405244158 4288769757 : 4099
1497711414 7142795082 9588702992 7924683321 3370515267 5643942311 3502628776
8903446466 3632184445 9217157587 : 4100
9241131996 3298754131 2018325222 6786967899 6413293411 3176366538 8968320511
9163622399 6373640065 0624218691 : 4101
9822306441 9813515321 9731985910 1563625698 6218174854 7088883782 2021617101
4912432492 1653238655 7690852747 : 4102
2547859682 9812494806 8606644449 3519183037 4836655081 7554225733 5268514038
9878650300 7040289933 4381972301 : 4103
9614734086 4828347612 6073019822 2686144117 9898436755 8389159008 4699913140
5413831939 1816564308 8434978829 : 4104
9151717429 7486490963 8389673430 6517121736 0275453757 8343113521 7215018269
5929149322 7874737425 7245721360 : 4105
2566263841 3895262627 9302130009 9661963003 2325220131 3821884482 2215385253
1276767630 4855187006 8314684039 : 4106
9268185487 6538405638 4831921002 4722319166 1009134395 0767855513 8370482142
8491510169 8975339078 9756323399 : 4107
2177890388 0476336813 7484651689 2263571623 0718406415 6632492410 8669239676
0121601081 4456092321 3374291457 : 4108
8448806124 7863773882 6410208618 0249513057 3388369415 8508782319 7098151586
7117095173 8802867958 0151067880 : 4109
4493390248 0689099052 9195328446 6968288254 5529207870 8090501661 4853675330
8133690700 4801338828 5854616540 : 4110
6413320250 6938355963 1742436588 4064726157 5760099347 8411408406 2998236648
2357485543 5335905053 6126274282 : 4111
0018784805 2953044769 8632263662 7829632741 6370115311 1823408178 6739876610
7281273257 7851392113 8076815418 : 4112
9444041763 2946304900 6186478075 9891264283 2572998735 2871612774 1833680517
5637941952 4402321288 8549117741 : 4113
5065311168 1836226989 5319004959 2292508376 2608050033 1743338563 7848674958
2231058631 8894073980 7614496920 : 4114
1791751393 3532988588 5343364499 7913001657 1286809999 5155763688 3579690344
9984723426 0419431859 9122046582 : 4115

7495644137 6367770216 1143127001 4347716120 1646483213 2927118257 1328791058
4135786193 1189374595 3236310239 : 4116
1270889013 9129091665 2719237745 8686417036 4801203295 3287516120 1291706095
9270907773 5616740193 9117441244 : 4117
7124601417 8496797282 4936614589 9072550082 4349970890 9680963641 6891569620
8984519256 2671934304 7171456304 : 4118
4323998155 6886935433 7262302614 9800352837 1665135912 1693178382 3097964852
2206285418 8473486939 3594384325 : 4119
2998753765 1192492335 0991966689 3106839343 0992917742 9112608797 2830433166
3875840237 0220112172 3945611447 : 4120
3365412763 3402705845 4177857748 5248631649 9991704854 7694843205 3120929273
9986610752 6313197643 3765380295 : 4121
2214163742 3602372218 5067791138 8725805767 7755437425 3574423899 7963358197
1403227793 5613974470 7119461141 : 4122
6517615151 2388236279 0564886358 9472686057 3344797283 0925709439 1377951656
3058538904 1681689876 9258083650 : 4123
6882500936 1192610789 1124270988 1222693468 5319851706 6371742046 8096376655
7289364171 3249386443 3405288735 : 4124
2790255086 8995093761 5120464984 5093084820 9464606417 7940775927 2735187506
1493452818 1767517108 4502365204 : 4125
4236776815 1326743193 2510951920 0587674918 4930277695 5965409812 2539635771
1694671126 0602360694 3945721364 : 4126
8076464990 1663784374 8410973573 0098749733 8721557269 5976033113 7128831583
8030624903 2383304861 9521149826 : 4127
2358667333 6359436081 5330962043 5231806990 5867253166 7967198977 5739671985
0563320391 6276929612 7845043250 : 4128
9302784936 5575704663 6650050423 5380700021 0433790543 6545267691 5632116230
7081538667 9328752803 9918102228 : 4129
7966754927 4141381460 0656548508 7779489944 7855050889 4914805288 2658768844
4566272939 0819614400 6839830805 : 4130
2403725695 0641143899 3318116637 7016307519 3044500215 6616091239 7787650073
8743398612 1377676316 3799499160 : 4131
3580029425 3941509361 1828779256 4890199706 3611511833 4377322878 0513781700
6854618939 7707800275 4504705746 : 4132
5744096115 1865016887 2167815808 0161854864 1080898632 2233409912 4749225808
1118532699 8795736203 0363601202 : 4133
4863397052 9467124019 2398688862 6998354310 9200456227 9169984416 9883212018
0955945505 3488532617 5425495363 : 4134
8515063118 9625309217 6652916582 4315900458 3496939706 8765865424 3281945647
6478537355 2515306898 9109796668 : 4135
8781002569 8390687113 1921425419 8866419286 6687545377 2443176144 6354156657
6287140553 5364287851 8017662196 : 4136
4712668996 2431948273 1009397417 1042590552 4374968733 3036596872 1460188999
6692802582 3050002504 9474319578 : 4137
8736177439 8118194129 4800460295 0542736866 9231007938 3355277975 0038359156
2081647286 2995161814 1511824304 : 4138
8649558704 7642038715 7706452758 7236770808 1805904084 2352377577 5754008846
8775611166 5813925119 3567319094 : 4139
0209614289 0110964899 5715039720 7159050424 7857829641 9139818646 4569806900
3883646798 3679810123 7694632288 : 4140
8360915854 4301049426 1487603503 7990344177 6859995675 9933050225 3234765254
6889955258 0628931781 1721823163 : 4141
7950877845 9343728786 4151446387 3034437300 9824804924 9554003322 3434378058
8846442656 5167111725 4089024251 : 4142
6568074345 7602957148 1282240947 8460558543 2107534563 8218480483 7562589157
5170601371 4681964216 8971776054 : 4143

9530199246 9530239019 9614826260 1706284818 7963579912 3969700159 4441468677
5318564583 1272547174 3944500821 : 4144
6298283049 3756959752 1339743912 0310652126 1696229281 1787214991 4975372547
2129306870 3850875650 6150275226 : 4145
4203373041 2161623496 3978809943 5270804166 9332272183 5932242791 1165736309
2546649671 2499296143 9907500009 : 4146
7635710205 0913521721 0267487838 1800459683 3106999955 9077825546 4891283674
3394515824 7815280574 6103415113 : 4147
4356381056 6377035487 9809403265 2665484582 4868458290 6755643586 3245918075
3460775801 6958399758 0648813770 : 4148
4343413010 5968843929 9179317417 4742867125 1433132551 7759566739 1206113546
7364831151 9793542086 1547222832 : 4149
7585604017 7338917321 1141862365 2027644917 6308259943 0939167130 1603555506
6396640063 7699177448 2383984045 : 4150
2727764172 0112291229 0611855951 2750665064 9835460296 1029657475 4375906955
5507510185 9350758837 8946923408 : 4151
0884424210 4044351784 5101844946 9776602243 2572757633 7213826673 2831848510
4191372257 2799003023 0195198814 : 4152
5702121716 5722765189 1902737558 0323980856 0285417910 8696330380 5028305825
4155207922 1546755099 8816607126 : 4153
3796669062 6696222880 4105219437 5553599347 8632239333 0880742944 0316363392
9743184574 2947964530 4848126672 : 4154
4296055477 8937241659 2547542572 7294818303 5852407989 7060138339 0192188164
7311557850 5264328106 7107830425 : 4155
3827862550 7357514417 8094379451 5208769644 1803929450 5337106958 0028806009
5926155424 0429538493 9223669282 : 4156
5186545787 1541550543 7681721619 4941624398 0236697017 6458516822 4184823494
9856122612 2052594069 4686843355 : 4157
8288000423 6042671649 2051983003 0400639082 1604948418 2231793780 0187897147
3265413916 5969414662 8544607201 : 4158
1663385275 5083190003 2819349165 0079157574 2541572642 2107319213 1424033453
2607660005 4088324224 3039536428 : 4159
5170101907 1959689962 2216857242 0582700804 4884586616 0085246854 7117664403
3745417169 7257316643 5712993493 : 4160
8871819929 1759466318 1328648661 8479719449 3997567376 8896889421 8741250518
1286522654 3689028478 9523945318 : 4161
9075978183 7842937386 9271123324 5224027048 5501136807 1301499127 5390637656
7761950408 2472945779 5161472517 : 4162
8334058293 6314389374 7277449678 9651364811 4070023132 7461989829 2467466885
5898433555 4557604277 5737627609 : 4163
1205804848 0271999849 5539526723 4262697175 1222462169 6901278008 1098444425
5359151750 8360950391 5428095603 : 4164
2294024501 2534448001 3590713294 3742644076 5715671941 4139579192 2713654100
9016170299 1031998657 0525840925 : 4165
7998434141 5907909404 7612707383 0478178955 1094730906 6257504993 6351422786
2650746766 5960409187 2737225477 : 4166
9472975212 1567356857 2609788566 4645754101 3819301847 8212336515 6997722636
1516999711 6230814735 3868744559 : 4167
5345401386 6559999562 5362468346 2963168340 9797550456 4050102772 6108378785
1820349370 5332792138 9434083704 : 4168
1728605351 6274318097 1964185158 1334638617 8640580852 8993261626 7479202822
9007798061 4873787741 1733583027 : 4169
6131207960 2560784881 8904686228 9627843902 0499837036 8536881027 7112776491
6480789399 4443904092 5075904833 : 4170
2890872832 4142449335 2364630816 7981840710 1984769366 3055389540 2873674490
3373984749 0758595035 6060720035 : 4171

8784504301 6881102144 2649884729 7177449594 1058452003 8230125316 6078708859
9066303712 8041215289 0785515342 : 4172
2131215471 1394784386 3137802569 3727576221 5857679289 1521263000 6697169938
4726344308 2846306827 8319532124 : 4173
7975797640 3014368143 7346152646 9198950210 3421817625 9365978453 2667602506
6744884671 2460966866 1730097066 : 4174
2524501769 2278852056 9067359008 4316041245 9284748056 6894697818 2767794755
8847010378 8134571631 8817494428 : 4175
9989298716 7278685725 4665536773 7283112444 6176730408 7523506505 4391728319
6198305056 3809001407 1189077122 : 4176
1329209083 7662204930 4967945786 6788418100 3586373834 2709135352 8093942551
8198079349 7854644802 4964273686 : 4177
6843355216 7080896583 6820495433 3674108689 9344362297 0414443439 6109886127
0196212361 6829423893 5804471970 : 4178
2097389199 2082302323 1464235056 7106499113 6035773195 6388471918 1976988071
0058067038 7057250153 0190615358 : 4179
5559014641 2669669192 3629059503 8548687337 6121913721 7442714461 5555267452
2825740005 8034707483 5030814855 : 4180
1071539938 9168383731 1092750205 8170195123 1311776778 5184487776 8726804213
9392860342 1323899581 8513277666 : 4181
9365598180 6455715585 4038012159 2971267768 6438428537 1888091790 9957308068
0102254116 5164298547 9859780120 : 4182
0530325322 5752499741 0354310883 8019843220 0235756740 8273306083 9664255593
6595175830 5161534971 3443826367 : 4183
9902877179 4289082660 4259749491 1477071422 9702525112 5860603900 5296024025
4582624832 5755757561 2161312795 : 4184
8128121568 5384085591 6278057092 9372466124 3672058886 8157679076 9303505178
9575639340 0311125929 7972535777 : 4185
5442005699 6640220213 8347356603 7691695441 3189061916 0614468251 8131601766
0986167723 2066349745 6046509728 : 4186
8265951275 3972733368 6941755084 8721788938 8640618398 4600371686 8968509185
2298344657 7551641562 9814905434 : 4187
6762684211 4452483994 1312303852 5805184868 0195708892 2618832522 3553643687
3645834791 4690356787 2259825575 : 4188
9755960216 8145222994 9197379268 9788065727 7499602061 8312361912 5984229997
4459179319 1907004469 9554058436 : 4189
0693267316 7089261261 7013398426 7188893421 2498755699 0243059505 9536766540
2753307550 3094906893 3593376555 : 4190
7321753833 5664890410 7287803731 7426730961 8140745156 2072125983 1472457483
0121106609 4797879629 4904381332 : 4191
4270611655 2697331798 1222046734 2712122664 1381929147 3278943660 9182787888
2764146146 9764222050 2911444841 : 4192
8441381849 2376352771 4914690697 4336080814 5042927657 6158754217 0524939409
2838637329 4735778423 4240779548 : 4193
8209531262 3534027550 3505710289 0394313368 1481995153 5661092447 7427046999
1167237898 9551639046 3749839660 : 4194
3242741993 1139442903 9057605907 4155553325 0655482151 7929225476 4254508718
9622131135 1669933045 4312000074 : 4195
7182980065 0638738989 2625648905 4397396866 7943212705 4939232744 4279957173
3635910633 3484418514 6953429812 : 4196
8089686874 0832375408 0262605943 2986205629 1817544122 2290018402 1005925843
5570500116 2633413891 1164722410 : 4197
3293543067 9924686315 5390027951 3923299722 2766212995 1309940979 5053020739
0559581191 5124333040 4078852497 : 4198
1092537241 7474301388 3031797018 4410857045 1357681512 9153624429 4925037526
1611011837 3210046518 9614678269 : 4199

7244426178 0434649844 0708181946 4885701556 6472912494 0018323157 4748921227
2150548567 6173310551 7328675555 : 4200
5513727525 7228070158 4444306909 1168420794 4852719275 1675238846 9452014058
4365412441 9006882995 7459005435 : 4201
7430805615 5946524228 8193127203 2923409249 0339765471 4651811131 4592519057
2580493515 1124366891 6540226006 : 4202
2755458017 6117435281 4314894999 1614671893 5238144336 8464042767 2953871675
2608133950 9875962077 8572778924 : 4203
9855982834 8232891772 0811777733 8346633424 1884922791 9805296830 9263756731
8470970487 2233707624 7583698147 : 4204
1774211480 4321363094 1487254781 4926308746 6784789524 5556100534 9890788889
8459211841 0223597827 3731652128 : 4205
0197485413 4198697705 4395388749 7390145741 2211904805 3900855254 7517769182
7060709796 7712659724 8844196823 : 4206
3810582599 4352982382 6723631734 2426715578 2964601050 6831004613 7927489065
6030763632 5981027936 6112357062 : 4207
2546093038 4592230956 9944746489 9594352803 5957281220 7359002148 4674876096
2849701879 8980716170 8718601131 : 4208
7039698435 4371096843 5176649247 9554420274 2247063771 6000357522 9427613743
2881027737 4243763384 6548182324 : 4209
2545858652 2993701908 8847477674 0026800100 9673197266 8495586454 4676707987
8771751353 9808839832 0732770178 : 4210
0462499327 8618880767 1330925433 8928428954 7339980467 8267914598 1967469019
8306839892 2634392903 5718573309 : 4211
5966285388 4503112265 8632570149 5178443681 3918535839 2042964337 5838949238
4812217565 5021035540 6710582772 : 4212
6887575137 8435979790 4449145269 1790592703 5088146718 7781681401 4900991554
6216901425 6978035035 9587247391 : 4213
4976161903 3480456499 1698043894 8284871605 7330970807 2050466548 0348755712
3331222248 6247330163 9986713795 : 4214
1278867986 4381025542 5604253579 2751624131 6245495529 7310236459 1993011349
6142952218 5316982957 1040685148 : 4215
0382213988 8376963907 5814255195 7119935917 1118755087 7596254773 7751359233
8703229940 1391763658 0370678440 : 4216
0859562468 7639951401 4714572246 8543401280 7858564304 3939470699 7121975940
6459214429 0107129319 1405742653 : 4217
3471341641 2866451075 8564588123 9511401177 9550807321 6367810160 3734337601
5731556349 2556593937 3671656193 : 4218
5500458810 7322335593 0244825696 9965558388 3053413186 6761689980 8566682827
7132356870 6812262548 4629821031 : 4219
3176077180 1239058725 5347242047 4152001617 6660218058 8246194966 4874606456
3878996315 0712442915 3884042324 : 4220
5075603214 0477624258 0365260920 5191482910 1027157745 9241426287 1044559729
2699956501 1286066885 4627487571 : 4221
8766506696 9779602823 0413811084 6930878716 8483579092 5462780349 4426425458
6120459971 9960800316 6303479589 : 4222
9289456325 5312542534 4317399853 6945980583 6028674518 5081053313 0476475285
3762874570 9777767541 3077142430 : 4223
0232534740 4093030694 2182911681 6439039165 5747003216 5881100062 4207185247
5797469805 2765170972 7451530250 : 4224
9461865993 7285040116 8149577814 2596361240 1478096837 8688851125 1471276223
1791533148 7044871025 7937765503 : 4225
0401664297 0150767388 5504773832 8881787312 1246007452 7612354176 6657688170
1011494289 9257349101 2356466763 : 4226
9625806511 2713396498 4228245027 3056693708 9237358164 6095356116 4343750565
0299631516 7974534093 8235328999 : 4227

7025627180 2165624362 5511798697 2162498323 0956587196 8296025466 8065000046
7162224023 9665382418 5505730658 : 4228
1446063590 5595549641 9820113969 6528443930 1573931052 0628308914 9218063428
7155353700 5900350470 8646096354 : 4229
1097884106 0965660343 6535444906 1700707899 5831805614 5339065047 7052743156
6417609721 5179194892 8336481286 : 4230
6046083530 7188194805 3442177304 2360408411 9157616446 1309136759 3152863992
3396387054 0720547988 4795798861 : 4231
6996218064 4201535353 1790044765 2558227276 7064593635 0572878042 7573485589
8768296689 2335724288 0868062463 : 4232
2489791895 8416944579 0029592863 2288829382 8009160324 6063532023 8223124737
7367874061 7940901041 3815330576 : 4233
1028300884 8864941592 2557543809 4606302586 2676989173 1846156393 6799577053
8251493415 3048349312 2434806333 : 4234
3168840269 9767024473 2749061897 3633454332 7828082077 4402667097 7817820783
1208572634 4560947085 5485214265 : 4235
8489101849 5735031866 4217482298 5673406328 5256420887 4664253405 3395045023
1027544934 2901916000 8441503984 : 4236
9520215325 6001639276 6936647780 9584126560 4290645256 8061709558 5204187128
1481347434 4515393746 4754963442 : 4237
2053892610 9954944289 1463675389 5787605584 2484190258 3118172885 0215885837
8806898930 5945215392 8137465408 : 4238
8777536100 4235101493 1084607654 3290946086 7969100188 4038416225 0090888837
4112448306 4612377523 3176545420 : 4239
1486843335 3152580752 6673468582 1776462554 7987715800 3780737152 4124840869
2569070492 8900666781 1402797916 : 4240
3653743339 5145161411 1072264561 7628225991 2478849099 2848729888 7052980830
6524599355 1737408241 1336775797 : 4241
2723356826 8033781163 5754998347 6669908238 3781455439 1621551285 6385576818
8044803432 1321029394 2162408135 : 4242
4802623088 2224191231 1925661205 2550161349 8997753721 1303331839 8096362166
2471863300 4541360033 8330530579 : 4243
3860790211 2932032887 1749951452 7204506239 5800542689 3173481835 4080848734
7093887388 6190102136 2175650259 : 4244
5911351442 1270210116 4809309303 7494167291 5936245687 8600346602 2400339535
2251424491 5160604299 7647942423 : 4245
4983779611 3498665910 2717829299 3124652842 5630398244 0607897986 9734206039
5170075318 7578630616 1264714500 : 4246
1003208115 9416960435 7882471847 6564441351 3385091862 0897857462 1805394727
0574914758 8858463426 6518241468 : 4247
3879858080 4824882080 5502019288 7763490205 3551585624 0018934643 3242686366
3411740303 1223423717 2519433745 : 4248
1401469092 8714729058 9015061523 8956228461 5203146170 2617566758 5862095154
7768283562 5450643162 6437458076 : 4249
0714178578 0027587608 0483325967 5887885318 0959596581 4816008065 2129817919
5843985925 2951844584 6927246647 : 4250
7076091516 8517463856 4718762214 9193215623 0101271310 5284450748 1070026502
4464178587 2435677992 2130399652 : 4251
1838279141 8482652816 2510461472 1163047907 8224202348 4217623506 4391415298
2300554362 5701476369 9588784501 : 4252
4405130574 8180118673 0872375965 2465204643 5093431303 9666558562 9392515681
0381694666 6203136394 3991734489 : 4253
6713982981 5627411988 2627235109 5251221821 7047506862 5339926753 7797846680
9995420421 4991134354 0701022056 : 4254
9268199940 4251665893 3928498565 6673473325 7949343731 1852048961 4803989074
9110350313 9468334077 7866936873 : 4255

5475491107 7611269470 2987821125 6476568149 1566691909 5529367320 5595772792
9512982090 7151577300 9178847796 : 4256
3374300214 5803603144 7789062007 7155490476 4805424270 5131677668 9353266943
7359097224 0083159402 3758388314 : 4257
7635421404 4086042129 9577016536 7296599020 4836305777 9854577067 0281905871
8648306138 2527527860 0758790702 : 4258
4358743812 1184097439 2908362239 8912538931 6731842659 5596595484 9588145573
5273119107 4691798283 4574376785 : 4259
8841443980 9369945323 2606572599 6975882421 9293587914 5436934910 4390422637
8176755732 2498731475 6957013416 : 4260
0414988729 7960683857 4148390973 8944823408 1301095658 3016594628 0719178447
9826332845 5536359745 1218726033 : 4261
2535781792 4068700106 8616506202 7823681063 4292874145 0813730609 5489089247
4450662277 3308385268 1090209081 : 4262
3243131511 8495335969 6898903910 3170864371 4328426824 9064812666 2406926704
2133671604 7057426553 1382424342 : 4263
2085609793 5006501412 6838867933 3024186546 2230747202 5320717893 8050317199
1787349112 2677621899 7084782360 : 4264
8111600798 0195606821 3562664092 4595140594 1919888659 6052780459 6189879889
7812460445 0752043911 9510874257 : 4265
6253503374 2853443599 6680254877 2112938561 9101220742 9651120888 4295468554
4392519090 4093459699 7623225136 : 4266
6844939593 9143105822 0054977616 5119323633 5578447738 7007275167 0887701833
6089731053 3306167400 9407487608 : 4267
4857287050 2540396522 0624984387 8079142123 2920380618 8310481723 2109300172
0191485125 5372112309 1515719016 : 4268
1271374065 3683546403 4590901751 6919077376 3122126257 2265866454 9644959123
7496183719 3955151966 5045731490 : 4269
3486981404 5381745736 9758982275 1137745630 7867235621 6976682523 9245298159
0467562185 2858455891 4530148222 : 4270
6584053595 2639098539 3717661971 0158783873 2782383755 8472442882 5363726210
8186657407 4402013077 2139829403 : 4271
4324598455 0348898469 1608935046 0604602972 0752430522 2505088484 0981344559
9990758681 3154752220 4656145766 : 4272
2331404526 3269653527 9292379792 9373929190 3748696719 6463071806 0724784747
3965505170 9747509660 6260234290 : 4273
3764684445 0582247222 0213238638 8557145132 3474982034 9966040662 1177777297
8606443414 6736621110 6480533161 : 4274
4307054651 6482946603 3764356209 2912703240 9021043510 2759540397 0654729979
2721089618 3967315084 7030362204 : 4275
9840208056 6869922524 6595358373 7496013314 1790035370 0524645605 8694776746
8897309441 3691010744 2020272238 : 4276
5751420522 5470819089 6349914219 4431740411 2374851728 8954308018 4574091031
9994611280 5564334422 6228957934 : 4277
0517365029 5313360283 1432970249 3656220118 8073837965 9686693856 3040365542
4493757197 4210249374 0570193694 : 4278
5138482569 8600195321 5162568971 4018351165 2549600636 0110312028 1995349670
4616097462 0829177365 7299785645 : 4279
7130696516 5001622777 8852738340 7259835596 7397082404 6315259176 7380422974
3178528315 6494449545 0369564011 : 4280
0936985581 7985115027 2119159176 3800482965 8130789859 4703973120 6537802032
2760344162 9732969824 2059066824 : 4281
6901430294 9399525015 8989047710 5080253835 1571584280 5178152642 6400657458
4027917036 5562834115 5199917788 : 4282
4995168857 8980881405 0968676482 5608157587 1689213785 2614348240 1986700859
5949183437 9968727493 8039243990 : 4283

0124089292 6764545234 2219220757 8413785334 2487883353 5474932468 5978019743
0738916781 4296105044 4496628579 : 4284
7198684931 7044974915 5067552127 9111615838 0570696819 6572594815 2986672133
3525808676 7899959649 5159606109 : 4285
8889306228 8696613263 1217557575 8648326279 6857114153 3502647214 0795023598
5958527648 2031363986 5562813744 : 4286
7612938148 4774020225 0445085912 1825095624 7790738896 5494526502 0370785100
5311569656 2202984993 7475086136 : 4287
1904397629 9599031311 9008043197 5245809400 0091455821 7454526625 8052740399
8131693257 0018276842 1722318051 : 4288
4068200502 3092966177 2701854460 0370433014 3234525164 6866456105 2172069290
6036720273 3353961577 0828488823 : 4289
7288135889 4045344902 2699453876 8784657248 5781928755 8201702633 4879541782
5355455575 4906825006 9979739505 : 4290
9504613420 5343092698 1905725531 2303387133 0291285031 9228135696 3960128385
3455211594 3569140977 4601581987 : 4291
7741238198 8958667271 1142729583 1082708986 3920462180 3453794556 2433971856
2497127140 9579259619 2307818840 : 4292
4535552979 7774221068 8936733885 5485881072 2367687848 5938225154 0354409948
2252905928 3484780136 0135009151 : 4293
5811453292 1442353962 9875548306 8015588531 6596656339 4478186222 3936580697
3126128512 0024220474 1143375961 : 4294
7769910633 3914637580 5203581547 0306229527 7217035148 2123853049 3265705844
6113196562 2545238528 6064449330 : 4295
9181674712 7236972720 2950026266 9493867606 7390723574 4132686071 4781306327
9625225213 5834514617 5319707352 : 4296
8090113543 1467773945 3471024006 0895178155 5830757211 8215079950 0558837744
5473192612 9253830499 8174304023 : 4297
0702238909 8160761531 1099288865 2872560217 9498866339 6420618209 2131604317
6801177815 4296636735 0787241918 : 4298
3405805978 0194136413 8016285792 9818320868 4244469747 6095946754 6593513927
1920270860 6667009644 6912642578 : 4299
9697600503 7528190966 1537566185 5458062643 5229492165 7908349843 7925743197
1587706468 6820018182 3204868998 : 4300
2564563340 5013849665 5764029708 3448061878 6696893121 6351339668 7831362351
1774979419 9305482289 8661904006 : 4301
4715435959 5792275442 7777943966 7263372974 6627797753 5731960843 4724918119
0210119294 3925903802 6026484174 : 4302
8444782005 1685684303 4661441250 0612254411 8553603669 6829948065 7213953513
3407886924 5327059129 1498280174 : 4303
1121071884 1342687878 8829800210 7119318415 4769063232 1330356647 0428019983
4162572610 5167041311 6849386770 : 4304
0277509498 8441085136 9316956444 8607593170 8354676736 9017773894 2973154551
1459227701 1103608430 5577182412 : 4305
1223403292 8229874439 8644640191 9560923000 1439499345 3060442579 9693849177
2397816149 4511312042 0486863791 : 4306
6752530634 9006652395 8044028984 3539255578 4845807220 0332029250 3465974481
3261401733 7334841522 0872649858 : 4307
3672364880 5643312830 4693053048 7353905968 4897769410 6624899681 6465510182
5562769089 2330654374 7477325157 : 4308
4823464207 6182693720 2001112884 9083740841 5666378790 4917715791 6261744725
3356921102 7963136363 9619333830 : 4309
3169096058 5634786515 8364104095 2185421892 5393845365 1900094568 2188235121
9678534912 9074727334 5761908795 : 4310
2770071453 4296428857 7789197970 0517737331 8942564746 7787059514 1670950151
2543632545 8585059092 7777223574 : 4311

4136906107 0592541796 5794073644 8940133684 6212597403 7769436292 6710786480
6916569414 4947649627 5547975269 : 4312
9750611239 2906590555 6029980618 2775792321 1986904515 9059424907 6760144944
3302144753 8110788616 8394173626 : 4313
8247379536 2048578667 3661943401 8375399507 8873570769 5697363348 9060966234
1520330327 3664416840 9155972675 : 4314
0606818691 9542897295 5496780074 2088808731 9998422933 1801642263 9183011407
9597049126 7195672661 9387623534 : 4315
2306778374 5037399215 5604973161 9654537918 4136237601 3666098734 3740561564
6163459852 3847828523 3197307913 : 4316
7019825090 5853269294 2864012889 6615562366 5336680867 9676269021 9338587009
4706204085 0270178945 0516817868 : 4317
2770319342 7843070164 5193131391 1485790961 6968441606 6209283732 0833387867
6414883913 5298925848 1845308669 : 4318
9758841288 9658670242 8755687731 2359003496 1649957608 2923775226 8936555707
6354134082 6557724889 0243575485 : 4319
3975257909 1134201798 3026115347 4517489394 2282388277 1044974234 4359228203
6621472973 9913674036 7101215970 : 4320
9430824875 3447698010 6697699031 4194078502 0801000638 4516220354 2748953285
6955258016 6987140127 9094554658 : 4321
4468531729 7663885922 3272280239 2295725516 2170439537 7986809188 7085119555
0148345006 5354205895 8817281907 : 4322
1594632777 0613634760 9047316518 4177320017 7627496686 1929830048 4784222251
6625268124 1060317143 6519456728 : 4323
3488928109 5890446951 0765410361 8988534832 6694340218 4793134763 8061335551
5202360217 6365618271 1315453253 : 4324
1524831850 1600255035 3002350998 1187456840 1397841324 5041292489 9510635618
8398860593 9985186066 2669837430 : 4325
6821560893 5364080372 2105692217 0621065402 9033468957 1523900667 9969843981
9719944948 8473637992 6562713791 : 4326
4408554512 6277376803 3692487909 6474511063 0943048104 7440825975 2902764930
1909961828 6720668008 3812477082 : 4327
8042534854 5154944826 7335177099 1586513972 0744453559 6290620297 8965148227
9964382284 6241004949 2538096631 : 4328
7158494746 4968773242 9417148601 1775792546 4809222939 2563484734 4849734476
8767897255 1867684457 8041930104 : 4329
3588384787 4498471915 7546612527 7421065198 3403688768 2177098564 7989749664
1796375853 2760889483 3993789803 : 4330
8693590500 3885915004 1822476926 2139163222 1151170732 9740757299 9505921614
1953417954 5395648258 0695755819 : 4331
1410105474 0858366976 3889748544 3567038088 7776225342 3725236685 8625286860
7011120737 6644417703 4759239022 : 4332
0540292118 3363592076 8287468191 6357344362 1225846855 1849117378 2814989331
7329432868 7866734127 7094195061 : 4333
4067843095 5963466118 3009377235 5931550084 0818820429 9011125362 5495158658
7987779332 0160602302 5399639582 : 4334
0888578524 6406838930 6031488155 1188185106 3392801380 6882947755 3386857687
0062873818 7175509620 2301678908 : 4335
2957729937 0394812355 3225117730 6514137747 9705343893 7964519477 6233680444
5661037728 2423774640 6531747191 : 4336
2285508752 5701248555 3034958425 4775511192 3104174126 0896041764 5304738448
6184959876 8824455494 9742237079 : 4337
6274252261 3785956150 8152707173 5522514060 9247146628 7707657592 3280000636
2366178185 1820146612 9968973204 : 4338
5067019387 9122339028 0196792848 5764215175 3605032428 0495374126 0170275740
3030534227 1641863949 5786000064 : 4339

5995239276 6917288951 0347832507 3178136744 2373727643 8574252177 5361860499
6157784516 4056212512 0075711268 : 4340
6925439127 0548041741 3062908526 5480196487 9811114373 4157554754 9991708223
6619650715 2797220508 9698520513 : 4341
6490535547 2784482972 0710781457 4514568460 8611157295 6596317579 3886474842
4633163765 6494481161 6192160462 : 4342
8737029040 4097536728 8613066251 3809093509 8491430915 2013685940 3499036297
9313644033 1259011869 6488012087 : 4343
8666141208 9289781050 7017559263 1593289475 9333535558 5396153573 7486473277
8846929005 1483756269 6456927138 : 4344
3010871929 8422825611 4413262879 6930855435 1482195929 4806708059 2226824590
6695987049 8056520226 3218860004 : 4345
5813384821 3383801071 4011937050 3199986558 9124946066 1314089799 4650450291
6697723288 3060195184 9785565324 : 4346
3552234794 7617679257 6544582052 6605167994 4760722705 5026040253 8069305498
8610650921 7465565888 8730178883 : 4347
8412895999 5965915932 2793045738 0841437898 2357797812 2166391540 0107412133
6095716896 3667005748 4589520547 : 4348
9261619617 3666179668 9247300318 1633077684 2912398177 4002938069 0464820050
5930567210 9704707235 8576276559 : 4349
7152866857 4058099115 3560586905 7244722420 8349859427 0765145780 4339879341
6795768137 9636208339 0395083293 : 4350
4495580595 3637604854 6223113625 1679236438 1354247748 4194804358 9332145988
1942136057 9294167412 2615999836 : 4351
1085501672 4402649352 9026929476 2413425825 6387230319 7435078616 0646176951
0121854106 3220308171 6611241488 : 4352
6764403388 7297576584 4395220221 9610073743 4541485089 1687133744 2678359527
0561315212 5262073869 3321830593 : 4353
5658930621 0304929209 5355381454 9511601214 6419843979 3903718964 3986934878
4158909220 9093907042 7578219059 : 4354
3570943076 7237024389 0053453120 9670035089 6192224298 8160876864 3298293971
4882496041 2446513280 8211248922 : 4355
4188331444 7492688036 3423918296 6671662822 4156781839 6752437966 5967458116
9914928136 9014502279 7801359769 : 4356
3138653945 5472068457 7771745913 8536178479 2669537103 6950896377 2279061281
2365479157 0888690766 7689081937 : 4357
4934061036 8410673860 0541100262 7870087470 5947106586 4514391969 8055970695
0556505015 1233936665 8905717331 : 4358
3364476421 3070656757 3199190393 8811893825 3075210882 8595161475 0680946883
9245740011 1354702747 8698216862 : 4359
1274321497 6651883003 1960333456 2235534421 4173681170 0595766349 3676223290
6918488032 3734519243 4912465665 : 4360
3297182384 1739691986 5871333134 1207105170 7368341724 1234472371 8672494151
0842704961 5549503795 5015287382 : 4361
2486053846 0676720392 8758686262 6169843822 8056015875 7866830925 1223040159
9897538489 2283600959 8075215949 : 4362
0258074717 1577353748 0669005001 5349853590 3697672297 1043715921 0984693912
0490542624 6339608550 8249182328 : 4363
6584393402 6293826856 3715259623 8947447178 3369996618 0201154788 2499133236
5221595665 3404857011 2325827708 : 4364
1886215013 0715779346 0027439516 8927524355 1823962408 3915012347 3924668651
0022276735 1541533981 7443806293 : 4365
6818843187 0219539467 4578806838 7330450266 9934820474 0930850952 9400887069
5181863254 8354824966 2607066502 : 4366
4995646819 4106470407 1227311004 2154418554 9123163407 3401961809 0749872367
0338997499 3439243991 6588065128 : 4367

3667905641 6957365216 0502822448 8521757361 1331742947 3563577483 0847833098
4929574305 7306045041 2840297114 : 4368
8973655233 3702529333 2859341544 8831374005 8107262424 6477515535 6118977342
7429425968 4019406813 3338141610 : 4369
7490916404 4307805277 7297700938 2768736682 6928083628 3467725910 0328412812
8935787611 8865330379 9439157109 : 4370
9173130876 8414829814 7316743892 4150708123 3304663866 6526885171 5228773495
8086507090 2110271308 0114880705 : 4371
2510861641 8161525567 4817104786 2194001056 1280013464 7100048960 0816181363
3986131437 5953783573 4463470097 : 4372
3802239706 6733317884 7121742166 2379783395 0508494584 0564804300 5911201841
7300502241 9629811376 4325550862 : 4373
5993667297 9212123968 3362625433 0792019745 7555379857 7860333993 6113973147
9735858804 6748266319 0555031714 : 4374
7510740826 5754553845 2600524391 1716394798 5454385464 0477445274 4423208201
1580589401 1041094538 3309204755 : 4375
9831301278 0340587673 3811345178 4704239620 8849358293 3994762948 9274926307
6151959475 8086352305 0039063526 : 4376
9755089737 1963214320 2797580886 9585297731 6219835944 2566894666 3498655664
1828527840 5307103946 0383705531 : 4377
9530705781 4332061654 8372087637 3054215104 9819639823 1622394615 5540162448
1922748898 8991548181 6161014129 : 4378
1114223327 4248071051 2027849723 8490440264 2968076980 3440316010 1365988085
6422960754 0695695571 3350800535 : 4379
6595188841 4562653198 3654599814 9354703533 3965788049 4761273209 0048553113
5908697471 4535368901 5718969841 : 4380
5775481177 7941076041 9019145652 7288840405 1882795084 9008264536 0132429152
2888893797 5199955998 5968288936 : 4381
0891127109 1030997306 7570621865 9729673463 9843061928 4295504292 1054669160
9154182803 5178323684 1552181865 : 4382
5679634071 1124306438 5515415624 8975176397 7188126209 8408736098 2992383637
3030275059 4187706262 6357378481 : 4383
3688683682 9333969889 2774446869 6966294996 8966027691 6003698605 9689047184
1716000197 1841236265 1997930127 : 4384
8741913927 2279869802 2724595229 4015243220 8404817539 8124008257 6371360670
3033447765 4114042743 6996205412 : 4385
5145048270 0210515174 1189088254 9260977472 1462055366 7790075520 0879989232
3465359252 6127377276 9545671215 : 4386
4449995014 8452685432 5974087574 4450235595 0317588809 4678670624 2795049681
2231738195 3193469955 6151004209 : 4387
2153608326 0321530691 8572722599 2986003953 9254734318 0646477064 4415440963
0859833209 8942811738 9795068274 : 4388
9552487066 2827328053 6479247410 7235461194 5944265379 7874986979 5964322144
9501435206 1663273608 3387789574 : 4389
6269008896 0941621373 4426382002 6765753391 0341892476 1056359888 1700835396
4495585219 2240769813 0525217319 : 4390
5013237419 6453428976 1658135407 3313026303 0285478022 4851925520 7832323745
4832679632 8907125627 4752461229 : 4391
2929641409 4248502905 9474155759 9275124250 8506996634 7353764294 0738391820
7419091625 4160972844 5916256144 : 4392
7272170590 8293857676 7130454384 3359908261 6086284671 9632875160 2860804035
3472085362 1937494941 3665519915 : 4393
0853689237 3512748507 4294814451 9654169382 2917931530 9180378204 7287223894
5587726485 8456583583 6888354176 : 4394
1056472125 5643864015 1856876957 7988275174 2224347111 1552623436 9721170763
3450466532 7247663342 3782119587 : 4395

9117615783 3972287470 8451279886 5992451832 7916414459 8280214437 2701894626
6803579233 3375174806 4104993185 : 4396
2188455068 2118360856 8975632511 8138750460 9424195574 4326397198 4310789277
5247235667 2288395527 3552360662 : 4397
2598788598 5396328772 8445468369 4923676064 8441227353 6829219132 4554704074
4216668206 7461265670 7964301200 : 4398
5535159904 7417085461 6373362027 0386595227 3806423126 2187406268 9090399816
7108615021 9582650064 4977817357 : 4399
3185052157 7151608476 3131429710 0682482491 3916249289 2556126384 7673862182
3334522830 2857013815 5437476505 : 4400
7846569058 8394825313 4199813962 9573625019 1985710580 8466792364 9110592908
8055068338 2177183709 4406269993 : 4401
8269197068 2447053200 5794540835 1861003661 1615609450 8242398650 4199873162
3838357466 6454521887 3590261219 : 4402
3453160485 8339212666 3772506208 5190546236 2373373281 1425816643 8424988979
8596679541 7592003691 3004537037 : 4403
1962753606 8718301376 4812127410 5614824487 9598635108 5005487349 0298447752
9757534939 6655307150 5515441039 : 4404
0938653741 6665215918 3354997382 5668932645 2686208235 9621402396 3409712242
6826870025 3395582693 2243805998 : 4405
3242690845 3047814199 1498931107 1015716947 4955576352 1945874576 8629630183
8990582420 1278786755 7969347231 : 4406
8503074461 0603483914 5987949779 4871139135 4629025486 2822249743 8944743868
2661643606 8318124885 9872326879 : 4407
1076616230 5380086029 2158338144 3284532240 7462355418 7998881747 2381285359
6838289500 6202252464 6357316108 : 4408
2663603643 1485635531 6050499154 1441308872 1541443317 4525373210 6513961197
3306948781 3913096873 3186136669 : 4409
6193940257 2004993080 9999534821 9427655878 1132352187 7812457183 4239764680
6973069793 4060813830 1890778537 : 4410
9227621548 4664088231 2642981642 1239417456 9713566615 3982555543 5294981722
3901452223 5919257419 4314642182 : 4411
6647768886 6772502171 2329415228 9784461629 1056628542 0071453883 4167814891
3456674450 6929129063 1913561846 : 4412
9153315855 8135076487 5413218594 5574577015 6486652616 7466208580 1070023556
8168758105 9596769989 0198547270 : 4413
6417531288 4874941390 7086642080 8062769445 0131967019 7033551810 0417192425
1364427448 5219781537 0357770961 : 4414
7978036554 7150100641 8180502754 0735836312 1007584869 0420360453 0637430343
8876152382 8983557372 8526027901 : 4415
0965811358 3990082025 1064150068 4886037785 1451703993 6269383226 1888230189
9591271000 8790754166 2874832239 : 4416
4452285023 9184867913 8156920568 6354321704 1633586370 3532435303 8603138245
1788944252 0080485301 4804232640 : 4417
0652099629 6009664177 6937613082 0806870201 3088347209 9991664558 0574702972
6506424859 0310071846 9895310690 : 4418
1743228378 4757153675 0984970434 8348205993 1223224751 9854853545 5451980842
2814507464 1693251741 7116603293 : 4419
2667762278 1908348972 2751003080 8975252050 3024664935 1264612784 8908741183
0385877998 5696639063 4505200221 : 4420
2393452668 6579920442 3861484757 2428901051 1432888381 4583700148 6782283330
1640727611 4036941362 1157371699 : 4421
8560584173 5445608033 6888906925 2435338105 3971593150 4525920940 1288682676
1957851131 3238397636 1566212857 : 4422
6482640972 3566892206 5045948831 7985864101 0564381869 8894289927 6314816131
1411978394 8514389640 2016161447 : 4423

0282221756 4827342006 4615301617 0156730451 8618437705 2127062573 4225871506
2895985488 6176205229 6168865652 : 4424
1583778424 7149479866 4877070673 2481799084 2497441413 0307727812 2172757039
3898531076 6265694827 6197633287 : 4425
4465960455 9350180212 3231361682 9469105165 3679961314 9656188142 3079790028
1970207530 7204137744 9667453062 : 4426
5110784565 7924638003 8273856860 2245895406 1439361499 4048484286 9698064832
6210846438 0743336558 3986880899 : 4427
8847151429 5701014512 0549668428 9667486610 1866876512 9731396283 2144102683
4165860359 9114389318 0762564434 : 4428
4609274750 2825373247 2243963582 3144186618 9835337323 6916808695 0292204814
3623720481 2347239217 4822684291 : 4429
0666117849 4353167259 2385041053 7793110490 8157295800 8218904109 9653740522
9404620198 4820594719 5654684300 : 4430
7682203732 8399061467 9207880313 0621080742 8878267456 5222795103 9753018549
2450108066 7143565220 3409852097 : 4431
1712751588 3904861311 2946667339 6409924349 2647162312 2346056816 0044054572
1462865662 5616892869 9746838216 : 4432
3390433176 2863708095 8285081909 6974176212 0740550851 1001815330 2081718136
7176854348 7272089172 6067431810 : 4433
3868754990 9642874280 4106528574 4478313994 8749789244 3435373201 8837509779
5346044005 2811978526 9275442489 : 4434
6325741629 8794288255 0607639516 5108387117 6646737663 8752697508 8379398903
2883574021 1229563543 2087437414 : 4435
1624949105 1576822711 7417711932 2329453919 7409213659 0860000476 2024328188
8116116364 4928715559 9082998145 : 4436
4358403717 0956530527 5679006103 8582200502 5774224292 5697737322 9660938546
7336929449 6815432605 6858234085 : 4437
0739091504 5923150813 2267678106 3632505958 6274551489 2282856540 6945222105
6355802862 2416764055 7695260322 : 4438
1833562396 3739880801 4246118755 0546095550 9412272002 0016673290 7897800070
9638482831 2446295596 5450221207 : 4439
4332643657 1869713451 4468968889 2295108020 0162568079 9512184131 6098330970
2178276860 2448714433 6131445923 : 4440
2060147349 7481486913 2077424597 3643394920 0730885748 2067224118 4792682405
3304598692 7545897072 1315458415 : 4441
4047723222 2083920803 6324127217 6311628918 8164339403 6869317135 5826800063
5173209745 0524378305 2633429820 : 4442
5158572929 3729065144 0822520931 4003833442 8600943717 7696508292 9411189712
0873707840 9355275475 9466760770 : 4443
0739582996 3258388610 6737450792 6180433067 2514053521 1613376789 6036538579
8503221845 0837234572 5893974482 : 4444
4492032873 3860279920 4201279950 6284586176 9293493474 8645046491 9686353383
9144956932 9353499139 9224536641 : 4445
8810063103 0710950568 7734454230 9645895436 1418630876 2414944474 1986669834
7713405300 9577978720 7929146238 : 4446
9626390884 6392996906 3931016383 3638321319 5393793042 8049848759 6279495977
9336483584 8003784161 0186331741 : 4447
4981336434 2738798030 2577970218 3088650364 1810622157 1029282021 5871814456
2664355191 2892390746 4494910427 : 4448
3596927665 2311469566 5844581811 4476377375 2350128593 1265925760 7505565019
8433567754 3209175069 0113587867 : 4449
1619549365 5202913909 4377173320 1558080722 7333191001 7793165196 8154228088
3705765075 9883759702 1754117738 : 4450
0871379853 3896289683 0262981628 0553011075 5775897853 4823854812 9828105014
9628858871 8232218048 8826018274 : 4451

6164487902 9172841840 0029367469 7066526008 6282843883 0881335633 5802435105
7763528593 3515444380 1010244994 : 4452
2174781896 5531884708 7378155223 6440714542 8295483213 1189537540 8182160748
6597977762 2791023082 3364843617 : 4453
2336602917 1933992616 7882868980 0578911911 5696128611 3730404222 9962444456
4522881179 7471536635 7914615049 : 4454
4381217969 9911719553 3929154679 8588856908 2819110314 7183811299 6786565145
3059679813 1908004221 8270900569 : 4455
3044807483 0083456425 1502360936 3752556383 1935174270 3436292684 0234776413
8990390433 6235185135 0263930095 : 4456
9664448987 7604174378 6038172342 0607907143 7119444753 9522843551 6778885709
0934059098 4262800755 5002487258 : 4457
3065055751 1259353896 3151718260 6584600128 9326829742 1617913081 1687821272
1039217579 6983370070 5560047926 : 4458
5997432364 9525447559 8623484474 0409617147 5662882753 2509173759 1546032508
0960114432 7041198731 5969680079 : 4459
6893604075 9775018037 4094171469 9818850466 2898263050 3406281730 2823197991
2553091836 2807799251 1330655577 : 4460
7258958054 9721937547 6854503466 0721841155 7053096747 4247232869 9207294954
1768648905 1896687507 7362184297 : 4461
9151157585 1590669815 4334699735 6951876553 2300615728 1995740879 8433268786
5483877037 0483886796 9634471044 : 4462
8810366625 5295862834 3438480462 6781221476 4251660197 6422702036 2945087987
9092306899 7762622023 3634117711 : 4463
8201539905 0173071493 9961155484 0371522826 8577998385 1093840045 2826367576
1260563984 3913529408 7394557427 : 4464
7648499241 4146858242 9138579177 5623295402 0168247866 7600210215 1382666411
0130042178 0062344321 7919523686 : 4465
1776961388 0500659250 9070252900 1557499487 5260137231 9426962837 6917390123
2800260299 2780683106 3027349010 : 4466
1100314067 0779004579 2769827931 3684945163 0895091829 0483029650 0190892833
6498837214 3961807932 0874232876 : 4467
8068649099 2044495073 5986052895 7211699519 0404083389 8408356330 6643854128
8900714754 3866567070 8501846953 : 4468
7900325716 4362715713 5734722510 5929586511 0984068456 8971565825 4557256333
5559457657 7747618020 1568378523 : 4469
6403149785 3809831566 2531858094 2578510581 3904615465 2480727198 3290957729
5805530088 3576439354 6571457913 : 4470
5927674664 3254483845 4558273091 3986167177 7995943499 4550317477 1436597966
8345645423 4578234953 0220337939 : 4471
2308962273 4428667174 6679865568 2537325145 3774300922 3721202753 1693856659
7761078235 9801886307 5075604083 : 4472
6771274187 1458966983 4570275384 8703603042 5842802127 7883107135 1132772777
4992046483 0373404864 2790225836 : 4473
7600773900 8365615912 2196628426 9036366602 9854323536 3179945147 0339639496
1386871671 7763056546 0918856551 : 4474
8443445168 9033462458 4682912414 3175358657 4989124759 5560896040 2921553939
7744642887 7279516223 0612464779 : 4475
3088265423 5748010717 8091527740 1654871046 8265535703 0989813108 3104309486
6753941511 6316766582 8140370086 : 4476
6865547643 8339770427 5712504474 9221422969 0806227876 0854440485 8813995712
7475613019 0263751507 3878311885 : 4477
1441005371 0314183163 4743367511 7492320531 7682583915 0342354505 8602263058
6870277924 3229691617 5459534404 : 4478
5744739971 1712119207 7899158121 8062470944 1550694727 3514012345 5020462906
7783378176 2141918901 4690880707 : 4479

9818100673 4859396797 2669348769 6523832201 9108208731 3696579981 3776062143
3456083913 0799199491 6188426151 : 4480
7662011516 3301336524 0268650603 9217254034 3455286751 8082580104 6290582083
6468110735 1833729751 9614561225 : 4481
2537135677 1188784251 9940235439 6224247751 7373345353 3891507262 8823213219
6100844063 1515479985 9952158888 : 4482
4711599783 7510010625 9724395444 2905647837 2428271127 6886130872 9717667450
3877844706 0504981792 6326301121 : 4483
7932857796 3021341729 0584506938 4231278574 4709828199 0368933332 3225245658
7848649909 7175032022 1672852149 : 4484
9882775465 8306559015 9584351618 5552054897 8267361357 9627957493 3052201119
1610842080 2369961389 0043993283 : 4485
5066345181 6407683899 8792241323 0641588672 5004798621 9148727609 1383655534
5671757845 4745567604 1291332244 : 4486
6095514831 2535361128 7145210743 0436355478 9821369772 0092793013 7261704890
8280393909 9983557408 2898005613 : 4487
0326328048 7840702588 0190853877 3031881871 2730746136 6420509108 7496124190
1769134499 7077457528 3633978323 : 4488
1561103564 8212470903 5953681610 3938413756 8833771578 2793829413 5462391627
0576056022 7177983532 1108182513 : 4489
4419902677 2861049687 0350607337 2089355308 6400814265 9916508722 3597269462
4860483666 5891130395 8507510474 : 4490
4947269939 2645264605 5226651892 4976453080 7213109352 0962315760 8488311382
6036479166 8246800841 4223588777 : 4491
4104611556 9322707887 9326874361 2837957576 8538184074 6778209846 7762736724
4691118916 5465543048 4797104215 : 4492
2397792892 0676389663 5896383566 4180589442 1474539622 6231753265 5797252681
4663117697 8012993711 0982453405 : 4493
2292709330 0399629513 7144383195 1350573178 8056663961 0421225778 7252801325
2838483308 2997288182 5074988518 : 4494
3147176842 0564258524 8475129554 5343386763 1293546440 7773405000 7219968380
1405226959 4709199118 9551884053 : 4495
7007723165 8195210530 4731149144 9137585879 1465623887 1946479845 2652848026
8964713172 7151501040 1260230712 : 4496
1222879224 6888113632 8743346672 3782055811 1794202721 6479225653 7042244877
6329697012 9276913150 0425202259 : 4497
8108009997 9885590235 6890432902 3147853873 8724020947 4483853941 7783679432
3325613093 8177205017 3430979624 : 4498
5324951033 9055541937 3115567810 9600573604 6904087935 0621298748 8240528497
4393964694 6584677743 6237028014 : 4499
1430030440 1661790116 5797602697 9051136930 9091941300 4043992505 6649562424
6838020340 5089480290 0026376220 : 4500
0578640552 1562274015 5388028003 4689559175 4312762186 1314760525 5689989509
4989323834 7287908253 9984234639 : 4501
7579120047 0117076564 7978008566 2647461192 1493201815 2685676334 8584378214
7331176775 3820642489 4580965820 : 4502
0413472695 7162370383 7207793565 9762005227 8022518225 2963200325 3692765319
8445965838 1037550792 8342527858 : 4503
2591745643 5006260843 4416311297 0195537547 4851856666 6109082130 1768897917
6081186745 2945298815 7107661286 : 4504
0214031901 9653862395 6291513139 1540587849 6630720194 2185802070 5480546447
7547695567 8720896015 9717907361 : 4505
7241406114 1977276502 5290138599 2813531379 7092560361 2068498388 8215192251
8132179980 9333597821 3175107048 : 4506
3806021942 2357710899 4591747051 8247086038 1957578800 2816155616 1166585331
6284726372 8589024423 3722087246 : 4507

2933270283 5317492841 0044347263 0417990086 9416928737 2227838080 2882663780
1392053292 8677292804 2365659125 : 4508
6451489454 9289320084 0060968418 3178390593 4902376205 0002243391 2426191828
3291801084 7706957415 5867364119 : 4509
2801283496 4582068331 1575462623 3916641813 9061988522 4573436830 6773983610
1046423164 6413313553 2352273752 : 4510
6644533820 3004095510 3219288976 1040287882 5716543401 8178388741 0155412948
9492109750 7774609559 7715247869 : 4511
1288112448 2928425719 7468820488 5654909561 5768512861 0593526780 2656143938
3358592178 8673566005 9653009853 : 4512
6546206425 4346379106 2988857389 8318354369 6792192647 7578037976 0991519977
6315697301 9113881696 7770383136 : 4513
1743164453 7854570073 1936052197 0015173541 0076686960 1305437275 8816070499
8778936501 7422246617 4381932691 : 4514
9286242930 1061984254 1634604853 3061312244 4410053811 9042170529 5021020394
9284993280 2291875208 8000318240 : 4515
8337953638 6422378318 2669354546 7637857036 0640399221 3306997783 8159195259
5555888781 9155273311 5500131708 : 4516
7949016677 2728426303 4494974713 5658767281 4394265492 0526300619 0800059109
4495621854 5217579084 8418298466 : 4517
9384194955 8812074700 1060376835 6344103320 5450016151 6572407201 2529860495
8894707972 1834259701 6438475984 : 4518
8658280980 8690549036 7726146202 1539863629 4163771793 0476272183 1365968096
8500291803 1141079704 3113811864 : 4519
1849141586 9758498660 2245464927 6513102872 7586293867 0400213000 5951781635
8644077874 8117806522 5685065307 : 4520
1350734245 8944602683 4620834580 3133921453 5422338754 4486303860 7087278502
7777775086 7762992022 2402710743 : 4521
2459134004 0289287209 0265865447 3562108425 2719222866 3131418011 2202868765
8119592727 1283513242 1508816450 : 4522
1912089966 9021973736 7720181554 6709893992 7354219911 6273165262 4506949920
0284852366 3471619210 1353179446 : 4523
0181932739 6072124887 3981575039 3461817020 8223027956 3933543167 6555278850
2935163273 4594726579 5104011499 : 4524
7344823774 2065875730 3463602505 1615613927 2689820660 4832108554 2322084072
0240030913 7515825417 4604210574 : 4525
5596153391 9845890027 3971574375 3802242641 5567778759 3079348042 4536014455
0080463441 2405440897 2659399663 : 4526
3007997972 5921292231 0941947078 3842041830 7596625539 4324456096 3589541422
5661367379 6928542371 3764749526 : 4527
3375569771 6260219922 6734913942 7236091784 6761965872 0575970336 7639965915
7563639934 2505878894 3766830142 : 4528
8916417573 8503388810 0786002186 2503880088 1754282047 6784272595 6719604840
6057121007 9656164926 9169969392 : 4529
0855774499 4249775621 1414133332 4323795150 9364096641 3408084475 2861415797
1242508505 9258050810 6462051245 : 4530
7221688482 0197934411 5322991552 4820485989 7513010075 5988479695 8676495248
8027942323 1241980738 8735906666 : 4531
4964915113 9629244887 1057045643 4199781743 7109316921 6647620690 9405665634
7912220876 6735316201 4395794445 : 4532
7621663868 4660165500 5339479315 8087747107 6476445478 3961099927 2348900285
1457979744 3446604405 1117604558 : 4533
6177008433 8760531480 0885367357 0211565036 1045487992 8403753217 5914820188
2993997086 5269064553 8159173954 : 4534
1840830826 2969421422 9603619265 6880385459 9914531553 1953051939 1834550185
8845493495 5788237465 0985608212 : 4535

9842400795 1804176372 6731295371 1599532987 1379480968 1866455468 8514436258
3503666244 0852558593 0624414700 : 4536
2018821220 4642192591 7234593398 9619257901 6317752923 8655519765 6249237162
8465685389 3447270690 8824411028 : 4537
1325841910 7757550548 1596675431 1956994397 8415163324 0688940704 3926608112
1561861902 5303220713 9064676977 : 4538
3268368150 6611237692 8002489683 5575713723 1128842582 7689287823 1095678118
9966969774 0948934872 4146614973 : 4539
9727949412 6610763694 0295823628 3034824223 7176331997 1843239990 3981284939
3985868272 4464605407 1699259408 : 4540
4518183571 8891094351 2641768469 1477909225 2837837532 3956019474 5349090551
0164233807 8115996634 4826207141 : 4541
5872381354 2032404931 7575753370 7935096946 1348285274 6513771468 3420562623
6321971729 6199917086 6638304381 : 4542
5049988726 0671726765 7248389966 7394934781 8640599731 1990052966 0035329941
3932648526 3280930082 2108367705 : 4543
0197128707 6699002478 1300851305 2729668440 6061016437 8230719136 3070494220
4938341960 0953509993 9871351587 : 4544
9252197394 2806151356 5643134081 6156888490 1747824857 8366068004 0392675996
2908630937 9031115067 2767194187 : 4545
6882462830 7518879506 8973905489 7988939998 8362663176 2238054216 5500973438
4360758921 4242046806 8102248730 : 4546
7387780947 8772352739 0164557430 6789841755 8609780598 0115965586 6608180878
3531930027 1259737913 3589578973 : 4547
3400876344 3118861486 9768378811 2151087712 6677572310 1833251113 9198516665
0796809644 8532556638 4175831166 : 4548
9493885921 6141667456 7555538237 8604424455 7126333966 7721462244 6741586901
2956205456 5276810473 6368260978 : 4549
9864963005 6827907376 9198631560 1291614256 9306478100 6989197020 2265386741
6260304963 3997127366 7679574289 : 4550
5556631401 4881184615 4458200759 3167883566 8267089919 4556985802 4090677654
2744450798 5684535490 5475878108 : 4551
2742772021 9796600382 0209978659 5196994492 9335431694 9164917055 4412944820
9296760925 1479903225 8890049539 : 4552
5499572299 3018062179 2621124071 1016220957 8552704888 6053819284 2360852957
0140657674 2213483133 2602634217 : 4553
7054373095 3570108907 8073022135 4982636285 2887826558 6915685634 6625947313
2585195659 1384689244 6244583440 : 4554
2277049670 6653499387 2354752090 9770648817 0900011063 9125571874 0682470471
3672257092 1712228672 1119172532 : 4555
2046914596 9221561229 7524374555 0688205617 3695494066 6273325260 4137044629
0865446151 0442560076 3291794861 : 4556
4510692258 8794437481 5778912508 8591113417 2233031567 4973819208 6397220651
3520158280 0869824687 6537375323 : 4557
3446669783 7732810111 4526195995 8866650108 2168815975 9495666029 0352758332
1493728643 5874599676 4765258208 : 4558
6055524473 3912117129 7976729814 0379890972 8650697649 2032209927 9240402066
2929795608 3333483709 7343711038 : 4559
3136407411 6484760050 6985201567 4472277835 8999608425 5788427247 3530611705
1602021864 5681801484 2377128587 : 4560
6616591199 0118688140 9516094024 5806826199 0511296112 1275457411 9275346382
2939947534 9910748595 6141079594 : 4561
0104716129 6588915916 5679925544 4422507796 9298705769 7058479143 0401182545
4535611172 4273371399 9994326721 : 4562
2771932319 3391300068 9026267852 3004204477 5981367284 7437901286 7170965176
4782559460 0907601267 7038666595 : 4563

5450974046 8485891271 3323990972 9337983586 5350809726 7094853161 6454530790
8715735299 5507516924 2502752012 : 4564
5482063398 8339034478 2848948625 3263843703 9404450543 4363120240 4956819485
0386290285 6933719155 4779296289 : 4565
8845325057 6597682076 4344629468 9814524433 4205580449 9465858237 5053659570
5362244038 1842993593 5441506813 : 4566
5070920577 4449799905 1667690538 0209621766 3656068357 8158732037 9231026760
5286259507 7654529057 9527210420 : 4567
4138524615 6860576562 8218747553 8902520578 5080074069 3372942987 7374630579
2569937750 4026856615 8661680407 : 4568
6154221619 3889794638 5363383112 5958244187 9709719549 5224074795 9539519422
9768582513 9007695483 6547899338 : 4569
6356880618 2905679340 7651975802 4674217910 2503322169 6437760044 1282077990
4903233081 2317611237 9972572146 : 4570
9283312184 7175212958 0119166422 7393514437 6698375199 8453064578 7366407773
8069965291 8703123194 5950041645 : 4571
1412022582 6972469814 6148818462 8904987272 8361923812 7204197791 6155713134
4799687575 7033935401 9325461289 : 4572
3602565440 3122892178 2203409543 4438836935 4604089544 5802479018 0847629703
4197962302 1602945161 1247691650 : 4573
1860059811 6417274382 6673072895 2978409244 0165695776 5725555729 5761302425
3354790950 6761981919 7014241516 : 4574
6593574460 3959673618 5325510610 0792443466 5329441818 0701214764 6030124862
9639975333 4724745983 4339008685 : 4575
1604672402 2521613626 3343752004 0663234254 9930842141 8588260982 0943615743
2408456471 0825443101 8830907726 : 4576
2585702511 2104655164 4620072039 1537464739 3215292063 3797970985 1707477303
8060536283 8981511969 5460644201 : 4577
7113929566 8853118832 8062643343 1701717020 5170264852 8131841251 6343924730
7171335549 5060515867 1361101058 : 4578
0504697835 8806115072 2926127576 9395214080 3029828397 0431417265 7091329508
2911253431 7694308075 4686551329 : 4579
2224276499 0417404748 1341076368 1949573977 4643796865 1830444656 6839831848
4194824417 6059863821 3837823531 : 4580
3730452285 9498310101 7622494394 0539018649 4271323242 7248943202 2720850526
6704016110 6549298027 2688629950 : 4581
3201407835 5006817832 6612442947 3379347087 0403693394 7444883640 1463014307
6654888691 9995190654 0631166476 : 4582
7819404973 0100375198 5705416725 2764702308 5919922739 9842341493 3499839097
8640343907 6922427293 3766892660 : 4583
4934944169 4715697155 4705990421 7421925882 5753425929 9956598642 6768451410
6123846878 8065445791 7717504027 : 4584
7628236063 3007943758 4074366948 1319018301 5709126766 6207989331 7001720205
3954906864 0945149774 0977410943 : 4585
7345210942 8596847172 7023406450 6710071428 7458635586 8678236996 3390799448
8074376088 3533682031 2113732445 : 4586
1571040418 6429258794 4963777514 5772351175 8660881749 3878267538 7764130377
9519060504 0647630268 0647426906 : 4587
8794144744 8269351953 9380970178 4967319464 5289143892 2238632801 0121834314
1180980750 4797024389 9193572725 : 4588
7052706054 7047281474 6474462082 2245407471 4169406520 6969516760 0105591770
1364897132 2784201700 3356643205 : 4589
5511462217 9331323222 1711932737 3477715778 8658376969 9079117113 1728353740
1935719863 1824447047 6154481845 : 4590
7702523441 2484550968 1281166649 7090604322 7471833668 0981167836 5157004680
1876817095 6889324184 4073623106 : 4591

0256627475 7479697637 4720906354 8402949173 4727487308 1201950527 0963758360
7462545479 3529542341 7127268198 : 4592
1339902641 9025763373 6116569738 5508744130 2411046998 3643222100 1267728682
1809498907 6236868926 3384417518 : 4593
8562628321 2965994121 3751796989 0520924751 9987894668 4512057837 3584504314
8908445688 1613840158 8039698729 : 4594
2009854086 2969471322 5986323131 7127321942 8914135494 3233591229 3724230966
0154616499 6985579627 6327555457 : 4595
7984454403 2230409049 3465934666 3308857369 5960232857 5018026850 2120007127
1420765556 5772640798 3020729194 : 4596
8065129360 2177611360 9391135939 3530812177 3628431894 4173962135 5453605002
4613878796 0676040805 3169300250 : 4597
9344182571 3421613614 4401191522 8953065300 7779053972 0033966289 4179465964
1777983559 2534222877 7465640003 : 4598
4393526477 3517835159 3305571994 7159812125 0380175595 4757774595 1520713908
4336100115 3790490150 7890705670 : 4599
8818445741 1450883051 2299700209 9696113097 1218937285 3308359428 8099049270
0746796012 8096971829 2839412819 : 4600
8960521323 3610705214 4317601499 5019958953 5897183369 0782831386 3423685795
6796563419 2930497698 1520997452 : 4601
8878312145 9736846075 6747663606 0039687917 4354213850 3489532629 5820544748
7692413303 6611832868 9112778317 : 4602
5460291279 1295567600 7418733225 5429082447 5967304300 8045896634 5337819875
3699486335 0537722445 1590105416 : 4603
5665887069 8989013753 8880243275 8515141185 8454353652 9874915727 1886535640
8268327225 1101470090 9898627167 : 4604
7224152877 6898622677 0369839699 5978281655 7940888253 7183764537 4022181884
1187448133 8287947196 4454937570 : 4605
0207234696 2533601592 2182028081 7972604912 8292420726 7129800242 3760022464
8500879889 8857231588 4221469461 : 4606
4749102789 1554652123 0594248610 2726047126 9813241493 8149901767 3574896094
1182680504 7717301316 4589468234 : 4607
5409075051 8616610154 1070179554 1759759828 0422938627 1555107777 9967459122
4661638474 7714601117 8964858677 : 4608
2191861044 1753073190 6039973098 1271821910 6724424419 4814711527 8343039206
5368122142 4655022693 0206567588 : 4609
1276475434 3855734742 6328154927 7056266006 6546711848 7652072673 3565473166
9662177069 8336458358 2683020776 : 4610
6337778647 4895873125 5219478599 0526293246 9843638246 8573872328 4277789839
9257375979 0403828529 7787397684 : 4611
6638096241 7547759148 3096429915 5281451658 0372824423 1957520927 7265826932
3059752461 2450564355 7559140534 : 4612
0142976750 8244334746 6243985834 3761487938 9964144402 5141130024 6886937195
2157830448 6934438407 3455908669 : 4613
0946448138 2353532042 1925347916 8899801439 1786232376 0175521459 4654274459
1797476063 6166524018 6206188847 : 4614
5586487380 5830893796 0029471475 5490167949 5904281561 8428725158 5293982983
6320669197 9956738947 8800199587 : 4615
9803482831 2599253297 5991050207 3507362660 2542431051 3286294034 9554176399
1094763294 4854290444 8102689152 : 4616
5958952145 7320122663 9910332168 0006863504 8620073570 3485896106 1790277660
7181677068 1793663180 4538237919 : 4617
1259561984 3885760951 8028943215 7965212056 5761109398 9915496884 0313775493
9354085695 3449908009 0127038401 : 4618
7262618018 2046673795 9946829543 6971942325 7922064747 0975895251 0700203774
1719426389 3464967571 5427250404 : 4619

6938718835 7277563444 8769445950 8498638877 6168355058 8497002868 5726928054
2129179585 8131914228 3803871372 : 4620
8275528683 0645433314 8688092151 6972378760 1375527981 1045966988 2917779624
1097070913 7037883930 4506450124 : 4621
5988678589 5788637091 0482165298 1152058332 2780817799 1560826384 3346554197
6031504888 4191986084 8406264900 : 4622
3995878169 3294270619 6229745312 7136532694 4415146362 5697342752 4160873440
2697978790 0214636269 6262012037 : 4623
7533883677 5496915720 8241546392 4241350575 7494323179 5379554632 3852419930
8940605113 4109851878 0558840927 : 4624
3559116716 1670187118 1553505823 6609856919 3922700064 6822883589 9448657522
6714800463 1656752194 2019661830 : 4625
7034521639 2823182511 2778342832 8954142472 1726008101 4787580302 1229475153
4403712723 3216708657 2434181080 : 4626
4779578414 4265189977 8860114020 2712329482 3762338268 9209641111 6301103221
7548508094 3353504340 7762834159 : 4627
3003000533 9134180446 4235262400 3807395109 5137011158 5095347324 8223743803
5089352893 5216587780 9760032571 : 4628
2128354195 3122550820 4126987603 5418900771 3651108604 7181941080 8843111074
2536391713 8988135975 3193233465 : 4629
1579864775 0468457586 9893919631 1882411441 5430357040 0082640126 7779538411
0666509640 7904987522 1717226415 : 4630
6490434959 6267065632 8990457837 6011368513 2624683460 2134845575 6462607773
5218836021 6622168327 3267524660 : 4631
0907786809 3484338791 9815646612 0824848678 0506442969 5846021029 3045372463
3487602599 6224270099 3707358710 : 4632
7459968463 3237610502 9378740807 9067362688 3143222400 2634254183 4946497494
3378764251 0786294049 4355970889 : 4633
5096801337 6240248190 9558113491 3297913613 7997670206 1469804350 1490706733
1506378762 4020662390 7027183279 : 4634
0480285148 2956061154 7406959972 5619492966 8302102119 3050250622 6826388089
6540629845 5619793389 1284197588 : 4635
1135820072 5743561685 7677236659 3657751853 6374076870 1008248672 7325827055
9473587594 9915271835 4765991563 : 4636
0391378571 6729854404 0275171738 8265873849 9177463098 8127769334 1113336407
0227666786 1050716885 0450924177 : 4637
6817981250 7691095090 9144115833 7030312240 8506723054 5812454372 7125939201
3793712989 5049768896 4104658348 : 4638
5074819454 0118577023 5917246901 2966511482 1500743422 9928281222 7997873756
6992007129 6284554728 3485924965 : 4639
7682030788 4679506470 2194730980 1494977443 0994977910 4146628500 9066700904
2196029866 3311030110 0111327823 : 4640
0180588984 5558966393 4608753728 5190743600 2284230389 6215171619 5641806562
7000999956 0134896928 5573932626 : 4641
5725163640 0204079984 7141233603 8886746834 0872957614 6976265971 5075739705
1452167403 8808385420 1531984940 : 4642
9356208349 4180844821 6518439071 8645485793 5411988652 6224032719 1774938527
5782250382 2693597054 0054945456 : 4643
3248668816 6937021367 0549848865 2755267963 7592542953 1152739433 8996216801
5529068835 1370329024 8458258150 : 4644
7127636005 4835685965 4294106517 0419630389 6699098713 0227585345 9041651175
0974861330 2711360435 3170371718 : 4645
4168098731 4181958503 1564870772 1494788546 2800392627 7518456650 4309126034
1422184230 4210133140 2856915028 : 4646
9035417750 9096542281 4083329824 7325956328 2470188681 3189069773 4122400982
4572047781 0212442578 6794196217 : 4647

9708245494 7151657388 0791213055 9425579829 3659101859 2952145889 1989736410
5748908948 8774613063 8684259775 : 4648
6423252686 3463439677 9838391211 9284815590 4632572684 4298069038 0184893738
7866267760 1466755693 0214317529 : 4649
0992680576 9995789731 7628916600 9235193383 2885941822 4344448373 3745857062
2675049768 2736139292 1256590388 : 4650
5590939178 9629362867 9219569856 7220771145 1676871151 7036384009 9018030189
2877956685 2973917705 4160961453 : 4651
0699375915 9622783391 7112798101 2984300277 4890935874 8945398117 3533815768
4613950841 8563804583 2986728493 : 4652
6286928780 1275472398 6638924135 2820450578 0353885364 8470973275 4492568009
4275596516 0291034577 5124513594 : 4653
1932152689 5090145445 9524169611 3698064708 1102315333 3882766622 9870284100
7248866052 8760375404 3021857842 : 4654
9642231635 3133503406 8040278444 1475200826 8900378635 4784156536 3689567311
6883950433 5456317801 0166155103 : 4655
6438448861 4982713956 0795729139 0579008646 6129824415 2281347786 4422645334
9015863717 4488626108 7774283733 : 4656
7586277208 1968306963 8318058811 0789389572 8376778301 1869997008 9433656062
4571898381 5017674655 1248089349 : 4657
4477000837 3453061948 2002243872 5236049547 5711739466 2913618252 4460576260
0948866902 5658132931 1350674437 : 4658
7932320250 3802054350 1852994980 7781052599 8530448875 5072493125 5065100388
5719082485 4783899326 6028603373 : 4659
9356873644 5574675696 6011919160 1438807752 9876643945 1357947607 1447098889
3277660530 7411631153 0139110492 : 4660
3282193110 5880973670 4743234405 8908828564 0026797594 3534642742 1078414906
1540929624 0833879108 1565789909 : 4661
4988157126 9023662004 3280215289 5896157752 2206062145 7988284049 1823383657
3512543126 9368037987 2686847552 : 4662
3692007777 2138117964 9240036159 0234577003 4941153406 8335573824 8739131353
9191475815 8527625413 3758998420 : 4663
3618879551 3807317346 2244771235 8536727905 3008355143 6917470851 6100914754
4822406092 2455757610 3986096635 : 4664
6788289455 0033360433 3720846556 7251648946 2327278693 0989509455 8630983011
4378782588 3212299067 1329181939 : 4665
7652202572 4558400814 0241309321 3342095068 9694190778 7020261535 7580218210
5123290813 5283195091 5725992555 : 4666
0067198796 8751463023 4571315295 3633128866 9612988751 0461508593 5633839150
7026224045 0567951168 8694605110 : 4667
9466276723 7607237505 2884555332 3504747748 9980158288 0457530050 4609169021
5964523830 3826292106 3032258517 : 4668
3215675280 8913584883 5485487124 1265327424 7165752201 5793560433 8198464883
7766055270 2655267847 0835601024 : 4669
3568608832 6611788097 7031360613 7790936091 1308723407 7209618102 3901578532
9203547127 1969105674 7947044146 : 4670
4331213143 9560967516 6261128944 4050366388 1934028642 0485038072 8609879198
2980487313 3499004845 5219406573 : 4671
4654155537 8014662305 7839211851 1432796651 2426186044 7837066569 4246510969
7462111066 2557267176 4171963200 : 4672
0060679461 5444473830 0958784936 3123252352 8518998571 9809160006 8141919186
1683890151 8124804643 4719012840 : 4673
0070704117 3800540248 4159936710 2849240755 9303963487 2403013753 5199115751
1804441626 1522200880 8978968333 : 4674
3245220004 4042973126 1225161765 5694246030 2075359589 3240208528 9249243527
0739656513 8857291831 6494763448 : 4675

8104460321 4390876483 2655167298 7621999736 8370945854 9393433833 2588462059
4015729844 6429786277 8628233290 : 4676
6904492327 6593292449 9124046133 8339690152 4758560101 9682761237 2034305510
8979231752 8677738782 0132668450 : 4677
9880718026 8203922028 0484921574 3424256804 6730697177 1218734560 8571520356
7062170083 8936982753 6191719737 : 4678
1079407399 9060368395 2061279228 5497138600 2256544673 9856276967 5644776992
5630094957 0927013119 1929824955 : 4679
7757637508 5942369144 9068347634 5460439443 7131867956 3025883826 8179289757
9073867577 4381973520 8131268068 : 4680
5284194436 5315510971 3356812574 8283469790 1789014318 2775261808 8331363997
3602829599 0100005197 2437701525 : 4681
5864819106 1236176764 1145832369 3479550647 3752503406 0992789723 0719508277
8497001196 0313162475 5510087288 : 4682
5217381291 8560876774 7856196063 8935124092 5078580628 6682155371 2362464896
9325920348 9219130671 3471710997 : 4683
7033736289 5959783509 5325235982 7128964919 8711044815 6130108291 8581273929
2211017926 9486428196 8328624692 : 4684
9210846438 9599692881 6550769582 2898733723 8615783294 2848807562 1932309574
0630838089 1462836413 8116929229 : 4685
0832336756 9040678765 3576163961 7542666459 3802436019 7471995403 4836508568
8712885529 2429351867 9946556844 : 4686
5132086630 2074899531 6987888798 3908989545 6546759232 0605819226 1778837661
2225335473 3863709677 2959302399 : 4687
1180556622 8092649500 4498775277 3346252173 3263893894 9491465403 8212432180
7361584536 1140460215 5525975366 : 4688
7202239198 8105028468 1073770437 7232931215 9753622750 5157238764 5294363483
8791352898 8787905821 5501111042 : 4689
3358451831 0227611334 3358047897 1414977152 5157789848 2989361644 9182897931
7790326467 2287494186 7545326918 : 4690
3640879828 2666191059 5314252866 3092670701 7324059507 0622738667 0579287338
8271849518 7976068532 2806148088 : 4691
1864659328 7994277974 6124529540 9526712496 9044104613 1073909192 1123969406
8351844696 9930471686 6174240505 : 4692
1922297675 2985296684 8673483681 1625766352 5545752244 0318439224 6233255616
5225141042 3592327320 1888336360 : 4693
8901360504 7263374962 0582370618 4846895299 8652674663 6302195871 6136936786
7248353561 0994037381 3996989004 : 4694
4051538895 9039103282 7945728151 1397587762 7474453184 2930536613 6379516981
7274605208 8546469232 0417591143 : 4695
5001350395 0753484896 1569249509 0020755605 5754384170 2901460781 8186992492
3939119438 2411002570 5182138823 : 4696
0098035458 5878116400 4553031127 7179266614 7437748373 6623746020 2073538403
4823096876 7789362305 6167893806 : 4697
7317856631 3716353870 4715874017 2890527249 1252082716 8930437398 2296588377
6322658705 8276813473 5329947863 : 4698
8582774381 6547793609 8561315677 9341800020 1475804245 5937737830 3603956313
4913576340 1350303484 0739927578 : 4699
6595343912 7551602075 3086523653 8527450378 8859712486 6716500398 7741926537
4145011160 4950710018 6493015358 : 4700
2757306955 3026842287 5848340321 0477930534 8522095490 7925735639 7702757373
2045507991 0158184367 9809773130 : 4701
3516449897 8068477009 2412465890 1717949986 3901523452 4231446407 3314623970
4547452605 0430187421 2994382125 : 4702
7064491507 3546929709 3206380290 3775911881 8739015791 0382794684 9262860649
3557705792 7535699984 4748784544 : 4703

8455800506 3460310194 3277182722 0312508097 7502995191 9768802465 8963981141 5337135068 4361001131 0996298251 : 4704
0923400768 8626817381 4285820414 1106078957 0114950689 4308627965 2602887953 5397611684 5807304352 9965627122 : 4705
0921968552 5228849169 7344907527 8234090379 3436643175 6882408319 2949783543 1473144939 9148449444 6342896571 : 4706
7965533285 0400946759 3996359499 0733864266 5621874810 7263247872 4304062904 9467920253 8485915398 0266086302 : 4707
8206837106 8192586367 5616167488 3003149343 0128105299 5998880618 8629972368 9641652045 6806077951 0100917746 : 4708
5730814546 9192581327 3129673025 5920587165 6588042033 9131539744 1591728719 4875701426 2221471927 8911066027 : 4709
5076128944 2732242913 6699466592 7065725104 1936310238 5981754907 9869943892 4388900893 9346341213 8281362194 : 4710
7180798111 4503000204 3390157920 4393125539 9519222609 0889997125 6309233027 1429125014 4039818705 0042025960 : 4711
8780870813 5868664901 7724447369 5219446703 4906502238 4096930518 2593996803 9421038583 2160164007 6947789192 : 4712
4212148954 3593744009 9937964372 8561303628 9431167437 0894674473 7971676182 0864159316 8356184240 0484542707 : 4713
6218560664 3959788021 4216431665 1900960728 1746064137 6846555288 9186181960 3488258520 7484555661 9089693189 : 4714
0551159916 2690872569 8821763884 6374595506 4737773914 1580667407 6650097783 4149348421 6184403838 8310937601 : 4715
5393889363 1063154871 3459725290 4837036883 4150168607 5158579902 5411531953 5607077038 1497122744 6960968940 : 4716
9530526009 8081749270 0921933267 4213887448 9748595826 0981627811 2393479338 2770561974 0666378971 3662110111 : 4717
5062064178 3273090943 8642104434 1001568487 9462446967 5999352470 4930849927 0573111248 2397592335 9898645282 : 4718
7704154827 7629085165 0379608150 5783509823 0275874215 7795977900 4592060888 0043548743 4735298281 4703676657 : 4719
8453926047 8341670459 4176917489 4680825377 2836216066 8568691135 1945238338 9089189111 0912798514 5874140765 : 4720
0285259067 2091111527 9925914212 8435543975 7589985220 7175909946 2414469284 6326862634 0215826602 4982921158 : 4721
6948010765 7666054916 1908778645 2758667194 2368344343 1434193812 3350021289 7771314250 0052922497 1673107771 : 4722
0017168511 8882299889 2084729467 5210622424 5534716079 9120832204 7262682159 6886645081 6735096164 3933992447 : 4723
7511462369 6703082302 0942646253 6749134703 4074245876 6524088261 9103362519 8046411371 1612273934 9796447567 : 4724
0145458128 8427010677 1962632936 9178482286 2791205618 4982809090 0732388615 4644578857 8285840970 9604774411 : 4725
2659161889 5177662916 6252716872 1718762516 5136317089 7961764648 4309698655 0203832749 9036901395 6616772453 : 4726
3477735942 8410507910 7689335238 7935549818 7846781230 8125958197 6513263542 5639402346 1702890360 7081664241 : 4727
7750144015 1789871339 4071954068 0368401437 7753823023 2741686512 7331446717 9277067541 5259373709 3421319759 : 4728
7040931481 4919948898 7228678226 7469651931 5663052914 5957136183 9749552474 8019433927 9493656351 1497990843 : 4729
5426571310 2019714434 5952011778 7594580375 0787462518 0499505221 3989814784 5950331786 2578489997 7510091836 : 4730
7587992192 2826055038 0285078100 1785441580 7312470071 3812448131 9901891828 7915172974 8063153541 8402724973 : 4731

1986676050 3046989214 0012207784 9186541670 6288946143 7398736216 9714665740
3395403546 5704207132 5868663336 : 4732
2792832121 5614881054 5073264464 3989838002 4170684921 9912974800 9428508166
9435649295 7834344123 6526169848 : 4733
2303194094 8633410216 6069882846 2310983387 0141892807 8207374832 0543074954
1642896324 8139509455 8993370449 : 4734
1826186642 6541508190 0582707235 8841898280 9980191510 3517706388 1643966427
9347043773 6480269578 5368103679 : 4735
8665319390 3768776582 6608503687 1810357283 6664336505 7717227866 6728793090
8099722192 2188813671 3093480501 : 4736
0525726415 7381340641 1781621048 0678714302 3891668115 6631072876 0536945753
7784780650 5850299526 9117032333 : 4737
0371671960 1123523444 0747526812 8928355868 2868402632 5716056974 5001944449
5170180661 8108190644 1054552466 : 4738
2199746055 1083766582 8858129814 4515267490 3802269567 8115007030 5237616097
6160751647 4358393341 0407798055 : 4739
7722934775 9913433136 7215978661 0550734227 8371167761 8319948593 0275857261
3970586584 7098686679 5339609462 : 4740
9188678055 7171136878 6100785500 7438818571 0641269205 0966991492 8942274974
2545108502 9268848939 5447482286 : 4741
3067499394 2734032486 8374491455 6473583037 8788605984 7569183700 4542330128
6627585792 7095106789 6905333597 : 4742
3429361420 0328824671 4880458353 1441386307 9762659740 4893087110 7688512323
0786079342 4512201715 4558018416 : 4743
9216407608 9577132875 6451994313 8023272075 0051287556 3762033975 3300605391
5208115880 1054294431 7855589339 : 4744
8246356458 3104692415 8286611788 4901004634 1407609499 8214465414 6039283751
0247350589 4407349196 4954037027 : 4745
2245345722 0124463687 0031384833 9842309183 2402490157 9518659337 8599282947
0159857959 1968598386 8598194490 : 4746
5441298097 4841995536 8963699635 4717206181 8135855259 2623299539 9169375917
0924286136 6421855846 7667720623 : 4747
0544414885 8262537053 2082454374 4822903027 8027065751 0380017108 8178328186
5145986558 3258719554 6458525341 : 4748
9757769782 1772923612 0713336412 4028163497 0014373150 0885527129 1758919608
2897388980 2021760821 0487754500 : 4749
1333613191 7131976208 5641491481 3796406630 4065403542 9084875144 6093045428
7978881572 4796980056 5660799594 : 4750
3883648202 8316033894 6679044511 3655307314 2332760578 7108207926 7604395353
5466422536 7818502682 9228847438 : 4751
6393039583 7400973858 7014265108 6344620470 5175847605 4990457272 5750315085
0585856819 7468429290 4170671661 : 4752
4133188504 3846980797 3560177710 4811768366 4107793967 9436202684 8191177314
8240899769 2555128826 3145278012 : 4753
6438139137 5696394891 1084647998 4997677906 2995454495 1573178700 8040712480
2833371117 0264272301 5442391360 : 4754
2066185340 6307113082 8152074975 3719314684 0097106225 0371959216 3865824282
2894083649 2726282455 9605155252 : 4755
3505912078 4146707605 6758736733 6189411174 2761679155 5301719943 6756847928
5326732265 7804559329 7864425666 : 4756
9444338925 4568991712 2537729840 4289632547 2825553367 9890350772 2310316422
2255936866 2240293289 7079007749 : 4757
6213095713 4962928964 9293878759 7564476663 3100100120 8538049136 5778048694
7710046538 4219708755 5332495654 : 4758
3022798404 9186231967 9062531769 9314735845 2802777820 5886769223 2477965323
9355985991 7992633825 6860793129 : 4759

7153065955 3948783277 7388807136 9814974640 5066251751 4314106466 9754807888
2506210803 6930301495 2360247668 : 4760
7170177830 4594493982 1354666812 0189155589 5109323779 1066397432 1717152861
0129007333 5180113311 1413566846 : 4761
8007910399 6224531505 9620716207 0402855157 0261433581 0300780277 8165023775
1720096785 6240169078 8151274926 : 4762
7410160630 2319578673 3309667419 5332631791 5486900877 2125173072 3578980925
2253022563 2265419023 3991410380 : 4763
3099343538 4580270680 3344427165 1927725823 4253690592 4927564449 9599318933
7302024156 1339217286 8821116886 : 4764
2565852709 9872906016 5657108807 3894875836 1695217062 0956326824 6090399618
3788978720 1822687023 7853568344 : 4765
1810012149 3461723067 6752863599 0112808891 0089647377 9245979568 8250073985
0238077509 5912981555 3066924895 : 4766
3573727641 3238566846 7817781405 7638772348 7604032512 9467647135 9436659413
4553106314 0695856263 4633689731 : 4767
0651638074 3553431261 0069096745 3765014405 1981183492 3531887194 0799399909
5538928957 7987504772 7780327084 : 4768
6609361548 8559657996 9702595623 2802846174 7441648263 2048724128 2989494781
0298822837 7461856185 8232334667 : 4769
1858688349 5818421919 4388322204 1292566218 2136162779 4490041022 3801402421
6582366043 6391323122 9544346569 : 4770
4982722206 7908232880 7024151373 5241623122 7477573032 5612470426 8544338532
2470127977 7859978351 8199543021 : 4771
4875947781 6076188527 0838202084 8349974712 7552820257 9694699665 5381939593
7982896137 7844300795 3018540036 : 4772
9661661157 6286812079 5797161535 6066202735 7626724712 0993482629 1145872585
6610030202 6165460068 4814387146 : 4773
0837279792 5961459932 3921101703 9733769873 4954390474 5166541954 7377441161
8809167007 2852497347 2353480092 : 4774
4594736914 4102322834 3634384103 7207710013 4620579466 4067075309 8240855503
5768869970 0784814367 5510401545 : 4775
3365221491 8883832021 3279173749 5225085910 4094126462 6158726646 1619176963
1733141663 3744493848 8514859513 : 4776
5867901870 0795817364 7585040709 6563444530 0631086513 4015456864 5460985779
3768228464 2303310173 2799271214 : 4777
4695553139 6650548172 7640197265 7906219538 8132535096 3250947445 9898910878
5901696922 5819860487 3251616314 : 4778
8722533208 6811525038 7783903092 1096397314 0870146326 2774873619 9925160369
0351640181 3228638410 1577891383 : 4779
4548208695 3949114165 4192122441 0372588235 3735283296 6225404053 9599551254
1880469553 4701692796 5818084221 : 4780
6911467794 9577090318 1471995224 2004890588 5593746441 6153490934 6821095273
8190692493 4012585401 6212983888 : 4781
2360671129 9047278538 2510932212 6760086356 7887299111 0606474438 5406313070
2691579329 1147168357 4893086073 : 4782
4176203484 2422557974 3251001003 8643688160 5524668277 3280148166 9787349633
2099634172 3779237883 0351660548 : 4783
7167179287 4001119144 7262456712 4700249762 2458224027 3977070270 2350323771
2416913149 1308448027 2640099449 : 4784
7259572092 3593023069 9273000522 4907365141 9777811827 2030525906 0505366479
3101848708 2839762435 9765410245 : 4785
4719021296 4624407218 7868542913 0719911560 2213435981 0926127809 9244929883
5321421151 6880443052 4223133517 : 4786
2036687591 0920612181 7157215013 2304915038 5243127601 6026780710 2779059878
3630284909 9513323332 5655424283 : 4787

2331269708 2604828410 9116462935 5307970134 7159992856 4268898897 0076744233
4648965004 5114824894 4884381190 : 4788
5203162023 9561251550 8011429345 5938402258 9044523549 1606770575 2641775197
3609044020 4483367881 8915689702 : 4789
7789366044 9983167550 3334609974 3453906679 6812379833 1363984458 6220491826
9859801850 6280836505 8175856328 : 4790
2867239702 1647037926 1157543687 8345664046 5906078885 8593615726 0733764794
7362839418 9492835760 5048336449 : 4791
4011349865 4912082100 4813255068 3756281970 2568696995 4918762197 6397204502
7900446675 6395119376 1315600645 : 4792
4486485525 0749799420 8500289544 4499533574 5046836622 7668720824 8316413599
4803070601 6118223091 5617525952 : 4793
8490028995 2934287376 1735102674 2418815937 1599489096 9792225772 1440139091
2722461788 8387436080 5196751530 : 4794
3147911414 3357320736 5859049304 2773379844 7129195444 6454304400 9597518309
7418233761 8633781152 9128015178 : 4795
6636009016 4769745449 5895732314 6795549938 9337513949 5643626014 5495926467
3734721602 1885313265 4468600885 : 4796
3720772213 4527510310 5952625307 1110235537 8856916499 5911692208 3888770780
5173543839 8567867015 0963780886 : 4797
8647575766 9825323540 4424542840 0882687477 7703285382 6199762542 5819929342
0591217977 8082778705 1185845230 : 4798
8972985638 7686511275 0727634100 3599414660 1222794895 7495592036 0994603780
4848385525 9599108171 2362827442 : 4799
0401784980 2110317627 8778875030 3630226361 6097666758 0106039555 4799787456
9981579733 4233997424 4475884531 : 4800
3933453664 5917552581 3475504634 4267161094 8908179968 9592267046 4402169176
8051005907 1844735263 1235416442 : 4801
4864777438 7877651738 5347897501 4025204069 3299113532 5561481360 4353296831
2928991295 3561529040 2759131767 : 4802
7341277704 6365308522 1325754863 0793354788 2996383406 9993715154 2249438088
2406473326 3123350461 3882159479 : 4803
5169175925 4509848097 9110892331 1377895396 6407460834 5730097651 1706075241
4362883466 0918003849 0635692685 : 4804
3296400551 6535997857 9130606640 4745571908 4866250414 6276335720 4425087660
3320764120 0276837147 2025839577 : 4805
5725483081 7635228170 6577594153 2708326625 5391096897 3058504562 2593689849
8975622702 1582652652 8062002518 : 4806
4416489819 1969095582 1207898972 6717646133 8008395664 8772201932 0463671881
7239547049 3020927986 6104118469 : 4807
5704868470 0496386412 5953067376 6660389421 7618938754 2375224018 7458159722
8442522907 9773722925 5101808867 : 4808
3990839885 4921449138 6356253886 3789161591 8884240512 9981936517 1369259169
5779199884 9494149771 1519431575 : 4809
5826307059 9485758635 3474963975 6855970386 2678054007 2008974502 5270519397
9698125296 8831191645 9045209756 : 4810
3052833730 9486023832 9627213225 6900737675 3391116829 4719812770 5742624375
2213758250 3200873637 5320464500 : 4811
0573891793 4659235577 0836220414 5285013907 9464067246 6736018275 3798544781
4076921256 0856944810 5416619567 : 4812
5264750745 2390254517 0531094066 2636682474 5757346070 4065275753 5777432010
2391334113 8135775033 2256114390 : 4813
0976099469 5213981770 2841461088 4136085659 1839329939 3135808195 2707069270
9276077217 1777909876 3854634460 : 4814
2405169049 9497747577 4828730093 3978940641 4736945719 8504829905 1348428670
7099150933 0445796639 1953558914 : 4815

7801094434 5098270173 6772409790 4948059288 3846575269 0821406840 9367522488
8424661256 0526950978 0101260577 : 4816
2822873599 3798499770 4573176117 5869419846 7474290486 4012319967 4462226863
6332600764 1107029348 8972360126 : 4817
2149845968 4160318742 4531584352 0548918590 4535694196 4460633379 8849315311
6954758361 1157497667 7407031544 : 4818
8057897817 0504573192 2588154943 1143902579 3450499895 5003727042 3618264568
0415899700 1369770936 4618431829 : 4819
6660690731 3547636451 2034680880 4485446394 7988105468 0984967013 1795494286
6813882765 8444650585 1797712368 : 4820
1426745854 7557138229 0726634603 1643813750 1465429194 5359934360 8206282790
7253188657 5179734245 7466122502 : 4821
7443019092 2429627769 3665315871 6809444242 7862498407 4946655635 2680450276
8435782113 6273406943 5616425153 : 4822
7923224666 4582371079 3214590444 6111092210 7039760456 5101012869 7605935565
7979972303 9393868396 1799189869 : 4823
1799159386 9470863242 4160101610 3098873785 4395677314 8297234896 5947671427
2134128400 4376210660 2005505662 : 4824
0233395195 7558645128 3024177142 8237157769 3287679784 5227571042 3728172468
4454616224 4727036661 8640546724 : 4825
9250144671 2045654783 5726921447 0538798054 2744365066 5491900369 7852300703
7200614183 7257113091 1168102172 : 4826
2867212589 5948574600 4353295033 1146957965 6490062446 2269539125 1125286803
7826375208 3462010919 3920529940 : 4827
6735316501 7583782632 7641004899 4464890719 5082147841 0208366699 6441555489
9731185444 5953123278 8198843519 : 4828
3646690756 1917443638 7489862636 2765030272 0686092518 8097236088 4793801625
2950392255 2102831859 1195209521 : 4829
6030879709 2230631824 3049051361 2517852667 8309896484 0762618711 2119385645
2332588015 6358456632 5681536597 : 4830
4140063516 6538527833 0279933276 1818731642 9284343663 5161566325 5280877405
4766967000 1881483129 9264949756 : 4831
0617449945 5604840565 1692066062 9441907471 1647405755 6195518345 3874064046
4663446523 3320351037 7844767588 : 4832
5666660901 7287520982 2447115645 0455672431 1091957346 6626779195 0351111912
4849140645 5543525772 5496566392 : 4833
6785221917 6238547538 1971083198 3228483946 9222949537 3749317257 7554258096
4794985891 0268635945 3576135514 : 4834
6603751391 4968451229 6745547842 4717793060 7499992487 1503917371 1331059841
8240045774 2575917866 7121950517 : 4835
4289611967 3898889045 7931260778 3631612978 2569411368 4507888434 4497866404
4958957817 7977537633 1750386865 : 4836
6921432845 6590763877 0350854549 2288177459 2134644790 3640967193 7884800156
5990852627 6175097739 4016437406 : 4837
4232153788 3413005026 2017131975 9124864587 9230068500 2533813548 9151319984
7109339798 8436544604 8272891549 : 4838
7107603731 7445126097 1717062154 9152309355 8078456283 9217995093 6649904409
6091569942 1493871124 2914663120 : 4839
5469132238 6063352687 4046418697 6497513460 0318428982 5747512025 8445983103
9820140609 4846870769 1618383086 : 4840
2387891061 4682419728 3200526150 3851106819 9058351769 5632365696 9414236261
0152155381 7250369391 9882029271 : 4841
3685444010 6294367264 5120333442 4480829528 5077865022 3588668479 8729233633
4526758265 4613647644 8809927763 : 4842
3165502130 0744945298 9543396175 2661174901 6912130058 1095542449 1226738528
5122343836 9598397804 8397524645 : 4843

1012058826 7428306399 9608907655 7344363448 0149048829 8357189816 4096451011
9289853987 6088229692 2643542082 : 4844
4456841236 8752332589 7783852438 4542275920 5676092607 9584672793 8663976401
3337529817 4103800839 7136698750 : 4845
1792518889 0050961627 2530290638 5122448156 2947348667 1807921675 7439502335
4187971463 7034135950 7068879603 : 4846
1015016464 4742239232 8672923717 2565384290 1470659068 8672940694 8425427159
0520534093 3901432108 1313854582 : 4847
5970434858 0931485506 3394187054 3754820191 7022688231 7510901329 4136171877
0058091133 4579616422 0301259802 : 4848
2043107382 9668649070 3750445386 8442844087 9094580713 8389620954 4920759961
5536655713 0684404695 2712994878 : 4849
3097134507 8827572484 4899897349 7039622465 1104987770 0417937126 3198665504
2482645042 7132291612 3093246164 : 4850
1353303916 7689451483 5856927021 6603190656 8999303527 2966942540 9329858504
7142854083 8671088927 3443108736 : 4851
3457002691 1192432496 3772982294 0094402075 2024662066 4473839172 0442575483
4014539408 0354627340 9990964069 : 4852
0720756229 0737468413 1198586578 7988559142 5214708035 1784967358 9369335350
0754467232 4613125258 2687046375 : 4853
6343964639 0594790703 9289612597 6486670569 0718158460 2993604285 6641652378
2711376077 4562109031 7980865717 : 4854
3199934312 8179691294 2962045569 5201243315 4460835957 6565078324 4672112585
0077609968 9071429906 2146372250 : 4855
1837031998 5169220050 8139102053 7898391562 2992500646 8735679795 4066278295
0224745926 6561768688 6293211625 : 4856
6560500845 4294559032 9173720098 2085881735 3746878318 2064470226 8612764976
0906343394 1566490264 2621944959 : 4857
9972627879 8874206865 3774840012 4027120252 7473344361 6681302272 0475458702
7046409864 3467573641 4944556040 : 4858
1804690256 4908532516 7271670951 2790052393 4217027430 3288614132 3309616964
7556020183 8621599787 2360962122 : 4859
7570421099 6873700071 2257994295 1869630041 2954188779 6287240932 7846179804
8647907699 8905777338 8605583824 : 4860
9114228316 3748692878 5381481431 4622428398 4962337855 7735086051 1010509261
2932309949 2474189267 5131618818 : 4861
6207532574 0386848069 6490469827 9991221611 3866566382 0326019135 9684025156
4924790644 3983330031 8078952369 : 4862
6165184056 1410338434 7993761781 8210047435 1909070654 8064032056 9117166432
2099215471 2404616942 4710431522 : 4863
2205104885 3663827047 2083522207 2281792307 7243786214 6298606683 0039443184
0318971195 9387588071 9811504839 : 4864
7750862492 8452053766 0993570351 1384695978 4295954406 4857515050 6208497122
8123360639 5414804523 1830375903 : 4865
9230419312 9358047025 3223963911 5127937593 6015140870 5351460629 8995194074
5755354886 8619822693 2469553821 : 4866
0219472021 2488490442 6365539530 5394353300 4050613359 9683201666 9879477207
9525866914 3318990921 1865226058 : 4867
4960881496 2683654672 3498404814 3810943198 5740368377 4663709365 8063733929
4587946644 5168261148 3334598089 : 4868
4813230142 3449736300 0817700302 9015416740 1795744361 6201842207 6053577654
2316284564 2764409375 5689715537 : 4869
8728695987 2245740725 5862889162 7542274001 3939248909 3720555456 1607842415
3956188399 9055824797 4029707414 : 4870
7881435727 0791492210 4355774546 0934900573 7681596828 1968019771 5906457966
0575494254 1814453299 2479819966 : 4871

5711964551 8856556568 1215133490 9804658037 6847618266 0137371756 8372741306
1704644380 8292439165 3713296382 : 4872
6165307231 0606823338 8999951025 5307327441 2094269941 3792011010 4663208749
7154105874 4582611599 5906877240 : 4873
9428697875 9462694298 8054778607 7645800522 9407570308 4063851604 3605931255
5658553312 3116554638 6554103007 : 4874
2699344773 1181690786 6280119553 2248099561 7127881556 3935190125 6077112884
6947322636 3577830259 6564232697 : 4875
3847446967 4793817191 8349844120 1028423519 2195947941 1888645207 7960231343
0632171743 5406924886 2375881350 : 4876
7950130195 4212568538 9606693125 4279329444 5046041439 9751105930 6830758986
6631208970 9178673814 4310626732 : 4877
8572913524 7335737334 1399649816 2256056419 7417879844 1560213301 2302960042
1489114784 8575264386 2518142961 : 4878
9136898382 2274941558 6395787408 2378090383 4140879297 6445118888 0233720098
8640122361 7417643050 6370381076 : 4879
9278341189 0172940132 9462385639 3085315322 4422276881 8571728893 8851686500
7590484415 7097366365 3939411381 : 4880
4143300708 4179831172 3439743212 8982693088 2817908454 6083522761 6883923670
6701819377 9063875186 8898267898 : 4881
5926122487 7072553707 7468090917 3812195568 1542467817 5824986359 0845775282
2108733709 4710053447 0431626453 : 4882
2650987541 8704826040 8504943289 4496711511 5564045037 5992218315 3150621894
1768079740 0402678059 1637359492 : 4883
3058021891 0561624590 1301913073 4230922011 8774890541 7634752347 8720509575
0830124580 0976089449 0201422832 : 4884
1517081786 3829715178 7541023277 9377854072 6836395180 3722727399 0724613281
1667262023 8536295804 4768744894 : 4885
5589355612 9416678771 1514445952 1507280112 4528978389 8892439805 7926113395
1171633727 2476322035 6134137665 : 4886
5493609107 6077542886 3941537509 1128355095 3891017569 9273481058 0205244749
8860656941 4068805981 5714553496 : 4887
1021164041 2992071191 5738223964 9687900615 7717959575 1167559084 8192969043
5593444902 8936964861 9008661184 : 4888
8364084454 7092298808 2018696435 0102416589 2806728934 6189212883 9980588675
4723382266 2874674898 2760072139 : 4889
6917908199 4180723645 1438298943 0173553032 2511478911 8467993569 4376925654
5351140882 4626441787 8175405852 : 4890
0462752091 1945517150 7153223229 0031783786 3929972645 7950413622 4307388747
4295884203 8653796300 8878284974 : 4891
9101536714 0448727060 5223983274 6544345293 5652807696 0477221503 0240603299
6082014428 7406646309 0243695582 : 4892
2017514819 0212769591 9485480650 0251596627 6672115626 5827038707 2145811474
4697943698 5067657688 2350350632 : 4893
7908552602 7524578452 1393118971 2195956022 2778010521 7056312396 3444611270
0212586723 5737090090 2467406949 : 4894
2576669653 0479737142 7626570535 9574487908 7619929979 7549154709 5674568662
6669331781 6026949924 5637861334 : 4895
5040128648 2682564546 7814425734 3138486006 6826797672 7082780868 9640327625
9416788405 7145941148 0462916488 : 4896
1285200102 6105619845 2785431676 7343019418 0028739280 9461519481 8546866310
9907951504 4881336501 2546122141 : 4897
3542791228 3913036947 2625897800 2733168003 0312219807 9222723393 0233396990
2493238136 3017812136 0113136831 : 4898
0388677339 0012252524 1037628839 2001146874 7397830196 8151623448 4584417084
7115405630 6712215025 2400600956 : 4899

6845020996 5442357785 5478800658 4453063354 5865197485 1727110468 6923708210
3742383445 1621446359 1642683264 : 4900
6976748511 9161178393 4195142162 6343457280 6939369671 6440345231 1179813313
6747255810 3023300335 6470919313 : 4901
1184154932 7154187214 5395500896 2151995577 6480953224 7281705684 6283525646
2754305762 0000172149 9028964513 : 4902
8438331171 1658584177 6451061858 2508210472 4546989085 5260571543 3473640776
6059802184 4622302303 5764928698 : 4903
3535477228 6691167255 0929264234 9535911297 7808670676 4009141760 4725110184
2295428946 9724271931 3957825111 : 4904
7713736466 1288610993 8596154463 6685556640 6324240965 5115271244 8740813863
3958403290 6830166375 0515937889 : 4905
4470032446 7944420500 6232380567 8075861405 1617949156 7587848547 5379803613
3204588672 0131374684 2227773933 : 4906
8301015927 8738266403 5207938573 7273318787 2211600697 8868491056 8876737354
6820717934 4205169186 9120733906 : 4907
0230403595 3144024848 2755249884 9668639056 1518129135 4913330689 1980509583
1954745243 4768855877 3831041434 : 4908
4539933978 2643790560 0793257790 3268772389 3359704089 2906379862 3591326882
7691204823 5322685331 1130949276 : 4909
6145457891 1147071816 7829879548 7397796649 3933404999 5804451114 4263787805
5006181201 9428565801 1511604965 : 4910
9751470936 0848244784 0715101656 8819151547 6308537089 9177749500 3154677919
9405543023 8455568432 9587122807 : 4911
0339303655 0805203590 2690222574 3269417733 7833020778 5088164599 7749984355
1470304189 9310927886 3628555954 : 4912
9866645215 1460832243 2702939865 3547117146 1120978644 6464432561 8002352000
4864169280 9012519555 8796777707 : 4913
1201415854 0324292727 9651399821 4262542785 2383618248 9744008376 4657199183
3956603365 0949707852 3288974588 : 4914
1493347730 9948639738 3182019324 4687919192 2759819287 0591900590 7540673571
8523481186 8346434778 5054157828 : 4915
2244907985 8408700586 1205284471 6335877723 3043808270 9137396889 5056845553
5232618221 1102373127 1368814205 : 4916
8398385474 7049000536 0938657617 7773048783 0785597217 5388186172 7200525683
6873008672 1769154412 6287754950 : 4917
3239222116 4872217861 5226209112 4259237746 7055016125 5823546154 4208744584
2149647760 2559572740 2918471645 : 4918
3158885258 8278842868 5588949650 8427039429 8993544274 7557848840 9253424367
5964612654 3810920255 6299820515 : 4919
4148019374 7004803579 6677458741 0088413181 1872582385 7854484888 1668277133
0320509148 0499027063 3866946909 : 4920
9351248792 2710050927 7461414952 9146261051 7648599069 2992381759 9024416187
9497023565 7180951442 5854995765 : 4921
1792962105 6744600221 4133673754 8079860359 6959468569 8337977267 2852875925
0604441846 4357782403 6237766702 : 4922
9872832370 6045274425 5506815180 9555888683 5891475330 4570116921 7308190413
6821961979 2444604264 9934384845 : 4923
2732374619 1137110824 4643350169 4801550253 2841846287 3349226613 4568944139
0024706633 5547078814 1605485679 : 4924
6686564326 6981303178 3621646465 8820830050 3644548129 2288496446 1641012502
3757682821 8945511845 8059573209 : 4925
0939462064 8677509380 2437912896 0180082794 0474817422 0633279147 1284209513
7010561943 2717629286 8579090949 : 4926
3218055568 9790662862 8915473754 8673198693 7384664195 8561889977 0827990016
9246494251 9034737770 3988630149 : 4927

2011183528 3723279199 6507771558 1634061761 9698407222 3186782701 2354034994
3842916855 6505151743 8377353920 : 4928
3225611599 2975687965 5748563713 9984984188 9818723863 4477477355 1501353507
9191081808 2698953601 2524782547 : 4929
0432140977 8469323294 3801472551 2584376086 0998146021 9969863154 8473576292
0809967022 3123077999 9766892614 : 4930
9370948131 3117612672 5029020251 1250176953 8583175793 3248237475 8716019598
9309689954 2945715238 0277892235 : 4931
6850487641 8241391023 2691491655 7444484512 4608305145 7874175073 8717674417
3551101307 6455378978 8752222185 : 4932
2058613301 0759127201 2841581609 9012918291 1856667157 3929980929 1799149161
0004660333 2922577608 6756566635 : 4933
9653418185 4225605988 4318442721 8109423181 0906310605 9677332784 0395059606
6977210277 7361411827 2064342985 : 4934
3844246587 7284718517 8836337528 1932542583 0054996146 1483735041 4191861619
8629106706 3879119517 6205055641 : 4935
0548148530 7823643720 1653999652 0950423890 4116749764 3602902434 1956762244
6851420894 5656803232 7777818280 : 4936
4023531717 2716161384 7452643425 6170300403 8807168888 8421475795 7281324163
0693977190 4869321017 7484934395 : 4937
2208897797 7460641321 6065912354 2873066349 8564951384 2991511937 5415682688
0464034407 4003191648 7524228128 : 4938
9908496049 1088752481 8976629544 3946375362 1983060853 0266297677 6229728237
0884106403 0679839570 4550798642 : 4939
5562413295 6970690312 6069372722 7049738584 9063548119 4665353366 0264315355
4561276200 2245436501 4092870701 : 4940
0384502494 2299682469 8948193110 6380584896 3834966842 3154689585 7394178100
4780274310 2434361706 3937322667 : 4941
1414047645 0553206852 5452777271 3379959897 1912933510 9533521837 6237814482
8212389833 1839226954 0362944194 : 4942
4477939338 6294782085 6539160276 8654742216 0219745416 2503479486 5733087486
9636214605 0796657211 5585682647 : 4943
1256401983 2368722180 1627370274 0854612331 5314874653 1939362613 4387278498
4098267899 8619691220 2669754590 : 4944
1203347487 1715272034 2378485320 1997739410 8939918323 9935343608 3831944835
0407170160 5001971940 7955692916 : 4945
9441248268 4602905546 0248055663 7492567690 2358565496 3056305499 0947383386
8200225327 2751014765 3820157189 : 4946
6891764386 1760384669 1471270502 0901016679 3143538898 1799261389 4958205684
3152542717 3339845849 6778421955 : 4947
4228496937 7538214198 7539836280 4151385327 0951507067 7831761849 2686749245
1878878256 5238807875 5933987418 : 4948
9923941064 6608428415 9517680029 1899674460 4562347684 1746284361 4905240996
1460913299 0782507767 1729948737 : 4949
9930049706 0952190291 5918788156 5502948692 4994826388 6396823849 5608685005
7999208125 9169809576 2500632522 : 4950
7429775722 3532446463 1299191602 9527828387 8086374379 4848781600 7986691844
9449520338 9091181996 8400143601 : 4951
9370933054 7221904717 8616199521 1092911279 1700197667 2020216396 6918422241
3495352640 5415034347 8493730094 : 4952
8107500099 2303709153 8820155082 0033012007 6040367400 4750282381 2172353310
5469837101 0096575548 8116614281 : 4953
7419714224 1111465396 4940881068 7135316799 6748974279 3722635135 8103899882
5451605536 7251021364 1692900321 : 4954
0731223430 0723995569 1422524643 0126273566 6817822859 0169033995 2558864708
5175428439 7934232749 2533998925 : 4955

8403923207 9835993252 1574358425 2303743666 1389938632 0008006457 4750880709
3799846274 7096657080 9436929936 : 4956
1070027345 3815684801 4398180022 6497961654 9839242472 1495385516 6490613388
6994794487 6407062566 0160551717 : 4957
8911310515 7898124840 6744154043 8634321808 0496035776 3693369650 7502496754
6596535171 5008599750 7640004559 : 4958
5426370119 6268335042 3969409324 7325407321 7465365771 2189786335 4556824170
3910378182 4265672441 5781843849 : 4959
4538256203 4978117494 7104658950 8232140820 4782053999 2217083096 3792471914
3570526892 7378829630 1720459841 : 4960
6396765979 3992468451 2021673155 7594061085 0110840150 1493958481 3243143264
8317063835 2293389835 7328629625 : 4961
0064539653 2323409016 6345534976 1453977754 3545510180 0227298781 6661057242
3124306235 0399126692 7255939838 : 4962
7044682244 0569021752 7208905973 1403171949 9393757606 5170443081 7843584689
0232264090 6702558256 3156527103 : 4963
9919878744 9960056696 5311694201 7890333193 0791287640 4500245292 6077757355
4483085149 9121604626 0407966357 : 4964
0042929414 1521078517 9395124892 9311310872 3403687549 3332119971 6941558224
2253234526 9916514842 7080749649 : 4965
8243209108 7091302719 2207360528 2398890333 7764824402 4821643674 4892838932
7178724630 1295213777 5840656766 : 4966
5034225484 4795273438 9296263521 7069248295 7223372372 6052121486 7559012437
5106886361 6862068481 0753252551 : 4967
9080870082 3937566799 3000525640 0410568687 3213457742 0110043021 2747964046
2677207960 2886807545 3328446116 : 4968
3963670296 1676361061 2095640915 9039226759 7725612770 8233691017 9793240276
0094779050 4939059499 0355097623 : 4969
2855245692 0149233803 8955511453 6945379896 4243907753 1543866107 9617254935
7971644803 4461266623 5380414555 : 4970
7367642625 1445905719 2580222293 0640330494 3177399110 7745994805 1848434169
0301247105 2840011453 0117015926 : 4971
4176031004 6687984340 0676366135 7541593810 7394902338 4595997856 6490063310
0192580761 7965927489 0217308818 : 4972
6545124915 6100845649 2191733984 1849364007 8924253400 5288512740 9782607281
8449936233 4439677783 4164303286 : 4973
1707465557 4470958871 0661228597 9843832898 8827786089 9498259344 4570625520
8466693362 0736456132 9951753464 : 4974
9909660709 9343125635 9749029656 7184680351 8878764437 1927343228 4967575348
7438060586 8393873208 7107123411 : 4975
9603308930 6023521935 0237964753 0151415937 2862118229 5259067018 5758559048
6981033613 1061953704 4107720860 : 4976
2330006694 3559822989 9720038942 0507124130 9633012473 9898898650 1613446041
6369764129 9185513985 6413348024 : 4977
4010903820 4209805098 1881637076 5602535422 8852064250 4748958680 8991794668
6117193250 3324823024 1059805584 : 4978
7663804552 1378932305 7235020097 1557476025 9377207677 6087468148 2134522531
6302087888 2398535566 8408462018 : 4979
8776333889 3823940059 6938234755 5811966044 0536601708 5155431409 2863357092
1594481160 1753296583 4133347177 : 4980
3027110597 0905178115 8901708660 2990160512 4794507024 3012323067 0262179701
1415106820 0226813999 7507258321 : 4981
3035616679 4912610054 2012864532 2980067268 9000948209 7075854102 1988488529
5459601974 7306361328 4296985538 : 4982
5226523816 1308896659 1450912488 6813125953 5362960576 6031975042 9504118843
9397247053 6057894798 6283171400 : 4983

3968480764 2119094142 7568120273 2454233195 9311195239 5056290622 6110099643
9894838164 6448745866 8307548577 : 4984
8532874081 9937572748 5219741371 8094296777 7411722239 3641356033 2119093344
0755678783 8113045199 8451486289 : 4985
8000608483 8694206218 5271928018 7780424866 8080299512 8970347329 4463170946
0038595125 4533868355 7968905846 : 4986
5172300670 4488896840 6108630406 1213515520 3874213928 4496202225 7546585820
8669864060 4986554258 8590814553 : 4987
0994843493 8427338421 7864505139 8542739742 9095857008 5614625618 3495270022
8141732536 7653979469 1275297470 : 4988
1317006383 5415965446 3424496835 2635059485 3447447210 7805610781 0829649426
4788100259 7931877563 9239043291 : 4989
7853276342 0375229756 5752743408 2950845479 4701524526 0899313885 7831239117
5126922556 6757288513 3404397696 : 4990
2540393117 4933713994 4952935680 1060379694 4595685975 2498772673 4807907326
7618245233 5521216214 9680234492 : 4991
9254288655 1457337565 5765945557 0923953342 8142462903 1727815403 9983415564
1983771801 8982112476 0855955518 : 4992
9995062073 0071403452 0815503329 8149750702 4426772643 6033873753 9731484313
7407092665 4492295204 2319900734 : 4993
5946393119 9653505680 7332981486 5841109199 4439462723 2845367711 2848473622
4606331360 2859105963 5237193871 : 4994
6345986963 6443906854 0532231931 5241354693 2487576730 4633817030 2944798352
2602051814 9444585049 6120326909 : 4995
2337527162 3551335262 3432072194 3009358815 0339359897 4493352695 7874572783
1403967039 6910017073 4149325310 : 4996
2206326301 6925237018 0120244226 8849290981 9555117195 6120838155 0144858365
7166510269 0866487173 2381901486 : 4997
0992469913 1546082001 9927050473 0568876141 8932981083 1023526482 8108102485
4502208757 2212834413 4379484999 : 4998
7279202583 5434172044 2598469327 4091417821 4394928017 9974565987 3698287428
2682674844 2121371546 8232751128 : 4999
5341652370 3165307043 2582833721 1237137609 6959399375 4953622322 2219746596
1933252907 4042487602 5138195242 : 5000
6973910175 6371975343 0044796178 2504311533 1506758256 2735343476 2525391425
1527570478 7843767885 2418719663 : 5001
4624199270 0802576108 3927497622 6365490018 6532064549 9515580290 8398513272
6273219672 8478302538 5221904792 : 5002
7868938953 8780368699 3188466031 0436335243 7327154699 8988111864 4367011402
8426202615 0473882358 9974728149 : 5003
3343257065 4746370224 5188728906 1255030273 7919026403 9657741760 6898976983
4566464705 2047163592 1448307099 : 5004
5843064717 2750763371 8020714597 4456525141 8850503716 3773818902 9696852844
0925819317 4410055576 0898502923 : 5005
2710160089 8257223817 3945436985 2965476949 0187384046 4656437307 1319590114
0761317428 2203883313 6029708526 : 5006
1451234907 3071476245 3405024542 3763666685 5757401206 0405988695 5630114415
4431416969 8607403220 7886003132 : 5007
7905539178 6967416355 5240250767 6530856322 4273784714 9746037784 0646093468
1299187190 2892759763 0701598408 : 5008
8778174519 2690147691 0300302349 7945851100 0180886606 2162868011 1091511610
4098327308 8089954337 5511871831 : 5009
7378655877 6487358854 4903668790 7416918253 8313306362 3382058201 5488544982
7883821742 4375805492 3815972964 : 5010
0619631058 2151670919 0326318730 9338138501 1309133277 0930342511 2210972155
0561049138 7050426219 8052602476 : 5011

1160750597 1037943664 7715293534 9171861250 6611636738 3304995648 7877936569
7911293195 8782777740 6010753337 : 5012
2432790009 7324072560 7038880087 1137309659 6344570960 4117508944 4791191621
7455169709 6484476204 6174128154 : 5013
5361284226 0155130158 9525862806 9570639244 4354780239 0516861338 5498619266
5676227603 5823498556 9192248906 : 5014
9011645986 6096156795 5415492157 5812835409 2025393170 8073781465 0788320352
6651345707 0655368569 9281124265 : 5015
6708448017 7864614974 0454921184 3122762184 5224410884 7150757019 0883495022
4375698850 4945424105 7266246094 : 5016
5832095962 5728720309 9477654010 7398558069 9661980195 3450050651 4501054924
0188610891 3417306075 9238439461 : 5017
2465698616 0560909454 1772907364 0093913219 5676836433 3299619997 9654243480
2656940683 6986706175 8741316765 : 5018
0506027135 8825744337 0721645881 9522613397 0545205234 4128092317 3065051989
9545028585 3835228737 8728698291 : 5019
7275807824 2098261060 6090150925 2110908989 9643892978 6294301411 1400676287
7499207817 2379474620 9168998389 : 5020
6189486307 7306027491 0267038839 4335247424 3421236712 1770246101 1602406111
8571687050 8244049162 3894343504 : 5021
7025081364 9155155104 3272507479 4102473747 8192265205 9335513625 5301812196
4249924882 1299260177 4080103719 : 5022
8842125474 4721602969 5002774268 5077521758 6691799380 1362584224 1739856076
6991654327 5706123045 2867280707 : 5023
6318947058 9544061884 2131571380 0339848740 8680941446 1845854230 3449514485
2105399124 8932455486 6593053355 : 5024
8457827714 4237743385 3392082412 7763216596 7538326650 6430638806 9556915620
2622259469 4142900799 8769834420 : 5025
9146999897 9568326670 9041413980 6863332515 6525696878 7899225743 2796139643
7026543144 6393337990 0081985753 : 5026
0797881761 3374360948 3928302233 8032797320 3865364624 2498055114 0224692264
7893159294 4918040382 9836490348 : 5027
8863869756 4414855608 5605371772 2873714822 8236864278 1136572039 4749690693
7239915610 0368280751 4117847378 : 5028
2562212775 9959167758 1403853298 6888513655 2323139384 3174225853 5628070191
5440077630 1634764778 1261324380 : 5029
2484218656 6985527400 7641051190 1095452898 3825634840 7045580809 0522147697
2204287382 2020644511 1655802021 : 5030
5813722046 6347663335 1756179905 0089778088 5600932081 4541764439 0323524193
9731547218 9209202024 7427040585 : 5031
7913543792 6009768076 3629875661 4782644338 9261212134 5659445680 1042227616
5241837640 1867345339 1488079001 : 5032
3636035284 3217478040 1683146726 1880679364 9304686587 3646311483 2826980492
0227876255 4540178509 1774977282 : 5033
9467569293 5451666098 6419481458 3644478909 6038304582 8536989560 8066566619
0360310045 3047200242 2639388157 : 5034
2068674690 0619298730 8224849001 6816348921 2554337544 4758038873 6158658859
2485975587 4278385999 2954170991 : 5035
7225115340 2542222969 2234366177 7819296533 1349674775 7220978430 1938184450
5985075080 5649483189 6792205834 : 5036
3414789051 0257617490 9975820012 3472441028 5079412318 4376014714 2137829511
0218735899 3917081686 8055988133 : 5037
5469913794 5378313771 1413549197 1243465810 9311278332 6621759222 7589369934
7704984352 5192107351 7573702005 : 5038
4573451305 8731322143 8462087602 7751840534 8937132376 6290711205 5269490530
0388822804 2048082108 5192876107 : 5039

6772814548 9279440322 7236284124 7943192644 1924307968 1370382948 2005869742
0150132353 1987205199 2035387731 : 5040
4215882211 4859743182 3879098919 9325393415 0378768995 9695804327 2517776898
6612588939 5442013917 1306418862 : 5041
9510665268 9606241452 0844803403 4741133148 8004313864 7317158446 8157548365
3167051260 1746640760 7280959672 : 5042
1327294224 5361042667 9280946977 3583638349 8228114766 9169596955 7327702827
9120997926 0863817017 6542101511 : 5043
1581109268 3287485950 6462623629 9259249694 3781822291 6923432548 5361604152
5142063692 5214774598 0941988926 : 5044
8147764439 0537611191 7463026074 1704144922 4194980017 0623866816 9466079892
4751697185 0009428744 4466241005 : 5045
8638035586 3065063609 5727097973 4930709648 8907606301 9079233461 7019745665
6248712884 9867382678 5462689451 : 5046
2094622905 4827542330 2517321328 2785316517 5869340541 1184571510 4834632832
0081011525 2541311936 7954561261 : 5047
2161031974 2783210387 9548253917 4314098770 6525700603 7671968383 0478692910
0326748344 2066840667 3515088586 : 5048
0916629765 7227799692 6003868273 3496418758 3321180809 1686185171 3459115685
0893149404 4819961072 5023378967 : 5049
7989918872 5826867053 7757435124 6508137986 2813483506 6431324662 7092452365
4698351740 7042651369 8824143308 : 5050
8355263194 5475425324 8458840939 9378463186 7338133661 0310058114 6455151877
3050141308 1566601806 1132268352 : 5051
9639711462 4010498314 8464604395 2006163735 4258469826 0521254538 5915053646
2014165930 6152413774 3303724644 : 5052
1039759850 0156892102 7909378678 6031851767 2404695991 5723405290 0381676072
4247701697 8152148615 1947462705 : 5053
4614221125 6912722538 1590502039 5830514685 8500360242 0054909455 0268261850
9587542621 0973991322 0950295489 : 5054
5828917423 2981343154 0692575705 8801573654 5351789274 1527128941 3143182694
7690258885 4786599196 8988359990 : 5055
4388542274 8469473118 7519748877 1215019159 1908885813 2637699812 1590808969
1823350590 7971487461 3336803970 : 5056
6350036955 5394225907 3453693326 3896961623 8425410479 4629483793 7463813246
1240825306 9026914650 2151719655 : 5057
2425678847 1272297526 6792028038 4129661575 0218468110 8869393992 5268794395
0326728700 8758188610 4930085975 : 5058
0196928095 2048237571 3894543884 4241621756 6263735180 3364327740 1734125942
7455820267 5440278812 1409286880 : 5059
4120300022 7063180825 9318676648 1490447257 9944670459 4238281114 8786117712
4712994023 4353193476 3897789484 : 5060
4625604174 5478020529 5530815025 4298015709 0392382147 2145748055 0269684443
8096038016 4437261953 0044039908 : 5061
1842658748 7953263601 7477343963 4741333029 2755818200 0846488609 8183291707
8621243703 4594042815 0160414770 : 5062
5818564413 2223188779 3921639889 6127369924 0022516053 5128293723 6557373193
1938983417 5384397083 0903585333 : 5063
0857183693 1136503700 8190565450 4329842209 9381490045 4454004486 1822140550
6052871683 1341426278 9644169533 : 5064
3980296879 5175366678 4755096725 6739077347 1816933997 5900118989 1113962465
2061607188 5649119266 4269401606 : 5065
9590616104 4378014986 6699828433 2465257608 8257101089 9195911118 0835030365
0797428123 0709395198 4025480169 : 5066
4462620592 3636351071 9901481567 7440001231 3072541025 6056015931 6840532816
9973390707 1537208809 0132654154 : 5067

0840848694 6485133762 2748284991 6127474705 3222550579 1833719782 9980217375
9142434471 4866343808 4599101392 : 5068
5549446596 6742373011 9121569104 8473306980 5081712972 7313806629 2296784931
6570700310 8305567889 7912329813 : 5069
0375103174 6783401120 8135309146 6073367511 8762187223 2478968370 3827950123
9544495396 7588537384 4664134720 : 5070
2470015885 2017613121 5481437213 3479265672 2446917956 2586330030 6994583628
9103812648 9523388807 6384381880 : 5071
9211807397 9276831412 3086880946 9171765526 7197134112 9027158146 7529442762
5213295015 4301153369 0389221384 : 5072
1013378832 7539359202 0905194209 3938979412 8938076595 3087027355 0773042219
1143004325 6684624072 2890340217 : 5073
1151273168 9053394678 5187382278 4851582247 0484127739 2682805651 7260376445
8440068284 6673735448 2592496916 : 5074
2142390338 8931811701 1389131501 9222854107 3818145229 2592185148 5752524761
8384807758 3829866667 5567628883 : 5075
1008601526 3589358635 7127421777 2895154458 3129347098 0084717828 9672734129
0370760715 9708978389 9342792625 : 5076
5738188956 3194050727 5208732599 9456545608 5868553582 6313370849 3406413934
9137491790 7721822853 1160094982 : 5077
7036679278 9235878843 1608491838 9533623615 7657683342 2289592702 3939153868
1195815299 6066452081 8861843896 : 5078
9720018935 6785489494 2216512361 0171847376 0460603829 6164427331 1730402586
9985215372 1647753632 0660720612 : 5079
3292932396 2407800474 2806780812 4542992795 0935051439 7602119291 1303780217
0395082950 9906994439 3911955003 : 5080
4072157414 8017759229 9719328271 6303971339 3800349267 9133416508 9113946424
7966132094 7250898195 7237504911 : 5081
3312916698 4273711489 5866286493 3613397030 3624716301 2832100708 4152026276
8996306818 9695266669 4405167511 : 5082
3293786905 0034120042 8417394295 4144584447 5412467718 5087103019 4976373064
6456618634 9406566159 5559022203 : 5083
8813571361 7682347784 6310988547 4803141607 5183130544 0342811270 3507195990
1420305881 4923214884 1293569445 : 5084
2964317031 3190357793 7739190165 6912577178 0698512180 1152036378 4718232894
9534446551 4626784682 0263039666 : 5085
3019906922 6106232408 6257493667 4402658139 7192480133 1606041280 4309403011
7818395338 4569734777 8755496505 : 5086
2267198920 3980646508 8098765959 9040268354 5829084048 8743590121 0248570719
0569558518 2299648163 2457015703 : 5087
7483350716 2561778784 0260248079 5436431427 7409752046 2960077059 5232159378
3926560584 7631831645 3946086912 : 5088
9139029707 5293245752 8311220631 6530779455 9970935880 1910179547 8468222029
8137949673 3900870993 4835236592 : 5089
5215317478 0914929635 9576803641 6453157419 8408294970 4240390235 3402659221
3525985308 0332431785 8577406056 : 5090
9003951749 6624783114 6154707471 0152410927 0990190694 6055592396 8305614807
7265373998 4996188802 0452613695 : 5091
7223089073 6822583697 2534673293 0434907962 3158386261 2634442408 6314103499
7403655086 9216050046 2809793942 : 5092
5957370775 2242127755 9627928584 1773136930 9036459436 3844939823 3929185489
5536764735 6087898997 1977807038 : 5093
6620927463 1851985085 1926852624 5382587024 9573805010 9938163238 7082022856
4474021890 3503702051 5868245711 : 5094
2552023153 0878090161 9437053871 7681644829 3944911865 8976812285 8228286305
8898610296 0528325443 8604524159 : 5095

8879043704 7737737136 8502135393 8226539490 2342171088 9783710051 7498175970
2068682570 2725971239 9904777313 : 5096
3208906742 1121200821 8398677787 0565629424 6938200423 7813491306 6630287587
5209256477 3273167463 6966474625 : 5097
7061200346 7156409034 7896314818 6821345980 6127296825 6545576749 8591972867
3797538106 7421738580 1948693932 : 5098
9232978545 5610079000 5472533051 6708185495 5082957331 6453038966 7280573878
2805825883 3867439567 2132016122 : 5099
3642719023 9990318506 9780948472 9803770517 2000913056 2696720528 4316649649
0463110313 4009037815 2174924340 : 5100
6714466245 2624788444 7970326708 0920479660 2152536229 7779841303 5291824916
6775473338 6109757717 5208925700 : 5101
6270953984 1927672468 1021271464 4651636182 9038678557 3635507001 1836724154
2895404774 8343284348 3747564304 : 5102
5274450680 8442943288 0032635348 1326604760 1721422694 0034962856 3970257256
2837188280 0895476023 8666306549 : 5103
7470453498 4376640385 9894645814 4805319509 1992594514 9823361365 3904412316
7407129663 2618704241 9884078656 : 5104
0346889440 9807356511 1719088416 2131337038 3284752715 8014023337 4087460550
9019733708 3604720090 2016509576 : 5105
6376037863 7218832003 8639008631 8700252115 9432783932 8479264705 6206906139
4036488309 4552023943 7276001155 : 5106
6448767844 7540835616 0399848851 3772923034 3235300979 3967833698 3009127779
9794971704 6285314100 4353493338 : 5107
2267484965 8177523531 2719601590 6172828462 1371401030 5334392027 0345130194
9107034461 7456577179 2191324143 : 5108
7364969396 9048965732 8257566913 2764531083 4456871594 4990050920 2240475799
4214851413 9245457253 2607827164 : 5109
7130569958 1372348962 2724101513 5814633935 9857391762 4057344627 1578414318
9580687674 4880080349 0104731595 : 5110
8190724146 4172911598 2896970259 3777767500 6732038336 7599920898 1411198985
7577054569 0088066735 1053470372 : 5111
1329348905 5455497205 3530105573 2469661053 8066099500 7027547522 6267638316
5272527915 4833145361 9391671955 : 5112
1030251519 4171781239 1244827611 1453221886 6777176734 1740645237 8993015037
1042110232 2388477693 7314255347 : 5113
9838888837 4132460169 4849452299 0510710895 5879321620 5632208515 6530074079
5055924944 5974351034 9190441133 : 5114
7915663666 1772461772 3425057713 5160442663 0343126850 8796541039 7347392745
2905140551 2410302983 6258933623 : 5115
5834267865 5320477807 5390909094 6014043806 7643829114 7830296465 1821538618
9201400711 4945518626 9299132473 : 5116
7572992206 1152524782 9166545756 9566383251 1478672525 6097578274 0644380460
7071542461 7738733768 0719306797 : 5117
1913031072 5974717095 1488737474 6950349152 0242571874 3866194207 9828728831
0571028157 9364027146 3453279847 : 5118
3494139024 2082297386 6014373548 5609080295 7134692265 3311826406 7965254434
9899969780 8629347038 5536157001 : 5119
0379699864 6180680289 3285729725 2901428290 1970405011 1709396040 1799400772
7907300595 7406453592 3916663881 : 5120
1445976418 7426920093 7829705225 6405569829 9485145116 2539665867 4223630839
3508148778 2114697145 1562642183 : 5121
9723245197 2563939261 8648161829 7743853241 0643492525 4168626435 2270431372
3804662598 5568603693 9297507794 : 5122
0069921092 7268702532 6458806166 6922557296 3522442048 5867181204 5093427166
6318905192 6245014908 4600688052 : 5123

2350944434 6582740225 8290512950 3049799614 0166077255 6448746146 2709786889
7325672101 9292300982 4547654380 : 5124
0864374001 0975723666 0924137231 8680077447 6268629452 6391724897 0267674567
9273486954 2632962933 0779938306 : 5125
0695174258 9565375330 3319825137 4669746258 7161719676 7115785959 1293894641
9005195942 3116255426 5589036040 : 5126
4261172499 8909306405 0953199964 6107199705 1693828158 8424916149 2363960011
1384532591 7165402764 9547661729 : 5127
5659031279 1649517360 2942136281 8726459267 7678276896 0296649722 4763374234
5885325543 2357319176 3216539724 : 5128
3501468576 2391656504 3387680021 0156944358 2360863484 5735010253 1746380709
1205815681 8796838583 9451042367 : 5129
9587670566 8608026594 8295232559 4202921026 8875291154 3692228947 1789836330
6284089773 0393136992 1911471481 : 5130
6289438323 4242910144 3255465123 7764299212 0827639297 3771594064 1397405209
2305309640 7841631655 4355424023 : 5131
7060821781 8229921582 8215654017 6766976612 0899144806 0059529906 5437623636
1202630598 8951007723 5755651659 : 5132
2197145167 8367435512 9084511136 2955595308 7586708377 8001919961 8165568310
0455702282 7891916130 2403618351 : 5133
6412231627 0904542939 7967412823 5084330945 2202160626 6842689197 1808149654
8307692214 6809272437 7218327397 : 5134
0929550675 2479298131 1566988231 5933109698 9939122543 4479357745 6066112690
4642965640 7405419288 2264785479 : 5135
7979744559 2299821353 0029315531 2717617221 8631970898 2235311610 6940684881
5755289481 6208209181 6672481312 : 5136
8908104623 7699070132 2935328445 0081408941 5189311098 7965407214 6182757480
5585802435 3816151391 8772004445 : 5137
8506579847 9202545694 4114779663 9922979053 2027713023 4997864403 4421824961
6320189181 1804326478 3111515946 : 5138
8111481647 2645477617 6349787130 8668875695 2976260303 1666841214 6343026244
0794686978 2859874183 9503171394 : 5139
2900135590 8772255770 5373858205 4889531696 0180686323 2587915107 0588932716
8594220715 6076389078 5976077655 : 5140
0843037319 0771766931 4552391351 8622363114 2711608544 5804084750 0024799875
8230058708 8265491462 2078726726 : 5141
8063637453 7598067416 4794484814 9738923875 4728459343 8752792867 6644327256
4795561569 8313724625 1045428885 : 5142
0073348951 3025043675 0092419290 3435436010 8566920829 4261850388 3972921380
3592097852 6334813384 9959272530 : 5143
1325821086 0871562799 4119122426 5330135185 4683014758 8361727164 8821814983
5028778855 7539659788 9061068322 : 5144
2861861952 9827640257 7225660572 4744655934 0325286581 6014386518 5373616114
1630755729 7037999446 6511411405 : 5145
4079427147 5377811114 2551339728 3496115287 5330388644 3281608116 5190097196
0558504031 9934565279 8222234280 : 5146
7299181713 0543227748 7592302825 2948026550 5717408619 9252370629 6814981242
0426377685 9881953959 9684750680 : 5147
1203170391 3507600700 1949208483 6887963851 4240583538 6068968037 7544207064
2716519419 0884481774 9095380525 : 5148
8104482473 9134996228 2964378356 1309374571 1476150935 8959457926 5351844458
2440734844 8910093955 7560229105 : 5149
6062448456 1256046382 1227900342 7907039318 7107523808 5638921136 4039358354
0105820737 1156598072 5630307728 : 5150
2358517313 6193707794 9085709984 7401336685 0020841298 1911223595 9696822259
6454588321 4339357119 4834778926 : 5151

8345889740 0760306939 4540213572 7476634284 8666575966 7909377976 4676816301
5848816770 6460389706 5832759236 : 5152
3081409299 5503945837 8138514377 2011353236 2892642037 7834841213 5950821407
2712089533 7331687881 0000342084 : 5153
0576180389 0377556602 6792357942 6458250463 4285826405 8246647041 3641994747
0546611833 5438791076 3314524200 : 5154
0537808999 5503474173 9824797089 6323065778 8066150878 8209327159 5756544842
6168832058 9402879672 1171980537 : 5155
2836240656 1189079991 3802883209 4343311246 9126814143 2182067023 5390665254
9656361761 5132401567 9568910354 : 5156
8807955825 8253280000 3740724316 5059153059 0694165524 4026442226 5534657082
7097143842 8418562650 3226274931 : 5157
9629589024 7541653457 6128240935 3540627014 4007091148 2095483336 4938657662
8566250273 0215442597 9845382948 : 5158
1913014999 3816839032 6408029823 4887741471 6824958784 8181800555 2066910156
1370253273 6823984640 2589188012 : 5159
0337650092 0647729115 8684977059 1281425393 6463325614 9381380988 1934729589
2191223730 2958434157 5051021773 : 5160
8760027425 7006713072 1327155850 2718326316 5673229741 6556993878 9693238288
8666604753 4639873633 2850561625 : 5161
4877385435 6883005726 2497407546 8515505447 7920664951 5963229258 0229326617
2296221390 7725474952 2646217975 : 5162
4608682099 3904542065 7122232019 3765629382 9780886180 3061952849 3132084967
3872332603 9474869730 7938567850 : 5163
0759409394 7867425498 8202420415 4706671982 4068073770 8036607512 5464173710
4593569506 1956103385 5527322098 : 5164
1093912275 4665443071 5285653966 0824821860 9920130748 2570769885 3489469154
3197165697 2247661616 1766706519 : 5165
8873444896 6245862862 5152225384 2901560874 0899543421 7695175410 3353363403
7347503819 6456961620 8605048641 : 5166
4145849222 4453185569 3921046091 7094993512 3097068124 3367449929 8196274387
3931320970 4016695375 7317336392 : 5167
4066206733 6606643546 0951047803 3905319817 0587577123 8273780252 4734990313
3500461177 8877727476 0720342573 : 5168
1481597071 1634540909 1842317519 8274427763 7794025829 4921372461 6142441229
9249358946 1434008809 3891446700 : 5169
3203465218 9837272897 4239743797 1531983072 9612428502 6965615884 6639448405
5677766860 5819505314 8337097686 : 5170
8517186306 6764959789 0678887068 3150525108 8021562700 1700423292 5652577553
5714748807 1703303840 6465942131 : 5171
3222560288 1837518818 0027781993 0167409986 4410627199 8252745569 5968061642
3014905837 1294203283 5260868134 : 5172
6582068661 1208819994 2560871923 9598657559 9068847944 7989572847 1897115760
1126635233 2118719974 6658400358 : 5173
0218134043 6535578951 0224096323 0644670362 3964969102 7401415384 0261860400
2521040701 0782659915 0449718939 : 5174
4637653897 4233959444 7809412596 4467938810 5954017706 1710349317 9060102858
4817942869 2776802516 0634698186 : 5175
4876158623 3701525160 2363718178 7999034379 5955115086 5145402587 9726423200
5725044441 9433171201 3746825419 : 5176
0708531072 1520851269 3984662801 7711102642 4456380791 5882495667 4349287553
4221533673 9800701841 4596890640 : 5177
6936381641 5346937335 8041348621 6823628124 8551962820 7372517116 4471004843
3926771290 4709544982 1771245997 : 5178
3537949895 9250181926 6700991923 1981513968 7203924078 1317332882 2740618348
9936891596 2588277943 4181259782 : 5179

2374682335 6556966161 7070377394 2455538602 1304334936 4007531953 3984910869
9633356133 0849820046 2205237942 : 5180
1261127386 4697734352 9840614120 2933574769 8542837985 0731423029 6086006458
3902032668 3297106647 4859280280 : 5181
7062228941 1984685237 1184628817 8252337801 8175736273 8656338576 5251145897
6914251738 6579768133 4553312596 : 5182
1089216796 7599216200 4777773660 1739698157 0051893055 6320934768 0575208239
2470930750 5648653540 1470969258 : 5183
6332447535 2400235795 1420808860 9935236904 9792716495 2973307312 1762702277
5828058433 0992778199 3804392172 : 5184
6811176593 6516774634 8586811523 9136038172 3993804361 9137087041 7832458368
7073268792 7591512421 3279306924 : 5185
9368067256 7164093795 8115437410 0844743180 7984798184 5622953378 3159209437
0585987930 6743149584 2128091771 : 5186
4693715998 8393383659 6763332855 0845214487 1734987298 2802287224 5972111003
3706788519 5708636469 3267471590 : 5187
6123201811 2861922070 3781668982 5406153183 0765384398 3956690719 8048513952
9940333118 6528088123 2170777351 : 5188
3270097093 4377286450 6905524832 0188640544 5912131117 9044127994 9017804606
5119346213 8351820079 1941711869 : 5189
4779020864 7877508811 8722447763 4397287600 5334911972 3576540686 8201936808
6771886415 4368080099 0685123604 : 5190
6326960994 0850990396 9483621715 7071869815 6808599431 9275507836 1763106352
2862443582 9757772397 1617124439 : 5191
0007852725 7874425455 3882651665 8015450494 4151640169 2449890426 8345734562
6173402057 1401179382 5315539669 : 5192
1786508532 8616022779 7761828448 2128672946 2937149110 9116351044 9320107800
7237074496 3492569121 8904931357 : 5193
3617672737 0757948988 1987081480 3294560802 7014245740 2799289156 9507497718
7763259012 1855832116 8332456383 : 5194
2463542641 3663047442 0695339068 7461804874 2695730825 2892854977 5570446553
0725372653 5449091655 0585449050 : 5195
0908073608 1330077728 8059176211 3407812702 2071577217 9918469199 9647647655
0010097766 0172796469 6655224452 : 5196
6699337699 3129003168 3521179052 8327287495 4475129813 4025062913 8109549668
5728387795 2905044288 9661515542 : 5197
2823760174 4913735138 6273975389 8374287836 4570857778 7917772282 9418378625
0926281394 5141402838 1810712499 : 5198
3708570349 6006797685 6336968647 6720810794 0134332689 9229203210 7971528543
2230936878 3898249784 6997755897 : 5199
7758416212 8896661183 0971893956 1805589888 7347882238 5862983606 1482819706
3885376638 7023699719 0568879142 : 5200
0015005893 2781566679 5541265615 3240777511 8370166626 8023336345 0787841366
9795089586 4028584925 0404849337 : 5201
9617118032 8045570796 9321708308 8711762591 9959435644 1397149528 2622098711
7939305641 2254938187 4381430808 : 5202
4727553083 8917269346 2948996786 3754936115 4624333630 9926769946 8701922308
8148390348 1460666095 4452470388 : 5203
3761208316 5142010076 3183375893 4939996200 8248009986 0462830624 3249049275
7335719180 5826053249 3551274110 : 5204
7358603022 5305769820 4994600214 5986397578 5058321391 0161582089 3810253664
0432838562 2156050481 8745659654 : 5205
5074877137 0455025814 5886878709 4158414236 3695825390 5004410260 6727943451
0615024772 2806223509 6633157361 : 5206
2514266794 3414074665 2063534546 5839222611 4632355783 9399370970 2620191622
1347388562 4539830201 6554714620 : 5207

8476297265 2574642687 2338636231 5313553556 8166360519 0357189524 4664237586
6198041144 8025219901 3025958934 : 5208
9990701834 6443127068 0599004325 6138711210 4506890557 5067907676 2563050384
2626225137 2906522565 4942654827 : 5209
3213499848 4160539500 4718249856 5601847314 4907426919 6457452658 1135702651
0716238773 3468833725 8997786436 : 5210
7259174115 5701655292 4842063152 1873791595 2464483195 2515980633 9460374218
6735024198 0390342917 2882058569 : 5211
3526786658 1582011958 7606982523 9649292348 9146794430 8478657497 4508366962
3296079018 9307962445 8748437251 : 5212
7492298595 7167812557 4842331673 0358061373 2387207900 9336336925 7424578919
5386758667 6113412582 4716030708 : 5213
9431687496 4912571672 7634443030 5543587943 9672798391 6055021351 4833324054
4829591240 1218783134 0304592944 : 5214
1621292750 4998450167 0946484953 4409152471 8377825576 0029397942 5014893421
4633867588 3458963870 2611272977 : 5215
9481845303 4442314713 1244078498 2148467358 6273265362 8625501039 7360301794
6321255354 3423298346 5042878299 : 5216
2697952507 6211766041 6274969836 3594996032 4349579121 0300046219 2014621172
2288584413 3017429091 5199745617 : 5217
4122142548 1510127488 2630094697 4835123007 2170223802 5163626542 4573166829
1547444218 1002777612 7607297892 : 5218
3682586116 2066670221 0658425698 1663821798 4120387841 0264487735 5661996291
9187799243 4897623591 8131207934 : 5219
2126285808 5245471431 8097767637 2136458404 3359927672 5551587668 7292681898
6466457737 1815671372 7127327069 : 5220
1744607683 7023027879 2424382140 3830439878 1254700621 3796556007 4412906471
0835313042 0289291215 2146345225 : 5221
6753566839 6813627818 9995449676 2854591729 5133246231 2686271207 3070316867
4086601595 7099166315 8646522388 : 5222
7456234496 4515990749 7920994590 3431392149 4067984830 2722545724 3051435304
5490785504 6857565598 5031418630 : 5223
8483486501 0575124492 8756718289 9127645724 6072592552 9643384645 4729452076
4118443817 9028342584 6409298456 : 5224
8121756169 3893596597 5740429295 5715291507 6829707802 0239789169 6925219956
9167435795 1072811013 5844878034 : 5225
7612045865 1798666327 8274801506 3649022752 3192057452 1535572880 2481958656
5693666305 4699706295 1144220612 : 5226
5550456869 1480327448 6028172682 8102742488 6402616018 9341813654 0881702751
9036327982 2038270382 0898351245 : 5227
5779035361 6534469844 2851334548 3864352548 1083128499 8328941817 2916120839
0713906848 6878921381 0103679650 : 5228
4435251204 2221465614 3647220149 8942002143 8344201240 9701466236 3438582726
9296108625 2552511892 1290714836 : 5229
0456679559 5168653305 3105451019 7832165741 7626828559 3469187149 2472425479
1232715290 9308767435 3283875056 : 5230
3162595363 9199685925 8022570941 3045416467 8572397376 2668711162 2820934741
3738395309 7453856228 3954012484 : 5231
7808174876 6221827151 5701356954 6601451452 7686148821 9451341105 9801959086
3768978510 2904745476 6574382151 : 5232
8113451025 0246512821 3078825731 1518251357 3208273422 0695504162 0563142444
9297757854 1810479961 9444293639 : 5233
4669739813 5936691118 4515777799 2643506435 4447789078 6688566763 6555675101
3302039206 4109944674 8697238001 : 5234
8901857712 3739333851 8911517615 2917903357 0357835979 5955444897 8594989542
4664715432 7706708272 6518943783 : 5235

7948064033 4221433772 4047867869 4063066207 2056502773 7002585237 4393455794
9437175759 4822239484 8213090905 : 5236
3923116986 6409648373 5068307724 9445046879 2906337053 8796371011 7044090375
2038526402 9625703004 4175089847 : 5237
8007785474 0069098315 9410928873 5871895339 4735828847 3512024777 5144152864
1433301523 7602025716 5466393135 : 5238
1516958028 8664731538 3404234437 0579637648 6698042917 4989729433 0983715121
5686192965 4197679087 5435900658 : 5239
8519119491 0142607559 6946536884 4561880990 1450714928 3559365332 8658780734
4656348810 3298867967 1646781357 : 5240
3843886039 3567548126 7316383463 2608152226 7743361034 2562799044 5807102437
2908605563 9419844097 5135534819 : 5241
3689894438 5856295365 4998625886 0485713492 0887479532 6091097703 0346501294
0402286028 3613342235 0547868311 : 5242
7531048043 7887068288 2815000408 2704666900 7244607660 8132834861 6471638169
6784460515 5176165371 1887635152 : 5243
3592785897 5553158127 3539168302 8019688722 0255186609 0543478414 5591634664
5871317346 7524042320 5137846651 : 5244
4548485210 8274337280 0134867245 0682567941 2532619761 6598876523 9668887392
5240799849 3528141270 9665420506 : 5245
9769479109 0919691843 1017573132 2507064339 0879078608 6732900306 3983535165
1598000740 5308268960 2417037023 : 5246
2670697644 7029760260 4698157613 9529220964 3118412199 0812443297 4534969714
0355250428 6138984205 9873557626 : 5247
5543534721 2310996418 0755739430 4984758358 9778675340 8277752559 4534690720
2949290138 6136911719 3813173601 : 5248
6464766254 9743551192 3723190337 5664039955 4244778394 4298351828 7742464123
2791662248 7261166153 1195565153 : 5249
1837930374 4830710564 9190863769 2464303925 3332818232 1026516298 4382479889
4081615315 2609817835 8205420475 : 5250
7574239190 7906610890 6172633363 5275358836 7285253235 6699956862 7018308121
6740528018 0002553689 9293369386 : 5251
7881167440 7772991654 2788044678 4135620093 6297554569 1518076713 3129637553
1481657990 9072931041 3796284759 : 5252
9417790724 8909988399 4658129887 0509883672 6884172624 5600654681 1448063717
4295084587 9829312533 8228957064 : 5253
2635558951 9574792100 4748780099 0674713490 8177116597 0278740048 4855846184
4362188045 5975536139 7818771100 : 5254
1600120973 8065852060 2274673984 3219801950 6909533162 3045898291 7616258601
1360218289 3530594673 6287128555 : 5255
7042048740 3580738017 5224160105 3644920727 0031358736 2744654707 7793526648
4464080671 8320237279 4201434472 : 5256
3474168049 8214353219 0661854299 2546902783 9023946936 7565855047 1520114175
0008863744 3752809863 3553566630 : 5257
5314474278 0108559946 1612399029 5628653163 8308517479 7943117702 6166211075
0667259113 6795657312 6107906874 : 5258
2527132102 0893420430 6864426562 1908898010 2687881677 5863339198 7936068177
9680015207 3864581118 6433222300 : 5259
2760660979 2372849188 1652219262 7682090188 5478038944 3560320424 3307256873
4606173702 3245268043 6617589619 : 5260
9744461169 1103048605 9055319970 5636008633 9357467682 5493273026 1029297265
2219997013 7847456683 3692781960 : 5261
3268412053 8725039677 3387410094 3023310659 4985975756 2894408874 9450232447
1045141157 5928385431 9043656099 : 5262
7944177869 4565059086 7372794050 1810358913 3007802251 8367047129 4785839325
8945203373 7115409652 5172614324 : 5263

5821365109 4928649637 2790675667 7570290788 1082452198 7372794038 1999502492
8527363407 4361722349 1460131337 : 5264
4188261577 7539449981 7582444937 3817062049 5051672294 2428537471 6677617880
6443475674 2240965920 8038034162 : 5265
7847256942 6829022925 2125238485 8533734791 3671592469 4997360835 0100908415
9991613780 4841580776 6179309191 : 5266
5947188569 0996102325 6006367965 0977588295 4357381862 9851803285 9284770413
7243664895 1461450501 9206944691 : 5267
0055451753 1628065330 5226400767 7708933108 9867475923 6311424921 1964737983
8595425577 8544792305 7506806386 : 5268
1269020901 3250630793 9519222133 8609185950 6512594966 2111567524 7703172613
2363270363 4237172046 6923617527 : 5269
2964371717 9484510462 3856042582 2273820574 6694163997 9217811594 1355964646
9298483067 0920142577 9742380093 : 5270
5295307642 1311879630 3250033849 8464286036 3407249831 2855593980 9627486824
4319578185 9038887550 0902591367 : 5271
5504375891 4720260583 6213776664 2009090107 0593053387 1909593483 3833029991
2816614493 3023488324 2862780945 : 5272
0374459419 9622771925 9121297396 1871592020 8473155346 9808229179 4557411069
2956227707 4646829225 0640769884 : 5273
4047590191 7347741466 4989263523 6134717021 6506656059 0473280364 9682439836
8514816737 2969698615 7231072880 : 5274
5030865542 0546534599 8321799681 3769385218 7938647137 4152993484 8299188089
5775760669 7549607425 7348997204 : 5275
9994459247 4776550655 8813098857 7915322125 3482542661 8429008335 8153300425
5332405321 4212483604 3612819221 : 5276
7295783844 4800154600 2989505011 8238964630 9785607091 0788010264 9379376565
0947187932 2583388446 0889554615 : 5277
2946266400 1960838911 3091285689 0749524410 9929559185 0870862987 7492462935
0984305416 3189206518 9010221083 : 5278
6873850768 6044955836 7008844971 9941471180 7644132023 1950290439 8729819740
5881795755 5924943246 9415479940 : 5279
5009578634 6491078935 9582772786 6007415601 1277589271 5697466918 5672088046
1585956897 3929234646 0865182597 : 5280
3480987627 8032735783 1228242397 1491807934 9952189648 9149874891 1954466018
4648597939 3343279816 9521734810 : 5281
4730074648 3125306807 5519706380 7910667955 8987664530 0045068524 3345844501
4194380760 5732159724 2893329409 : 5282
5357769028 9281327309 5037334413 7444842404 9652672649 1230745760 2800339029
7467260071 9069651789 3056846085 : 5283
7048624192 1921895529 5012064993 8297904522 1890711544 9969937914 4340897775
1170261408 2584014441 5357391257 : 5284
5268782112 0477956764 3328875396 9726244731 8790329525 0147526864 2593745614
3548799640 2355502556 2412865641 : 5285
8887114956 3905477942 7685872504 1127991197 8525278635 5553260549 0775416372
8187879421 6675674857 8561416230 : 5286
4336382407 6503788015 6019880957 7238048798 8087768681 8853155071 5521356754
5824862107 4536528904 2299172215 : 5287
6412432965 4859604047 6034010896 0793000197 0188160340 1870668767 3493011680
6762537583 4101944939 5995975179 : 5288
9294529013 8196724876 4294353034 9403840454 8902069081 1451703580 1938191590
0539925385 4990428116 4432095981 : 5289
0429725600 9893982978 1428018158 6790336160 1288340891 8242656076 9657275476
3199672245 3307756635 6514988924 : 5290
0332801785 2909671591 2155922079 6605920051 6647151130 8401574719 8573315595
0814931327 4438609677 7289684562 : 5291

8669479813 6503135660 9445293868 5876216472 3841947266 7526482797 8036461664
6165460093 8432009638 6651058793 : 5292
8358408749 6361419706 3437324407 4406175963 9504054302 3084204531 1689261869
0547745512 3081429286 7131454787 : 5293
2493672262 7140194805 8072089560 7295402660 3573000165 9184622869 4429707353
5927977913 5141545723 6366805613 : 5294
5103373594 1605798693 5094459307 6459253643 5012949183 4074656822 1068024329
5447211988 6042088970 4487725978 : 5295
5222935514 4143595866 6171084057 6191296480 2496895729 6336190810 6473429291
5801249233 4159507279 0860874734 : 5296
7297779283 1022117036 0078554576 9509313647 8690591301 9289750813 6820693747
6624409797 3906085819 5703694795 : 5297
6217339092 2423568787 1954488870 7655404175 3864894947 0012505326 5954906591
1585793127 0481659345 6876477106 : 5298
1383451657 2863565673 0814481072 2413625202 9771512301 1521636643 6175554153
7773135312 6073673845 1161060841 : 5299
5135837496 4999144671 8312384781 7266127832 2910119023 0569342678 8974768845
3059278879 0534903105 0776142554 : 5300
2996293875 4841751963 9979120671 4877105669 9673222538 9139322340 1564458501
5038358971 6631779696 5033760872 : 5301
6334162566 5152168570 7737530402 2479440398 7545693099 7611847595 0363021128
8883923449 5926700379 0422257102 : 5302
8065129805 4546955321 1495375827 6649718670 6631970791 6922605079 0892188740
7617766347 2692934300 0028544451 : 5303
7296016471 8901564120 6994309986 1693668518 9833401054 8662959489 9465690353
6055415702 9370869503 6832258132 : 5304
3917400113 3578912283 8643818662 1427639112 0167628509 1701580132 3940797165
8678751993 3592260677 4097310151 : 5305
1742268902 3963220013 0165011170 0896408566 0915964199 6020206039 9859516759
9336164608 7251610537 3830600830 : 5306
3854701146 9178633538 0234107417 7249771296 9498238933 4746097814 6663308890
6239419319 7079768488 6220394489 : 5307
6652185530 8553719676 6750963598 8051779747 2617930326 9128318267 4862057809
2652565290 3428019088 2705901405 : 5308
4663747001 3833025889 3616017150 1548886893 5910569440 4497217202 1139245648
4749888583 6600119303 0933405042 : 5309
4819094808 4678987600 3827669218 1719086769 6601829782 0355992071 8977804292
3618366306 6794572605 1889605382 : 5310
3472875932 8230119515 8655518017 8283133565 8309861373 1652593503 0805823962
5590828657 2921386547 2195779312 : 5311
2433591473 2375240378 6908253602 1646294813 2048845739 8705136724 6382465059
9873114603 7622534977 8221030101 : 5312
4542987511 3822448213 6735267372 1070466677 7974242250 8385846488 7294909600
5218614774 7715199660 2465522606 : 5313
9621270620 8541687872 3951599782 6861207922 3288649559 9471294392 0009967608
4626144529 6688181911 1809448899 : 5314
1955881471 3799314815 2932958251 6076010711 6624951556 5937202893 0345139657
7612021582 5001792716 5821147493 : 5315
9898252164 1313983316 2921591167 8786311490 3503708046 7291134675 6645736776
0316704471 1872286977 7749372049 : 5316
8411552534 2283689364 6994865992 8163890293 3332959670 5023331063 1864547159
4619892117 4696040074 7930112114 : 5317
1664974922 7696638793 4136055436 2801356521 8859169191 6012650170 3473492198
5779121559 9655450788 4590280024 : 5318
1979283614 6886166968 7294067326 5794485514 1210293316 4767805043 2030238929
1620949699 4094626811 2688477619 : 5319

4192988872 8302045325 4835559564 5610497507 5261052442 5580938627 3420635892
7768284422 8129581464 7347367073 : 5320
6225597282 9437263774 8776839928 6376323453 9366344505 8464032220 2749734150
2487576599 1021519544 8391414894 : 5321
5282238326 8148398051 9970765592 6889831544 7157695118 3155510618 7300676856
8040571230 7832433568 6278130316 : 5322
9868434085 5208154148 5255677208 8289519403 5878583720 2454478168 6346084119
7036059946 9204205022 3895050999 : 5323
0547462252 4732812866 7550756059 2665702381 9679854436 9738430363 4755135301
0142231318 5966598355 0172909122 : 5324
6635606839 6272443084 0465236286 5284232381 4234990867 6609158081 1529985673
1882700013 1093821576 1913296593 : 5325
1457997811 9982342630 5914119833 8941256260 2152147455 6185978394 0816914555
5130001751 3102083825 5631736162 : 5326
1545903512 9106180015 3799212994 8114092826 4215227003 1879995958 3363724143
2647124378 0519752445 2586298660 : 5327
9139765363 1205034855 1968701426 1642737394 4255387041 0409340117 8690284831
7324446696 4383927413 2175880980 : 5328
0049498861 5024556839 3213699722 6039513159 9909527837 0145480127 5952792565
7066876033 9430121291 1978505936 : 5329
5743941306 5417846820 2190033810 5036071495 2969682719 3408249954 0824216163
9655390093 1007144638 8762031349 : 5330
1340775627 5562058777 4807984394 5811949518 8200713205 2594033442 1340012436
8874611015 1026668901 0672516105 : 5331
8423245264 8551398239 2400435721 9566830736 5786346362 8471394771 8760514239
8707775353 7119885446 7657293463 : 5332
5846988478 1033914667 8478547087 3151504290 8742459239 4613261089 2919503710
1182549231 8420721043 7942006267 : 5333
2244936264 3509595503 3758381719 5105443116 8732796056 7532982916 1312575435
6825245646 5008122805 2263681461 : 5334
1597532119 9170660364 3597448082 8856219322 6739368656 0625429382 2490061995
6077533306 5246484430 7273487208 : 5335
8797672926 9681479748 5326374608 2115171227 2174344612 1372624336 3240118432
9188986371 1234758307 3897410591 : 5336
6818188166 8955953718 2399583109 1589558651 5510793911 0515371592 8239197272
3851893459 5024177790 5515151572 : 5337
6497570872 4287679440 9731453020 4095907691 7387500961 3264837074 5595341535
1331344900 0387528031 0133785644 : 5338
1911474235 0224523620 0886553420 5354859562 0977763485 9173787806 2122554402
2592527384 8307765179 9963738230 : 5339
0909619759 3801747525 7830796156 3233626646 3730838957 3846711167 0927506441
5747632824 2109681866 7021420776 : 5340
3268375526 0776111089 8894496467 3277534390 1491089616 3618414435 1402764581
2228960506 0381554653 1573231035 : 5341
9157359227 5891134925 6900935607 1479776887 0073195410 2283427174 8757490709
1871630476 2388723409 6963025340 : 5342
7824746509 7250027224 1452603350 8279170509 5244089375 7333312319 8203021541
6350786778 2655093217 1378171231 : 5343
3716114212 3181244080 6330981536 0743763950 2654255747 2387777479 5160370760
2548634894 8281530335 2021941346 : 5344
4669637514 3271528080 2910028162 9344182641 0827591952 4951817369 3711365151
4537697574 6303550396 8815577039 : 5345
3898348715 4988401323 8769153330 8860839615 2038792659 8342627243 1774927686
2696354131 0684656244 8434931165 : 5346
0896687484 4693471034 0534822954 8440424944 5480290150 0871209116 7917658606
5278724812 5797753474 7880316689 : 5347

0035108745 8568174549 7070497359 9571133983 4554368894 8216100537 1526145300
5639912424 4539962580 3132802285 : 5348
7815556753 3705179614 3359354831 2607199002 5851110866 7542907999 1703537006
0643683886 5760343284 2134674936 : 5349
3284798134 5995095244 5941366688 5886365358 0165643967 2455889752 1380576361
5902148421 5833508687 1769121384 : 5350
5760044375 6081874548 4730537879 1606854309 3840810194 9720610599 3813537730
8743093608 0254374618 3604369148 : 5351
7411272220 9890114767 2877947895 3951733778 9411265620 4674800771 2938053458
4076556161 6367140328 1364506738 : 5352
9621785209 0789222464 3916798604 3804019848 7605953575 7297848080 5364654179
6474341632 0801881522 6299727917 : 5353
5361841901 0572539102 3526100666 1285793861 2848287211 5114433121 7481644040
0561836018 6728632793 2539692823 : 5354
2426014000 7259899509 7051264681 1985138070 2881169297 4703651388 2781618018
1143146872 7933233145 4315102504 : 5355
8027376973 5906929697 1888893454 5315353695 1518987596 8892467578 8779434750
6970959483 8918709199 9856782139 : 5356
0784326370 3165798784 0807613470 0595423153 3063135629 1972347631 1123701263
7437169883 6456993918 0204023601 : 5357
7065484741 0448271684 9915119737 6191570800 6875431889 6722949153 5191956193
1247877010 4883649003 4307122021 : 5358
6932716448 7378907486 9716889055 6697893495 5748268599 4538815934 2379901066
9245538086 6035693093 4748516271 : 5359
9970008372 6662594609 1636911918 6683408760 3981605345 7187304488 3787844599
4564106619 4649282790 2987144851 : 5360
3829662277 0697710186 7972627483 6234505524 9868433462 1758253519 9489358389
2640008216 0425785815 0101486886 : 5361
9194413913 5668696883 1445986211 9923349722 5849337410 5532462372 2317841154
0422579197 8377626057 2497686118 : 5362
0888568657 9798004500 7436559961 6849105084 9723073534 5599811561 6755131668
4195733447 5132922964 4291365157 : 5363
9966325538 3479450766 1032896848 6987628172 4653418103 8300169434 5217588085
4960747849 0075673039 7047497994 : 5364
1243512618 4213715027 7561668273 6902616881 5088921349 3507321826 9322806605
1823157630 5416072303 6713898748 : 5365
3779334050 1584198071 0583153109 1345171439 5671880125 0305988695 0631165440
6196180647 0807361237 6753683950 : 5366
2282439875 2860510511 3518141919 7379569469 3638681274 5316528106 3503834682
8357384405 5142293480 0799366821 : 5367
0804493606 8621646006 7663448303 1922495893 1559540458 1694972413 6805749636
5679330791 9636715256 9807186187 : 5368
4731197459 6214344478 7453875567 6792302136 1485972863 0338941823 7450498917
6974007142 8842397668 0628317290 : 5369
1190857403 2503385452 2235403375 9894400897 9329603741 2054603543 8418200763
2588093628 6978935955 8085094975 : 5370
6506799535 0334583706 8005723967 2681329462 1407540131 2828636882 3074537737
5496644876 7066237058 6745285605 : 5371
2628768437 1031208597 8307077429 8894233319 6659933706 7247096895 2524985445
8209814394 9212079393 4365568025 : 5372
6945592260 8937043796 8088305921 7354990280 8187239423 5008316672 5862989891
8865217070 4858091843 9451741645 : 5373
6009022597 3841145776 2210305768 8690926001 9619975636 4983572315 5371164800
7804190340 7349958308 4414551229 : 5374
4281776021 5210493581 3182396627 6787157371 4998284140 8036881971 4227388302
0874501573 9858523972 5106689569 : 5375

9623496283 5492727345 0826891202 7147525704 2705061118 4594647182 8532054542
7233451594 0869862411 9906327074 : 5376
6182667956 0117012752 5835086473 8849204602 7857128502 3478467533 9031318786
2966495520 7064550768 9584383918 : 5377
6767906678 5694775621 3003776080 2451808273 4525214320 5516063315 0403699415
4860613067 5953365971 6800235921 : 5378
6089478140 4681853314 7106095468 3369162184 8065587071 3410551936 1745103710
1599216136 1995097086 5013778053 : 5379
3897593988 3669378739 5482278968 3336138162 8700251093 9802307692 1640166695
9917739312 2483908264 4791280998 : 5380
5780640532 4340913718 6218853230 4910289011 3716108308 6808748672 3811585974
2181565370 9296744655 1849292875 : 5381
0040651580 7192028280 5299224441 6273543223 8256495251 7424671577 4826896121
6185256399 8936569445 0558351032 : 5382
8185664131 9976266974 3911116778 0948717936 9657369037 6143192238 3476655463
7308885177 7088449683 2380581866 : 5383
1049089021 4787876907 3115712614 1234536496 3900842699 4470802551 4624680875
0232039897 7365460713 3043284170 : 5384
5373441390 3893234842 5031816883 0269138738 8458847942 0355798916 5117728792
0689338631 6827004321 4700842467 : 5385
5785294166 0469640375 3841191237 6432743868 4183406337 6908656334 2469199412
8446600486 5121778043 2768985518 : 5386
4022775180 9879011069 6764581909 5323132112 8789090149 7686946981 2633598854
1954556717 5712676678 8371557631 : 5387
2670276152 5977769925 6164372103 0108499313 0354371898 9924700939 5752824298
7296922410 7111180096 7264157307 : 5388
2618623927 1371578280 6114143002 0171114264 8258488927 2072257212 2032771300
9982949992 0651648922 4551883661 : 5389
4215959589 1282375248 3606217204 4495902485 7568099691 5586234438 1651671463
1633100487 3922932262 2194319536 : 5390
5465767781 2098448625 4203960830 1127767452 8313406762 2943688413 5422093949
6071825983 2600135575 3099690969 : 5391
1637349665 5974503738 0554178792 9382030075 6984307669 5439031187 2706744220
8400259867 9309241436 6290522730 : 5392
5487332037 4993181703 8992564161 6744964935 6926652153 4815077105 7177730318
8284522175 3638550909 0287600520 : 5393
7644455852 1723466926 5970493470 6045532756 2960679060 6632310267 2442059559
4961738355 2205881466 7667226152 : 5394
5390990922 3734700140 6561397500 6335694487 5096877152 0848323518 9794212322
5929201006 9507200894 2754098085 : 5395
7268094505 9064154821 8430923520 5988084931 5619983713 4863086316 0817753099
3431356826 1193911313 1975531178 : 5396
7891831559 9227750248 4615967066 4664052848 8092119285 3256995495 7327013366
2891020185 2959715102 0987318184 : 5397
6394428005 2695190904 3300223556 7601977870 5309066426 4844723116 5417685422
2119167486 2350821405 5524163362 : 5398
8561937291 5602656303 7485196337 9831249371 2512584668 5449441793 9893950361
3056358538 1465349742 9452024282 : 5399
8047303931 9639196656 8679509781 2119450371 5184948395 5346185707 6075329697
6206042126 4450505727 2373623975 : 5400
6574400005 1885064382 5547281839 6964755953 9802630009 7329759096 7668112501
9203508629 1396430856 3802687805 : 5401
4242269120 6708466695 6677938084 2887975906 6333385191 8316580733 3128864107
7969980277 3881216321 6461819722 : 5402
9799382327 9225034392 3207155706 5563693152 9218829355 4619023650 7169076916
6585341141 6202786881 4520633343 : 5403

5182444586 9277959804 6627970127 4348819999 8367141615 9300403854 9093477123
2162308539 2496688136 7722657236 : 5404
6504800742 8664506672 3197281356 5442043144 4220544279 5569289248 1651067906
3605431422 7622982051 0603946839 : 5405
0735987427 4419025274 2980580057 3784246772 1405075868 9327578238 9649330295
3931675436 5582636829 3342278667 : 5406
9969086201 8808955961 6889819483 3665683085 0488836176 4603121668 7630865949
1983675260 9718937985 0864296154 : 5407
2780131573 8842262690 9720754930 1238347693 3259946850 6448850358 4484386598
3493006017 9435497954 6847118805 : 5408
8153183483 8854670090 0036239016 5675090985 1974382608 9176297051 7516264463
2104682240 0453548606 3995436349 : 5409
9876179329 5439245093 2753221167 1851255361 7107202565 8333543526 9680085114
8963267718 4139489710 1579682237 : 5410
1232377799 8170453134 9779534409 7553782240 4146878310 3118732043 0287713984
1266742843 0830099637 5651519942 : 5411
5640823532 4995905280 9366080177 0815480297 7468046987 3018442817 2248913336
6907642573 4983149530 1549179718 : 5412
6038684886 4164136682 0315278415 0730160382 8760904785 5070467970 5102972376
2400497326 7965404437 5743461492 : 5413
6309232993 9531021385 7334039465 4643138039 3284754601 6682894570 3985202898
9705619905 0058566331 6810254247 : 5414
9612249799 4161923808 0817186956 1768335559 3082034329 1711186119 4066142459
3023416858 9283254022 8112237514 : 5415
4878383804 3753338600 5245983771 5219808574 8474091159 6560510126 1021357390
8775874485 2230707278 6751026934 : 5416
0595258400 4438769166 0888998376 9297115767 5464446866 8682258216 3159384153
3402184543 0200445567 8786165353 : 5417
8779006847 9648161154 6147742567 6643466678 6165714321 3956058051 0685249293
0203486311 1876979067 5737306130 : 5418
8658049805 3996249830 8851629744 8323420918 7093248655 6631685998 8821807634
2633618038 0330937080 6856552116 : 5419
3083570426 5987849396 2774257427 2212686526 0875537703 4324279721 3162471865
6218150212 7210015338 1435276404 : 5420
4674384081 7921398524 7176560645 3271666305 3443606587 8850200216 2008322243
3975474388 4656908084 8229761095 : 5421
9890936151 2918143246 9419017754 2972016826 0292606936 2867759394 4130399807
0048562472 2707461638 1708027260 : 5422
9324491040 7161694311 2721693245 9613466992 5072649358 3327369223 6372077527
0567953185 9869893932 9745192340 : 5423
6580320289 7787347160 2317472274 6080557162 9448861637 6900498802 8768178344
6200622032 7101913512 3357703649 : 5424
0080845554 4255768267 1091481559 6736553858 2594382219 5135801563 7711971679
3670206222 4568634854 6905987780 : 5425
3465897456 5844785240 8819268830 9227770546 0494635561 5841409745 3554239477
6734275305 0083936054 2653385834 : 5426
2924088679 1409368144 2328587900 9223052915 2799915905 0090651924 7441417091
0047708499 8890303385 6339419441 : 5427
0486439219 0202256628 7739846266 9712962868 7375722161 8817127821 2356334904
7154994707 5886804059 4632498061 : 5428
3245053389 5452386272 5524564400 8130368634 6872684137 9219379578 5077498902
7824315838 2732350331 0967089933 : 5429
0157066419 5240164048 0259861958 2946661146 5708789699 3126661948 6819119516
8899785713 8260889720 1650221151 : 5430
5000045916 7591831413 7952842924 0162630759 0692810259 1504769068 7227211554
5836998472 4422211012 7830084214 : 5431

9489854275 9892875796 2681408284 2462027344 8115637761 5885856666 0628707597
9845389858 8853973591 5067846667 : 5432
7440735588 9621176490 2638877304 6312288150 0824148640 3525126070 0750590426
8551545873 0819944836 1425175009 : 5433
6919881214 7177694290 5584636475 5781593888 0662395420 1893749084 0322751020
6348923283 0934102823 7894088989 : 5434
5317257072 7814669839 2546761505 0901133856 2734748228 2653535901 0089528178
6892385130 5483370801 2163757061 : 5435
8687637859 5480214881 0067207114 0270517244 6818076076 0129942225 7578378515
2623729804 4375489744 6037591002 : 5436
9399148776 7534772424 1123099838 1469118795 7254447449 6042009487 5863210496
7562261710 9806570375 4579778452 : 5437
8755506706 8133301194 0370283684 6318300506 6807530624 7064752243 6185743826
9778398335 4212993579 3422513520 : 5438
3492355031 0322632422 4496188486 7197753558 0316521471 0619958410 4062211908
3596690789 7963380812 1363989284 : 5439
3731074315 8230935296 9442335189 3330134370 8243264385 3278990826 1497650024
6622336013 4797174646 4208079836 : 5440
9546513329 4537755766 2272524462 0507851699 6720198999 9340101333 5072075807
7104038265 5836481700 0188252048 : 5441
7386269472 7977686819 8412449713 0143674446 1773836037 8450588677 3104155213
8029551148 2941381968 1116530162 : 5442
3559891014 7759480759 4117715701 1650869218 5411471004 9853076615 4450326362
4343420505 2785612868 3711788698 : 5443
6242284537 1122780475 3731921404 8956991086 1835758095 4983844456 6013084746
5276512015 2251640460 0863882540 : 5444
8921010246 1583536189 8855615679 5455594341 5393774502 9100661215 5870640166
5298168145 4915048709 1045363602 : 5445
3369626790 0742045245 1078747671 1108581603 8833504907 3019845459 6575431151
6272250132 8826413273 2720045864 : 5446
7504003597 5388411508 5871236676 7330536716 5176672359 2347410822 0556620738
9714589673 6124428601 3184480990 : 5447
0996619418 5512659140 3124482682 1505050968 2117223862 1231152652 3007131658
6542736092 4785213735 0152630873 : 5448
6450420970 8686975277 5065195387 3468375231 6717031793 8647078333 8125470305
6742160675 9571812448 1724154900 : 5449
6779553950 3394974344 0651101402 6688323398 1384072995 2577942686 0509803857
1003884722 4848237875 8702492339 : 5450
2172716067 2323783759 6682806558 7790533662 1109097334 3419979784 4334852653
2435072253 3603987194 6098706430 : 5451
1678895409 0843383183 2738009554 9056808509 2791321896 1619966362 6200962269
6371104592 3058598579 3321394571 : 5452
8121849704 6923974684 7119409031 6286548267 2781602334 6641245867 5531537220
6986570758 8845615920 0392763767 : 5453
6159553914 3811410641 0712803664 1827384857 0213166476 6755240750 0949113768
3189196875 9459457677 9755050543 : 5454
5913958847 6862792044 1607389943 8660597048 7557360320 7618939929 0784732570
1218938635 1243450402 5611153110 : 5455
6159816041 0629347212 5903381033 9622107691 0135920641 8442727572 5636244992
1884192969 2845048865 5716808147 : 5456
6678966093 6346427193 7555630095 8026571667 4060696704 5070055976 4523975309
3566926721 3475656322 3105217097 : 5457
9726788201 1756733844 2025858585 6309958277 2237670124 2619501402 1412415170
1706257482 3919937573 1539805651 : 5458
8447478149 7885071568 7141423536 2443327302 1226810854 7082779756 8225700932
5701405883 6963664043 6028018657 : 5459

3558394491 8479033626 3501554324 0087944455 2885841494 5115640942 7092240658
8405815702 7469906282 6291235403 : 5460
8024579975 7493270182 3914476262 8671972800 6741510646 1578426087 0892100884
3375777396 8560088108 1369285756 : 5461
5080184359 3099633922 4874822734 6979480784 0647767757 7435081064 8131138541
2026681801 4621696063 4863869351 : 5462
8228333021 3466133693 2791685950 3079924691 1478332513 0820766910 1517857971
6958660185 7927299671 9400552366 : 5463
9049880576 5228834054 0382957232 9495666610 6181631097 8922639940 7301157002
4576283747 7357630818 3798161638 : 5464
1190797360 3158721219 8313819620 3449769816 2064239850 7522832773 3732583324
3728821605 9788610987 3551913778 : 5465
5588140372 9865083817 5116326674 5475940834 5296967092 6281699808 4488901044
3639929417 1335504917 2185304441 : 5466
3612755372 7404380196 6167062333 8297380240 4906298258 6095139350 2749589875
0528072067 3183786473 0248913057 : 5467
9054815627 3669342031 5336593480 3545221237 5932319892 8884374803 7324107862
0860562246 2175566914 0664965603 : 5468
2562701524 6939470098 5147116009 9408912835 1947568274 8329672272 1766829934
9444534067 9073655254 1717488658 : 5469
3901039319 2874133368 7055929138 0027727314 7754869775 5411984013 8896045288
5541646155 2959673062 5328830411 : 5470
8555011888 2984158691 1577081853 7363775060 7133505432 6871353825 9132850979
5856142302 7436039576 4960678598 : 5471
0108937955 6543087594 5538732132 3500760300 7730107917 8964083799 1887086196
3208679051 1588913741 2620483854 : 5472
1157848822 0395923594 3327045245 9141147350 7805455040 3358190338 4706481014
2486891972 5638116863 1933216497 : 5473
6298148495 3964608387 4849543969 4585152579 5180911626 6545236733 5174501636
0833352973 9655892210 9211569119 : 5474
7831672918 4425792575 7550040307 4904666422 7090664322 4285132454 4080760503
0997258302 1790257921 8216908237 : 5475
7943883480 8656545165 8032249266 4139624733 9605115247 5958393563 6607443829
9366773118 9712438143 4507144263 : 5476
4906683367 4626796514 4875760421 6490046578 0456681089 7623735246 6304486513
9464823196 6040726712 6392740537 : 5477
1480600202 8814119479 1417421096 4506313055 9424100563 9877803076 8315454955
0715730058 2014792515 9590460246 : 5478
6115178393 6785601612 5324062752 4204371192 5061379648 5089198190 9587770580
9924908420 8560010388 6023943155 : 5479
7064092622 9658546194 0599098696 4765074089 0903710204 1073773228 3300019434
4131318989 8291146076 8793479475 : 5480
6379260766 0871483258 6479309319 4260650376 5031220789 6059737211 9426458074
3339322946 4562171529 9286987757 : 5481
2374421745 2435105937 0152244869 7613048297 7756125599 0175145475 9543095729
7867274007 2177410442 8762419527 : 5482
0575130818 8966699039 0251341804 6187913963 9175199961 0688039579 4578140869
9435579293 9886140071 1417249894 : 5483
6051202315 2862732692 3240980270 3440393572 1351903458 4028877988 2338142266
0989331849 6674788926 6988742745 : 5484
1477269584 9546580773 5894966320 4319684227 2939054023 4206993659 3465517594
6440193569 6808875736 5135655273 : 5485
7827558456 0031781257 8547257973 9016890446 2967021262 4861217876 6264775483
7065335028 5918705939 3901778046 : 5486
4717594243 2362408204 7979583793 4985204559 6799025405 1594049454 7744176083
9577973686 7376905855 6479182794 : 5487

8871367856 2976470361 9719277273 6875817422 0619342121 5839221471 6939705250
7247831658 3686909212 0660602977 : 5488
8144229145 1540611950 5865265141 2881871106 4918796979 4316047996 5128847403
6540839996 7083236314 6792721324 : 5489
4936269857 0740704713 0584348263 2931884544 0510586619 4037558732 5025731566
5275392902 1523622074 1530144657 : 5490
5930722750 8883267063 7058214884 3196156531 0418901791 2554845787 4634053818
7505614044 2314611978 4513097056 : 5491
5453691268 2557944487 8042588295 0076968232 1898626238 1111072368 7449921424
6915040404 0031332186 6823059154 : 5492
3595496718 9013559031 4696913736 2240776144 3103122648 6037647919 3926771749
3540753416 2571025588 7112283009 : 5493
8045539403 3370487328 2358839765 0909521672 6564747258 0205480906 3989279095
9497565422 9061846067 9476258129 : 5494
1478097037 5574292480 8981654174 4855096727 0889203351 4269360692 2157665634
6844171425 1418841300 7346342117 : 5495
7091338849 7281568269 7556483084 0913326336 7172477250 0819038332 5739818737
4366036429 3416815416 0852874645 : 5496
8725565350 7621150730 8850688034 5696823291 5354060692 7242413612 6383496243
0252227348 2207667432 8244625195 : 5497
3438168663 2983333512 3189264153 7539026898 0209230690 2469188086 6165342597
1370319964 6563655178 4185225207 : 5498
7400377550 1336093586 1416370449 0921998912 9808172359 6202953847 7316594515
9325457512 7621826588 2696502308 : 5499
3274877952 1264161760 1548390384 5869845831 9371172093 6819908328 8932591601
6831523789 8417509554 2145657354 : 5500
2447703079 8689073885 4236551194 0355299976 8615121850 3691185811 8061965525
7629645824 7439549463 8849523052 : 5501
2939043192 0834700408 5959062477 5869267390 0383856751 9596357188 3814773408
1913671101 3050785355 5897601972 : 5502
4131868873 5176938889 6092097081 1996086393 2376114367 6107035675 6751954509
8566772609 2830783033 3384947896 : 5503
3026102904 9827785167 7349656630 1792404581 4188890168 6110371492 1870755368
5442128423 0029436184 6299784470 : 5504
9902389209 0870282586 7759979329 0024068740 4128958781 8001551262 3850669835
2076101525 8102922206 9185058987 : 5505
9780761031 7527440586 0231492711 1485381252 5022120659 0865901782 0258780104
2247426440 3050814934 0088935536 : 5506
9942155412 4279402428 8330163556 0084548061 1425197360 0079468367 6743928913
6803754522 5852503564 3374923040 : 5507
6118813478 9864950179 1203952139 2725893518 8503227479 7283045553 7783064856
3922152548 1817233208 0667252766 : 5508
5964316205 4180594480 7093733762 7385191779 3186994730 2459081165 3170846477
8489793524 4335720380 6146987009 : 5509
2327006538 7351237811 8935558807 3827614313 8710477321 3306585242 9217695200
9865221558 3227076240 4786747201 : 5510
8538357834 3770223504 2508142457 1284375657 1479689310 3634180893 7329972201
9185677528 3873360133 9441772638 : 5511
1562202849 0159604474 2284106914 6815777966 0193578485 3607566769 1643346140
2954821501 7627807940 2653275401 : 5512
6988343543 4195316982 5972572703 2376723271 0438850214 5755116078 1078490367
2734137568 5054751283 3318281681 : 5513
8068792348 3942357902 8304653257 0909969893 3929319568 5942215104 2753161377
8589133026 2383995724 9246256520 : 5514
9238098216 3100220099 3851783878 6291284754 6474997380 8460988491 5717838740
4461981374 4839068734 0591284238 : 5515

3139922683 5506173095 3776007733 4415519711 3447780113 6367129304 9539990019
8370576056 1472107745 8334412357 : 5516
5225879819 7320154370 8493212917 6882040450 7449491093 0517869705 6711770114
1965616954 8746956995 5457946649 : 5517
0245988058 1820522678 2805544006 3055171620 1091706454 9665772809 1231282730
3711793448 8881927575 2509942568 : 5518
0572045715 6003088351 2548354531 8216700312 0209888175 6656024815 6433398561
2103352180 3122203872 4546176492 : 5519
7719760756 1639899925 5588448247 1253975444 1995752875 5338160752 3828907495
7943408478 4098190504 9954722349 : 5520
1767808998 9735554281 6911986602 9105054819 0046634137 1555736993 0816878792
7857366696 4628266282 6784498376 : 5521
7811270499 1616800588 0869983507 6429727402 9251889063 9492218318 7369232560
3991587304 5117744833 4267912516 : 5522
8538532935 3340557406 9546616393 9922312566 2373328913 0824466051 5960756818
6413772904 9551578853 2889906154 : 5523
1234378136 1699480463 9083110947 3541619189 4425279561 0239155036 3062024648
5820967519 5205668630 9608898274 : 5524
2416026648 6843101853 5792010934 1489959132 9401297008 0979209338 7285245068
3677063522 7193756121 0705529957 : 5525
7005685216 5473956309 3401459950 7090684943 9995261099 5037011491 1485835658
4035694439 2440181375 8729506682 : 5526
0507129554 2099185087 6507111812 7238305255 0942709701 8083673324 6047964391
6371196248 1671287816 8808667228 : 5527
6327543572 1997602795 8550212499 6042934207 2378556610 3352134288 3540324720
8840753622 7477076722 6570322161 : 5528
4598530848 0701432711 3323279558 9627033533 1133019138 9309874263 3572074810
1442606041 2219499873 9520982043 : 5529
1409400117 3960150011 6374649373 9333742500 9359679835 4959808168 1749426411
6175890029 7445379024 6451001157 : 5530
3256811536 6055049218 2290718208 4906673068 3734741354 0876864688 0362810534
6366063639 6370508400 6032259444 : 5531
2368068407 0109041479 1408074372 4325375420 4529454923 2054885409 4472777330
8677289572 8448351993 0624062333 : 5532
6014894657 0102953727 6931923980 0791374246 1289880620 9404251071 8216870035
4182717824 6935885289 6936838712 : 5533
3395081807 1126520323 1606942139 4350441642 2118045865 9728470198 7470587704
9174523614 0621664718 2684227650 : 5534
4309872213 0253292080 6244101356 5190192429 7683194184 2247953230 4887222761
9999813059 3434354096 3408734149 : 5535
4009836113 7992901040 4902165198 0016965345 1050932011 8652889500 1349855481
5797648691 9975355400 1996239218 : 5536
3037155704 9566111407 4906989061 4688031627 4565043092 9296856492 0493914266
3256004909 5467246245 0603867027 : 5537
7906597782 5588642987 2457951551 8278895492 9379136186 4335494956 0747960629
9394332898 9560710185 0074129740 : 5538
2790275983 2985344536 6547373825 4430506492 1875584905 4725366313 7453707765
8929988754 5318170758 0339014469 : 5539
7612612080 1219236484 9601319145 3758738420 8870327736 3104544410 3175204216
8170202492 4246854339 3848183239 : 5540
7201173367 2714375914 0357324974 2190766283 9518726209 4877364228 9035555070
9956806448 1385391213 9494407699 : 5541
8176571221 7675877697 7413747693 0080385309 2414702592 6730972434 2514738979
4333082207 0919844429 6349368273 : 5542
4555999275 6543084921 4500589594 9858388972 1949080480 0211031010 7746945750
3027986720 4560296682 9024330049 : 5543

0195865656 1523385066 0108608524 7051259251 0723689039 1442604480 4191068856
9171156055 4676557754 1313378467 : 5544
3435728191 3557921521 2524416342 6728793514 5798726236 0684768794 2449245432
9593759322 5680382412 4308040441 : 5545
5170323303 5468059302 5545980840 1814193883 9130991313 7803951658 6688564053
4425045976 8630684647 0385293042 : 5546
2812537128 8812406383 7191968251 3155064045 8658212202 7033445251 5002455569
7185204494 2712661755 7097790765 : 5547
2201631409 9243456249 6582346327 4199966965 9190963031 5077216229 5973794133
0449069123 5466289997 6787063964 : 5548
4275439705 3072010156 7866765146 2562096649 8520697707 1028331992 5355222101
7907154255 4910668909 8901571435 : 5549
2252320882 4935483264 0582544332 2899338818 1433246077 6202119276 3535560554
0181246640 1180644907 4865679249 : 5550
3045753030 5406358998 5810364305 0669178836 0446337600 7252105930 7210192292
2418953223 8291589093 4128228750 : 5551
5109801463 1623957367 1434576541 1805422274 0156504411 1311050586 3376808693
7833038419 5969453861 3373289247 : 5552
1602532229 3697313224 9379972928 0059926755 6724797579 0953496341 4170020386
6341961453 4418285905 4952805340 : 5553
9997763087 9343399759 7891056860 4133901260 6507861256 0823204380 4545448631
3325106637 2391013324 3023185466 : 5554
5265092939 1180543257 4169027595 4071908417 6288677352 9998642546 0585144807
0324558508 8002374921 9853889982 : 5555
5763569046 9402036938 8274466818 4013998417 7397803802 2542914925 9195749277
3178379574 5010729161 8969993288 : 5556
3137541980 7672649180 0130643899 0549637448 4069145828 4642612420 4961678749
1900287079 9697918853 4126248108 : 5557
7950554207 4352644179 4719404034 4796510910 2724481791 9032055597 7587738427
2738131058 1436032811 9623482496 : 5558
8770668087 1902988723 4172895279 3641870640 6948463980 1089564315 3523342290
0314798101 6684313694 7919614720 : 5559
6097308580 1361507055 1665133391 4433244858 6946901204 9247732557 1823058188
5747169914 6360318779 3301384759 : 5560
7598611209 6170716267 5773157297 6133468570 4814097969 1548600125 1242060298
7885535337 6478465112 3129289985 : 5561
2004061054 0065834250 3451634685 6303094050 9789836252 7739341235 4469101293
6261701989 9713956710 3939774531 : 5562
0096305490 1054483657 0216719918 7220346357 5636382441 1161709378 8893854939
3556125009 0479363717 1542240806 : 5563
1034381188 6033057556 1248673326 8456069417 5279344094 9702599774 3501467019
9502710791 4478321097 8457155621 : 5564
8883274107 1097653063 4168129793 3459346742 7567072441 3469431451 6216704733
7673582685 3171960542 8171285887 : 5565
0830159316 0488414921 9032436305 8050330721 9660182809 0094043271 7917990576
9932354438 8105032406 7491860691 : 5566
5784062898 7344737670 9234225782 2449454348 2808126567 7173212580 4023869289
3611255653 0820450653 0663405614 : 5567
9010369686 5856206381 0884162112 5072437468 7218924209 2630265334 7648510064
8824018753 1107752387 9905191475 : 5568
1301701155 5125936509 5027766386 5604759932 6777269553 4780632704 9303948934
8979598091 1778925653 7443265439 : 5569
3742278506 2716532617 1004913012 7647658588 7813004808 4071184414 2690290625
4200678976 4961699662 0372407499 : 5570
9018383924 9623080167 9713342360 6997570874 9062665937 3346349331 1747238915
6734441286 7677628500 7328674289 : 5571

3474298435 4169140694 8981446418 5413445247 2851022266 0007961380 9607527010
4077275968 9266151674 4436606657 : 5572
9171195189 2709120693 1157006078 4613104860 0959027972 9514654677 2338319753
0039299820 2306775086 1479378310 : 5573
3409232516 6793058858 0944497178 0201240607 5320584269 0453320997 3279358065
6207789144 5512444048 2552693035 : 5574
3303513359 0148144451 6470171740 9678054134 2209529918 0906029126 0715683392
7661689267 4456115532 0928000933 : 5575
5523419347 6248116875 0783337505 2144864123 0046935912 7245516840 9493435643
5920208400 8663972887 4526447641 : 5576
6812224319 7900574037 6752041131 4593569082 9486365028 8665138667 1870984019
1265208719 3821461049 6291771605 : 5577
6112413262 2984729181 9735019232 6469347606 7359179204 7346019502 1468502042
2272549906 0500390527 1739830882 : 5578
3934696132 9546058235 5961688596 1443855177 3255682572 0040864066 7157261428
7421586562 9365346667 6503053994 : 5579
3726433777 1175524863 3346546866 1747102947 4257074711 4524084069 3574593305
8106647910 5872577087 0394121595 : 5580
8349720801 4347320216 6732002959 2177831146 5479415676 3235809015 9344983934
3603113947 0196023680 6436471015 : 5581
5265483303 3229032494 8488740174 8716255541 7843234598 3583131736 8567411948
2164882832 1983039293 7820890066 : 5582
6864165635 6329990025 8912425367 4659874275 7842445313 5056811633 3665037981
3616103342 8769139977 9093956537 : 5583
5874699178 2052951266 0654358748 0249531054 6207908286 9238253173 4870934885
0922029899 7492167707 5930466511 : 5584
5811333830 4783246584 5377999596 4224511055 2052822598 5513579954 3314078046
8738288309 1817168865 0474647342 : 5585
0060161594 4581759276 4878317510 0615405715 3340802777 0017339348 6915972544
8358499570 3116908905 0234790041 : 5586
8269611277 4389101113 6824983651 1243522173 2941277060 3813026341 8165235751
4835007798 6826106893 5781083578 : 5587
6811158166 3802542863 9454882447 4315217144 6531882122 6560337168 6438855279
4908372296 7271505859 9839020073 : 5588
7043520451 9621300668 2618638971 1245753979 8316748358 0286609024 3753365970
3795530174 2860170982 2852543466 : 5589
2822605029 7822819229 6797495070 6846947014 1114718271 2377179454 3954247545
5823176370 7200209395 2480305502 : 5590
4861529194 2583807464 4561247566 3032936211 1940643853 5126173363 8375785199
0930337789 5635070998 4872647563 : 5591
2881577185 8778297353 7054845621 8406166516 3805211633 5535956157 1365548830
4023083904 8499053464 0227536350 : 5592
5322131262 8579847488 7120879742 3919712980 6515715622 5145357376 2296496995
7851618947 2586019304 8018878841 : 5593
4077064277 8982111550 3893404172 9907842887 7913960319 0909475642 7746282346
4504961855 8957186727 0893050509 : 5594
9175082590 6016230356 0854100261 5065995829 4094188223 1711766856 1034310940
9015195560 3594197629 5271519144 : 5595
6546570114 6273564760 3416640733 5530107840 6612170688 0487767658 2960344592
6238647585 5747734122 8559099455 : 5596
1261972615 0335698046 7429465446 5241074799 8946799784 6489274548 1446292704
6102427724 9451728542 7202517075 : 5597
1372587973 5089028835 6512373075 1402162131 1702640482 5133197504 2822098951
3845528136 2688417732 6700882525 : 5598
4317243579 8898927526 9871603988 4633307640 2741020725 4286075391 4632056334
1900517801 6954816417 9493173488 : 5599

7812244472 5186119410 0883513137 5554904941 8167286424 4172188026 3881656275
3985733360 0411059599 4336004451 : 5600
0938252578 8027664814 4257909554 8425763256 6676861865 9127051483 8980441597
5502020223 0844231648 1714578201 : 5601
0230876136 8940062171 5918636966 4756680115 0589395069 1794861793 8811069135
7391119291 1947645716 3842439850 : 5602
6767609270 1560138762 5473877555 1308216149 1786331107 5676996932 6398363601
9984305639 8867930350 3631101462 : 5603
1259261823 2432920230 5048739735 5510388061 8396303383 9202244502 1877806341
8013902925 0800165476 5599060390 : 5604
8806917718 5244075096 3515195819 3085485349 4363752693 1428347260 1286321556
9558934759 0375217333 9569860535 : 5605
5818133404 0468345871 2039407449 2363546970 8013539670 2968920564 1532705761
7850743694 1042162002 8138597409 : 5606
9445803948 4371712237 8085916106 2547291289 4185014337 3320113941 9237769927
2849879216 7957048572 0848262117 : 5607
8953745099 5901605731 9145103301 5921950477 9986163997 3513634301 8127076636
1962564218 2961355714 7414196825 : 5608
1977845129 2399518094 8027615779 0515052996 2456467685 9410800319 6552477778
4431018487 5368377693 0481208594 : 5609
9347514957 5363519083 6645103834 8117838304 0200872495 4329435980 1839909611
6144252080 8102461548 3437721590 : 5610
0747465469 7078956825 5917810215 3564440601 3968931598 4451593321 2019827378
7780746052 7984806318 8533935925 : 5611
5326405680 4758713927 3617127439 0494442064 0017604659 6009726741 9950111801
9442199470 3076386808 2018915210 : 5612
6960803373 1147927545 0469180708 6633068304 7177127699 3888434975 2036150838
6403092367 7079665962 4074665303 : 5613
2058855795 4353589859 4777542084 6585446621 3117417321 3638193761 1174526719
8453771073 6595659498 1659688759 : 5614
5352725655 3052258677 7874361969 9554527008 8885043127 5909414330 2364295198
3092817046 3636710577 0040823556 : 5615
2634578787 4785234482 3998469437 8073980038 8210355197 1467793674 3948439795
0422615555 3063937295 7759761213 : 5616
9082829477 6160906986 6159952443 4740720825 4872747416 3790541203 2043763264
9767578839 4491159485 2617550814 : 5617
8236435201 4490068449 0375525495 0452871127 7590244026 8289660239 9138122666
8728716943 9142423356 9071672197 : 5618
4852985490 6502866938 5313900278 0062228105 4659491949 6576773408 7173852622
2585319584 7657410136 5001688354 : 5619
8356236992 3544012235 9693600605 1229704848 0867065598 2833625461 6249888405
8080568387 4768059241 6721489925 : 5620
4667697070 4279759470 7443992401 9217135876 8929455771 4824440483 7002827609
3844436672 6557959205 3332863638 : 5621
2118343620 2774646087 1764560182 3692049975 2142614111 9495091413 6593993959
8884968732 5390545616 8986309629 : 5622
6277455693 1596271112 9138906875 3371428581 6832655822 9153167341 2889027734
3355493447 8868355341 0612823002 : 5623
1846623652 6025203082 9905573599 6294121284 0361584876 9828447672 1665060508
4309332357 7916341259 8672524107 : 5624
4116285556 0887417648 3498207142 0906963904 0582853918 2621622899 8268695975
9493805904 8857536815 2351745149 : 5625
6466142696 5879562019 9766438100 5061504180 0687076584 7045347714 7005963307
2335779079 4376706421 1961192058 : 5626
2425444418 6413088962 9668960333 9150013243 2796099227 7835339589 1846625759
9319452669 0242146365 9868461586 : 5627

5059340714 8400860403 0338552638 2246381589 1581183633 5966437381 8562104058
2013281656 9854031673 5563816301 : 5628
9680564587 3480396751 6057164490 4016838278 2016031003 2606803266 8396046855
8981291340 3117536801 2912557689 : 5629
0009703604 9925914526 5139775772 9834685305 8553693635 1824757233 3780440075
0475514350 9075612721 9522846296 : 5630
0672210621 6074612377 1515371186 8850400371 4786281788 4264613905 8053647502
8946907239 2890947226 3625662125 : 5631
7205691977 3693290313 9341358756 9782287912 4283350725 0272859563 2347802504
0789612019 7892164132 3874369299 : 5632
1691397743 4727149780 0996496729 7895391487 2704895812 2750145899 0446238905
8696429492 7230354129 3353238761 : 5633
8921156458 8764429713 6389781641 3221384394 5580346265 5791314402 9141250116
8851998922 8707998820 3332745885 : 5634
0878739620 1958428491 6999880962 5666397846 1402160950 5972997287 0961243304
5762531292 6815643291 8037383948 : 5635
1915146495 2919885361 9766896498 7775347004 0989333379 7271594905 1939180303
1244093812 1636064272 0597499374 : 5636
3009579616 2204706746 1174085734 1097442874 9024072224 0719200849 1185818151
8124276338 5231140880 9193386990 : 5637
5247375517 9697915334 8369860778 8473417923 7590002069 6454778980 4654420961
6558245456 5757260109 8292794621 : 5638
2016035864 5909800121 4611081297 4865267664 9377548555 0163800936 3914403874
7044068074 1730711149 1203955955 : 5639
6476378636 8725212586 6419965181 5527268261 0249104716 1897279219 9637288140
5772954371 8948300129 2061255825 : 5640
0088095864 8234350311 5842725044 7144179924 0885831604 4363542631 3119988381
5034474732 7397732657 2582918374 : 5641
2486825322 1336201914 8473697626 7555076004 7847475071 3026331527 9144246484
5831054261 7927325595 9789950216 : 5642
3649805680 1672170239 8636422151 3849136789 4696651895 9963698189 5289292091
0915814558 0415830296 3877917869 : 5643
3541218300 4099868888 8707650560 6757845234 8837144892 9958031397 2269250026
3442393372 9377836121 9989460046 : 5644
0805192918 1573650714 0605213243 6657117486 5186510958 6655317669 9331817383
0344832523 7239280960 6769052368 : 5645
5146455827 2384358920 9066695738 3546278011 2429104142 0564745807 1394447904
8166588098 1587834729 9839103102 : 5646
7522874694 7404696773 8211615109 7247127560 9181821603 2132711544 8287990220
9158099544 6717910239 8577577600 : 5647
7593706623 6993152851 0617800162 2280013068 9503482824 3805988974 2807809786
3373237536 7387515639 9625002026 : 5648
8891715608 7205681980 3813215927 1334649860 7978324698 8263250521 7246773232
1585052767 7276908073 9518020633 : 5649
2392022289 3513074342 6597860593 7025106926 3878950489 3955632192 1166611355
1555981326 9057575409 4401663689 : 5650
4260092675 5204065333 6553951459 5944303364 7298697252 2461302873 9834973048
3019618694 5556575297 9106778727 : 5651
7547211347 2308106665 1220266183 7023659008 3531181275 2978241047 4176812005
4732854088 2448388546 6837414233 : 5652
6505912599 4228687922 9483507726 2714575470 4620061650 9400348912 9260399554
3195783268 3200403542 6871828068 : 5653
2549652538 3158353257 7307988741 4298463873 9305884324 1116758545 3287548999
7195502300 3383521326 4235652711 : 5654
0701750793 7488068307 8560332541 4601943320 9677063749 3574153953 3003747883
9909900702 5314629659 8041526455 : 5655

8977993948 7647541072 4850931927 6032948979 1717413621 3784198103 5068496164
0393871356 1098187853 3506494822 : 5656
5067534562 6451525297 7403298927 5375616918 1748537555 0733716370 4805113108
2092768493 5994530695 5812100822 : 5657
8531454181 7055339623 7627676852 3646265893 6773733428 0355878578 1280821115
7430619791 5537124356 4354768881 : 5658
1631808683 3937758278 9315224641 9954930016 9784479090 0079766476 1987833614
6456619219 7575452830 2389984112 : 5659
8019862103 8498830157 7437087384 1082808014 4737287666 8190323709 6742894197
0934024336 4458161318 0747722821 : 5660
3377537599 2468949488 5688725904 8714181460 2376469599 5080138660 4347059435
1749860090 5231831220 1394591848 : 5661
8907530401 7368699612 5439466721 3996723140 3034936228 6270110183 0211066751
1115697441 3093694485 0884308639 : 5662
2094696380 0556700634 0478765610 3708240980 4867884265 8505599647 7627529334
5172179481 9545507384 9381133042 : 5663
3859464446 3901683723 4401990718 8086077474 5846502332 4552057248 9711651503
7354612483 9533550370 7166335469 : 5664
5583359220 8900331481 1093105035 6252415751 5546073932 4444620243 8951629450
7183976761 6987097469 7327731850 : 5665
0836328590 6286338132 5773471767 9708600828 6365784771 0142436557 0873713729
4057536068 5199619901 4232615351 : 5666
9121878183 2403826601 0409932276 8038702518 2826899050 1392874943 3754762826
8055926443 8064463585 2915698379 : 5667
7510240859 9405715559 6201690611 8060638530 4794627810 1163688371 1150185564
2083240988 1625698054 5241961108 : 5668
0501075913 4257423116 2743886126 4992086892 6439355212 1508479061 6735964953
4179203357 2993192298 7009457311 : 5669
9991169784 2268853665 1053937230 7341483362 7765946108 2027507201 3548479905
3771977521 1020802148 8139107284 : 5670
4348389583 3745239607 9131264461 6573885318 2117046599 3665343126 4959034724
1970089105 7207310514 0310031420 : 5671
0160783683 4277549263 8478125557 2681147907 9790178690 7065870634 7495144162
5253213465 9135416115 9377354271 : 5672
1274878442 6401032091 3869535451 4175104568 3594010162 2677546837 0908677917
6383299513 4146804688 9569352868 : 5673
0453620097 5579858801 0754417592 8524296410 2754439417 4983197584 5436916715
4537583187 9858306467 1534276462 : 5674
6016617073 6520150241 2509413289 1717472435 7727936423 0528420491 5384313671
8688623786 7006886699 0269549824 : 5675
2234826535 5688667764 3797575821 7353681724 1785261396 2129235281 4651019033
0402029598 0863199433 2812220298 : 5676
9858917413 3129412548 2553096868 7233116292 1846782131 0026202656 8569686333
8986031149 0682515184 0653582620 : 5677
2849203691 1080130045 1065828997 6889398622 3020029873 0202666823 9598343372
1483435941 1418680094 4102423948 : 5678
0597129516 2152859580 3182583624 5884073891 9247171307 5627136269 7428833359
5200543374 0229716897 7565143850 : 5679
0239796312 2083229688 6854415180 7687575048 5099198641 6003851929 0649018781
8432826073 8036579415 3750889224 : 5680
7333091289 0232978391 5701654709 8990259096 3377562583 2771152197 6990127206
5427673634 3144359633 8669837899 : 5681
0691427314 2987712102 8098135403 8990518196 5902575287 1711017725 5359810918
8971805969 0665346225 2559961087 : 5682
1060290385 0682610373 6595190365 9809459038 7568023489 5812098378 1845663284
7510122625 5811761539 1139727878 : 5683

6965663647 6038783309 5845869521 2974136021 2303926230 7275831620 1715327098
0917602947 0213889754 4744047645 : 5684
3541813844 0232395192 7105008365 4112614498 7477629576 6461315292 7304082624
6467017087 9217673162 1559023521 : 5685
0339715859 5470580242 2838270279 7149401860 2228872477 4495150192 0484063908
9778470639 3683763842 4702769184 : 5686
3714011326 3995349055 3916092843 6499378627 0814923084 8515856910 4536572034
2141118382 7241925996 0984403071 : 5687
5132883908 4613953670 7141210527 2205061025 3405101940 2940749759 5745271749
2953907938 5860638632 2716975883 : 5688
0913157754 8083427308 4500345820 9437567851 1762382918 1332285007 2395652673
2881809023 8219283414 9414495655 : 5689
4284260221 3790588610 2004188339 1973178632 5472260696 7863498146 8979548112
9245649195 6275748589 9108511676 : 5690
6023520108 6703572062 4104191113 9896508056 3101776254 4678994028 2116489206
2993099395 0416269193 6328525056 : 5691
5907122368 2642913459 7500011438 1266244639 6194029226 1249313966 4600821783
8602422263 4029098826 0707141310 : 5692
1340225182 2925181145 0745324961 1798278098 0909040598 6688873946 5434533741
5292835273 2068452037 4228670618 : 5693
0187577441 9308457568 4590083048 6689521818 5054620583 6400727652 0648231602
4479229457 6503502716 1024023604 : 5694
8276091892 9259141865 4431079730 6158572168 9758130145 9977941667 1685835670
1456279748 1377628779 1201997077 : 5695
3376009154 8850548543 7349191072 4448878268 5079767274 2474988775 0371695099
6456850662 1052359813 3155973577 : 5696
0965590640 4999570137 6219792921 4384231902 1934015133 7337146388 5699756025
7526096919 9204167969 8230878351 : 5697
3389340972 1274136179 6713331802 1610655335 1478401227 1805000560 5899625441
0874291771 0596386148 8871216534 : 5698
2027420219 4001089823 4916321433 4109664552 3645641574 4254761628 0614994862
2628197947 1209953326 5692883575 : 5699
7076874231 4825654762 1396657615 8701886088 3087352063 4213818055 0809538710
6264331097 9218340123 9101558732 : 5700
3449789928 6404340085 6643324403 5520634294 5708350867 4597822201 9072043491
8209816527 4154755619 2053287163 : 5701
7706698839 1265389325 8830090785 9330973252 7980300713 9032546111 6679061262
2091484958 6424631374 6047429285 : 5702
1212258409 0588471531 9438431133 1074768044 6329529101 4411788533 6084147241
8307882287 9553889265 4286664484 : 5703
3467401260 1752783005 3237795047 1739461989 4984126586 1788389973 2766773092
5977236372 5112409369 3571530993 : 5704
4453343631 5957211000 4778061319 5625664941 9026610029 2052756670 2498156483
7479664097 2093861428 7428218067 : 5705
1772944466 8642296989 8060104500 5527182047 4193533036 5947648428 6197418817
3599121091 8110517831 7173557233 : 5706
6204876797 7349797951 6425829722 8610893435 0157998396 3113356714 4207775122
4522159445 8881235393 1831789842 : 5707
7767907761 9574751252 0272576345 9241059992 6915418595 0946053770 9471536644
2336816034 5377494478 2038031479 : 5708
9452485419 0241582254 7307801051 0922138304 3888730097 4159589762 4392851682
7241735402 4953352564 9788361744 : 5709
7651981462 1487379737 3350201389 9631749840 4803141747 3311253576 8108772820
5440275301 5794992122 4822818831 : 5710
5859903217 6421808576 1179589830 5076310457 9394151675 4013599164 5960889661
1203563607 2409926071 3876870353 : 5711

5308360231 3716182758 7949437078 8026235451 3499947100 5751616584 0831401814 1609641484 8555695573 0484032393 : 5712
2205248542 0840917721 4991579667 0505394094 9709130942 6035844241 0735665967 5150594129 7650572681 4953177565 : 5713
4706723150 3130463608 4548358457 2144624672 0883776265 1946049230 7291085755 1718087040 1192629859 9674373996 : 5714
6703984299 7856292449 1578367945 6050193823 2289199784 2022914384 6192877103 3981179532 7919640087 0648499992 : 5715
7364161019 2982828364 4198702283 1823536960 1337295269 6400314320 5504271571 6563003478 0171924642 0651854607 : 5716
5681103879 4804264588 6919236548 5930336260 6440276948 2209740683 5423424397 8019485317 1920260633 6030218984 : 5717
9987739570 5143192427 9415742837 1466917177 5653622153 8380395561 2588336253 2556198988 8138394131 5190594078 : 5718
3614415697 8797339022 0266643667 6056612603 4177238527 3381717007 4654328762 2673577991 7344206401 4595759856 : 5719
0581198520 4360990748 7862010633 0950503989 4971353174 7581834943 6118333585 2575639212 4646558514 6177331430 : 5720
0998747082 9349366305 0146531674 5749214912 7422582208 8849460920 9423211433 4628251716 0783182427 4822368063 : 5721
1197587626 8107227796 3874119144 8120760796 1353984499 8783245877 8085584707 9140358040 3227933215 7013895936 : 5722
5817735396 7847577538 5919860590 7702571498 5199792918 8620717554 0665044143 6740619597 5690246107 5245136349 : 5723
6607249358 2493815286 2368659264 1392363275 8445954235 1653026603 3702306645 5584086230 6562445697 1108791978 : 5724
3006102976 4884611057 4242652954 7417648662 5207870400 4909017904 6710359849 6470060348 6476171102 9493672651 : 5725
4970098727 0328479905 9993478928 1851306023 6900749309 5737937181 3869516821 3954681295 9146498623 4149183262 : 5726
0755026387 6824895095 6748676320 2646934551 7551029281 8249839119 6467909182 3935241871 5552522863 2683189420 : 5727
8769977596 7873611749 8348588993 0089824631 1854478422 4101131019 1145821330 6528058112 4123005358 9649036369 : 5728
2652436919 3640694048 6516075632 8368948571 9246133771 9895892533 6526525704 8202672064 7698022098 3714151087 : 5729
4808272712 1455265654 0049463226 1371175565 2255785578 5438620484 3972745128 1124698930 3953851327 5572087385 : 5730
8613633284 5154980999 1216221760 8194229832 9537528843 0849748152 6598950959 6031707675 4986645374 1376304678 : 5731
3260728838 5165158982 8190598366 2442409841 2397675433 8199564138 8773390255 6191040434 0709254058 7331227195 : 5732
1500439073 3257007402 2910892710 6398570264 2339450723 0166256217 8032650525 0808879203 9039830239 0563040930 : 5733
8301813017 2614570730 8395001842 8619529012 5738124421 8064366115 9699702227 6933679377 0489676516 0022948925 : 5734
5184171690 3012990721 2012965013 3350627007 1422766354 9741111999 2198196646 9870956664 0066532421 0039471451 : 5735
7812910000 1780324540 6453689450 1473949749 0056690622 4257146068 0569254946 2264794670 4886636289 3504625320 : 5736
9784701286 8109030596 2783791319 6010909078 1603725759 8889091566 8049431931 9589059697 3623783181 0429437253 : 5737
3961007287 2574632977 6748022624 4825115785 5302750058 6014154190 8753722113 1528876724 4349548893 9371268118 : 5738
2357650797 5737559186 2260954758 7939006855 0537922635 2071301751 9988485814 1373912082 3909552910 4948088632 : 5739

0773452653 4495606973 7731565388 5478357543 0682309858 0903306345 1846343524
2119359009 9177251932 7329122989 : 5740
2982399848 0331430713 4208898676 8649183176 6482764551 6485097831 8312757196
6685940965 4673991686 6673803114 : 5741
2877256054 7672156667 6445897568 2178499580 3697938800 3509182753 5854837351
0238035096 6032255255 6599141555 : 5742
4441736919 4496215692 4331126508 1247949877 5233971600 9896404320 4516324156
6124325014 5503431660 5675360644 : 5743
3540198147 1072977478 0115502323 0507765864 2923557297 9795505513 9760232195
0701458779 2644147392 1211871557 : 5744
5931178810 8567349467 4367757908 6970048686 0076104855 3967400939 6682669252
9948537691 3467099834 0658310623 : 5745
2213642074 9971036766 4880906366 5818289088 6783654476 5605239961 1687466435
0388545496 5793933667 8299942212 : 5746
3905754675 7896113200 2146388875 1427704284 8516141037 9085362672 8543299282
6190091240 0426930018 4230897419 : 5747
4723371882 7707653645 9963443767 5073059724 4890946843 7335025368 6017508317
2039515236 0017879073 2272888524 : 5748
3637033300 4444092781 2905934536 8663141470 1046593418 8347468928 2629988236
3013060137 6692698821 7798851721 : 5749
2454145733 7848823038 2467191665 9511051746 3243127903 1560874148 8607081554
8311021325 4013356868 5405588343 : 5750
1018870889 3876139373 2502340880 7965938201 4804830316 4481123176 2015402434
5025897217 7670052598 7685752911 : 5751
0799488761 7033468123 2019932311 3219287434 1846612599 8701817465 6117914611
8689268370 2520165299 1198988874 : 5752
9488292420 6169649654 3089442346 3417530646 2620663204 1270524790 4652222594
7485262988 2180166510 3773915209 : 5753
5692571767 6051391512 9079083306 3089131384 6707678071 3608298991 8994490539
9843274940 2438897106 0176275164 : 5754
8654324350 4174682174 0477205357 9072978819 0300647621 7956560515 9378531746
9975436785 0429962280 6859383605 : 5755
8350652163 7181437581 2035946389 8013538578 9008753863 7799944252 7513971642
8576455853 8815095998 6542599611 : 5756
1121263525 2183537375 4089383829 9407147671 9479556565 3338103356 0920916513
5879604317 5645214900 2108737452 : 5757
1940701660 7907421171 4638920928 7184760160 9024923191 1042267151 0290601789
5674642383 4095198359 1142408642 : 5758
6457110707 4853007624 9802206736 3837798445 9884147751 5071622932 1920310260
5005515090 7697894319 4378348211 : 5759
2231317197 6968730832 8746838329 3986801931 9165370266 3820034824 6498888280
0995308021 9176380419 7594627304 : 5760
3423705049 8168626631 4633138199 2449951350 4093368521 3264862216 6261430456
3801554167 0299756701 8107991459 : 5761
8371430134 0032034976 5295216438 5778342024 8049746048 1356556278 7670014116
7645327657 0915946987 8574710951 : 5762
7077561758 9719547014 6914052898 7623863446 6607521691 8405152920 3706434167
1434458101 4881245904 1088366769 : 5763
3696301612 2140430307 9623341879 2780707414 5543096121 9509880330 7323271225
1430746743 7929490847 0001118157 : 5764
8721760472 5628436874 4402999903 4907235233 6477956148 2607275430 4750733835
7941695208 5411858141 4211633663 : 5765
3188436139 3046086404 4381203050 0873747407 4303519812 5879556512 1015437961
8540181768 3516395531 4297889210 : 5766
9793350644 2189220638 2792601708 0859661513 4092310144 5509598050 0497093334
1826034628 2226613652 4578624368 : 5767

9338287481 8080831166 3214088601 8962793337 9691796702 3892600395 1088492322
2262487914 6995246944 8221322207 : 5768
1622818763 3754117440 7176440825 6359777491 0049844113 1586645655 2169347946
9938534589 5276480298 6158402264 : 5769
0999942100 0433420644 9394164465 1586082274 9727905680 4659105802 3199814041
8166646897 1070381589 9178259905 : 5770
2443794164 7676653136 3703816495 5688078417 1970666908 8818711189 2963554097
0894493508 3808672087 4087385891 : 5771
6782805784 6463873013 3563290056 0817556570 5186898351 8288538558 1894187618
4643188554 1883532205 5865514919 : 5772
6084013505 1091304296 4386737267 0176920946 2568404821 6955942243 8162836317
6054907299 3983829018 7707137864 : 5773
8219596279 5827372843 8493021076 5170111412 0971271895 1367781133 6345225119
4325640609 2909203989 2030311428 : 5774
6931102996 1628971574 1651353122 6509765663 8725415021 8818945769 6063382654
0252017462 7484331378 6593668353 : 5775
5892728894 4137222712 2323730318 9976271218 7563590305 5240593344 0680406716
5885499108 9223395103 1852280400 : 5776
3163077793 1393887881 2426373994 5766173505 7804548647 0971336561 2269105452
6803233517 0946578223 5132634711 : 5777
9775415664 8001216476 1891538394 4543827074 1357110988 0273825243 5819294527
0638724302 8983887862 3739726994 : 5778
8101909956 4763398726 7797438188 2407864696 3720613457 5020404386 5140481454
0848637229 4808918749 5333684538 : 5779
3329185692 6116001360 9052698074 8507788080 9719922079 0549384644 9129981160
4451051248 2015768134 2303697584 : 5780
5979313525 0764991726 7189783592 0462441355 8353968020 4390112988 0820848792
7059939845 2081514555 2771604555 : 5781
4509166391 0861465981 0109436485 5299583415 8940501132 2175918918 2274078858
5450573707 5417198079 3576571347 : 5782
6422566400 7855202712 3596498418 1478091852 4754051782 9859835871 9450900926
4562032214 5679360032 0098036589 : 5783
1403659248 0297059704 2389340141 7849408983 4058894208 2813754108 4532719476
5940157849 1808798841 2786712883 : 5784
4697304445 3463300113 4078424469 7467610052 1632523146 9607461797 2352275188
8911366108 4728250447 3338769880 : 5785
8997882496 1745714326 5389593198 9193809453 7356200697 7956600732 9207375987
8773953340 1224262824 6381176046 : 5786
6552954932 7760151654 4433987977 9657596421 3036484953 8029733629 7340540953
6566027156 6620956242 0401897254 : 5787
1002690308 8730688596 7583236348 4860800313 6749337880 4698810817 9243487055
5858612604 4335111341 5506834721 : 5788
0280388630 7988424864 7959934426 9107098070 5308289506 5139289872 4560947408
9911504991 5932660761 2639813504 : 5789
1864212683 9879243828 1063901902 4427167350 7646240245 7681752412 9773437704
7211534086 1680417829 4996765068 : 5790
5806251274 7529950655 9532498818 6611187221 6992147209 5560654755 2197614550
4910999069 7568423675 2153392973 : 5791
5597122527 5715087665 6596645027 1917178205 2938851093 6944721032 7912997299
7894959537 2179654148 2204684847 : 5792
1079713315 2924225658 1065964907 6885751212 8315157570 0891568839 0775159233
9497055571 5543961203 4287580617 : 5793
5188670398 3086783340 8101348168 3943703392 1934197424 3163346877 1675540102
8790595185 5469702441 0748369099 : 5794
8853159223 5768360501 8587567855 7363745857 7148410634 0133489759 0877749058
3345539770 5795352135 9016826647 : 5795

7382725085 5655781354 8876359883 2002785770 6316422404 6839516165 7169656331
1771164541 2249718086 2165255308 : 5796
4508026356 1891326043 5962920096 4032384354 6372129515 9475370293 4135578205
6091034650 3148279361 4106566034 : 5797
5442082853 7160023113 6681319091 9641028730 4920850041 7437038337 4462810464
6209427769 6378558093 4275787598 : 5798
7841833340 3996601935 4202671488 2612819488 6255395043 8151533608 8819835281
7494735425 2961305205 8898947452 : 5799
9789819276 5362302146 4927164086 3202923565 9299419175 2454776144 0843060223
7931856760 6483039434 1621875362 : 5800
7041474976 1396384298 6391528708 3114593851 7668485369 2245247913 3970185679
6618981007 0204722123 3180454192 : 5801
3079943922 1508991833 9722129466 4854266918 2478057987 8782653881 3387791747
9992986271 6454333930 4246091128 : 5802
4741410076 1105420089 7125853667 2836314860 8986343464 5693410241 7486756764
8864999320 1676076913 9511745616 : 5803
3032737449 8044609078 0906403046 7634944431 5588698973 7215023060 2240876896
2808996777 2008295400 9728621969 : 5804
3697999085 6286378189 2068734312 4342519125 7166586084 8853313223 4261843260
5583673517 3575594624 7449424091 : 5805
8913520207 4248917184 4022318266 7020736467 6861018624 7364849275 8014735888
1296157116 4530777730 9918790610 : 5806
2988862871 4930204679 2275251671 0370807163 9439123743 1679286682 1934446224
7657260407 0545998596 8287895948 : 5807
1812296099 6644984189 5435505126 9746222228 4055821601 7815638489 3241562942
9410235472 4474406529 8275956508 : 5808
5230803988 1041917531 0953945082 9566866700 9805968039 7238783071 0888730991
6708399098 6667030216 1465717224 : 5809
7840852262 3333842572 0816810073 3965346032 1498432069 7266393091 8651492548
0137010383 8705478495 8056923908 : 5810
0907147014 6803194411 8829167741 0010867607 1463670346 0970165877 4793861986
5572514916 0321261997 1997380349 : 5811
0164842267 5449125967 3931239799 0074831055 3850686618 3048290644 3355681392
5304490175 5675497722 4586553701 : 5812
3114885452 1455752765 0034001289 4742742237 5583403216 7742658602 9415028540
5959573417 8734907098 0159085826 : 5813
5302204657 8069213686 3441823833 5855058044 0690789048 7694695230 1682422689
5303019503 8490457409 4772378584 : 5814
1308094244 8126386762 5452617907 1856678494 9159447575 2589043298 5971556253
9168706640 5003386911 4702527528 : 5815
7746323076 3947736622 0502124317 1119766975 5407073331 1267595581 1430766435
0837766138 3937418821 1987281402 : 5816
4301959257 7233992497 7456535991 7373704823 4552569017 4683861816 0590685025
2368717229 2558204547 1781431991 : 5817
8580749491 6821191010 6141017546 6753076202 8915463213 4291872260 1569145323
3924467835 3609292392 5956317992 : 5818
4773642655 8854142993 0289457142 9764367323 2226292360 2401555030 5643202837
0518644027 0320700941 3308930740 : 5819
7897145934 1135466306 2636587285 7188977005 5691796392 0940895404 9496757766
9166831282 6151980538 6857951638 : 5820
8745693396 1269736698 7222044985 7426520785 7339345005 5218249597 3648387278
1039461205 4451563797 9612030291 : 5821
6594765746 9934154327 1014074745 7728926544 2299660080 2191430751 6320121147
1223362886 8911003141 9826976208 : 5822
1161023720 0462099132 1164326070 6919886802 8640972266 7809023807 4035935421
4499157461 9796835571 4813677142 : 5823

0102843682 7004103443 1879942143 6138119770 5387057025 1577675008 7453539287
7472019654 5049062159 4472377056 : 5824
5106196759 9990856948 7775939149 1159420150 5099136774 1964053191 2235392749
7551027522 6212593290 3159292020 : 5825
6322743156 3163988355 9894769491 2780282598 4508358367 9986203533 5202068546
0559216786 5528357649 8156695323 : 5826
1585885723 8729888822 1915594480 3787090891 6485672990 7213738605 3604371214
3961691038 5695176160 2847570707 : 5827
4122088557 4454803861 5549299960 1110900895 2930561509 2834665028 8039831552
9188908659 0281766493 3855036021 : 5828
1301004261 4046121856 2027290863 5851705705 2077500603 3082951809 0619335033
6573369268 8723114598 6400466223 : 5829
7348473629 8028779881 0214710192 4585493748 7774531159 6289792540 5501780747
4919647784 0674655279 0393195565 : 5830
8138966925 4392861168 1270286078 0164924717 5794769004 0713838418 7102292173
3518989407 6408089714 3188308922 : 5831
1639365968 7537987014 2040037849 1301275010 0361893552 8646480423 8014072668
7789499470 2425251395 6832936672 : 5832
0126727746 8876032284 8694287301 3499735546 3449841082 9039902461 4311248528
8482555246 8148762739 9427149890 : 5833
8989640658 8465382777 4882015498 9400559486 5085108465 8197861933 0248608338
0072550353 7057526726 1626271208 : 5834
9574838570 7810716790 3963214061 1479857589 2731652066 5138741418 3990141524
0806942716 4153124841 4657507367 : 5835
1621014372 8566150672 8048482094 5901214115 3970570484 6221539045 5053205451
4086490834 8169336750 6628520708 : 5836
5044761687 0476424706 2925198421 8234056711 9317597738 5071213843 5661612005
4129148709 1099968133 1855034567 : 5837
5525027394 8056094553 3332426165 0049742736 9923689595 5712032345 8164445061
8398094463 6812010841 8926213314 : 5838
6656721599 4708198176 8665914883 2682318546 0165541728 8345341670 4493091663
7484656897 6763423120 1898326438 : 5839
3910341875 8413624196 7457994649 2022219798 3459305656 3692756849 3597776710
9310304141 1307312539 5642486385 : 5840
0145550075 7943604266 5449474702 2596898510 2663374383 0181532607 0463610412
0350698291 0077402475 2336575842 : 5841
4349259806 7819610676 6125498936 6947457932 0383480118 9180462399 3440204860
5474005397 2919887064 8908353273 : 5842
8462542597 8152377016 5393409066 3961614181 3699362622 7242206373 3819843067
7526480387 4177190613 4560708695 : 5843
1288294213 4188943261 4115598374 1984309650 6180799248 2485995574 7397586597
9178350016 2512479117 6820566112 : 5844
4568789795 4672289441 1612072462 2182150361 1187196038 6759404634 0815340520
9319548994 5280136392 3920455820 : 5845
7050232815 9177110790 8638599432 6625268337 0835162218 6279069635 1346100018
9272878972 2396733421 1224885525 : 5846
3794962334 8050174564 5714169688 6360100538 7174928821 4974692896 2534740324
9065911079 4774699550 1662902714 : 5847
2984650883 9179574390 1191544231 6633387279 0505489315 7337140084 3033387711
7939845502 8810515225 3878558588 : 5848
5276786724 6546822526 0139414212 6380025151 1052536202 0285088336 8116711791
3145351827 4582690793 6214338287 : 5849
3636714785 5025406183 1507426381 7135131076 7393576500 6518722579 6621355848
4525199814 0046504966 4429369446 : 5850
2643253534 2270481087 3584386515 3165747836 9349438175 6184393891 0192099339
2079359173 0235133613 4333617409 : 5851

3788943324 3636766210 2057520640 4986003394 7626117730 6597900717 3384350861
1904667283 0919191405 4876182490 : 5852
3540960361 1171758738 4282953107 1297887413 0067815729 0071872028 5253473736
8305268388 2088519006 5288899206 : 5853
7114141756 1482180485 9030161269 9363022004 2457303654 5063083444 5212718140
4811064626 5502183349 1808728134 : 5854
3170005938 9454647778 0717800755 4115944795 6636875231 3028096856 3849766467
4164239794 0380978024 0068223930 : 5855
4397514877 6185510146 8074924443 1304936842 4027979663 8069701072 1859444694
6675695263 1588382852 6261340027 : 5856
8056513954 1647267978 4720187392 8734317431 9563427146 8612868703 1868026805
1307783311 3336497051 4243458619 : 5857
4339937603 8313489195 3616522198 5717340060 2626816423 3315262753 2561526998
6604467428 2100016307 8713356756 : 5858
4176057061 0365397244 0343499640 7552391445 9700042488 2780700901 8247852047
6973060681 8272868950 1112304020 : 5859
2596546463 9168826534 4062451389 4380086858 2630992637 0738304783 6303898086
0109948994 1257512561 4015344638 : 5860
4423708749 0956244130 1959987563 8910465209 6675458776 6008659039 5215269307
2494759346 3765524999 5739813687 : 5861
0468238357 8222135022 7515627717 4392239955 4134549014 3078065888 8714513281
3370761485 0257685232 3638293314 : 5862
7428059668 8096462099 8422476207 4394269002 7942917237 5897478932 7985624247
2965908532 1594720533 2369490434 : 5863
0279662663 0740273131 6432230471 2428965781 6081090460 2256804488 1972470679
9349489374 3915075505 1735578827 : 5864
3674663011 3365128062 8067638738 9443510734 0477854284 4945810324 0215302688
9267092892 7343216222 8866530807 : 5865
9172552536 6482531922 2486046719 0401188149 7966918972 3839048992 1449906378
3422472582 9744875713 8716393766 : 5866
0383531958 2212583899 5005317567 0095529364 8507888404 2900036232 4607985108
0944704118 7766965698 5270022423 : 5867
6542148408 2307424965 9128990965 0888536308 7254327321 5141598918 1628756781
1307051625 6851055815 1267135934 : 5868
4832178026 7835089604 7258005426 1710332895 1883638910 3244737167 4832059178
7336509628 2974559694 3462409255 : 5869
6528166566 4281336902 5930758740 4400234673 1373767779 2486726102 6258403688
0816938609 4183043542 1605123289 : 5870
9431137753 3910651173 1742579190 3877442755 5774666030 4066200990 4063042605
1492029870 4318460132 7389509099 : 5871
8152703064 3369446904 1004457120 2235451171 0113287564 0395937024 2331710298
3934900820 7273903649 5979673246 : 5872
0701174416 5743432549 9611780691 7646759647 4687979151 5572781516 2473060583
3452636485 1289816778 4698088189 : 5873
9113210039 3955511186 9683602326 7657819460 8392777588 7735609407 5598291775
4280861145 4330139500 4552465512 : 5874
4291004911 3728859660 6867189535 5711890373 3300649089 7568335165 0049482437
5020133685 1572849963 6967464259 : 5875
1495360373 9411549609 8234431435 1093202218 0970935978 0329549759 5988950811
0435013606 2164200304 0542535251 : 5876
8200915587 6233217544 2175880859 4192994016 6160003634 3910153400 9403986138
1614185296 5918958274 6862217600 : 5877
4007540224 0523491448 7411541445 0603504256 3623296960 3659720823 6492559421
4765207713 7457479512 2002325330 : 5878
7577273544 0666725460 6385566002 0024685704 4600372754 0392329608 7432532813
9244892759 6263699974 6081980307 : 5879

6121586944 3681254346 4760058234 5170986588 6875789643 4602270548 0070837900
4133051417 2192659415 7615687911 : 5880
5019134029 7485850517 1486081731 5609739898 1871178896 3997543859 3851481271
2285659202 7869352860 7609610014 : 5881
5004686282 1433081002 8800342379 9080316038 8504060829 7629418230 8278380860
3522724981 0236770590 6046463477 : 5882
3095240249 0251187179 8642433919 0253045895 7320390858 5078719522 5501777037
6521626642 1852819817 4050734002 : 5883
6663725152 8093405208 1167101126 9698677937 2259856933 4951943269 3201259024
2307651827 7713527188 4472532778 : 5884
0205511448 3586444782 3011547118 4418352293 2511493257 2698861749 1226032840
2072777884 3300201824 3512889526 : 5885
2643485040 1801176692 1894003013 8462303925 5957312898 1537243816 9530773158
9478556460 2548901233 5984452603 : 5886
0584211078 3664177049 8438042272 7756181463 6149708220 5297894046 8419642105
1959529763 4427944938 0087623752 : 5887
7458736540 4368603239 2568120396 8153978062 0318441175 1734063549 6464494688
6431290056 5992397103 9802605527 : 5888
1913444121 7649315767 0125023282 1586829133 9917094347 2186019906 1499472704
1937223244 8113657773 6478433420 : 5889
2259996962 7985529882 3483581345 1982181425 6759243498 8631331576 4355498522
0161874700 9448486245 7290141545 : 5890
5918948887 0773043749 5867207924 8383857434 0109825006 2896166079 9710944183
6998747844 3956767929 2388862416 : 5891
0244369027 1546527600 2249393490 3690547167 4482965770 8307392421 0015283272
3379609356 9239903388 2446560129 : 5892
8007919176 4303142021 9423739939 6437444250 8813987203 1104733044 6839944062
9881969793 7197577353 2541936499 : 5893
9703329803 0950573019 4490517681 3411652445 3593299051 5291198614 7095703537
4526557874 2451856888 9601351304 : 5894
4654670270 7588099460 9033018356 9536601327 9171879444 9541015603 4369228648
0222247044 7675869609 0322096842 : 5895
2563613405 6348368297 1743439491 3450350154 5627121130 7069128196 8263867332
2131840444 4149770373 8450944461 : 5896
7548305453 6899360682 0580388987 7247411952 3892924216 3746784562 4985427985
0314493299 5331585543 0027667154 : 5897
0262962651 6695809146 0788101747 1430699174 4199865847 3290401655 3566585762
6308050241 4955884775 3348985236 : 5898
4672238934 1636565324 7943645100 5902522586 3213646412 5849984679 6161843552
3403523247 2111052122 6636091573 : 5899
6027130213 2944820897 6614103780 7091936558 0262218178 4957122075 8511904228
7800087459 2867736276 3323009690 : 5900
4378031370 8952520766 6717572718 2998614393 6555118371 6692237254 1946679808
2166668111 0395660439 3375037280 : 5901
7554514848 0681660436 7467894326 4045371156 6586375053 1512081271 3275492053
0682220005 2569298501 4308858791 : 5902
8383385888 2722616677 5683455460 0420387321 6650375630 8540835999 9738344203
1879253515 1098838338 5390032909 : 5903
6587405487 3988529729 6837997229 3660129231 2307160205 5097339309 3605034590
3955144350 5307799861 6792471614 : 5904
4327074762 4508513019 7897386992 7099332578 9524645547 5067636682 6464527152
5522543338 8053548362 7391626239 : 5905
2529667664 5875489467 3447577273 3560138382 7372900539 3896656592 2305985710
4848277439 8049720583 8211155382 : 5906
0098920966 1369468931 7719911147 4717037337 4826981059 6270612913 1399606088
2187721485 2557889824 9605715119 : 5907

7409955071 3992866920 1545658383 4310142603 0808586884 9327192298 4158950926
4357183140 9247104705 1845128758 : 5908
6998841092 8735902874 3120393437 6279851641 1032441226 2926311001 1096914955
4450309453 3576921409 8033156765 : 5909
4806421257 7277675625 2536621018 0850636818 2957928716 0839823402 1472035362
5982063645 5200852312 8058003267 : 5910
1686683448 1511046373 7048499734 8399072102 7211903580 0884324222 1164334445
0800225977 9528179717 2269973237 : 5911
4386451794 6984457648 0639489491 8334385251 8042878693 2632752902 4478904759
3794042859 8452749922 2779721000 : 5912
2389112154 8938382391 3828729899 3173119476 1739061150 4478279287 6911023764
7550252257 1732194818 1473706301 : 5913
3088417889 8195981629 9954108339 0244410692 7067375959 5699711953 5930938496
1102865740 7650636769 4490893018 : 5914
5586498703 7289727234 3345722492 7891532609 2232477022 8772629642 4917698080
3027823621 7239379885 4005036257 : 5915
1554887536 1008901145 6864982824 3767815051 2482820550 4920676147 2527146521
8966300496 8857959976 7752259397 : 5916
4060305110 2898580396 2621881971 2821705192 6322308951 7468158647 7249400663
4762523998 5417319602 6161036924 : 5917
1957159776 0197169490 2399328727 4397465880 4365659364 9688016852 8639775155
2247599976 4941859502 6804050064 : 5918
0969843511 3073797110 4411979180 0574646549 3078021521 2529810087 3140604694
7356590646 8924181483 9126360000 : 5919
7362471055 6481982589 3808874576 4536277429 9376813587 6541917973 5722961270
0089296847 1369649368 3678963525 : 5920
1823038913 1039926337 5859652579 6164964499 0890955243 5508658902 5530278599
0775532590 1273060023 5531124137 : 5921
2288339546 4048657778 3316157682 9861517865 0924137474 2372088701 3088054395
2259278853 0239430921 6595649098 : 5922
4077060959 4261296282 4796778811 1335332629 5287479754 0987883556 6787900429
1954515767 4414867840 4482363922 : 5923
3350956600 7275479391 4016971072 3185824412 7989233882 0023779406 3975753657
2516250133 5163672644 3591597747 : 5924
5061192571 3016230090 9373451004 7452761801 6380709677 3700943768 0596671422
9413589600 8247553832 4597480393 : 5925
2060796044 9050176920 7058512367 2619845895 6830937968 0625434025 0957462165
9518879755 0577965491 9550494928 : 5926
6712332513 3755673871 6057356380 0289429902 4885121880 1240568679 2361892475
5604824874 9553282638 7314646416 : 5927
4205988538 5147743343 3172591297 3171197400 0426498722 2438106142 1103274992
4136371337 5474324062 9667251815 : 5928
6579138643 7025620243 0396479048 9005044298 5262446575 6623621882 0854094942
3685057327 2737762283 6552938642 : 5929
1319461785 2606260499 9062547968 8474585304 4130593739 4727793077 5350781935
5762734410 6921558940 7275736285 : 5930
9694496638 8909215851 3270610171 0614979762 0538570852 8120957527 6329498576
6777194759 3521524216 7687786817 : 5931
3437055674 2374024365 0963517997 1533020571 4311146401 3586402829 0245151732
6107671692 0225250063 3762431074 : 5932
1617874762 4311018029 0133180972 2311238240 0446652702 5579134333 8648233847
8240836415 0914263032 1466554736 : 5933
6175962561 6966594331 2066598512 7670461450 4355650567 6323192723 8034514025
3542128096 1853640658 6065956865 : 5934
0090054298 4050046093 5485306206 2677047658 4564323035 5579621397 0401284145
0715513295 8915455166 9286582783 : 5935

8940339152 9922388233 2902538885 7260584924 3307420504 8077496581 8966106091 8105854541 9793248020 3796556828 : 5936
0399926145 9692054638 0587713649 0314877440 4809112742 8165482419 1745721110 2449742316 1561692475 4790847307 : 5937
5166326609 8195237956 7646387678 8253431508 8220817956 6747714706 8016351059 6475683188 9832497120 4609208556 : 5938
9971337574 1446504693 4783022432 7100384214 7687932214 2485713565 6462834010 3242412826 3276542081 6089448070 : 5939
1691549541 9078899085 8389973870 7067015416 6653384195 8350716971 4519341920 3744574382 0510407777 2973273608 : 5940
3932416374 5628589224 1337653863 6749550495 4305663770 8434508365 1770046466 4638153286 7444822629 0496018468 : 5941
6050368834 4077608448 2397002567 7621323572 1377269139 2393095925 2379422056 6769837043 9260789034 8267373475 : 5942
4528332857 6599177610 1256955535 0192620551 7993980215 7103124143 1145302306 9858987010 3035894212 8852315150 : 5943
6444142065 2849534936 2202442156 0328509445 4454628741 4074018450 8573337343 5077630594 2612250192 5255325129 : 5944
9186342147 6582140383 0797952738 7376105273 0263924182 2426415421 5090646009 8831844152 5643072600 1468614601 : 5945
1619491302 4036693824 7501714189 4225920208 0670774549 1575953845 4237813886 0870217866 4247860286 8245538257 : 5946
0607007852 8273322265 1056334456 6490874361 5822952264 5069096083 1695617260 5265253491 5020704138 0219034005 : 5947
7017883118 3123741998 1786872388 2510105974 7512023490 6541684015 7335014317 8373352481 9386198287 1799710861 : 5948
1704819560 7925864281 9561977024 9670042110 0095380047 3880392004 7245467873 0906292796 8600542682 0228388866 : 5949
8402908313 3520768865 0527791865 6290128921 3124031511 4784046500 0757126177 9711587696 0036259178 8995845532 : 5950
0352877641 8478397863 1665070737 5096690883 6131673914 7668310680 4830017611 3605941255 8390261849 7547666962 : 5951
1728534035 8592190345 2376715116 4313372600 6710559414 3593321358 0593431965 1546323178 3380908185 7823319571 : 5952
6802322563 6454354657 3965389158 5126961726 8356652954 5299336653 6165073980 2987340183 8864612441 6365174666 : 5953
6698938924 8273782645 4263142720 3865011755 3097076155 8733454310 2676089168 1516242126 4870580775 0635927882 : 5954
0073571778 0569088816 0798433456 5970610092 4240360984 1782625417 2021527883 0719157976 6742885145 0587738133 : 5955
7614484000 8391264395 6891713569 3227613352 8160479732 5611648002 4364781339 4194939199 8144463450 3389773048 : 5956
3079017221 8978761141 5267584913 7827671364 0481452224 1700976380 2459275416 6726985901 4203411158 8045151837 : 5957
9364707694 4899216519 5823326382 8168333632 5513023426 3516944400 8445773426 4891932074 1277155095 6432261038 : 5958
6891038570 0958521921 6284841848 9882732686 5547042366 7527507529 8431229087 3054198395 0440894202 1416678082 : 5959
1968098279 7670774928 9849712423 8809335414 4951082942 5629732782 6692300410 1161806478 6854216339 3012745589 : 5960
2232424786 7491607615 7695411688 3021345421 7159658460 9084801971 9649487228 5422924913 3226957718 9910652192 : 5961
6823520973 4285293627 9886092116 7917076294 8584749614 9978359834 3087200700 4771021856 8974412617 2310910355 : 5962
8622624994 6839024978 0248231061 0773890804 3031729059 8477045224 3032100330 4957569659 5575909808 9718773551 : 5963

3274829633 9886457187 7846910640 3556448961 2527351448 6823105300 2778188431
0676814363 4883686881 5197935919 : 5964
4805864518 3785865973 1027120780 5878176828 3476422045 8041748546 5272557925
9321275422 0935506709 1521746074 : 5965
1863450104 7954448472 8043228759 0427853279 8925864532 2429852338 6332572078
5494344100 7130491816 0075095719 : 5966
8178380956 0002874758 2755714595 9121423798 2410344201 1990429800 0834846679
8477917366 7633916755 9812330736 : 5967
0449981783 3000271462 0794715396 2607424019 0517782696 8287930733 4273726355
4559682051 3214757796 8851655215 : 5968
7856382150 6100375744 2106878698 1759087972 3105471878 5979450934 1635317309
7134275573 6848046549 3684608589 : 5969
3279519387 8054835351 8384579551 2788897107 5385264812 5918197952 2714467314
8897830668 1441294809 0438764754 : 5970
1720328836 7931539487 3192784282 0614083782 1111238551 8592573720 2642344646
6169852063 3845340600 8552668716 : 5971
9998825468 5318368450 1164335422 4246766319 7456134600 8496308560 5745373590
3033205858 4604742117 1983158007 : 5972
2893001356 1157562071 7423148933 0479644746 8064963812 9316429235 2358113902
9669949468 0014506883 8572950498 : 5973
8003174294 7556236767 4376499424 3612959018 8781636342 2319493407 2584973173
8971847387 4274935509 8506472696 : 5974
9684412652 0678050219 4204287610 7362888938 5885038732 4568556438 8165788462
8098866182 0320357823 0333800993 : 5975
0591300723 3341323450 9602597374 6052004357 0998600298 1455095784 6283200151
3573592546 0273515967 5644163653 : 5976
0112264712 7864032448 2400773799 6917646650 6023873396 6563567934 0398356568
0722196540 4885119322 4882054279 : 5977
8091297110 0770045002 2775461206 6171669155 9139809765 6598227169 6317371323
8023308189 4643812813 4866452495 : 5978
9954457359 9602734037 4953198103 4137354585 9961495498 3609176126 2853953078
7384570759 4632930371 4882251930 : 5979
3817175115 4383500267 0899582654 5263811037 2525488774 3926013540 6025221454
9198169957 9873716453 5132550990 : 5980
5208796779 9440782253 0807758169 9560271112 7758544868 4402760529 3945142888
0029095380 2848541101 2261577841 : 5981
4915814077 4999841496 2922401988 9130831785 9666915388 2290099469 4745024784
4902571367 3569726397 9283040328 : 5982
6063454681 9859014808 6774140892 1089040105 7657503110 4192216149 4187431458
7847613671 4773918305 3531438322 : 5983
6695453832 9922394045 6133606017 8214118865 0929220792 9496640912 1600359051
1538805649 2162705446 4191236518 : 5984
9082065327 7589199738 9222939012 0026823222 2369773672 3300393821 7236746530
5265074391 4068309474 7321260320 : 5985
8840098990 1480267801 9948268585 5351480657 0539140057 6934547136 7320387577
2451306975 9606056795 3900372658 : 5986
4611384511 3230645833 7250580531 6793447259 9430552175 0085317786 3633981947
2177438498 3941664621 4485505188 : 5987
7706616890 2788747419 7775072785 9461678481 9648879238 3924297012 3021952643
8487691711 6929419136 7645398975 : 5988
3022131894 4274689864 4511952336 1135808699 5256573849 9513227234 4858932311
3867978311 9517843877 1350648230 : 5989
7870482998 0344715507 0141882053 1041426682 2948160081 6095024682 3597889332
3946769501 5594757502 2359260204 : 5990
2472263849 4100311367 0440974536 5861030801 2059308927 5276107285 2639425752
9284362186 3776425354 2781899306 : 5991

4800665696 3672751616 9718199072 2601937571 1689259479 7447612487 6288821798
6501367475 0750063832 3479883964 : 5992
9774004884 1235756668 6571614215 8311084736 0913934500 0273200513 0798128157
0222561690 6552683303 0836656381 : 5993
4134700708 1942216648 4822104159 3434919082 0405640859 5224038800 3780734926
1650300231 7179993148 2592911800 : 5994
3774744659 5015659399 8138623869 2869062652 3820612336 2367459640 7209835077
0110829907 9028069034 1091750963 : 5995
5731456182 3190444770 4954866187 1606922803 0350137359 5224123316 9641834879
9080748080 4086899822 1727551316 : 5996
1958780967 7521653989 8309620348 9409368385 6539421196 1230810211 0347105174
2416434655 1719207792 7713852950 : 5997
6026751864 2439692655 3672334478 4100681455 9511490367 8283881757 0535380038
9460027690 7056312702 3230141413 : 5998
0663168017 4679733509 7254146260 9599578815 9410727806 5966542285 3016083098
0948279808 7779954151 3306341851 : 5999
9778723030 1266392253 9995594139 4962110041 9546078252 0674425080 3288180503
3938921875 6524445169 9554137647 : 6000
7841671637 3075584797 2338659392 6352240182 2608031692 7670846826 9071288406
1919749117 6562869996 9084970730 : 6001
8233756477 9768748466 7530526919 2985079280 3668182143 7679607305 0873808083
0144642975 9825417007 8643973049 : 6002
6108341861 9696619596 3220184035 9163563411 8435818598 2051413631 9153091251
7440662404 9390924513 5885190762 : 6003
7068893662 7099055946 4689376680 0692046828 3630462501 6402102743 7917854480
2485128618 2161251211 4570003573 : 6004
4674069253 6790368909 5025923989 1548171622 5418252452 0806003906 0040615540
5892907732 0687309580 0617499719 : 6005
2036471209 7884192446 6670920444 9749874824 0886590526 6935889487 7525751640
1354367423 7920145307 2202353576 : 6006
8345446812 0686595139 3272635592 7699659957 3777441103 7909107156 8358658465
6220858621 0713954935 4539122567 : 6007
3280629527 5190075494 0904896394 3888064254 5570726221 1593631243 9491645972
5649101842 5754082200 4722888846 : 6008
6345128030 4483190178 4007401167 6477395615 4364713952 3558199976 9359010841
7752197336 2030832625 7616596841 : 6009
1936614585 3311420753 2119529276 6971704206 7051859842 4997628346 0412316390
8122790890 0560239147 2762547230 : 6010
4465613738 9193482911 9845493060 9446294509 6159115367 5525286261 0591271212
2041446841 7749781863 0501112974 : 6011
0039411193 5081890835 7333290551 1440743044 4467585330 3908198677 4585870646
6753105873 3204448664 3819547370 : 6012
8480984019 0145711080 1511114446 6295074606 5233051734 5945257725 7589307863
7007195767 9284954220 2391372656 : 6013
8259931838 4963573717 4554050387 3578054083 2235428668 2509834074 2461917212
4106592840 5281116620 0923282960 : 6014
3017213638 4928510477 3585298392 0870989263 1698435885 7422063744 5799561054
1443705248 8223358026 7567460992 : 6015
5440227768 4009359318 1777507857 6733453207 3118530837 9736957382 0246047450
0960452405 5600641568 3554046864 : 6016
1810641559 1598692574 4890303471 4608636868 4207141529 5195399688 8639944162
9850126219 8265478995 0631292147 : 6017
9605647184 9993133924 4495297288 3378335522 5306656081 1391115575 9997907138
2892418373 5740905193 2411808327 : 6018
5321057583 4434078628 7664294881 1335953007 8115142595 7827964092 8378127631
6746885253 2329802856 7924732045 : 6019

3209385421 0158071474 0180947946 1160486277 6786734377 5751143759 2333049254
9945720627 6842336439 4693270173 : 6020
3610844401 8756535693 1607880312 7015677432 9211095460 3746698646 3058964329
9619579908 3916388510 7358365539 : 6021
7358685803 9475629404 0228635209 6347217039 4504703852 5710853133 6244754542
0010525967 1217835787 4633359416 : 6022
5923235625 7039331280 1881979934 8769808508 8538737901 5678885924 9599338041
0450709566 8197806890 9791304753 : 6023
1270144691 1990817138 0579382353 6727157978 7439956478 9154906407 6938192367
8366723218 1905821363 9903497314 : 6024
3981196742 1740486606 9196506586 8831513483 4018768134 6790426439 5573859006
5483758071 5281812895 1074144096 : 6025
0450170439 6548535905 3827804348 1358307724 4516003786 0973737431 4721794026
4953077294 2955247321 6428585864 : 6026
1931339046 2255573142 8767902253 3447878688 5637977093 2207047543 8244713707
2108172807 2621619220 3451676638 : 6027
5698542146 0029371066 1317844674 3494946034 2745909707 7940257119 8873753139
9123813260 0095632063 6823628583 : 6028
0789874153 2274462127 5917935463 1214921585 6310068890 9580778060 5937282837
4066045173 3814756940 6689687907 : 6029
4464372890 3240457174 6893162279 9152607670 0874957946 3655298108 0060563205
3593234614 9132211508 1869171155 : 6030
0065566655 4774549787 5592906074 2276192490 1312867775 8013421084 0162883087
2921226157 7650952210 1508044663 : 6031
7440329782 2650479584 8394929088 0913783491 1052701891 5866659781 5395224336
3020381943 0779722100 7492952919 : 6032
0141757752 5169929771 4793500137 1899644889 1152064736 2966761218 2183930848
9926024600 4189915466 9973852196 : 6033
7567293096 4342169889 8063419295 3311166015 2026890675 5263925108 1017259294
7411597057 2467020836 2379144576 : 6034
5730763105 5046947966 1341506294 9874761664 1845707823 8555743704 7474057201
8709533927 2312325000 3365765441 : 6035
8215016266 0235171847 2672153312 1075096401 5851018981 3774914265 4529986669
2087088903 6949102304 9304630341 : 6036
7508983514 6799025697 2876511504 4510267583 5608349627 7333454143 9538796196
8611228627 1837770262 6499995437 : 6037
8975661863 8452242447 3949249215 0548510122 1670755524 0510210387 3300284593
6131845844 2786733823 1426176973 : 6038
5636427084 2120283188 4367381928 3471319508 7173122191 0120316721 1411093958
9992288467 4801741656 7609937878 : 6039
1968770763 4475970187 8701153635 0704268062 0343222196 2481896791 0905627992
6872063157 3443595078 9785292306 : 6040
9671951104 3330556678 3849538409 6125277958 3891054879 6848486208 6797174930
0845214435 9425346200 1124108426 : 6041
6566758689 7808277627 6840134698 2941929580 2033057400 4749139789 7105912264
2210407325 5789131404 7746709521 : 6042
6337310954 6710714788 2434746973 2253620897 1843414016 8951520932 9372895796
1799009745 3132280763 1818289943 : 6043
3189695689 5304723703 9953890583 9657483550 1508194701 0033649460 7541568093
9094827544 9981181003 1151431124 : 6044
3716206028 5082116771 6052901503 0383998177 8749861963 5004890805 2208969068
2794915503 8157223974 6651144204 : 6045
0712132800 5606536244 6019685738 5732581380 9450794347 4066036054 3591168103
8547455413 9010052108 5682696417 : 6046
4364592697 5730111231 4276156916 4064382930 4144191258 0970015014 7626045084
3029947397 7704434060 2558483155 : 6047

1837098621 0437182444 9093244999 0941239696 8072735574 9909747543 9290255798
4797093482 1903280850 5910233185 : 6048
0565958856 9341036975 2178796616 7710423049 4235235108 6300728128 7132147932
7804020664 6142623007 8561408403 : 6049
2598348925 5712085118 9853822385 1362097287 9195187746 5064186101 0501100015
2392140198 8115501033 3190671539 : 6050
1496612736 3813534906 2018988011 8606264888 1416943529 2751302012 0744485069
3949715656 9637005281 0443645796 : 6051
5400855804 4162484257 1854483720 8664333866 5752522858 1094828921 7257839158
1914769136 4603268447 6202255833 : 6052
7884307066 2682013656 2567060429 1660967399 3739633372 5598175402 3690188353
5300799015 9396724928 7745723100 : 6053
1781338885 0629426776 8452361064 2620854720 7080605367 3762684766 8768462104
3656625525 4577155820 9684895512 : 6054
5604270948 3869900453 7060236388 6713679104 2491149199 6301475646 7260027940
6939362920 8526804159 3916556942 : 6055
8310137017 2150024613 4125553880 3212017480 2466199405 7160259811 4205384973
3099095858 6477131121 9005778516 : 6056
8213546567 6925436958 6839553959 2269791198 1510567862 4278738635 5969635159
6525780100 8877751613 9485947653 : 6057
0289336591 7624022970 6578369853 6071100495 3475675722 8407933874 6963969820
5275488541 3863809128 0465567957 : 6058
8673802477 9624558074 9357238874 9181720103 0089198899 3237953392 7562492951
4306391754 1756523620 5655375337 : 6059
4784035475 1434991801 6962421277 3057517531 7271408992 8417997105 4379976646
9304839985 7656970388 9161802688 : 6060
9482866498 3964738224 0305236838 5787917654 9873616284 7160152275 1105553564
2270930341 2906341214 0503747065 : 6061
3876110440 5763127767 7687955828 3969360687 9749299247 3055757014 5071286487
7603721671 3666399647 9516812181 : 6062
5089563593 2214508085 3486264524 4138042319 3763653355 2748353332 1558341288
8866477801 3962249460 2435843022 : 6063
3059175741 5527544778 4716651515 8060159683 1434699386 0224116703 3961033143
4441421523 7812127052 9004152968 : 6064
3283581427 4572054807 6341739976 8540321142 7870270994 6582145669 6142049358
6005178320 3074959984 9994536775 : 6065
9639015443 3298372959 8770215879 8404530424 1723688539 5654311324 9128001668
8614321335 9018145988 1534511564 : 6066
9693087226 8799815440 1637903625 8474494027 6762231405 8383024632 3278535589
7049122876 3755160993 5228638759 : 6067
4826470923 4548966040 4395528296 9349632732 9619453926 3412540443 5830649127
2796994144 2577153786 6021215962 : 6068
8384800807 7648600684 4211951284 2811118606 5633816275 8796685046 7909393030
2438194147 1345044461 0996238141 : 6069
7080458893 8597963438 2447612009 4314750139 1451102903 5345846423 3986653377
5034032887 5117844562 1701907008 : 6070
2687537123 4894254845 2679529059 6728991141 6216871720 7252789541 3036625313
1216168718 4002908491 4010882474 : 6071
1929033100 3958533280 9030568981 6195958414 6403500881 8383544776 6161764083
4335657628 2916036527 8550533429 : 6072
2017344423 9999129821 5606563923 3096831232 6061134984 7459047534 8175724793
5228998935 0094349507 5396373482 : 6073
8911547110 1729844079 0711638488 2298841792 1854283174 9857560164 4356222646
1225946402 8308647766 3873594598 : 6074
8424504709 9086787716 7500913930 0382117519 8111842564 9944996192 5019393804
7253373994 5933377312 5252463064 : 6075

0434299251 0063627726 4404252212 9335363983 8887125586 5028214839 3519537829
1219232513 2955047942 7077479817 : 6076
5730690981 3981758364 2674915675 6380340241 6350300899 7458846644 5595105263
7730388753 3487344021 7725854816 : 6077
5703260033 5620490417 7355790973 4759843947 5995845429 7654636741 2107535150
7013851212 6171017094 3881638681 : 6078
8003253445 6078011389 3154572325 8763168759 1414183933 6568222962 4660091462
0551459783 3791156464 7926663543 : 6079
6278233025 4858219782 0709974731 0916035110 6470097487 4000731522 2876647396
2912778621 8446835550 0202043071 : 6080
9142007284 6279018318 6397870257 0277226878 2391036972 4454864110 5888916691
1105922029 4449329436 2703541330 : 6081
9880526880 0879346170 9563048460 2882766011 9070894073 0028200664 3598669430
9912883866 9523792986 6628178898 : 6082
4272697048 8860447376 7609420261 5377177917 9096775127 8719747104 4080915599
0679190772 3417208080 8599042860 : 6083
0454545675 1422771384 7378234110 0531182443 0632388715 2844086268 7566050697
2347847736 2196202376 5844110337 : 6084
2159043811 8946982931 3069286115 9856453139 8931399489 9983340440 0924228379
1251755521 7462189128 7605139468 : 6085
9884726771 8746650478 5270664362 5748319169 0849153712 5880541454 0363267478
6953964910 3724005746 1302202931 : 6086
9950310198 7750602880 2379750025 5215749964 4642453349 8859159093 6954395808
4528045004 9366398305 6378225410 : 6087
5626241683 2173011323 7466365081 8321551390 4980193919 9625148203 4852336035
2029798922 4377311115 0091658570 : 6088
1032600353 6444475177 4246989158 9357347056 5875149762 5632680396 9581696949
0397599461 0639763432 3054227213 : 6089
0876246685 7346704606 2234937841 9919838013 0993928023 6522741919 8605426424
9711792282 0503705375 8742713666 : 6090
7271485530 9460807796 2908093583 8546546834 9840363555 2168457034 3035006341
0235028534 8776635304 7125068844 : 6091
0872326675 9056557933 4784591133 2126078901 9286980993 5963677578 3128957269
7042883799 3551303926 9512405891 : 6092
9984490604 6319277629 9056460394 7687565277 6188987807 5082021154 8536425379
1970754729 0711263442 8136059928 : 6093
1191735709 9215255551 9802760560 3718090518 9020718577 3505552327 1391396250
1594272539 3023718644 5017661783 : 6094
5950053674 2452835334 6296600400 8468072728 5331808352 7248634331 6020649687
3873921616 0955927770 7472091863 : 6095
8361912857 5571939484 4572279339 0984130659 4059996512 6384799973 3328982724
4713523630 0117319458 7972985469 : 6096
5574966414 8606783193 6412121572 6453407658 0706689602 5318540147 8248747972
8063120191 6673807223 7763920872 : 6097
3247542101 3217402193 1705246883 1195661303 6567070305 2191237861 7793925690
7627223847 7050523927 0622228374 : 6098
9423814306 1034403779 8238210887 7414396313 9015107082 0312754566 0795464337
1353459928 0628719694 6972555924 : 6099
8762340560 8599760254 2338053560 2919869909 5607613736 8277070442 8667046412
2474056996 7492098598 3836128293 : 6100
6506774498 0245221678 0957009937 9281101073 9323086789 3546477556 5148775074
7946650087 5692695049 1325561264 : 6101
2806005988 3949951551 6257640827 7981605727 5574439612 0181474978 0045132178
2129786374 3751099747 6733763131 : 6102
3444066932 2169897906 4814152245 9605796036 3749293853 9058045809 8260355681
9528939522 1669574156 4220430366 : 6103

4372299146 9676043864 4194213013 6759001693 2422693491 3049246702 7077824818
4552311346 1103453489 2733150600 : 6104
0123028538 3342303638 2471550255 5136874563 2166936656 0444146424 5556981823
1927411945 0827968870 4694176702 : 6105
9660195074 5024985165 2980618680 5463813475 2514384332 7899199609 2710858122
0089357662 2256979935 9869922149 : 6106
9254647176 8210127651 9959597824 7005014221 7475458619 4296039239 4590928828
4181468774 8419136141 8987382812 : 6107
6483553432 4101696493 4655262953 4556346170 8335109501 6806940228 6750567763
4457147413 2177751673 0620778218 : 6108
7706492244 4475208280 0098342557 0457784919 1916884176 7717868863 0332126895
4919764573 9407537009 8837140048 : 6109
7025429260 3296387792 8754377069 5604373399 9010029485 2638150032 6289972855
1130103698 5819326744 8850522284 : 6110
2191805455 4082277475 2760746538 9890506437 4798169834 7177749076 1037600628
2890619457 6396781299 2880200275 : 6111
9672944986 1547706520 4732195418 9029670858 3780605695 0268859915 2022861683
1779318138 1113370084 6648282316 : 6112
4294019403 5016674892 9939240870 5756479425 1319995861 2783309735 2653843359
3841675247 5309684246 9778191862 : 6113
4343771115 5992528282 7313295136 9787821407 4254516312 6864523275 7808217439
1542146889 4359097731 8156580220 : 6114
5796404419 4212253701 4799189278 8537903377 7432873409 5511741378 3520191979
1527965001 3938688485 5693747482 : 6115
1612927167 2957278563 8473863246 9385840529 2467490402 2413438951 8832380107
9314601834 2916216331 9657370015 : 6116
9794273900 4500063841 6513145176 5590859700 2647003221 3022198522 4974139157
7952987929 0963497289 8511760181 : 6117
1374469220 9425310313 1383449613 5599318178 8354416471 4503878554 7166597698
2467247974 4031166060 6198912250 : 6118
4156909044 7646624571 2836382061 6674275647 0277596897 4627841810 5147670135
8042590385 7530620365 7293784016 : 6119
4916694827 1359285692 7354306769 1788670004 9220273231 6264040702 5502795620
9349621622 7338619486 8110608449 : 6120
3589560178 7085883133 8441728876 3890931537 4072400072 8025325627 6428402648
6565019686 9797443042 5922584958 : 6121
0474179227 9253400552 5247449502 3408392656 1723909309 4230060936 6303234802
0210867886 8089659181 6847927368 : 6122
3301432714 6956844570 4936542127 3852364197 6274894176 0376029752 0161535938
9448762202 3573913546 8342725946 : 6123
2829509057 6514319420 9595516072 6127413535 9833191841 2357841964 2134288725
6687389708 4383114104 6585600376 : 6124
8822032463 0865651541 0799296469 0465777065 2379505345 9602461494 0206156054
4843064378 7299452582 2626360919 : 6125
7006342345 6958121081 0188042948 5136828673 9852132534 5198519868 0652792016
1753896561 8411825224 2529689346 : 6126
3498323878 6265738322 4882146718 2212392161 4521633252 7567001704 2893990524
6548258778 5124125185 6125788689 : 6127
4555316654 9754643047 5350319035 5902321438 1285817927 5339840124 6082389071
7054605835 9605867719 9021834652 : 6128
8305718682 7751076250 6653709452 9888302119 6273029318 5889270847 5701485689
9852966505 7338471038 6205996389 : 6129
4320941335 9577964476 9922141537 8655112464 8537943925 4073621927 5246848238
2849973125 7186457655 5150869582 : 6130
4151349798 1570174378 2733663799 3430650906 0649809298 3863033353 9425021822
5663812732 0974354662 2445887643 : 6131

4994073553 8863587706 7206336811 1132294229 8365405268 8215612702 4596288572
3548264218 3145461433 1912545833 : 6132
1181255979 1473648401 2914686221 9867375818 9771951882 3278520809 3327828052
8503438813 8019528455 4650513932 : 6133
4690269115 6026768435 8544393507 6285667261 2665083945 3589830932 0803700107
8932436582 9155080132 2381298871 : 6134
4648091356 4402924712 5244100245 2525445080 2324615782 2063586871 6711055693
2854380162 4683461967 4923755227 : 6135
7513105100 1267760576 4387991571 9948597065 1760213891 4640631735 0223384643
4583948354 3502798190 2728797303 : 6136
2028508468 8401987594 9870378146 1796686462 8754667039 9896304248 3342254904
9244670132 9392472583 2362315311 : 6137
9971239894 4621765884 2719338254 6662161038 2140069902 3027742644 3857141757
4587943978 9795948158 0497290597 : 6138
7772621874 8279191542 1390156710 4042498796 0383839873 0807155042 5303932011
3817262336 6914341884 7662675503 : 6139
2586344926 7294163554 4061641605 8126006897 8504890246 5409536738 4485080441
9948123152 2643777835 8928010770 : 6140
5287235798 1319176422 5440790262 9775223299 4316244056 8228240489 7958586204
2095903016 7530770098 4125504143 : 6141
9517370577 2057555508 1755126017 9018120073 5133417723 7247622081 2000860440
7951239514 2659896434 0376424506 : 6142
0829599666 1560889038 5710684640 2914127173 7657151348 8794464268 9107694108
9531011909 9929995630 9309050352 : 6143
2277723326 2914701401 7864451463 5311873837 8495543882 5008569273 0878394745
2874920176 8864473117 8310411019 : 6144
9160063149 8818929990 6101527781 6870842162 1381839557 0791840511 9806759776
9599875313 7757726887 8910886459 : 6145
1654468983 1334742354 7929805191 0921514683 0716323855 3510387271 8754467670
8295297490 5345953765 2593192516 : 6146
5945147933 6850638167 9734786636 8832707895 4395966772 9832709668 0062790539
5999829457 7731682383 2607388018 : 6147
0654102514 6172612886 7883587066 1909367729 7966422255 9336908245 8671032121
4530157614 0656384883 2046204655 : 6148
1157310033 1062717763 6632725355 1051140113 7294797424 2341799659 5373489421
4214002365 8440811338 8319761752 : 6149
5505890092 4545313775 6058842247 6286523876 0627246990 3021267047 0780945124
1471629495 5702704018 9986666320 : 6150
1798423005 5070084407 5332796256 9991771876 5426525703 3125495139 7086494471
9145272944 8830509460 1841529556 : 6151
2514740409 5257980099 0146338379 7769021293 9408531024 8856156735 0606338634
9236844895 0752823340 1007520258 : 6152
2830620711 3719059426 7815521410 9218605705 4209610307 1329372555 3682257947
3558746256 7776516453 3109298228 : 6153
7602837922 5930251318 5165813377 0605209210 8657561743 0123342890 8469922349
7351511631 4217452542 6713978924 : 6154
8050025172 2320908212 4574110776 1163535916 8604652376 4118520831 0455600513
9095894987 3097070872 3111425470 : 6155
2312167332 0381085480 9201787391 4879344888 3726854568 9214877830 3900165477
4176228126 0580728355 4153313690 : 6156
0793139630 0063769702 0076253505 0726123344 1510111428 0709368194 0223698999
1308247424 6540127019 4011322299 : 6157
9932048332 8746713553 8349457963 5836899288 6232904397 2258449381 7107725905
8039497162 5950663691 6042428812 : 6158
8254838697 1596653055 4742543545 5973433201 6501747169 4261408641 3803804665
9532238806 0995968930 4939813989 : 6159

1441778108 0440177680 4126311873 0703803284 0781365152 3786595055 1008740358
3849737817 2321001662 3052721994 : 6160
7879907436 0574231409 9283345866 1530302659 1088028489 4388262719 2860592688
5462526118 1150655431 4391860473 : 6161
8638320149 5201419924 0165101739 7674092260 4325484294 5659258581 7768997716
5202674986 4198907493 3642588243 : 6162
0300822991 4088423037 0334920003 2109476423 5749370825 1538835961 2855402857
1511999684 1213095132 9760106062 : 6163
2384467853 3043036052 8332459477 1517521109 1321846929 6890135992 0399067517
4666377175 4089316263 5269159223 : 6164
1667585283 8151330957 3351829442 3401948575 9992887571 5896113735 2500733529
9446864517 7277810729 3555066200 : 6165
1116627864 0684583474 2122015354 6184274562 7781395631 0035038009 0185222039
9726275905 4682726991 4375360065 : 6166
8655126345 3165342239 9403325698 7619903270 0182932290 4538021646 9805315530
9882953376 1896730953 4457130377 : 6167
1285992545 8180227261 3746556905 8225957869 2098980461 1674009391 7323357544
5142418155 9427904164 8405012175 : 6168
2751116222 4841376487 9395289487 6891106208 3467875763 2368819950 6508172349
3681850049 2013953969 3115045084 : 6169
0631833169 7956500115 1633008378 2711074977 2860464151 9331149777 1862005817
2118357176 5889164635 5701844887 : 6170
3306567412 1671104599 1852850612 2196801107 3225482951 8774076669 9796023038
4720072533 2760059467 8695267905 : 6171
1431952573 5477141111 5730628379 4871723879 9010110737 1970337951 1138790244
2285766119 5134709382 4055168672 : 6172
9869870945 8855280989 6555090500 5839479776 8163621359 9589645466 9367741167
9523655933 0196254317 1459828163 : 6173
7637734830 4158535288 7106282009 2867345131 7867057905 5862422877 6977038033
5867189664 4007604521 0507780109 : 6174
0263740143 6327800462 8628932431 2169848956 9692681269 9655700961 1629781048
8083332264 0115844498 6578869198 : 6175
9155116498 7759500820 1165471079 4954761627 2535974431 4069895014 3479155214
8701805244 0688805318 2445054861 : 6176
5105575082 4583348306 0153051527 1410340134 6158717620 4932737682 2811793638
2237726367 6950899606 0057645760 : 6177
7434908380 8672495330 3401193647 3642216403 1877350174 2628383091 8160337130
5308194700 5481456663 3422929439 : 6178
4379129613 6117974299 7959789822 2018382043 3937515139 0081879567 5780849881
9671169957 7981480046 8611110202 : 6179
9985597696 2841938868 7612327451 5246277330 8044657336 9546365493 8404008197
7760970663 9132376542 5391868682 : 6180
0356685427 6619326843 9028859199 6788147248 3502319505 8877475641 5910641899
1240691253 0941631256 1954109543 : 6181
5308814642 3434083316 0970495044 9309811673 5398312937 3553934118 7320088670
8671067629 2802662313 1366609838 : 6182
3643075615 6824337100 3247612866 0874213918 9356752130 5950626336 2049826465
5008206650 1877463331 8404810965 : 6183
3726939935 4992508460 9322236389 1818790058 7249238610 7832157797 9026003556
2226643917 2544446062 8932945945 : 6184
4295831001 5673005507 5437247426 2118465163 7120770245 9968277475 8902127077
4608232810 8777465643 7622050892 : 6185
2117628625 9492333732 2306799176 1502464359 9135638162 0607408584 3974251331
5938986338 3102724114 3850753208 : 6186
0538973380 1159125087 9562340729 1394530386 2707068178 0146819477 2402893961
7221644175 8486302045 1648879583 : 6187

7610929850 6760537167 7640104122 7878179550 0182331972 6057461761 1883779468
4547320398 9183811701 9778662208 : 6188
0801810164 8347143140 3292545031 4249522008 2111433074 4664013624 2253192598
7509157512 1739132432 9653494012 : 6189
0953928653 4708463158 8215049551 6801442870 6494848315 6384372726 3048169479
5792035566 8445778638 2972288953 : 6190
5344118520 6100695450 4177044547 4492597086 6988636099 3447006199 3886472734
4992791272 2231658528 3623292536 : 6191
4825934210 7355249952 8548442312 7322046747 1078064243 6699584238 5286374322
7324420182 8343973400 0322418590 : 6192
1923803059 0058722292 8961055149 9388306141 3500649369 1047390212 9154397749
4536051080 6487208013 1190490223 : 6193
1107072307 7062428339 1952893722 0911487783 9087904495 9631522298 9682708220
5304896560 1639594558 6075535222 : 6194
2215957383 4959609286 4920413661 1204987681 6520941632 6912589404 8452822290
3607027727 5509104234 7607151026 : 6195
0847037204 9953307356 6165201608 0315883563 8796224312 0890070941 9217345047
7878774094 0714687067 9225942590 : 6196
5227518180 9492822953 3182148904 0420843933 3772858902 5366508426 3277258143
9486019593 7648754924 4711520859 : 6197
6616658830 8595533616 0717058520 4247977505 7905952120 4948991346 2737333935
1797353749 0955401805 0208624252 : 6198
9471556100 8799154147 0696535457 2999224070 9325803842 5538974677 6351480895
1876988364 6358942549 4284220731 : 6199
2036451005 0271607803 9833613170 0022763357 3220580504 7209990128 7776893533
7598574166 4585200763 9216878048 : 6200
5736753929 4950338409 8229397970 6583142555 5392829591 9229698068 7772279663
9729390777 9082178517 3247610873 : 6201
5564189670 8494182323 0292691324 9449134037 5769778800 0085212068 9948851950
1187042430 8197047767 7656477051 : 6202
6660073820 6498848571 7104832272 3457119259 1806527116 7048969229 0985807515
3627517095 5052842903 9224365004 : 6203
8248807448 1318574364 6566984452 1805366646 7548379873 5679164220 1329619035
1708641479 7337157754 1061516174 : 6204
2404449579 9580315561 1157910308 7313472099 0353018949 9946582192 9921964771
0568822829 8614101420 1946390542 : 6205
3285844381 7124083633 2265624325 1238385947 7635673012 0676441085 0147539814
4634285931 0494968693 6382640046 : 6206
4162596469 5149031961 1104854477 5919170658 4392706760 2400311452 1271764708
3332009417 5688738759 4777063240 : 6207
9980920684 6305347743 3241945220 0217630004 6622820238 0827780419 7794933893
9188985224 4085506866 6909872615 : 6208
0999343275 5942195136 1489460327 5485400282 4746381855 7430386722 4848145712
0412894022 0014152688 4770962461 : 6209
2221149992 2887643919 1910899400 0764500426 7363603359 5644644270 8189785274
1770774513 3958409576 0446211432 : 6210
7655989257 1212464070 4976060688 9475217788 8846753657 7313088848 3170413084
7083028117 7925946670 1208771841 : 6211
2865941990 1877875096 3200281102 3755143635 6123048655 4615329882 8299904617
4517748581 4776012313 3434173138 : 6212
7710905577 0936706573 6550203175 7900430672 2930314450 2419919774 2809676221
2425199286 3274025837 0400752972 : 6213
8174354804 1063934506 3732675068 4346881838 8748332354 1121663418 8042412330
3403490967 5779416537 7908412586 : 6214
8798288610 3235278857 3321553381 5198830588 0458531534 6904313089 8096369406
6418137041 5485931496 6671159813 : 6215

0899440825 4571535523 0065250822 8496172872 3967465082 5190045324 5582357486
8772006674 7949712163 6028208523 : 6216
5430278386 5361171124 5321486487 9424132133 1700852315 4337277460 7680663766
9961889512 2880491089 1117659551 : 6217
5736498473 8860695247 1668475237 5144641521 3365489246 7276122585 3936148416
5143858186 9173841675 4348278131 : 6218
7663142911 6937855646 1817166096 6340291272 0536530254 4476383065 3355045114
6415224708 6512121312 9009990019 : 6219
6815169592 1524303910 2294969643 9063552199 0651394321 6303653453 9747151573
5014459156 0970031479 5373825072 : 6220
2286432679 1180222854 5445051006 6868382649 7290748132 5848087102 0887495051
4269642937 3925813677 1841690654 : 6221
5215610876 1573780205 3527958004 4684913613 6917468253 7172803536 7843503618
9012457777 5833864677 0048718755 : 6222
1541811503 7141294549 1142726968 7720886195 2903110006 5214806047 9389430426
1211250474 6362225675 3934768192 : 6223
2221920063 5168766825 8215068279 8801607357 0611080555 7861686704 9474864042
0270006144 0097794418 7614785497 : 6224
6458239562 4980544495 5125709106 4027083239 0814460092 5117778765 2063803937
1357117644 7632922162 1412656483 : 6225
7394710745 1322905073 7205542332 6208635230 1211192300 9932821647 5364369237
9063250335 6825313554 3303478962 : 6226
9153304492 3115380919 9575553294 4987052801 9033511674 0752763665 5981472206
1218043857 3003072978 7921735685 : 6227
0001256233 1806742598 8724010996 8969813852 3973061919 5592833694 8611903239
4925359441 5836595816 1383912185 : 6228
4119515199 2655070437 2224511063 3671266896 2567275866 7738823879 0791336450
9386511722 0139628594 4786544296 : 6229
2183266784 5170020278 1884192400 9364903662 7235744372 8744856331 0872878958
4584823525 0820156274 2207923922 : 6230
0339204508 2819462261 5291844607 0619758228 5213387796 9632367864 0133130410
9955645374 7740645777 5768490351 : 6231
3791673253 2182650044 0150064624 1636404631 1782779660 5357566760 3716137442
0266916172 1891429916 3923048497 : 6232
3825428942 2199854547 8689482567 0545771208 3060696640 1517547011 3984382899
3193363522 8885729811 5482869135 : 6233
9603242851 1636219222 0766798190 0638840484 2715260865 9071553704 9718593352
2605711810 4150457934 7353963276 : 6234
3998872325 6063689608 4103154413 6421487826 1192854183 8499574304 4276868385
1459149189 1874240066 1902831447 : 6235
9859226445 9633479953 1028630278 0183115008 9300076576 2808870776 8885566711
3610618616 8996249963 8799370291 : 6236
1361950062 1550959100 2663942980 5842317789 6496506627 5652978344 1514973388
2552363364 4652036220 1321628032 : 6237
1350049361 9597507026 9172783237 0138371576 4323028881 0132963328 7393824573
8746245096 8950822383 3084417619 : 6238
2408476051 0272468601 9144743039 1510803777 4819238710 5291127959 0551749882
2939075512 7440803641 6932829212 : 6239
5537800884 9192870285 4675425466 6973573970 5365362454 0072223989 5620130676
0481133915 6349727760 5671449640 : 6240
9064045114 8094824868 5117962164 0442806897 1957629753 5622361816 8885002728
5694336528 8001318441 2121411238 : 6241
9838519527 8511948146 7901665284 0688382186 9586830661 2959039774 5990561487
0361228980 9841138200 6158591424 : 6242
7128622986 0041718906 4530100820 3279408858 0385760890 5122698760 0864246064
8269485048 6186296517 2218487518 : 6243

3556528814 6631275237 6870674667 5269172441 6729735456 9567331667 7184928439
4313859957 7385048506 1097313805 : 6244
7829209449 9444632394 3060687595 8119003860 2619049210 9839871969 9336474633
1140662945 1114715205 5694804027 : 6245
9873582430 9185973826 3977134041 1416016623 7752693577 2236147756 3479055527
5216648241 4609981486 2813287663 : 6246
1187524107 4174329874 3627538504 5777742542 0575766273 9315698404 3856772914
3837835901 8235173687 7088048034 : 6247
3742863236 6590745228 8539522835 7786813472 1953766050 0846861989 6263300503
3093604099 7822291481 5147525771 : 6248
8379528138 4891568049 1921812383 6041227135 8296116497 2471080261 2545920595
2486511422 7833911564 9754666274 : 6249
4867185218 7561816294 7119647138 0668777153 6085417869 4183861466 0748536539
5525809016 9662377800 0065583681 : 6250
8847197698 2444587322 9844530886 9902993378 8779526571 9728089915 9793941934
3675227186 6343782907 9368244403 : 6251
2062463960 1866990497 2319031502 4362504081 3055353383 0653161092 3789527333
9143697925 9369072869 7404262340 : 6252
4247587033 7917002214 9258343524 1015718645 3983478454 5175892241 2361367352
9136260171 2155441084 9303321644 : 6253
2300759697 1105885469 5869571172 6320379851 3719294014 4871195015 3715879163
3212538307 9389694412 7468922739 : 6254
8610118372 0851428693 1971502864 6909873284 8172073873 8152015911 6379451230
1010196666 2036445412 9562919035 : 6255
5481051912 5343871316 0015241247 8550452245 4804170858 0097441643 6084037596
3801883860 7489535266 2695035328 : 6256
1648096816 7944880176 1593992935 6306431457 1114835166 7475654627 7594216725
3782296133 7952004829 0422881659 : 6257
9956705076 0734870429 0859084996 8490952949 1046326851 7636522463 1701348799
8937688779 8420929485 1296278523 : 6258
0159883015 3361268342 9917766146 3925494770 5193020500 3105560549 3677633916
3283953895 5788699776 9714313544 : 6259
6101324961 8901917057 0120182066 7021165776 6541460513 6853434511 7330284374
1097526751 8355759247 1851518890 : 6260
9168989486 5760416453 3212417028 0811484675 9077301327 2547946094 1550972867
8967187961 8012204433 5029687962 : 6261
1964404832 7886379960 4098388253 6293823039 5839698373 9496101712 3582184217
7439742703 6469112581 0751594526 : 6262
6355464675 1436788687 9782290229 5599477157 6430671265 2597185561 5110357476
6104209641 7841247683 0158403397 : 6263
9360112118 7811200823 1745037140 7570040927 1083743540 1089199345 9498375670
6127409717 6992139541 1095210125 : 6264
0839813654 9562045150 2436684311 3398973858 7361130651 2474524232 1542571509
1709831141 4008602648 9053937071 : 6265
2774412440 6690768316 7085424057 3003611786 9052432320 5442682356 8650032703
0306501507 7480473870 0240267292 : 6266
4144800205 1530677327 0191114548 9873963492 9248920628 9712904783 0449268538
0028314875 3581059970 5614838069 : 6267
2737300968 6610988863 7890295573 2473773218 3929682372 6596409999 9476679557
6053071821 6186939459 9277816445 : 6268
2536969658 1224500935 6458994432 1917279168 6763985578 9253996966 0988248722
7058690249 2001784902 7121486353 : 6269
9553280594 6828846993 3578566896 8529914380 3410528336 9383898081 0654163125
0749494609 4474088864 6908365216 : 6270
5710685290 2937901140 0101710418 7582043926 1982433726 1561113568 5841730762
8652063100 3127497144 6997819103 : 6271

8819926090 1664617975 4069102972 5658474690 4507192524 4946583352 7641746399
5178861690 5673265931 6334124585 : 6272
4517825808 2449007188 5017851428 8717667208 3115995559 2926726211 7825611904
4595069348 9617572283 2507114752 : 6273
0441276776 5750508689 3670980336 6686786798 6680958545 1635645045 6700098282
1304671244 2375802553 3584949678 : 6274
3455027631 0756161856 7611024200 6229086221 7868011243 4591564775 6613627310
7961732523 4681409070 0110500979 : 6275
4586345724 4190266200 5773671790 4612327442 0853797609 7262686877 0094642272
5868500716 3645957236 0638163384 : 6276
7494349975 2065431904 8826875825 3505156910 7427134586 8417945695 5827177098
0275281686 2020313195 4435974916 : 6277
4686549228 7630804666 2143131853 4173986503 5226451902 5805451724 2379319713
3547989443 1301843024 0118980856 : 6278
8284287635 6115113592 5450517590 0660902931 7534197237 0373166267 6310455267
7057284108 2661953953 9676802350 : 6279
0467639232 2381988953 9040992691 6785915668 9219764620 5371386957 6979999088
8413176891 5113743462 7023755761 : 6280
3560235953 2129295133 9330688041 0662258095 5975301522 7590711430 7261209980
4061120495 9544702529 0224582981 : 6281
0326046036 6610790784 6145624771 8072122094 5378224379 6089208478 3698815339
9857843835 8476233111 1455294499 : 6282
3163589565 4517074349 4100864563 7568236524 5225528902 7971719998 6729963816
2137834713 2538829807 6612871462 : 6283
5534783529 7146301393 7978843229 8529584543 9519676803 8647708119 9962158170
8094439895 8038250707 1135827079 : 6284
8334780985 6610300809 2483633401 0664378517 1705008672 8560572566 5824900630
3816659250 2760359401 4207243253 : 6285
3503290715 3406437209 9104955441 7219296172 8518931878 7667540913 5877109975
3068394228 3296170848 0658343497 : 6286
1118140092 2782530261 5934813947 5517360355 8408942664 4749330958 4619986206
8129248429 9900690495 3095601991 : 6287
6735927003 4227705807 7772579429 8919248350 7500026253 5753826874 8323634227
6724808711 4413930325 7644516263 : 6288
6301415773 7299135855 2647618310 6154750550 5435003978 8791534532 7702159604
4566357030 6770661001 9201793214 : 6289
0171479697 1673846973 3339707056 0585989228 3092531295 2649427953 6183676079
2879940177 0860176084 7530393479 : 6290
1124788612 3969453298 2336275032 7417646243 2178205058 6312100328 0810253530
9052281121 3357690673 4827893771 : 6291
9290836686 4035028279 9470624862 4768867044 0208595385 3472413704 6928259372
2752895964 1559749916 7757872687 : 6292
0096143793 3909121938 6991136301 9731897109 4560373016 1110976662 4424400181
7806505557 2462339859 2568655386 : 6293
1168261270 4334070095 1800886897 1398949213 1948076545 6160954465 1226431496
6934969743 6169839616 8741124092 : 6294
6925087916 4951012251 8675248363 5616057123 4863846892 7964566676 4984846476
7165046561 2669908654 8540370528 : 6295
1050282325 4158231964 8245828614 9790041803 4575969586 5716578935 9912026924
0475446862 5626567071 4416277114 : 6296
3207057262 3204570586 4254864853 8642871832 5923588272 1005017819 2591032186
2102525429 0610641964 9321973848 : 6297
2292467214 5088027677 3631002510 6146589875 2818456725 9205007900 6099263317
9350293026 3391497547 8998055915 : 6298
9838740722 8201211160 3184744607 3130926422 3605720140 6831874074 1436847356
6930281385 9684496367 8163554669 : 6299

0457525318 5482659911 6179188647 5069922132 6838807888 0217815079 8752627095
9165282807 6767314368 7407626055 : 6300
4027715833 2846667905 6225244150 3160456864 8941811259 9997950353 0707283995
4188004062 1837690405 2860463782 : 6301
2068355374 4365548569 4778361506 2635999936 3478702790 9030997462 7721842411
0017648215 9012705671 1820876687 : 6302
8227576467 6994285411 3054242846 9796771293 6556371908 1124347524 9918894104
4238998876 5581490981 6313382834 : 6303
4398678830 5614220699 4865643705 6345681695 1020971434 2381265370 5290231148
9174162697 5984689067 5493815136 : 6304
8823125531 7855349374 6115050458 0356678194 4318476851 3482917267 9530465154
9598075641 1689982379 3686265452 : 6305
2544768231 9382165559 8816897565 4498984722 3360804023 6215212637 8985700273
2047970983 5073384755 8808852656 : 6306
0046011136 3664106953 5797344908 6862143285 7776929771 3883866875 4917368355
3591485305 7824571910 9983029831 : 6307
3951375705 2525620958 0954017897 6954139461 5172026176 4966079521 0633054864
5811890332 3277535560 8042092880 : 6308
7954813104 4308314254 1175646937 9644937008 8051788439 0646505986 9952993456
2288497813 6790056842 4690668982 : 6309
3480376722 8391414146 3938344197 0505255527 4566152430 7031168939 9586409521
8468006890 1136191300 9089342682 : 6310
7828837570 5639519533 0125180182 3500492931 0697272580 5703196643 1977564341
4186491957 0951944115 2022579601 : 6311
5794217433 2998712495 3984816432 1588401682 3180156768 2205889033 4434057690
6238372620 6054194708 3026988680 : 6312
8831940015 1779250677 5717517459 3722384717 7220508209 3070415931 1736220200
0388130607 8884009111 7739664188 : 6313
8367333204 4652964644 5934419768 5964281262 4451256257 7886323153 8319095654
2867923083 4512402761 6888359652 : 6314
5102882925 4701745885 0877854674 3233541314 3581398900 5234227038 8006077143
1783425266 8029966525 1595805267 : 6315
3968256629 7578541127 3245999963 4827193710 5702177276 7908830058 4907363364
0131647993 6837878094 2754761608 : 6316
2779063539 8026354770 8948992177 7344518972 9100846164 9056912644 5840049207
0830852606 4556605541 8879410176 : 6317
8916027728 3372571529 2633912480 9056000282 3379201775 2640689351 1804497791
9973237802 0380548451 5346421441 : 6318
1215740926 1171753177 5253425231 2656527895 6547994952 4999661341 8668561137
1726575357 6161246756 3936363465 : 6319
8529021988 3593583139 2192491393 4186424541 3593442816 6038405794 3034058583
0595161258 4120866417 9704045005 : 6320
5709015103 1427979014 5799585671 9745613645 3537244757 3259717624 1622166560
9815477651 0792433284 5973035022 : 6321
1418019104 3778487240 6174681283 7199614628 3916642534 8030966724 0241178483
7905118698 8338391790 2679307649 : 6322
5649132796 6578197716 9564574759 3331331342 6260774897 1367197005 8790516411
0575608680 3903926865 8263487063 : 6323
4540551576 3139861676 3810774144 5122594128 5507544942 1595294857 3989863056
8471553514 8771193322 7943103860 : 6324
0662876069 7072692238 8392110104 2205418231 4187838700 2847488883 8905667506
3312220920 5148078706 1361084283 : 6325
7440600890 4461466797 3715826272 0291116842 2932474824 7891789685 8777805977
6094181864 4316340028 8502645364 : 6326
4551355067 1213401188 6907855574 9941020501 2020584369 3594383384 3142118798
4966957966 7123182969 4191171815 : 6327

8049435257 9524060183 7585099793 4371130880 2640215428 8164434430 6720286303
0612449853 7156718090 9678367412 : 6328
7520201113 4541349983 9171117253 5385170214 2430673210 0031441372 8871055440
7895824702 3764904752 3203097059 : 6329
6062076120 2742331730 1765690320 0367749269 4227330322 7577627517 0079415069
1302352338 2295229380 4237429919 : 6330
5531100137 5700873574 0489604930 1491001310 5148285638 6998429294 1736475578
5529415333 7949392024 4023171942 : 6331
7160290231 2715943693 6461304778 0157046975 1026061543 3560235322 7271325523
7816494055 2536518894 6498390345 : 6332
1788574439 6354358013 4349760271 4738438551 1984781089 2866822945 7725753597
8429545434 9909526907 7618698050 : 6333
1260973242 5756673965 1664186094 3350384149 6183873703 5093807038 0101695366
3046160924 0729436221 1337355372 : 6334
2563179924 5208681827 1670641961 0045069000 1783517268 1539217865 8474812406
9882994394 4692847539 2120476967 : 6335
0400891751 6980044735 0134011378 0055210563 0498825434 9320679964 1734183811
3208226066 1908719833 6021714825 : 6336
6562393710 7277080044 2603025786 4375469141 4064673799 8813330627 4980434048
8444339372 8585142092 9071406931 : 6337
3278515053 4696812734 5232046363 5666300917 0263359763 2388614244 3801882404
9100851015 8252293256 5567279300 : 6338
9966171516 7571103702 2790900577 2432264519 3485395815 3367620180 4307906634
6938595228 2763485280 7373926695 : 6339
4153406812 8659434699 5691180472 4376608938 3156219243 8656410313 1340589115
0807219328 6739168832 3814944776 : 6340
0719920810 3547538842 3453896732 6548496844 0806351067 1575237120 3075032887
6853691661 2388601773 5354040091 : 6341
0880395892 7512100254 9716693707 9787186642 9220140145 5882456234 2414467031
3230350128 1032442016 3216305765 : 6342
3771387099 0527259684 9408782981 1861525888 4925272186 0328952218 2402628298
3232700817 6345561299 7145774658 : 6343
3785472429 6746182466 4384902978 6530077631 5937919644 2547839488 2826806550
1763316234 0101463270 9472597201 : 6344
8823335388 1335302545 6539046348 3105173008 4970426153 7367640620 3301379064
7873873712 1464525553 3658355828 : 6345
2531298690 8539600263 6689072550 5682714100 0417352289 8482171759 9940268074
6491418887 0306381471 1532964678 : 6346
9559318086 5122302663 6997204711 3174786349 0527766941 4273422723 2795808700
0259828052 5801378538 7832003118 : 6347
0872150984 6232270743 1627761529 4128040427 3173876699 7695548191 5380842773
5709328137 3760561766 9370728806 : 6348
1211955806 8900815993 9826487645 1200332178 5648698432 0906376025 6299992888
9730761247 7412283832 6961611075 : 6349
1648915228 2506445462 6830641722 7180338343 1737658197 1246395144 7878320009
3513333186 5522233895 5660251647 : 6350
0810190022 4467477939 8750108774 6162698894 0950281310 7485697512 8637090065
7719141437 7029675485 6231438325 : 6351
1595038525 1731366442 6550285981 0576418370 7048240607 3207871177 0552954530
9698183519 7292941154 1825195835 : 6352
3083363474 9559978987 6319942510 4174380877 3856423077 1733254051 9763611239
0635219894 9174390525 5002477559 : 6353
2393944610 7776141318 6765509240 6892223028 6443171562 3756205532 5359475888
3418841103 1832605661 6085707801 : 6354
2412132972 7491660000 4899274741 4021583437 0124815741 9129887709 4711693311
1041130464 6457319878 7106957011 : 6355

3548766016 8472395555 8088729089 7174691220 3692511897 2468059171 1257457394
3911456518 0610608056 3786530895 : 6356
7847735391 1887809522 4396978001 8458236443 9544824465 6556278922 9023722984
6083255470 5494068221 8788034731 : 6357
7899183416 0540268599 6748721800 8132077885 5338243052 7889525289 7097708026
0850788175 2684419747 4750300894 : 6358
4242700277 3032472938 1969579912 7666762690 2535976229 5246233268 7883138948
4003936817 0446872185 1013957132 : 6359
4767540822 3230437822 8175049812 1221849238 1106072004 4101737240 2920022571
9476280314 9945154783 3447030334 : 6360
6453163731 3152749162 6927728719 6580757976 5624902962 4121342739 4749494996
0580458288 7192221824 3736016280 : 6361
8616446829 4217844866 4330819416 5490980503 5061939353 7344184449 8848185595
0496502632 2264508622 5208709839 : 6362
4179516137 2926156316 6641163790 8625996619 5848326953 8206406102 6825170404
9489883857 9192421686 5098291704 : 6363
7583952932 6011944310 5729997009 4882000646 2825064142 9808857846 1198438509
3174349933 1587540568 4618443087 : 6364
2481690382 8496944549 1491211283 8991744269 8035544256 6720592375 9401508528
7584120562 3818670543 1189016681 : 6365
8097178919 0221229351 8874279092 1955155180 8886890313 4477084577 7714549283
5784529617 2487465733 3204316660 : 6366
6413022240 7809409924 0861486196 6164617915 7317812411 3520858151 6991824554
1054412567 6216433441 0701735120 : 6367
3236836922 4702394752 2319864680 9466576700 8441477627 1127818203 7067394773
0272527129 9203114384 5135201994 : 6368
6347089810 5814668173 2870787756 1024420480 7475308420 4104430163 4872633267
8834504450 2448423109 1775516736 : 6369
7076052841 0517521452 2934932480 2842648388 4846900209 9445286264 6418420347
2216570956 8643477981 1103662204 : 6370
2605284032 0341215341 8624039613 6792653976 3057239176 7808239419 7387937396
9931185790 4544687873 1500279916 : 6371
4768258851 7407844000 5612703428 0313124159 4006256008 0603168967 9607752678
0558515844 4773757320 4098686615 : 6372
0895683261 7959871934 7191082425 6221882689 0023341053 9718917603 6305647134
2188593599 1661004043 1195689986 : 6373
6854709529 6307607868 6347518088 7668213990 0467692737 0948474866 3262578526
3229965264 9029047557 2655642628 : 6374
1710673612 2779326818 8511621382 4241463851 4198006184 8256020216 2079647746
0122499966 9672286852 4506028513 : 6375
8006176482 6341859670 8376361601 7717878758 4787211352 4257127483 4527649231
7626462574 3670922662 1130773355 : 6376
5205683386 0543070541 5874024492 9003575611 1655557169 9857931310 0068193328
5380018287 7747232509 1411581985 : 6377
0576954509 5128707200 5765226634 2717714995 7535199423 7585201083 2755725591
0134198306 6117809208 4032701596 : 6378
3024197359 1496606810 9019295387 8864962912 0915479131 8409276234 6231311444
1025278015 8536443513 0263119592 : 6379
4384614550 7833436871 3210905114 8727123095 8757797122 2071836071 3602361416
2793363027 6200661513 0143175584 : 6380
2436447252 7128448286 0833476749 4112067999 0018473193 1904696130 1417860432
2552671008 3095029661 5161233914 : 6381
0072324812 7406943374 8481319458 5701844194 8519546090 7139625406 9592655653
6231923829 4985722128 6126450946 : 6382
3919495411 0726922190 6175681772 1293282395 0981632942 3697312472 4084346206
7641516583 7242952236 9301743268 : 6383

4138741020 9413223159 0431123090 0855917808 9809863981 1472423431 2597727307
2587496545 0798846085 0364940355 : 6384
6360642136 0246367250 2975825882 1423970906 9638947515 8521960100 5670875761
7434222006 8818401867 8289640797 : 6385
1342119879 8942420054 2623226391 0916080832 8221720623 8321815660 0956376613
1150703525 3943137043 8476406712 : 6386
5707365986 3704705747 2995577056 3329249287 0667671578 4263974841 6481898746
4420627326 2918091863 4569514082 : 6387
1112621118 3077842318 8055048390 2301823855 9619868972 5866375383 6854851880
8900275672 3914879675 7475871447 : 6388
7044963905 9666476388 4055513983 2085110518 0869467334 4214696438 8936519874
2929450079 6357933677 6306583547 : 6389
9134409437 4984874917 8110525952 9349460886 9603961792 3756363527 0568286632
3356938754 7828496049 1495594355 : 6390
8112337629 6279491108 9627284563 0669590312 9237389498 7390646454 8234526530
1124597093 6916636819 8429401975 : 6391
7039611050 1809396370 7677695746 1341673659 4918684071 5979940977 4921295144
7536435570 5104049071 8226804775 : 6392
3468921921 9223966559 8892640138 3385425649 4287750824 0456558180 3397987528
0750093213 2516595562 6489684326 : 6393
4720950845 7269426761 9620324652 4253611380 8128554108 0986138899 3231752761
0005936827 4818919305 7268792705 : 6394
2706681650 9473844112 4135225224 6216405896 6977377662 6694722306 0479259376
5890405451 0890872719 6692960261 : 6395
5646056922 3470830776 5972494222 5173449004 9588103258 7117349851 2933407482
8896282286 8451406587 0595822088 : 6396
5463566222 3796926845 7727080062 4286247083 9100012713 2274669329 1077595784
1160555232 5394275509 6006076083 : 6397
7053344808 0696135747 9866100345 6942905048 7428865415 8520571824 7493430292
6645020129 0152285085 7613745521 : 6398
0972739110 8825404901 9546790225 3732559437 0109330222 3343533679 2479155486
5089056103 1509201053 2977003311 : 6399
4909933191 4415825403 9376710385 6151258970 8413151515 2832179811 6250944078
3844635992 6982485147 7998262383 : 6400
6715422818 5069666716 2662017616 0970567094 8611250935 9420925767 2502737040
8358338931 6097505678 3035442840 : 6401
0003970020 2864233335 3238003083 6727576947 1670632072 7156328814 5135402665
4528053705 6163375657 5655615604 : 6402
5683973582 1127233493 0266571373 7615780888 1484169546 0695189245 0897761458
2773056447 1156036723 4013737512 : 6403
3911334222 1350995201 0356177643 0770908044 7304826899 9177976468 7600344803
6444148641 3463468999 5078455510 : 6404
2030298886 3338483281 8107269920 0088982857 1936841389 1539811167 6352370135
9960477679 4327352194 6249418339 : 6405
8303417727 5287197321 6535239747 8366615988 3000187013 5480072549 9677948156
4122250731 9820777374 9493693405 : 6406
1592615121 4725240913 3124328535 2266095099 1788624206 2147076181 4443651682
0590669370 2697284844 3035060252 : 6407
1914049775 5126145044 6572569712 3159579764 2958217968 3132072049 7667628657
0471311714 6597168994 1414194155 : 6408
8552792132 9165534108 0358645394 2360439763 4616953352 8984633901 4437097710
3171885263 3529786009 8669308669 : 6409
8435263934 1843697031 8857439062 8654710685 1908002382 4792265970 6695796912
9228277977 6008111989 6191705187 : 6410
6577047154 8040762342 0146901474 7012300723 6720063747 9414652488 4880021869
7254454637 0568600472 3267422019 : 6411

6980822142 2778472139 3055099639 3066658835 1205734209 5362327068 3351905374
3235096424 1692802434 9246375291 : 6412
3196824007 0383892099 4682079708 2050855302 6506084172 9064678943 2924890422
6662371493 9148718520 4533360509 : 6413
2819272200 2667835450 9006721929 3448357187 4004864525 8436195009 4552025533
8530150193 6082754550 8148367762 : 6414
9431734666 8701839896 6327368706 6717388978 3817058514 3556166265 7530589349
2837995688 3715661905 1174693401 : 6415
9853587525 0672746157 7175427929 7665411243 0661688579 2146630482 6537623629
1786363447 8732060481 1685644513 : 6416
3196337759 7452108914 2064262107 7337168716 2905791090 4737837639 9934031761
0132958656 6017868615 0841320259 : 6417
9439184688 0266009194 1207015673 6705275210 4460254246 4766222553 7968562211
2998222213 6619496927 8051234613 : 6418
5893880093 7870064458 8895530093 0967808022 0567122291 1732449621 9466633608
5015095949 1680911944 3188318875 : 6419
5727288072 6049204842 2572695023 4777273625 6681742643 0792407273 2430554923
8831405486 3945929521 6734612059 : 6420
3473310812 2670744358 5443063905 1354861245 8469235277 2955595950 8163391240
3448808461 5574831651 1028056596 : 6421
9126047382 2009624298 7861895952 2873019383 2928596279 9349844728 5095834161
8215438661 0869136544 0642590891 : 6422
1589698129 8974427147 0860637344 0889508112 0792563243 4391216048 6709803726
9298223248 5582499570 8943110863 : 6423
7654840373 7521383697 7540372535 0930868402 4083186592 1894754342 5465681993
3192028697 3641687638 0780559427 : 6424
2649094054 3186086883 5870623993 0380932489 3776063958 8088169278 8217232840
1008007787 4468552849 3009535616 : 6425
8133698782 1881319879 6091570780 0606587512 0413681055 1501389914 0721605454
0732098424 4571124078 6902752955 : 6426
6371764996 9227320973 4533903274 9384224697 1775604291 2196379400 4392913939
9369638343 1238105727 1231601654 : 6427
6238186080 3416937792 3236080648 4166851670 0884580070 7990539901 9138986928
8678868501 2546791682 5295729795 : 6428
0907781400 7758372919 5952592307 7898529009 5374969314 4648856042 0183072483
6681645348 8890061641 8487213140 : 6429
1761979206 7384253885 2829602398 4285540130 8933923814 1175701208 0830155418
8871910117 1220354396 0412588136 : 6430
8880489293 9409662947 6679405622 6564819392 2536371686 3010600798 2286989053
7184707195 5248636144 2452983986 : 6431
0549722667 3813992712 3232697913 5267819475 0815424751 5825957071 8215174779
3330838053 8542252593 5868790011 : 6432
2017697150 6948468723 2397756960 2190530277 2913441798 9533284571 7225695951
3939845008 0818550502 8461731209 : 6433
3463323897 6743966867 8411629825 3664412222 3505420632 3638997861 3011604606
4276425247 2119953879 7765254709 : 6434
8991387573 3370958754 4460688181 0333079159 2335169400 2680509969 0092016813
6950287589 3937714949 3311215724 : 6435
1590212362 2515497269 8785804236 2441274878 9986559314 5938269756 9314305549
8179506531 4418668327 3288195130 : 6436
2751995672 1753823705 7152252291 7889973898 3906301739 9172994785 9647328824
5148057315 7662397274 6812720692 : 6437
8354685997 2049158200 6549321426 3737853653 7776577631 0542564915 6377280018
9810994441 2679472769 2115095607 : 6438
2802362828 4880601982 0301770557 3603550343 1347690741 2760284240 7027856245
0186760493 6809217966 6630506285 : 6439

7919256903 2160921550 3829815738 4365049568 3338819914 2037563207 6289437684
6024766103 9435202294 4810101404 : 6440
1168409635 2222244652 6698404296 4025300170 0640703723 3171935220 7617831350
0357425684 5231020154 4861847411 : 6441
2946240496 3288983640 5515453802 3000657624 3147668415 3339805267 7981987237
5955284634 5594350875 9753081225 : 6442
4078311528 5645913826 6985406198 9075985920 2685372792 3008414753 0346162184
5748488155 5487518028 0750087011 : 6443
4860205663 0051107952 6262181650 9997046246 0938200305 6246103553 1104034228
7433668786 9698965895 7146052636 : 6444
0133947402 9550277428 8600482358 9976160326 0749857047 1221845586 6713227141
0069788469 5817126271 4993858982 : 6445
8585921462 5368691960 7865356133 5684500757 6646743331 0863247115 8007102508
6357042200 4275510279 6922229828 : 6446
8679505160 3903278422 7388738111 8303297732 9682416042 6832369253 3434394805
6151302774 7565564224 3538631283 : 6447
4139392655 9729662026 3849694533 7587293946 9025962873 8874014880 7094633806
5979969831 6511192908 8025119256 : 6448
0285817304 9910841216 4179968432 5236402041 2026660339 0613564414 0131389322
1202873062 9445326131 9831333565 : 6449
4125205821 1952929321 4940536488 2300337137 8130699337 5262675257 3427205471
8259851934 3971228475 4974232282 : 6450
5404076626 1730855917 9774871262 9887002266 7410474706 1146869802 8702735148
1988793306 9078040518 5216982872 : 6451
2319131755 8377555383 2930641934 3327670587 1185727668 7145649647 5004679237
0667719907 0691387000 8711630394 : 6452
4297482298 9518150941 0741915838 2849830008 9051563374 9970923445 8126841189
0557203901 3809374492 6726325991 : 6453
4733455221 6665140583 8474159317 9377470138 8310929207 2972574807 2689274863
6254501022 6180365457 9949694183 : 6454
6305185203 3485830613 8860891537 4757681450 6890483050 4743231041 7567254445
6786775405 6535324624 4263845011 : 6455
2540158811 3376287922 7914457699 9324549020 7100571938 9024332315 8180677444
4726672317 6645607682 3483862792 : 6456
1390923872 4304151108 8833740482 6020756447 6757768523 1345785784 3464984582
9336240746 4801415979 4645214350 : 6457
9285544414 6566236726 0070571074 4294714808 9930546252 4615053954 6672945745
4775223098 8593834922 7321181401 : 6458
2181993728 9727478363 9130242229 5422832265 8127299397 8869758174 2933644005
4623339798 4792366191 8522402262 : 6459
5456201012 6296227425 6238175300 5177632863 7553954276 8604195767 5848786829
3565801648 4751646810 8506725307 : 6460
3869350186 5443186067 8211748070 6710238606 1328973257 1244782397 9443239268
5146855870 7175932678 6933587685 : 6461
4716034489 6167761171 6341299667 3364505897 9211075460 1830123633 3606911449
6418838793 4121842194 9353787499 : 6462
6652379751 5391763059 1596461034 7152121465 6427472391 8539650213 1632591123
0816528350 2948681956 5713588747 : 6463
6772396290 1916931991 6776864587 3058755288 7475970901 5918885841 3402314944
5351637158 2046119392 9538650063 : 6464
0901968781 1487252002 4632500585 9550973265 6697594088 0924885276 6267444246
6726398494 5494096333 5885449443 : 6465
8550708277 8660432637 6695588990 9423392133 4532437458 6923695535 8408441578
5410486788 2308876591 4992585877 : 6466
6134937893 9338172395 6612673264 5860588609 9833130238 8177066929 3348340480
4894481441 1828434656 3752808166 : 6467

5602947298 8769848951 9112897279 9505976303 2773051776 9376782997 5992595734
4752814862 8394441893 6299209900 : 6468
2413580600 2423326855 2032534364 6493589775 2706903435 9880903887 7648352811
4172898522 0615766577 1897459969 : 6469
3126904994 2745958577 4890071501 7710280050 5102185087 4633374802 8182672386
6376110593 6721215117 8910066446 : 6470
0382809245 7244431417 3858083144 1736587686 3724975750 1724180505 5648156458
8826944241 3625697379 2922559459 : 6471
0205061321 0392317227 9468628689 5619967419 2340073901 3168793818 0418212831
7285099359 1980470599 8908531525 : 6472
5060611129 0296074552 7654160555 4323462610 1670881285 1520318516 3523305468
4501630978 7686886251 6095539218 : 6473
2019384304 3220857880 8474078482 3651151685 2457920537 6362858300 4724889499
0623220277 1501038935 4247920672 : 6474
7218039032 3180854549 3523002157 0322404830 3575168346 1419504518 3770461312
9450220640 4923254337 2082330956 : 6475
6895789820 8001861335 0050687537 8551738982 2960864579 2613460122 9336427894
1293472373 7724118347 1056719947 : 6476
4988132556 6302417485 5112544613 5307313825 0801775959 0020575525 6882935354
1033294058 2476334595 4891191480 : 6477
0263059411 2285629020 9919829708 9226752024 5118222001 1255715792 4073365057
4612758838 2381239480 2411040449 : 6478
0718897939 0179213491 7798575028 7517112124 4970710366 2889052674 5050625093
9260199653 9954670766 8561456593 : 6479
0046721731 0496756990 4556248663 3893772473 5088968724 5530867838 1939797414
9839080870 0712484440 6148488248 : 6480
9613102697 3307385912 4987903741 8706260010 5142158390 4088761237 0365371352
0292948787 6505488851 8285342824 : 6481
8410038398 5536623133 7645767092 3704477054 4344802824 9088623220 9658664498
5919433631 1920583162 0545310144 : 6482
5784822337 3995095568 1727341450 8503565028 0593064829 2718529747 2128532388
1753240007 7625654889 9011148468 : 6483
3462321804 6070717582 3781636211 5737939333 6563514361 2655788243 0018387059
7868274534 8622410331 2709824869 : 6484
4442254632 0518242977 3306319378 2880524736 2772790102 3657515838 7770445736
3802324310 1059133025 2239746032 : 6485
5442284253 0476392499 3687412259 4125253442 7457191435 7904261985 5980323554
1075551717 9602225731 0079403792 : 6486
9632241778 5536177805 0880494963 1587383212 8675554297 0688659515 5552168285
0849181846 7151197608 7907235697 : 6487
8603646204 1091972064 9300412315 0188837770 4583113976 2918700923 8052681820
9787516650 8952119937 8270494551 : 6488
3992044479 8943378848 4231222487 6272598929 9789548257 4664348575 5792646037
5862242308 5401606220 2312395837 : 6489
9396790618 6082020738 4862333850 0638603716 7953946412 8279817191 1311961478
2167000830 8089507854 4581326695 : 6490
4818166386 2569509252 3164821917 8221346241 4175913354 3797902834 1423778353
8889205956 6846197981 0523192825 : 6491
7665293832 8809119668 7610481858 8964544244 0220542748 9691120693 3353455235
1730720139 5279545216 1236905634 : 6492
5572702560 8582660648 5383918088 1791607076 0661386419 5339075346 3104231631
4346145514 9546839215 2545717652 : 6493
3017627907 1346702989 6411934810 4686248731 0912905882 8693056154 8550109865
7169943179 4832040020 4615028922 : 6494
4283253987 2040350919 3836454205 6806579819 3492377797 3992361643 3923284412
2291241613 1023681238 4123094008 : 6495

5955594392 1238902610 1480024118 4367325721 8566928685 2117117438 9577355671
3694403655 3418826630 3219673712 : 6496
2047784613 9672424239 3438206557 5710146521 0839041105 8025499057 4309270936
5971812132 5372945327 6129255716 : 6497
0338115993 2729512007 9560078112 0533146361 1272220893 1150147192 5179535242
8209704638 3536281323 7330503957 : 6498
9921425714 3099602033 9068800578 3518739818 8931411813 0306073718 9155369873
1169760498 5511407637 7510951304 : 6499
9911398393 2418267750 6786129309 3343419446 0047109297 8714189848 3801103110
2741884378 3292048283 9062201145 : 6500
9245419765 2212689145 3178721214 5133126779 8784556044 5615959524 9754529857
5352221546 8171489841 4511138425 : 6501
0814105501 3085489623 4243628497 9104194400 0953541984 7829409880 4365041833
7571950530 8169336254 6548668506 : 6502
7659486021 0844032088 8369791785 7456235888 1459948339 2712832581 7566478148
9307630183 1884503175 1877025433 : 6503
8151217217 4738086474 1836917805 1185427139 2830592551 5203696154 4265579274
3340951740 8790292846 8948476751 : 6504
2818289936 9369638005 2096625809 6309522447 8411416233 4706388675 1542015041
3175353850 0081328771 6280810470 : 6505
0188346896 5555544325 7656803765 6735020267 2652996826 6884467810 6657495635
0999859617 8894324430 7515831054 : 6506
9771763453 2380639058 2047731811 6208907400 1240149200 8019238695 4492021367
9302425802 4646517698 3105671485 : 6507
0197856958 7399755567 8750296617 3286707949 0827566132 1638302579 3652727117
2357649371 9772988183 8765016004 : 6508
6960198076 2607172973 3871958649 8709909968 7247174947 6111410147 8259018801
1824536102 1522042204 9819761816 : 6509
3530033988 5999787260 5603757975 9868030733 1475224756 4007571518 7922016448
5890554090 2200692157 1006776359 : 6510
2498572399 5527632014 4412168491 5435778005 5265954849 9985124797 8090199909
0881425545 3142011174 1169391548 : 6511
3314820573 8280414543 6271804646 8662629137 3582349664 8287522089 6995836290
0365962129 4068548508 8790079706 : 6512
0971536150 4930307097 1447935664 0149031120 5530808240 9502443841 9875824164
7086058648 9907394068 7544131101 : 6513
7445035194 6487233176 7462602490 0661157888 6109716268 8503239042 5642166579
3019714233 0777144535 5387862843 : 6514
2514336021 2501899094 5261445400 5268521377 1778766749 4322391925 9355906514
4012279953 2365738785 9633667188 : 6515
6798210051 1224848175 4029247501 3669500504 7596708441 8012637345 8801066225
3040018506 6981810971 9049088130 : 6516
2291872876 3660112598 1634073025 8132566262 0787942939 3773088340 5816982294
0480343932 4680675520 0085304321 : 6517
4053837036 8119674497 3642664329 9033781535 2550225246 8254277644 8272604906
4908269772 1388698375 5924420723 : 6518
7728578067 0194840754 8250245903 1366717778 7679458280 9712007470 9141033453
9792304070 2124704789 5972416306 : 6519
3167939042 6246365331 5382468027 8462999849 2955829706 7770367361 7809127448
6510961319 4783755421 4734726097 : 6520
7852214220 6977441733 3930237588 9154517446 2184146588 8174063321 9251066399
8626369368 8001561290 5397748917 : 6521
8692275227 1866000006 4868860694 3843341695 6897619104 7931550407 3790699480
5414239314 0342772120 5641761019 : 6522
1846519831 6227552484 7093788750 0888830317 8635730728 2918767303 5632713538
7492718627 8478368802 7045784857 : 6523

0870734728 1843655076 5235117062 5675594045 6147510817 4628886513 8894890196
1691873586 6440178911 3965033124 : 6524
8286282127 7323414956 6257038143 6670394964 9642201742 6965254125 0313866715
4925779242 5824718393 0406221193 : 6525
1552939524 7295659610 8834910773 1184650661 0853782237 2276507233 0072986433
6816764066 3482712659 6611335519 : 6526
4054621502 4279598826 0339343777 8513689665 8920777477 0380912306 1987976754
4853888493 4886151790 2298951335 : 6527
5151317816 9447938984 4805510160 4475387293 4602081056 4239999565 3132943818
1181794916 8206643422 9861417488 : 6528
8862468891 0910401578 3835789046 4703385062 4225450915 2617858172 0960154959
0419019808 1724986333 2365323996 : 6529
2035299833 5812881844 4003123166 8441282092 1571036500 3177932767 1655485926
7887642911 9443885218 2338965094 : 6530
4951082992 9260775654 3518139988 3335003165 9441874901 5738251515 3491173025
2921091190 7594922425 7483071597 : 6531
0710339463 9477290177 1111795483 8753245430 8219195710 9711631365 5851113361
6779926745 0022545674 6172702619 : 6532
2658228096 9301676526 3536797856 2181504406 8552425020 6327743120 7892459280
7308316095 4493228199 0578760422 : 6533
7141254698 9922037225 5809419313 2907397952 9846907496 6884973028 2861276834
2445094025 4197334278 3863713321 : 6534
6298271886 8028770488 5903981654 2665131221 7466506883 9034388054 0662876132
4151473649 8011322568 3524539644 : 6535
5917788087 7465457020 9538890575 5986962043 3725788582 8876643040 1486028813
4785667775 2231677008 5281567031 : 6536
3517086993 6872283959 7977614270 4304433875 3674046103 4096265981 4007959537
3380376608 2066807101 9298809834 : 6537
6461125352 1726264746 1480906245 0843560918 3266233286 1476571788 7367070432
6449642375 5831118108 6135320630 : 6538
4527313550 1338932584 6980396507 5894599527 8709868677 3661665652 7294096627
0953618992 3808343523 2506453540 : 6539
6536366700 8123808431 4212916084 8939260647 5750424360 7733700769 7801048622
6866973272 0613130877 1242883807 : 6540
9513682325 9604935462 6987547850 6976087090 6213944467 0987050435 3997170682
8557773245 5618338596 7258415295 : 6541
8438809541 0036379769 0748326392 2322696763 1086303998 1067968923 4160825946
6054119657 8250833791 0051764307 : 6542
1960066914 2267780803 1086358949 5491266601 9682168984 8727980997 3651882678
7773105328 5059191230 3608390383 : 6543
4063911136 5752938473 8732739647 5869511903 9096129214 4657800965 6278230854
8525764983 1923721384 1985818804 : 6544
9149140812 2029836664 7625638953 5915309542 4421023811 4009422010 5435873360
2176560751 5378419898 3618305840 : 6545
8095095792 3288723131 4482568811 5900293087 0419488419 5222515577 1425360196
9600071163 4744743935 0056845634 : 6546
4376900335 7417602825 5491088852 1532392506 0878448597 8816727896 6902631720
0106720647 2074045942 4890050807 : 6547
6282339720 3513834604 1011685589 5150189454 5922832586 9990993583 1848388833
3349590571 2543219713 9992855669 : 6548
8368262705 4269275987 8814989388 9555327752 1957161988 4675840926 6923712272
4294525243 9825238364 1763866889 : 6549
5023936487 6702931815 1507672585 9767848492 3374584494 3089231935 3106512504
7084459459 8229323390 4494552069 : 6550
1951947530 5035974627 8833569846 1320222129 9389749069 3428402335 5296263560
2769933229 2725089430 2020639727 : 6551

8510558905 8443007897 2050488044 1292348947 4672886783 8915026228 8852912874
2952750435 5650209629 4224736793 : 6552
4640352551 2695447933 1493143201 2374843865 2466578663 3810460290 7702516740
6752254703 5447178168 6148529224 : 6553
5879861896 1723244658 9528197151 4061410121 9273275694 8471047837 2857694609
2451816916 8799535580 4049841241 : 6554
1049197574 3929251100 9593142809 7659219158 8701463865 5140297759 6012953584
7007686116 8292264437 4888309035 : 6555
8961861126 6068498281 8072956009 4573218650 7395063439 5701253548 2591503536
8957714669 1650957644 5043362651 : 6556
2511991682 0408853488 2183992149 0374688320 6389219583 9677180721 2607327885
6195131355 7657524341 3453665708 : 6557
4865934159 8443546453 7694535909 0314079655 0810660494 6822472145 1838346650
1317890674 1103928036 1890014419 : 6558
2600451772 6990294651 9122625302 7545038060 3248851883 6047936463 2249905482
5804957760 1974085448 2383090288 : 6559
7041597312 0470316393 4836548926 5472302515 2908040772 7077303618 5182519612
0166242493 5513390875 8669104232 : 6560
9502521962 2274262425 6671592982 8930404311 9006795704 1201008409 4782659788
3650327603 1030284843 1325642110 : 6561
5347720758 3503146038 6175329382 7376171032 9521381275 6699397374 4418533716
3250635583 9050047644 0504318465 : 6562
5505073040 8990389269 9362886204 6940269584 4315063108 8978643382 5298791700
6595528198 7833755417 7791762887 : 6563
6197775025 2490672963 8062184794 5014014371 1027157594 3654040883 3825817964
4194308333 0859984775 9889176852 : 6564
2529027457 8421689690 0368437138 9240369032 5632174580 3965288188 2758186777
5090350407 7856772821 5280374417 : 6565
8285539246 3930301793 5575994696 1952814180 9981603585 6227245541 2759745369
6375361955 7098053523 4513266574 : 6566
6446826235 5483975934 2756845834 5430580915 9838250894 7389287692 1029768874
2998711895 4151732043 5806519158 : 6567
5613271735 1681114909 8354192211 8283470365 7288674488 1559980262 1633472873
4343000146 1760387865 9001368802 : 6568
5581277660 2634600144 9208170959 0733726661 0360736242 3085988438 5252469347
3596609937 9697569178 3492560369 : 6569
3389615646 0246531209 0800867027 2784774156 1036000878 9753367663 9449245622
9509234066 3833583613 2514639323 : 6570
9122021141 0475837363 8153882739 1201590659 4748848180 4948983636 5636223485
4214689440 0775250800 2307778637 : 6571
6424808628 1910281820 3471183174 3730349334 4261779754 3695377194 8128215642
9854650561 0964355938 0311477664 : 6572
3035513388 8829485169 8137863875 1307499107 4046317672 8759532365 5332912659
4764048735 6284411750 7971390474 : 6573
0055133239 3896195551 2856978598 0204776976 0052520606 8054835205 4715812429
4736280121 9088311512 0816936817 : 6574
0922796178 0504089455 7835991309 3476322808 0151499698 8913569688 1361110514
0234198784 1705950776 5640482463 : 6575
1188859014 8083377611 1552376819 3830635714 3194096425 7422671435 4981847395
4077948135 0877889548 8544130013 : 6576
3174612510 0043073325 4010753142 4253624326 7779801060 5783400730 4622567359
4886885619 1266923560 1232984355 : 6577
7726500006 4743197514 7022267052 8280388950 0570021520 5390805702 6851688179
0966535977 5812636888 5914769419 : 6578
7974829700 1132585765 8785726696 7996800320 8967718995 2205029203 4907180214
0169102005 6286547729 6008692638 : 6579

0272911146 9401471195 5067728439 6877248732 7158127480 4285978103 0245051328
9974410757 5415357075 2061036910 : 6580
2909431370 6923000275 8489532125 1197846884 6423010680 4490373289 1922605837
7388613091 2107117705 8775287031 : 6581
8746530455 2555815403 8878565241 7324957369 7748876076 7011950492 8941832459
1821807077 0062249502 8789984059 : 6582
9660972843 5314033200 7498193707 4803208342 4556361863 1659974150 0158189254
3723584147 5658435711 9349433485 : 6583
9756354649 1917525117 4560830819 1804386335 7646830719 2414107500 4094886850
1060599618 5014204586 1653689695 : 6584
5114758739 6066060761 9517162625 2502367814 3505929992 7432365421 7871595338
0317592362 7889878294 9132690189 : 6585
6017961588 7369958353 6315303050 3048325861 9209352221 4594208207 4130397798
7383790906 4062124779 7034395114 : 6586
1630488008 2178681941 3639283924 9637892044 2456710399 9952897567 0486179028
8062734238 4547397844 7281935512 : 6587
8116073381 2632791742 1992832097 2189396961 1128149769 7228426437 0275407503
4764079013 4533313313 2456496302 : 6588
8811571355 3811281549 9523882630 9068009825 7040325222 4821418954 5731194710
5049339274 5559351320 4206562153 : 6589
6892949366 4686409056 3109585365 9861276550 2965285049 0850254774 5982156692
9512737113 3217455907 5315718029 : 6590
0994325883 4053869948 1640996723 2531224608 3210899317 5248992344 3811948206
6431676943 2057280142 5335205262 : 6591
1878287214 8908350656 1087419272 1644374574 0883270167 5994309456 5014599538
8325564824 0406611283 3332293469 : 6592
4473822981 4758202131 7975355055 5457762429 6325512225 6361396446 8549978940
3443974936 8161016591 9412356501 : 6593
7327701212 2574966526 6463117307 1157348366 6997513141 8259115846 1432810857
8758134217 3761329838 2076751955 : 6594
5975541957 1723966422 2676454746 5200211986 2782287820 2403645259 9026354617
7059646470 4118323000 9657954902 : 6595
4871371179 1563189956 2108167542 8352688809 1179891832 7001510089 4309335022
6550988041 7268148908 8122912870 : 6596
8295210976 6415444761 8330981136 9127897747 7122980202 1332127465 8985263467
1926624555 3883192771 4499914083 : 6597
4101059524 8725690306 5476732185 5578141002 8384420588 9041468210 9669054337
5552065329 1342320411 1494441007 : 6598
7245985564 1528901216 2258426352 5511470785 9555321360 9880932895 0696448376
1877632525 9427108701 3146790767 : 6599
5027632598 7325766791 1904342826 2705147524 5364942311 6904215703 5922697585
3210231171 9445897478 6817232156 : 6600
1144409284 4987448122 5597231616 7944415034 4586844820 6178121426 8917234825
6874778766 2783616836 7743249408 : 6601
7732449489 8869106311 2600380478 8658181342 4621072084 5544736421 5184728768
7459371382 9842803509 2082752117 : 6602
6899384594 9109624624 3754536774 9182746541 8467347459 2704974484 2547339625
7541989940 8081139888 0961788701 : 6603
1452014324 4990958695 2926214021 9339730031 6065173591 8939332358 7541379546
5408213880 9127473065 4354517356 : 6604
6449093275 8490061726 3889764584 2184591345 9045197700 8403452259 9909618879
1933984028 8954323562 7906082493 : 6605
6894154632 3611677529 4038268346 2355519428 6330995651 2636968001 1564888048
9294626619 3646523184 3409883458 : 6606
6027305854 1333874462 7054535383 5020397571 5425221285 2521282183 0130299332
6617904122 2692387061 4682842645 : 6607

7480684458 5557582486 6887213502 5417309371 2604179691 2427648131 9974611402
6534138005 0202471813 9556939770 : 6608
2665136484 0094792620 1458975913 8181669048 1772316593 8055052706 9076839774
6184553099 1260231230 0839997984 : 6609
1076047829 3851415988 7872479842 2390914376 4543073187 6545189095 3265023109
4774263686 3639679565 1027995433 : 6610
0455019629 9510289841 4290991155 8001888557 6177703373 4674454678 4876147552
5096699445 0659033739 3015003131 : 6611
6083832059 9729844570 3580164953 9935756873 4065602219 9212181567 0455920865
8030544608 1036536888 0519721305 : 6612
7760336977 9128213097 6053685806 3007951517 6056476299 6410624739 8349695899
9491685187 0974989444 0934598359 : 6613
3523263468 1288025988 5425319651 3163595894 4314542006 8157243560 5398018596
8094371494 2865566944 1490034911 : 6614
1556979734 6153042368 6606894504 7000453891 9216435945 6291222042 0454155483
3978183611 5512898331 2621025076 : 6615
9674369298 5516832037 8541867742 2376337169 5145553084 7260914293 9192564006
8094877398 7875363904 3132845210 : 6616
9577870245 7216803676 5425369755 6693376916 8937215968 2642757537 4494974180
0541699442 7085677468 6279433457 : 6617
4127819345 9750892923 7014499253 4455143533 9662061684 6162869463 1883411638
1687326356 6096684426 6181685087 : 6618
8358302201 7268272795 1594919629 8894654714 1503006699 4178790856 9445027324
0803777842 7150341385 3735560505 : 6619
0774296153 8376582005 5942220959 4072775101 9318804644 2455891294 8593645383
7844050159 5920916918 0992093220 : 6620
8512267292 4749205144 0151722161 0979212210 2691572302 3656934985 5124309230
6603059365 7798502968 7173529133 : 6621
4499724162 9647348256 7376407878 3302942775 0736521959 6517539511 7328425384
8099934666 6970118541 3116405189 : 6622
4952655495 0081439314 8266488144 4646402094 0783235393 4236123495 3898624815
0164626287 2514068441 5403232410 : 6623
3038129212 0284342758 3407715871 4583810420 1688625456 1952959225 0289375037
9438292604 0209740029 7659130668 : 6624
6065952755 3275106987 9445414339 9655516194 9869209406 5915238201 1727678666
1628344383 1248166625 2038836576 : 6625
2541266998 4546456630 5626956300 5241453508 1644907959 1479909647 7820024481
3028767072 4944687100 7039715991 : 6626
1375934701 4971086244 0766515214 7198933063 0186021305 0480860703 4518380062
7956418712 2853624232 8066541232 : 6627
4259297091 0977002929 7212199474 4426228543 3086841273 3791162698 4752085482
3004156905 7210122026 5546980356 : 6628
7154815773 5251904691 4043093788 9334079783 5766389240 7945831053 8093398617
1461332131 5002679579 7258189622 : 6629
9057661511 9916240999 1932632150 5998246815 9612035086 1961621851 4095538350
7737097381 6551296513 5202631895 : 6630
2939243562 3951453903 1754422363 3130657390 1918412314 5894299056 0684292537
5989779182 4573698694 7330423936 : 6631
9233735211 1541335847 6203818142 6924586220 6816051705 7737223332 9961192988
8870521300 3969668000 4463367180 : 6632
9692370885 3050360875 8432040275 5756772967 0050070934 4562921987 5590098517
4161171010 4989587970 8446952195 : 6633
7848116947 1998915071 0844962635 2546703160 5791774211 6938220772 2093235260
0571823423 6239343941 1167924584 : 6634
0915260616 8641300255 9143056811 1117797082 7773615183 2307889385 6675090928
0970836933 1327427004 1040190294 : 6635

6144372591 0813483549 1707410473 2302975622 1693280169 2770491932 8931914111
4908869724 3890420492 2187582597 : 6636
9970675643 9265596940 2138114267 1289816729 5742804076 4685833698 7035423399
0641206190 8925873490 1166405703 : 6637
0824007649 1226128501 0187293510 0246046182 2256831532 3826198069 9821183840
1071948228 9989607372 0109075703 : 6638
4014732667 8500305625 8389980498 0719849773 4231901917 4724659837 2272074210
8557695432 5059034070 3340504822 : 6639
4226143087 6280888342 0167299418 4495616288 6958921347 2341857679 7210819805
4618158397 3331783147 9781392885 : 6640
0648258250 8818965761 6971791871 4150170609 1545003996 6398970448 4130921254
6912320047 5844245375 1004184464 : 6641
7710493477 3844112579 8496196654 9943416890 2643712951 2680562302 8208527087
8043999881 1778775264 3841823140 : 6642
5761652466 1931188606 9040513242 3239531733 1395144937 0956869460 9307469314
1642761616 6093185970 5889566352 : 6643
5960237753 0748304313 1652545728 1206579123 6728498218 5753474555 3875476826
0270022662 4748154671 0531782539 : 6644
1142737172 0745827165 0962764009 9158475043 3634318265 4635440264 0405973284
3711650722 5403265136 6553949818 : 6645
9050993479 3798198580 6978218655 0218132703 9315360049 8153697632 9865055365
8774363415 2179587756 1245185803 : 6646
4830994833 7088602548 9091525828 2377983668 3487395570 9870662763 2980572519
2187914536 0285571773 7671378299 : 6647
6157796051 8685217746 2472797744 5894561254 9743742831 9048153508 0953603345
1762018622 2460126046 2488022681 : 6648
4489030208 6995405340 6814154385 1849396599 0384591117 4586904095 7358146683
3603032345 6446280634 3052444122 : 6649
2099497111 3470241456 0239693118 1520775528 9287487893 6396189283 7294870079
3311171332 5796976022 8398317745 : 6650
5775458612 9931105181 0723969487 6579428203 7558844620 7770717323 4133890415
1955540464 7557632702 7653380107 : 6651
9265904501 9913999111 4863195571 3222124599 3272985911 8497350132 2417377527
6724983321 6157371180 0569092997 : 6652
0080181244 7186896657 8772680054 5551153681 9797198826 8480789452 2754243348
7036723188 4210352035 0485308724 : 6653
1154459384 7549658620 0922220756 9799320338 0369371970 1186206867 8541949525
5563324660 9115865504 2170617676 : 6654
5066648772 0230673352 1287122957 2864628593 2926867706 1307692555 1413724934
0019276522 4853943270 0595619641 : 6655
0907039162 0844032832 7989866193 9875924646 7689391149 5031828083 4793998604
6658372506 9502132569 2175649023 : 6656
8876561524 0185217012 2116212746 8589481789 8502864744 7787288052 2352356822
8520359372 2441279936 6379674282 : 6657
4483964378 0680849492 3017209468 8777742518 6607959324 5371729080 2613994109
6976765466 3409431271 4343688065 : 6658
8920699238 8966339923 7781474789 3502809437 3660275900 3646241613 0091873498
5959634280 4784736956 7350496512 : 6659
4278435885 4222294503 3800113262 0039715846 1145519604 1542091202 3544705338
2014780387 4100072252 4862318388 : 6660
8193780192 1658210374 4167097585 7353741134 0839395522 8578747068 2184294374
5044891824 7514388781 2334555863 : 6661
4507762627 0217716576 3086242091 7470500401 7086640127 5592307039 2599025709
6154954742 5743327576 5180189688 : 6662
3382854955 7132288485 9164029072 2819747439 0843291625 4992336031 3093772591
1624471648 7414226815 7619944324 : 6663

6160572956 0600777557 2712147755 5028219570 8859876103 0861652699 7434567736
8493453307 5310785509 5203683421 : 6664
5382026867 6367990480 8755129976 8255483326 9000521950 3130028482 5693688810
1150090070 8334010090 5416054397 : 6665
0501044880 1241231425 1727926697 8094246624 6160895311 3615416805 3097688007
9062255312 7495864955 6987999188 : 6666
8536255300 6113245833 5482857506 3694882255 7373040371 9252797800 9324019657
5453942601 9497499777 9469639167 : 6667
3674187369 8798782505 9437117506 1368451255 8358007146 5597991832 2786715428
3537193419 5490622248 9359562248 : 6668
7235001596 5515958203 6027828741 7452051357 4548409743 2753825755 5219568073
4777891272 4295252854 7537741631 : 6669
0543712273 9220305406 6316539460 7929542801 1949722859 8688626209 5970577136
5767457694 6422985856 7404085499 : 6670
3161468923 3548567863 3144181922 1221368862 9970654031 7111527979 2138934362
8299637884 1327782939 5098920549 : 6671
5568478674 7301183738 4550561314 7815705416 3011814605 1812583181 5266099326
6856564474 9486427236 1110063990 : 6672
2015319341 2882598321 0736949495 2916086270 2977939042 2363625159 1408247034
3247036819 2463982751 8952691022 : 6673
7950314933 7308998377 9927924543 9411604715 9119733154 1232099717 8133722888
6015363032 8005321283 4939228219 : 6674
4279859554 5541667925 8510853206 4032098488 5062781566 1621488128 4667672704
1620168984 3688069485 9910074525 : 6675
7530198237 6453842056 0472267979 8456326015 0554934161 8925533263 5667717529
4303411190 1878705682 2355471201 : 6676
5982950289 3609563368 3611856083 7692702315 0693340265 9423041956 2045967244
0770113647 3169496910 6130497283 : 6677
2176408469 5335392064 8150058350 1825585103 0854903488 0383374818 3194421881
3095847742 2037695226 4714441724 : 6678
5993959341 1907054416 3630797214 1768559382 8641120947 1656201941 0130765693
9546975599 6400829760 0535418866 : 6679
1788231110 2017271557 3022550595 2903350137 4025869285 4277197466 9844636030
2981411834 5183219329 3466211910 : 6680
4972061197 3683629567 0334194277 9167623834 1546392166 5589720671 0021103908
5968668390 5281737196 5278139213 : 6681
4501547568 7862913183 3284334547 5807220124 7546748768 2898189234 6880324971
2157862218 5846523818 1791603387 : 6682
6220544502 6105352773 0169177366 4780882417 3908844525 8913852404 1891705393
0136056205 0253228909 0913146599 : 6683
7522304654 6496264872 7050504279 8563552499 9568943548 4656290533 5283890524
0747682823 4694425124 2205243214 : 6684
6049691151 5995027451 1921156829 1361987775 7442997508 1994571497 8774047543
6646671218 3671863998 5938647758 : 6685
4804973795 7431031355 6106922648 8933237941 8862103430 3009967129 5756250603
3035903221 7667494155 3563372391 : 6686
1688903236 6927392379 6055594782 2231835025 7576956904 8499578355 3419927080
0453710290 9673080487 1931921873 : 6687
4905095783 2790929334 0979011230 5095355177 8787974257 3025912661 5147330971
9502128366 4439457963 9594272143 : 6688
4847918265 6385965253 2019504389 7241592694 6839252424 2186038072 8712661664
2373835217 6839344656 8942250005 : 6689
5758131518 4030850597 2561946962 3569650725 8485093481 3782928372 2170682289
7942641654 2578445420 0928474864 : 6690
5428513828 3727783813 3513581940 8279535792 2839889739 9113773593 1487315643
4126855773 8938282797 4403739403 : 6691

8580890547 5853764692 0265681816 6100037622 4592307825 3690283197 6059742661
3455826289 5200074717 5198661393   : 6692
4845228586 0670989229 1398600737 6721642745 0218418990 6875128210 5164071420
6606389587 8028324319 7758108844   : 6693
3726138355 4629612460 6128103381 3110128147 4830649250 0156551239 0649263760
5631557241 5059444644 7578971934   : 6694
9309910266 8095708148 4238104884 1857925500 5136768429 1351141125 1757939022
2625874008 8453641633 8712775753   : 6695
0258196237 2271590742 5554757340 1193630835 2925466276 9457148113 3866548463
9021230369 1690580495 4067841731   : 6696
0559465918 5983035030 6583139906 8316976483 8761011995 1936625361 3888976226
6165084667 3375947846 6304203938   : 6697
4465890384 6488347282 2845237953 1706994116 1364108194 4696996724 4407469283
9236413056 7933923015 2919108360   : 6698
7535780895 4407226578 4231508405 8834195478 2356879973 6621842186 8049687643
0753139516 3670366263 7544391351   : 6699
8542939738 6393035047 3224166952 8654384395 4922710470 0120921738 0103835760
9531156944 9746487595 6857723939   : 6700
5946125495 0830179421 8645124976 0690958553 2848141303 9828332195 7441569941
2308393266 3795175897 7091254917   : 6701
7973118459 3992615345 2213996141 0983491061 8405773674 1482444413 3572465291
5031005080 4462575025 3745275459   : 6702
8551539294 8604233021 4582807131 1737531913 9408935171 2155180430 1704622055
4479737669 6674789317 3096271481   : 6703
7804320524 1537398501 6954090511 6883068125 3148680904 3952323243 7602316225
2504388366 8986857066 1653066181   : 6704
9045685274 7466558716 1772658516 5073415506 5945175613 6678425907 1595185264
9232661198 9419927697 8330288377   : 6705
7941111750 1447410422 9080890236 6083391940 6521113932 3726761359 7885389234
2828890089 7730547336 3126803325   : 6706
1710912983 9261484970 0054268616 0299429609 6384886444 0600491029 4550396652
9170234236 3466065042 2229602109   : 6707
8785880069 4421569857 7053669844 8410868673 1050696302 9609061934 2316066387
4899001882 8949512561 5591102404   : 6708
4628205315 9377096842 6564803460 3378173148 1405384504 4990759203 5509064459
0990909144 5259725671 4953028272   : 6709
6452152739 6612218260 4134614750 1028649224 4572657657 5452903240 5471437060
1234431217 7520657776 4552719001   : 6710
1539533425 0541807539 1219150598 3798526157 3370516229 5441197030 7888743173
9989145794 8253499871 5896943787   : 6711
7343737675 1615663954 6476243520 8186531597 5103666897 0777583283 8650575235
8050171402 3236204182 0853877746   : 6712
7983029508 4423383311 0374580653 6419079447 4970628477 0654767948 2962188669
2590682176 0067313916 0665816078   : 6713
5832842513 8624683306 2603366795 4679124922 3032388929 5687728947 6121515363
9603094062 7653119766 9010911380   : 6714
4874054937 7875158608 2757323635 4566858067 4906271659 7215990299 2186708613
8968951373 7068317315 6800919554   : 6715
8277920465 0557777506 6888521186 6929765211 0390778631 8443369590 6723520283
9999643239 6387686732 0361379164   : 6716
6686795031 1217912679 1204942268 6319695226 4941526031 8384601653 7045092592
5788004244 0594434946 7622855508   : 6717
5452556602 0694226877 3218946333 9027671535 8202225470 3312234798 5668270282
5245988012 3468395377 6355174689   : 6718
1225367881 2381891995 4637844127 0332418737 3763325613 0422332711 8444507497
4242237826 6197046104 0111226532   : 6719

7588955723 8498505763 6480494092 4804011116 2770871555 2798840790 3612572788
3591241446 0810333685 6769783495 : 6720
8256973338 3995616923 9890008502 9353534299 9657406963 7795021408 5190300644
9547774735 4841168405 5102877611 : 6721
0864198567 2662267872 2878593933 5839702878 8359545126 3588076265 4146340514
8082978002 6991158114 1860547629 : 6722
5128819943 0838665320 4842674055 5551025332 2746134221 9931036194 7772049524
3388211785 0322504622 9221424569 : 6723
5555003313 7758428571 0265052016 8844871279 3185637412 6072774114 1786989397
1542267276 3228658486 1696775818 : 6724
7395294670 9294345253 7335065005 6960134607 3180257767 9099155134 5034841583
9846752650 5737423974 7147481254 : 6725
0093578195 3545973458 4707650074 4709732549 3931035952 4408780242 7666984630
8426444638 6736380162 8651193925 : 6726
9851334195 6303926559 5711671123 0449799232 1173606848 9759175582 6316223902
2622489846 2286579357 2796249103 : 6727
7640382310 4080481974 9428592702 4632127390 0690507498 8773173218 8546893642
4389420964 9285664424 6175264232 : 6728
7072727933 0815571558 8664841631 9531614117 3690908132 0190273895 2842515499
6722430348 9023763280 5437291613 : 6729
9822725299 0836908892 4973885229 9592377545 0126780887 3261195461 6362090049
0104565727 2381218607 3402918544 : 6730
7799535289 9592119455 1880521382 9562617740 8041273303 7942090367 8075311256
7600804260 4950740619 0564878080 : 6731
5023437355 8916155389 9274656330 7620080099 6912907610 0632053732 3126937279
6200739054 3343746221 0258995721 : 6732
3978891909 7031755019 7355937869 2404636111 6027206338 3384090322 3989418916
7224940658 4603868835 9912725214 : 6733
2008927927 4309866163 5079572837 9051438551 3872952711 2983667896 1619181394
8640736857 9422616671 1785990514 : 6734
0382347291 4009009792 2491302499 7870400380 5414564064 2906637903 9225230449
2171609589 0520281781 6159901170 : 6735
4358289134 0913588943 9304950965 9960511332 4294482509 8365107498 5148901512
8230584721 5258025185 0119996729 : 6736
0923019534 1591046634 9751480976 5758541954 4419386603 0102393289 4081909479
2784807830 4524045296 5721750860 : 6737
3472258594 1757724710 7124427077 9869072099 5580682225 1978998647 9329847328
3077934323 7408852481 8024125993 : 6738
6076280749 9552231046 3219910682 9980811820 4157702024 0378672352 3630415869
0964281330 3646625339 4949402426 : 6739
8686257659 1833522241 3541020048 8785056997 4716852406 7286575194 3425706257
3473667547 3650312557 2083747217 : 6740
0777317324 2611280307 7590654692 3950229067 5828206127 6734984710 6351976446
6442716720 9384615607 2556820379 : 6741
1375209490 3953881686 2720308421 9274933011 9689065021 8554242991 7425792226
2045088519 3664289287 7228775513 : 6742
0094773646 3000642410 9929900045 9324667735 3137444054 8668487587 7715788647
5428825132 5710249958 3552305099 : 6743
0406816463 4131457314 4849393869 8331347774 2735215011 6669767950 3234920870
2667063334 2677121526 4880879536 : 6744
5929937351 8992420672 2511948073 1326592334 3577207964 5297777548 6743842476
8926160681 7339169281 2463591354 : 6745
9460964501 1156588572 9253824469 1171975572 8510786552 9845427871 6583023412
1111553000 7466352555 0482452345 : 6746
4104857318 7791003476 2330588407 2578334984 4254988058 4587627134 8453269204
0225683831 6423222370 9764547140 : 6747

5091236413 4406361959 2205000038 7149001957 8335980781 0085988860 5597284847
9116532070 8745934356 3080218716 : 6748
2111489589 6268008973 3082142060 6706376919 3099058122 3716105016 3381359939
1835931871 3030449261 8018971089 : 6749
2820426100 3161499659 8585966795 9009433747 2123334438 9517882139 2131689613
7345600884 7211834055 3704173908 : 6750
8655022155 3897742480 5978954172 3435365861 4901584283 6947490060 4308895256
0611138755 4385849627 2691509108 : 6751
6239755480 5901362380 4035826260 0530735745 1248710774 1426347751 2509773977
3093907252 8233632100 2648028822 : 6752
9727064109 5201052883 8078676946 2037156395 7604766114 5005808047 7176427287
8001313127 7413244302 7234088512 : 6753
9896856038 3152465747 4367982370 4601279075 7806089242 2712832226 4403079900
6153231410 2079712380 0533820530 : 6754
5121911738 7093019838 5367908173 4700155973 4751328332 1602546122 5027248671
2583495315 7390056206 9353295402 : 6755
3597171539 4692464427 4388027596 1486494129 2475277337 4134950714 8780335377
8663631185 6230615348 6757767063 : 6756
2549018682 9580038258 7884851764 5563087087 7721920831 8388986225 3677915514
0403649128 1933272346 4641041042 : 6757
0881109312 6098396184 0185584637 0235435946 7248574915 3090804859 4183126096
7299406674 8441944008 4259832175 : 6758
4803345986 1740020043 1378508286 1832997568 1776571031 1963158186 5995049067
2379791787 3672190751 5640653437 : 6759
7644763062 1705766985 6708749227 0750280876 7246204225 3538944741 4241343631
1016643663 9699880878 5733504534 : 6760
6724540669 2181042688 1156695438 4513391960 5697691307 4521339412 1929508769
1112175255 5502502270 7891471738 : 6761
0068542365 0303237374 8677870255 8389059893 6305021008 7166469785 7118051661
7315670294 7958588442 6271984103 : 6762
5509462151 0005467265 3266686174 4293511621 0475673902 4197730966 1472422087
0452411170 7433388517 8666262283 : 6763
8975950520 3128872018 9523544821 7394782194 2443997589 9983500607 8041238792
1930656793 2572487287 0197685892 : 6764
4652325066 9611473973 0088004771 0443386543 7226884395 6825008857 1648130946
8446127095 6543139557 5475805079 : 6765
3514267563 1381936217 0242297318 7848122776 8318957407 0484623353 8362165958
6843308579 6285376160 3136764747 : 6766
8472565866 0926676629 2576087452 7312462223 5504415865 9706369989 2768170434
9941168127 9792821296 9226927665 : 6767
0708667244 8008802330 9789282035 5930828885 9785353411 9785021182 1400468425
0546740254 9771417333 3668952304 : 6768
9168314443 7629773482 7205059961 8427960138 2438264090 0540500505 4389956220
3257694031 1410729669 3855412386 : 6769
1845341651 3571633768 6204154882 2631196037 5395351579 6511561579 1687520081
6745362241 2708438608 2271549504 : 6770
4441606876 7350715278 6736164990 6840791020 5043844601 2308262194 9618692440
0308314293 0447840925 6311863207 : 6771
7751264305 9419959491 7717064130 8906867774 6385439177 9381975761 6446811106
6389323583 6184091595 3211635978 : 6772
8090633295 1732610281 7101486669 4864396113 1017175536 5403322789 0444701001
9763121234 1076463636 2458309773 : 6773
2689325932 7741919571 4992449855 9646976173 0467291526 9924740555 7698159349
6630773754 7040710403 4797497551 : 6774
7710849312 1353924293 2997367572 2029345975 7588025764 0391775690 5121552802
3145210293 1304996533 3937302320 : 6775

0920505409 6856615040 5868034163 7934252197 3861293658 9718569331 2537414578
4827827008 6059892165 4039560817 : 6776
3187429758 7191910693 2757121799 5208732082 6981863598 4600672764 3387648206
8156113679 1228409737 9440355613 : 6777
8714059739 0368679930 4742369235 0310653421 4666419163 6140197023 9342325752
6942990274 6933964408 1593015791 : 6778
8795154429 0533462614 1049155315 6884501255 8555000913 3220611127 0101844350
0695217461 7306595986 4168810938 : 6779
1630424976 0877061545 7839544937 9187191797 7821405022 5790496722 1220778160
2380149238 1294008546 3237140405 : 6780
2972776933 5214212268 4618597821 1744358524 2059109702 8941984624 6899952324
2239816555 9046677132 2876361960 : 6781
1838487431 9465372981 2296335274 0628643961 0417591624 7182063541 9743568571
3007167543 6810628538 3585873611 : 6782
4148832792 3310460061 3445141308 6470002471 9564721679 8552823676 9001963621
4043564985 3331976888 0648886200 : 6783
2046177902 5580227094 9683423461 0834925724 7535445213 3926887369 3931289850
4859588223 2814751038 1188809460 : 6784
4514397803 6923529233 6510900341 3934584678 5041229174 1350494370 6196058666
1948266564 3182166566 2964560637 : 6785
3347509328 9553432517 4972737913 6224270823 9356455824 4327306600 7731554713
0178084016 9319607522 3441839922 : 6786
7001189768 8984844635 0552660726 9164237837 9008047941 6176843343 8322845853
7684138118 2511677563 8388925839 : 6787
4020120613 5674469196 7075752780 4706934458 1827691577 9568602324 9588411602
8959837840 5115255979 7838818410 : 6788
2901478476 2345089670 2896672888 8964765797 7823069979 8819270883 5134933044
5898170864 6817971112 8381820020 : 6789
4419749911 7721541034 2051654270 2703199984 6634914695 2076159865 0684885220
3163287322 5903815064 8355010008 : 6790
4517784375 4507436145 0592219424 0241872628 2490356341 2652643119 2655030406
3202575366 0531509186 2694230400 : 6791
9333102482 9860919247 0044115776 9350573870 0428590486 6973642307 0558612735
6402054048 2677643880 9032580278 : 6792
8508769099 4809257329 0173311716 5305467839 2387839859 9208449494 4282354551
1110755671 4179415670 6197119864 : 6793
1995640898 1974696503 8359856000 1537763197 8659426847 0653207291 3018346587
8244925078 1987830112 3504261299 : 6794
1308041293 7786711714 4174364941 7041309599 8427749615 2385231195 2163541571
6204185542 6172577328 9754274896 : 6795
4252455836 9008052069 7198978340 5327952496 2832641569 3415613998 5728673098
7422621402 8851812860 0190846032 : 6796
3175013330 7807483123 6164178261 3805117793 3714470414 8121189959 1909802237
4678777281 2890453035 7995934005 : 6797
6720430564 4044935478 0269634646 3971915660 0092636864 2173474117 6193906414
5402317939 0631738285 1890365239 : 6798
8629007099 8309862522 4892343712 2606098581 0304095763 8851092438 9509763364
9477485841 6747664414 1210442699 : 6799
6776851984 9816506753 7953954532 8708054990 5298268091 6273410272 4026108419
5694627375 9163057030 6936049304 : 6800
9551760291 6771452567 3235033671 3095707817 6961089025 5690131338 3745115817
3426087208 0919768962 3761582472 : 6801
3279969676 2659491769 6890350001 8108711473 8574349836 4099092793 3694547181
5629034528 7008933177 8646260018 : 6802
4845835027 5093946565 7435197966 2469396278 9671332753 0010336216 7058028060
2032717455 3177354932 7571281289 : 6803

7223016943 1756595295 8213137736 4076503258 2412095839 4227661748 9875728184
8476302875 1860746198 1595309145 : 6804
1776753761 4228781538 3246915820 3968556471 3187068248 6414694355 4341864240
8986534022 6529793504 2531053074 : 6805
5564744585 9877651882 7085892094 2199651867 0330660554 0243235853 0574124988
9718839646 5826949814 1403272968 : 6806
0400985674 0944722029 4441441026 8192913411 0942187415 1328287331 0798232925
1450661078 2792101218 0103111970 : 6807
4277690769 4303544975 2830570950 8933635333 3318510969 0238906980 6293571634
3038815074 8514989827 9017106593 : 6808
6441270747 6112985186 8177731370 4723445403 7456002862 1916791322 1365782707
0240795099 6791418033 4821962558 : 6809
6266951974 3611248846 5946582728 7656449897 7159444421 1823492755 1749896827
4483846421 5544522543 3577710157 : 6810
1115312646 9088386801 0392137927 1888579610 9842454401 7603589273 6009257986
1974341024 6093628989 6955972601 : 6811
9190217380 7009755802 3753808355 0621119923 3147409088 4681435199 0335032977
6225541026 2247665024 4698571713 : 6812
5357265288 2185686112 3930221223 5302190516 2773569741 2863498354 0839032773
5481050807 1163896557 2944766845 : 6813
7921741120 9448401996 5561435530 6898308612 0849486986 4517106673 7076918157
2544604668 2028205264 8523358699 : 6814
2499908059 5519278288 1458694136 8229897692 9469666575 2303826034 3104885807
6055117083 0091971011 9849253648 : 6815
0728361569 7678187742 2150711037 3995369276 6152452449 6535091003 4528895977
7029742778 8541533978 5881066071 : 6816
5689492621 3082779492 6549436210 7345696787 8550687385 5611333520 6511000951
6722384444 3533158667 1798353595 : 6817
3576451196 6809574527 4000424340 1544004906 2666137962 9574706527 4075783714
6094002332 6556772078 3707945010 : 6818
2526980479 3497599536 3121472488 8120659995 0447324540 5295694916 1135437651
2759235712 6014280388 9143997132 : 6819
0628419546 8196885884 1752929226 6886442621 4340633843 2354490358 7183699050
5347094758 4495815394 0545298839 : 6820
1465186318 1509397362 3321405450 9690894535 3116235023 1626454242 8788803218
8619172536 2579809217 9692510697 : 6821
6631553486 6721289025 5353427663 2117995418 8944080231 1171241078 9106391179
1380111690 8415614710 4326412032 : 6822
4352019466 8781977732 7163070488 5109519004 3634508471 3867267715 3789281655
6212509319 9330846173 5522700918 : 6823
5624196261 0481528464 0455268181 7698323603 3294487179 9050874485 6244052584
6670570782 0296738438 6993693546 : 6824
5436456444 7230044536 2273752665 3145299219 6223098811 7645680676 7139785437
9165847320 9504694430 4384883707 : 6825
7186501843 2569270115 1713494961 7369807824 2694080406 4500922506 8239337956
8371761215 3905708238 5854829100 : 6826
4571562854 0523094477 0923367992 1158488718 8615244229 4590478039 6105235477
5294518432 9182177647 0401267419 : 6827
0255625784 7710396875 5033652996 2214488179 3526405867 8133326215 9209134904
4189412312 5521539959 7795265706 : 6828
8257462407 6495486611 3924814670 0657227719 6024957345 0805840870 2213078799
4543507255 6036021889 5109924837 : 6829
0331300766 1853085253 9127407692 0389698099 6769175607 0148475757 7914252713
8022094159 3877488494 0496468312 : 6830
0257215335 5580648888 1999490244 9124874438 7026964819 0456178567 1518215321
4214586889 6572634255 7795756265 : 6831

2700629297 8459645061 7489642530 0021092247 9818088894 6251675732 7350029677
1470503279 9875798996 2912792874 : 6832
6607578196 4794847523 9203522264 5594063126 2832484175 3975481745 7010657092
6022593172 3208740932 7491400979 : 6833
7818697026 0143742147 8631793853 1308155711 6718016184 1988500770 6082881144
6496753514 6217195521 3852392511 : 6834
0419841081 6414048220 4424941782 7953801735 7430963694 4364950635 1874820396
0509107089 2098443141 6784064461 : 6835
9109367713 6150577214 3004751642 9278462760 6125879326 4955586447 1865902984
8182201602 1004997719 8361803778 : 6836
2118959565 1689225393 4182635010 3713632115 7781214602 3709366910 6321911883
9169527053 6319994253 5844908602 : 6837
3980831429 8724363262 5841069153 6617915011 7100345827 3988450230 0190025577
4774312143 3421560910 2142563004 : 6838
7292952168 0512822156 4798791314 8361230334 6110277520 3382979698 4764875107
5308620962 8464767863 5312023068 : 6839
6050987180 8751282265 5052819891 6999445054 5825287427 4597063402 2960335685
3886949799 3828280147 1099860087 : 6840
0595684841 2151964733 4347663376 2743513560 0524136353 7190114791 2029922253
1533188660 2515377068 3310154889 : 6841
0999536990 1717369429 8447066677 0482793224 4823418206 5674258791 4678111518
9747749072 5584891554 6647004332 : 6842
9222039895 0649452770 7553293229 9426267433 2988281428 2556896573 3266786575
6193371959 3908066429 1914750311 : 6843
1328446751 4873963577 2879342989 7564133604 2658143050 9213481367 9852888292
5725159524 6368667052 6471898396 : 6844
1963598492 1134156953 1986380455 8148479071 7809078695 2927279184 0452294301
0294283671 1083950634 8425063820 : 6845
8555865392 3276545286 2272664883 4185074874 1393854930 7417630279 5550310290
8136770694 3028970120 2531770184 : 6846
9418209149 7923829631 6560222983 6198034820 4057952319 3813266546 0972135883
9304485216 6287214352 5751137236 : 6847
8450592929 8249561147 4226829218 3057942653 3469225932 5226372031 7019199984
4767571541 0489691608 2005560929 : 6848
3880438561 5393938861 0800038893 1415514251 9880728134 6757093933 1543005256
9037837144 5159973023 4429145056 : 6849
9986414675 9503031812 7152727894 2502383466 7263317333 7455880994 3544395003
6775670310 9190940716 1141154055 : 6850
7340884834 2833535993 1921443710 6006859366 0264699346 5720093603 1623444263
8215142442 8391715961 4658131742 : 6851
4731705156 3253482379 3147494492 0069795480 1646821898 8433275915 2807095398
8212683989 3642318106 5489044804 : 6852
8927684892 1854710676 9788613503 3352449811 3861637395 7310268025 1642313394
1154981062 6704975020 1361285327 : 6853
1679450641 3100084685 1005211213 6399583903 1402261910 8072899098 1630702063
5413111542 2666078695 4871773820 : 6854
2141054740 1402403263 1964563930 0962185517 1291432652 4427459441 4418275602
1445515344 0454002874 3249577705 : 6855
8449408098 4033027258 1809631796 5249173507 0149484668 4157254659 2899494762
3637005980 4890396576 3138240276 : 6856
9423666262 2690151478 9471169049 8025494902 5258782220 6853726045 7330359730
9077583539 3170308887 0854234953 : 6857
1223031802 2928795990 6405017737 3344213062 7431431286 1522090518 0899239090
9028791205 1223118684 6024133278 : 6858
6097500246 7507917126 0577404575 2925643648 2325915999 3166445098 9019146148
0286002548 2343871969 0337398705 : 6859

3600376311 0023850308 8551562739 4262315739 0714593247 3869850297 2134778226
0711680996 0060440749 9093418534 : 6860
0057217199 3409130820 8823412305 0761235269 6212454550 8422240149 8616621573
1602997904 0149669949 1727462378 : 6861
2673017894 0942495960 6884377389 4743154058 9473271470 1696198582 5091282157
1405308578 4661918613 2280610365 : 6862
0342003901 3081403792 8612843022 5378811284 4894688305 3710330312 3570043462
3165907687 6308980839 5415916852 : 6863
0545880947 6237119278 8990746785 7536699281 4235247044 1171665252 8910036053
7755973558 8435425423 7465176941 : 6864
8514470731 4525476876 9289970059 5461349680 3973823293 6238998019 8620565056
2460668131 0225203680 5483248433 : 6865
4969568965 3100385237 5466758785 6799400946 4253726661 5054675709 1298762361
9267008359 8762678555 0302863066 : 6866
8799392265 0993773988 4608390668 2082066230 1246129973 0378011574 8296449826
0524705900 2013885624 4146334105 : 6867
8145336313 2825605320 5200592926 4808303132 1203507877 6629584645 4418916953
7342428139 0351629291 4225048390 : 6868
3626144425 5735810748 6834892518 5396565314 9819630557 1241999503 5132382195
1186967238 1514583393 4908285975 : 6869
5991966972 1704534304 3273161935 7189676268 6397897886 8443886119 0418094418
5825869579 6026345630 3755247758 : 6870
3111543114 7620359386 9850737767 0742765108 2481928976 8101338596 9978614342
4430006768 4394207595 5495740760 : 6871
9325016782 3992960045 7403577583 3310658604 1155550814 6908508483 9959586167
6625129053 1530322439 4011724448 : 6872
8740345224 0718053522 1879714738 5313216889 6260285312 5348758585 5672513707
8186863521 1696331105 7203172237 : 6873
0604678119 5086114266 6656026580 6290848488 1723155659 7264624249 1374667624
4614371952 8456935221 5618440020 : 6874
8981301595 4860029902 5188585073 5730715293 2312318131 8499178936 3734145399
1341102203 5565410510 1077243598 : 6875
6055631748 2778472401 3798347463 0972453826 0700785102 5263816602 7436710934
9768677960 9742645392 3241963379 : 6876
8976824844 2209801913 3054859645 2868142462 9087353705 3411643476 9696472966
7525830078 6930930798 4984271168 : 6877
7904408338 1863267759 7338032526 8242243329 6807395064 1806715656 0271426821
1230312342 9485516536 5321526945 : 6878
7610646824 7007076793 2911101196 0165468779 9208741506 2186924352 5586056608
1435007054 6365728251 2899089865 : 6879
2988276265 6823007437 7036105806 2226645479 9657003762 7144257157 7258126868
2546629468 7239956406 5097283240 : 6880
7761438918 7704249767 6234221451 3091793687 7460629168 5183902468 0313643989
7416961358 8291859736 5538906457 : 6881
7513081193 4551982918 1491284576 5152299640 7262983389 6418739380 2463224707
3639276859 5129132896 2522675690 : 6882
3128989820 8892471349 5701429446 8682265559 5752774582 2828429972 8097773832
1273253051 4169924655 2250666146 : 6883
5591357598 6554748942 6015712380 8779428472 7911437223 6711153038 2747753341
4242407179 4886020283 4057759385 : 6884
8604743995 0943326511 3648775432 4221472663 1364859057 7719435372 7858104371
8337900951 7981796311 3508959960 : 6885
5570568421 7532144079 8955923184 8085248780 8367550068 3459483934 2478328565
6352476418 9493364922 4929770885 : 6886
8162423150 2049139804 3590449810 0431493233 0105944902 4747184763 9893163168
4443474047 5056811723 7464392161 : 6887

3431838126 8547094021 0927957187 9936931386 0000384685 1436845821 1668002182
8201299917 4809263989 0989566539 : 6888
4058295888 6002827448 0269991698 0106424049 2426581075 9212724690 6309924560
7421661169 4183897204 0284553170 : 6889
7331523151 1142056785 0949657561 2039007683 1567476272 1070457356 6171878402
9626248721 1591769263 2706359138 : 6890
7678465536 1999859831 8230451455 4009697426 8503775012 0083213320 6958860197
3051218708 5418438806 1088381433 : 6891
4713432566 6897926277 0244341992 1807686310 0164857338 4040824075 8539913655
7908117839 0854206293 3068122947 : 6892
2933559716 3936794620 3353915270 9063720694 6734233815 1383426947 1410243982
2438830866 7689896244 1571506447 : 6893
4790566932 3318085246 4295629231 0160051118 6409568800 6805225939 6549137978
1356366627 8658482540 4606873151 : 6894
6751685658 7130530521 1587055927 3837534786 8908081484 4722468403 3527700001
1912546712 3303888615 6296696015 : 6895
8721813302 4728247879 1621978491 1643328606 7222446352 8831053208 5836045805
1478498520 9442235498 4713141131 : 6896
8817352440 7691698381 7007803170 6833636151 0531335515 3309584662 3655392600
8021258336 2096679555 4446296223 : 6897
4869224431 9084746391 9351749601 5959286204 4537991279 2443954761 9127992304
1584432891 2730521691 2408131923 : 6898
4088661680 9185433824 2402945446 1017088275 1608810427 4721924920 9453946970
2686028526 3079403107 0690991309 : 6899
2204311596 4504298191 0857994426 0900152451 4242833415 8154882276 7582003641
5251050279 0358004805 3428092302 : 6900
3699820894 1459463316 9359389907 0015559835 0644690822 4449861971 7044307628
6932014459 5156484877 7012594850 : 6901
1151405397 9796339301 0043840531 9067907436 4712995847 1833213887 1109099823
5667221263 8054868387 4189401010 : 6902
7540013788 1018393159 1488087478 6587879500 1515539644 0761625775 0818635821
7931011726 1440466688 8481821945 : 6903
6044051379 2488319715 4571243841 9343896724 7551444236 6589364718 5182260718
8363875070 4592945198 1123641178 : 6904
0972307434 0955219414 0524464220 0902221666 8654835608 1018514431 7155508745
2030706681 3582038843 5215498794 : 6905
4960394947 2960654708 0639824562 7394706397 4406405137 8929533572 8586689990
0859908690 6703943434 8708008176 : 6906
0714050432 5690320088 1536947167 0760999678 0837875260 8107329924 7949817333
2139531838 7783282359 1673701258 : 6907
3124289064 7354742150 8878255149 8414142350 1301672675 0821791110 0255797251
7427600739 3731921421 0392603297 : 6908
0810769819 0589880389 1618138257 9914443432 2105291569 0578788474 5821110382
6605498946 6937643057 2820094574 : 6909
0752925336 1163657509 3600636734 9850284388 7265817107 2599410784 9303081301
1758648611 5832906605 7327427060 : 6910
3069996948 2400586086 4033250510 7457077136 2035206886 4946234411 9056316806
9892406954 1256556398 1942471888 : 6911
9220352243 2260972317 5488853172 8266773910 3757237101 0673933766 3808981515
8426760468 5203620367 2407891490 : 6912
1879273556 1807661755 8781815383 0714572203 8817729718 7593895662 9614975056
7853450813 4597659282 8371823671 : 6913
9243140097 4105437297 8445101625 7543958762 4579038208 9029349315 0148104836
4513538889 2803043860 3690164265 : 6914
8923162207 2189666519 3559162749 7955048009 2555159519 0461860679 1880818722
6970297962 2020888002 4877871779 : 6915

1258102117 8942507276 1319795431 0962466401 9772097232 2610026863 7455300186
1908859825 6565626527 9366085407 : 6916
8228115251 5511826170 4821756956 1541627851 3963778247 9952394359 3124697524
3690592186 7794263833 3073373920 : 6917
9231890347 3965575739 6649098205 2618777270 8461695587 3614559295 2198126893
3164433823 9731423847 2615945308 : 6918
8525408871 8865776402 2385105108 7973101722 5228452997 6947617375 8249952632
4714118173 0160201833 1119181122 : 6919
2813508856 9821004818 1611461676 0485187369 9956114717 1048969514 5284969675
8125834536 1297803477 3231329653 : 6920
5965223906 3357421420 2049904797 7949714036 8721532807 0643579877 1212837235
8671861232 8481014289 3866885703 : 6921
7620723484 4796759291 4421953864 2201210820 7798876551 1025031152 9371693291
9965153887 8042516256 0257463240 : 6922
8504228844 9977623894 6926963152 8657293741 3430169360 0760353293 6970641782
7787737793 0650105697 3945611621 : 6923
9240336676 4030907835 4145513831 6434889449 9792375578 1587501445 5206486934
7130864983 3007389808 8056894346 : 6924
0652073487 5952488738 1452856331 1886966371 7726065392 9012996245 1213621618
2124975855 1092460860 6703292781 : 6925
9860265252 3263228264 8357227323 9183834925 8202127593 8940571530 4112736281
8836102715 0253648154 9937009962 : 6926
4982612448 8262917986 7282497649 4874375237 7218823270 2328607878 6741446467
6075600182 0948906734 6574075767 : 6927
6455255636 4224302120 9167963726 1321178552 4325445167 2661784549 6108737903
7460663294 1769264634 5850239275 : 6928
5446300306 3866779511 5643109766 1700950080 5367466292 1490319228 9315461344
7952205175 6730507822 4066541216 : 6929
4309774725 1304495228 8499254238 1654379640 6991093037 2676864436 9877159674
5583878817 8236999390 7291362951 : 6930
1356585921 2624706051 2379920709 8445306806 7510219235 9762580382 0917959940
3622932241 6000214494 8306949888 : 6931
4798402571 1575886642 6014882617 2477609227 5527824931 8006124546 8044327036
9494280519 9693247405 6765621398 : 6932
6896794541 0457780679 3818986787 4389603154 9308087544 8508708413 9693776973
8770431774 2865816964 0187113187 : 6933
5555139837 4341494804 3673850691 0996892834 0845548658 8315078298 0444429217
8135196816 5206545962 8859248994 : 6934
7217333292 0607370072 5483852165 9865898890 7709883546 5950445851 4681125195
7872729831 3561434878 7393577890 : 6935
1650631075 8644069444 8397384650 8885441240 7993571718 9272387604 3392675414
2118240462 4139090542 9387061282 : 6936
3904674491 8290461763 8581776736 5973869331 0685073570 0688392261 4400496752
2441094393 8679474310 4730175735 : 6937
2214306870 7911389927 5198337431 7080328301 7996889374 4273125215 8213563170
6351219458 0243208618 5211252156 : 6938
3609484108 6274928276 8825625016 8628994059 8158930419 9588164286 7887107620
1109560759 7412872305 2213171260 : 6939
5071600536 8280281377 7234299991 7730050094 3273641748 7408931931 7228940480
3345998552 2641089298 2215449118 : 6940
1471363488 5180832952 4371156096 0510757651 0978703876 3128124395 9587830946
1734903983 4758336946 6057915452 : 6941
2058959144 0569186560 9642765085 2458957058 7748288120 4132612392 8589487352
3560338104 3140003572 8912378562 : 6942
7432502504 6365007438 9009557274 0636949446 1883652866 6054372615 2454172205
8298308509 6593648284 1958696169 : 6943

1157259324 1982735832 4585383144 0167068161 4873475086 1264875198 9780776187
9174968270 0117124818 9184661389 : 6944
8797807611 4957261378 2398415346 2245734739 1337669061 7788060546 3474924802
1501000162 4112121713 9454356212 : 6945
0000272927 3788090686 3891198335 1608116287 0910443237 3070159783 7680528428
1909521539 0089954681 4560480934 : 6946
7727238999 2107712708 8311999408 3118190220 4147888292 3964542865 0979881116
4285982169 5566288129 0310706500 : 6947
0093202373 6562567548 4437417993 1102181762 3100802014 0461666848 9913449894
2412244970 1917555764 9285423992 : 6948
3972482065 0415253703 9228810588 6341191972 0344394148 7016575998 4553384786
8765677734 7709176671 4680382367 : 6949
3134016398 2840788020 5349546832 9590014417 8046155394 0129223593 0446998544
5894149744 3909752401 2233423549 : 6950
6542515475 8972141848 9379143502 7789414016 8852816498 1439552982 9554011283
0729059554 7761225000 8820025122 : 6951
7256136937 3743769265 9490087374 6044344756 8747917170 2479007356 2716781833
7666364901 0405890049 0891903746 : 6952
4562017867 7383153699 1028400082 2875981974 8535247250 2014372421 0479284879
6759970557 7863311114 0447469807 : 6953
8827521309 5399080803 1170918739 2213784736 3226332007 6649516350 5108258978
8722053445 1471505163 6392345586 : 6954
2358747744 6622427362 1281109257 9582876119 1662541536 8465731854 0677262662
8180746456 1236527057 0261399929 : 6955
1616530551 9804652650 3055454849 9179403755 4110839891 3905394966 9149956241
3086576574 4020383965 1663557873 : 6956
0617890614 2079724617 5671478853 5806936170 0198445798 4350911655 1020022736
5198453484 7982496446 1874462583 : 6957
0476746482 6173612333 8632176788 8312504912 7792009266 8848423818 1854808972
7394461806 9227465490 4673210947 : 6958
4947459117 3096314265 5993105118 1446498491 6465734889 2990956236 0569180745
3782343016 6274316235 9500345076 : 6959
3718679986 9307186295 0698331108 7556768767 5209574031 4348590022 8811237823
6242261019 4674169308 4249295818 : 6960
8229097115 9753012490 4150583934 4052802235 0651030862 1844597274 0352939276
9896469538 5469623728 4285171887 : 6961
0062273433 7169764549 9882868235 5302327643 5037635204 4233031219 1859473589
4560426683 5449612689 9092401308 : 6962
9983378701 3512452394 8795384596 9039342470 6092469339 8335944514 2833816687
6839060145 4197079224 7310862821 : 6963
6859276547 2336421218 0787384299 7293257587 2522741969 2110088899 4834956422
7696402311 4275391244 7443241958 : 6964
9510468682 5138415711 7226630817 8034083469 4446916054 7679329388 9420433966
2086472722 7134058503 4694769837 : 6965
8481116592 5615771535 2284042764 8249729162 5573147864 0882004014 7197345096
5477703012 8475568141 0163286181 : 6966
3732440985 9068562663 7160588907 0719967656 9693695577 7186970008 3478267984
6572398312 8062485664 1195835594 : 6967
6686463368 2410147066 3496905455 5589265162 9622100768 4166189734 5442324855
8753157654 1157523508 2441977440 : 6968
5089409900 9451558682 9666106709 6709024218 8567263513 0220769461 6474196101
2483500488 3243990996 8585541868 : 6969
7667730961 3636285574 9905316728 3427905552 0874941700 6678209583 0556996249
4777334336 3507138316 6500916099 : 6970
4458603070 1991206362 0947261628 1628154305 0557973001 9787551337 8835228335
9301917015 0243774923 8293956379 : 6971

3589629549 5380106173 5070050621 1537941161 3692454017 7933110673 1368079062 9158577115 3435290634 1740273374 : 6972
6525121883 6939159355 2057152548 0391410141 1607607349 3255128653 2242006682 8708349329 4702711238 2304481910 : 6973
8063670170 5555168166 0516170852 5775389709 4270729668 6930226006 4896574146 6207261504 7086137372 3724645325 : 6974
4378255759 0092567538 4226193084 2019301417 0554457973 7519753004 7041686118 5897332422 4553893793 9958889725 : 6975
7396558578 6621204211 3770585010 0415363247 3380634849 3087817055 6606153341 1288607376 7434320714 0406875176 : 6976
8453010249 1894478822 8924670924 7645453578 1147272683 4119812515 7937479385 9013653631 9222853772 5907459494 : 6977
8916500890 6353103732 9346048208 1275476655 4480334689 7509941733 7722730080 0883513866 1216396087 6950803327 : 6978
7446327520 2112383366 3890757017 3658688340 7143249573 7637613189 0231795720 1193318346 8407571300 3556003589 : 6979
1019144267 1794848640 9750703489 1571557880 1216721948 1317424387 4698484019 3683550642 5753888595 3635406457 : 6980
8459525802 6348940536 8485245661 2071352028 2183609121 0728244860 3570984859 8621499110 7205603824 5919485982 : 6981
9823543664 7477244318 7443625938 5244319349 8859769508 4769983030 1144277916 6402353624 4069488936 4336665191 : 6982
2063300227 0696431428 9141781565 1576241366 0679408644 2773931508 7645327820 0005143114 4169384502 1360245385 : 6983
9527358241 2808228405 1480072499 5059504215 8198022032 1191170380 6022726228 5315277822 5079828160 1848480446 : 6984
3884542381 8778217104 5347616593 7626072970 8884217699 8044060036 0950716305 9379235337 2760325679 0842966377 : 6985
1490318446 9807592422 9306159288 7578809375 4556644160 9888103227 3324326116 0955215205 5865572880 8412282000 : 6986
0875492308 3828398024 9303701852 9206223877 0987704278 1012562268 2864582582 0748937450 6573468058 9573571270 : 6987
2446993304 0752354544 8638482611 7949774906 5229880663 9691549161 9456757341 7174146656 6677615589 1302285172 : 6988
4937498914 2315440420 3600837602 3193268910 9975546265 8514424413 0691699988 1628089074 8596904830 6877241500 : 6989
9109248226 0722119536 6278356268 8880666917 1455147204 0654698672 4461341839 2754666639 4924222757 8726053861 : 6990
9426620379 3909264926 9513080814 3167659282 5048764721 0486358393 8056093091 3730692084 8388137756 2730641491 : 6991
5510911084 4125268842 8224477997 6582804516 7088789610 0155962325 4578416946 0313922987 8054528721 5559674980 : 6992
4085046227 0826259860 9921560665 6776059196 4786619653 5011861033 8730120681 8626988983 1755538563 4933349620 : 6993
3625458890 5835397049 9641814143 4400220806 0299886141 3468584239 1393401535 6423247214 2530142859 1205657638 : 6994
5693562776 2470877172 6378207859 5318143314 6684288614 4057787948 4394185441 4553901189 6337566375 3705902427 : 6995
8787829869 8149149790 4899821810 2052906879 7424909942 8432273239 2927379229 3589493756 3460447117 4134618118 : 6996
7637755675 5317135332 8694467886 6336852529 9779133888 4062520535 5767274522 6875909425 0269880262 3029029591 : 6997
9800718727 1543905075 7092048947 4278808112 1617537057 0817778606 5723273944 3689675971 9613240934 9773380717 : 6998
7485067680 6611000249 7094831775 8504896131 6817104715 7601328115 9979796697 1179648937 4936580182 8293146885 : 6999

3266706863 5647384591 9381630061 4124738538 8455900458 2927311830 4531949237
5073949535 9713935327 3415735585 : 7000
1621085639 2947218980 9593934148 2083584956 9334691831 1894798014 4332181463
3528391077 6628884496 0868696378 : 7001
2651073755 0363616449 2166918087 9049103339 9484824422 0747450873 7919354745
6961564410 2072323320 0904265039 : 7002
4967045947 9467217232 2314104619 7781792322 7968911516 0333870859 6999475128
4280501000 7507936666 8644883362 : 7003
6725037759 0252110262 5881334848 9287313666 4049593470 2802062967 0055707734
0622029755 7733498918 4277808485 : 7004
0329492415 0006562102 2687299216 9408430237 8690711214 8745940890 5047676091
5559737701 4065431151 3641993910 : 7005
1532216931 9775060536 2646324562 9378601040 8510259722 5325173092 6509659234
7969499224 8331176811 6594806088 : 7006
0008261892 3704059498 0770400173 9674582189 2429548950 7168771331 0010397851
5335345314 7106859570 9445060537 : 7007
3359195365 8617196314 9700601821 6015116781 6445877799 3723274548 4370216719
6983240352 1589516853 5513924782 : 7008
0559128279 6153491663 0151458051 7888411480 9718670698 4272264732 5396278568
8562186013 5755148025 8411912578 : 7009
6169815402 7609127050 7419511403 9656379289 8288285227 4086014503 9535211265
3895388386 7793995172 9586292709 : 7010
4784127391 1524331129 2594640007 2613464924 9972681810 9451514403 3606300528
7128811763 4557474400 4968524233 : 7011
4339885902 9913064236 6501730923 1172549322 3569364627 0826473653 9614701936
1465648808 2366385687 1274311814 : 7012
6280242967 0841610218 7589986096 4969235031 2982446820 2286241015 8118338086
9587321577 6018223789 6180157677 : 7013
9989263352 7484089573 1040846106 8537718696 3984413185 8461150414 8287305023
1180669598 6341793389 5317469817 : 7014
7066866352 5117787844 5758531114 9522528920 9150428157 4922455591 3847113304
2715891355 3411142349 5667324043 : 7015
8530735724 6228684692 2284083793 7344847060 2916936013 9930409506 5459619003
0818380252 7282563082 2093726975 : 7016
2752989520 7663388958 1100393467 0501955027 2152540707 8422203158 6900159951
3082638484 1974877768 4991692392 : 7017
2196537231 6654475744 0764101009 9650617126 2795901780 9881925888 4779270810
1654593733 2535284128 0200757510 : 7018
4825911652 5137578340 1008951847 6852491018 6221173555 3486369698 4157401995
5602865124 2236187405 8340908447 : 7019
6878971542 1259530355 9132707207 6061508973 8197994811 8563835674 5062572946
9513436916 7891826306 6507952431 : 7020
1558263643 7199760377 4446201879 1016135424 4770062736 4765654919 6267616444
6807721909 2292908086 6228112959 : 7021
8849247811 4867153425 7155367603 1195245475 6156523696 7703853704 7688660469
8021385248 0305611763 8206875750 : 7022
6758136412 8628780137 7651406190 7267002330 9784587766 9182283339 1651290787
0760866937 5171168917 4162769080 : 7023
9551757279 6980520168 2102716262 1360697060 4160915617 8870530540 7209797684
9302663763 8968662736 0501139889 : 7024
8836733591 3513960315 4742889747 1385416969 3929318308 6800212328 9632959938
1271653722 1185295145 0167108432 : 7025
0746942049 0209766700 7422827466 7222156833 8634130232 8609110460 2401561230
4984559140 9523921046 0975654364 : 7026
8344461508 5503749970 5955336660 2213462604 8399601227 3273641910 6949439922
4652457000 5239136227 3333832849 : 7027

7275573933 5649594315 5230901233 2305634341 5235105696 2822867788 2109308461
2156866114 4438707609 5828406715 : 7028
1875349808 9714331027 4396702320 7286830105 3055854740 6112071090 3593522245
3189737748 9287212239 3887443165 : 7029
4802090390 0810810885 2669227724 6193143825 2909725863 1124615779 5502759822
0706166870 8048769500 5424198755 : 7030
6204669927 4663988639 3484868401 4223352872 8214440395 4784228917 4547658271
9454397372 1716746000 6085368620 : 7031
6659734305 1524283327 2893912510 7799228171 7140591722 3602073852 7892642080
3737929278 6435578021 1583575735 : 7032
8359060207 3781354710 8144950490 4591832117 5530154501 9836835859 8699232679
9363224360 4356246020 2847050065 : 7033
8345386692 6334314789 8299146785 3814916433 2937337706 4728561197 7754717802
1057206562 4167806204 9249605068 : 7034
1014505544 4000836915 8538639919 6405583027 2863321210 2715925478 0491989067
0890668715 1219434944 1699015818 : 7035
1216296916 4529656672 8882261734 6469987438 0606663650 1254919352 9951563426
7061483686 2978791948 2858285181 : 7036
2784137702 6940221675 5806828356 7621527125 1002722532 1755011753 9167618570
8244100056 9786355201 5457177509 : 7037
6098390350 0128251287 3029138851 7597460971 5463747851 5816291219 0028421493
7284094101 6515570616 8620646222 : 7038
0736689554 9393145435 1822662501 7737147979 9689206120 6532527541 8413914476
1139221521 3461406928 6419355821 : 7039
6544796702 5369448854 4417123943 3349986029 8027098502 7141815954 1678025432
4505451516 9222688322 2768070584 : 7040
1685710882 4894008108 0598243638 8069304940 1400360540 5882514727 4558825823
2852976555 4366006097 6141881745 : 7041
4395068351 8994306440 7677097341 5827497543 3460741103 2688819736 3251029598
6891271391 8327323940 6546776520 : 7042
5701268560 2158373854 6587562512 8673719730 5200835452 0571151156 3796715242
2854375612 3193671768 5670090063 : 7043
7354250309 4551019185 1196229951 4032597437 9209354374 2091184315 5116834985
1325491586 5438192839 9557383125 : 7044
9453672482 5067114738 9629672990 8788858488 8259970107 4227252504 0354818049
6285225705 2082363485 1226938315 : 7045
6280672836 7147435055 9839279732 6066842950 7263513827 5104197468 8976767259
1369017306 5123657317 1953243101 : 7046
2869561072 5805283363 1589009582 6698417353 9638487201 6520049064 9064035498
5449237229 4274183784 1103308077 : 7047
7386940393 4504663332 2001742224 0536339407 5275059829 3574910869 8818541052
2731299102 2233397661 2533246907 : 7048
3008027004 9025891197 8660954614 1221334121 3891962638 8266617865 5614634421
0813263307 7176078663 8063634243 : 7049
1909183254 1943227473 2411523770 4877561380 9279604835 7649969536 8748511308
8070624761 2732575037 6659107831 : 7050
6587857685 5066198839 7945830269 0860270196 4112202257 3660586407 3929143877
2259890740 7359876447 3315527068 : 7051
4661916708 1990942108 6085140332 8060178785 1430149967 9947363423 6622229182
0690324453 0152501557 8576015228 : 7052
8473671518 2349449265 4828857179 7699263163 5275384557 6557644067 7544890668
2489966134 5595218009 7814220867 : 7053
2688630564 2915401382 9653468573 4899671972 4201717604 1456299913 0251414447
5248848788 5952192226 8437236453 : 7054
2964338807 8851106993 3976961277 8269162546 3097323535 4514622641 3765213929
7506809924 7765973472 6126909397 : 7055

8722139280 7455856470 9998158942 9668515562 7040336605 8762305340 3109834593
2290012698 3226920502 2015814343 : 7056
0280800022 3658015463 2882552904 7068250736 3006307811 8956871980 9407671585
1261033440 1938296271 8808600861 : 7057
1697072396 7039431812 6130974247 1117179019 8574301239 9075403547 1650628805
1264920095 6055134533 3620292630 : 7058
6649809154 0875817634 0212996001 6194911103 4937515765 5887524827 2142376797
2642564269 7736819889 0882172269 : 7059
9265553049 8724790479 8100065046 8434265519 3059548646 3515965834 1932560199
5394272542 4637317151 5176352439 : 7060
5140380742 4423857724 7656596152 4567466993 1628602826 4855040685 9122463407
6966397694 2701366938 8291856076 : 7061
7582008669 4383375631 4593294171 4551996669 7728221814 7062090620 3497631712
0864660306 0567561493 3109167359 : 7062
6869262998 0336515985 2130088802 4046181683 3186087792 9665002779 2213476011
6174813520 0060259895 1502189632 : 7063
6415385131 1775210714 6410707257 7934971822 5272679005 9933701865 9615793278
0066176911 1831676147 1579646487 : 7064
4106952245 3761145688 7923115088 6303623840 7176919599 8900356580 7547270107
1991664477 5922773065 8896111939 : 7065
4275860246 6572916464 7782959517 7925217284 0786074432 9511315658 1166361862
9044736315 5515655823 3734102054 : 7066
1097055398 1055118587 7748472553 1770169456 1977327400 0762746629 1601557814
9893849915 4800670015 8802872713 : 7067
9647788555 3340501525 7888467196 4802052467 2424493955 1864372558 4409996006
3063002898 1857811273 5319400344 : 7068
4827534004 2960655950 8922535186 0155422289 1764294635 3097283905 7155251922
0067182252 0448345695 7283024858 : 7069
1587778354 3719512685 9654267081 6960798234 6616792592 6073561893 9895590289
5422982006 9011350119 9927833705 : 7070
7617540712 5834805827 8068123894 2856193111 1141325049 3316426052 8023303981
4169006798 8980373760 3209951643 : 7071
8489058699 3696068307 1012317961 5266924530 4649237463 2284627990 8242334798
3303740344 5017864480 3988127559 : 7072
2259020465 3660846910 2137948423 6888145262 9696866213 5664767020 1426725320
5644350313 1171444137 7135146736 : 7073
7169726545 2265465106 1440382027 4351610225 4601111621 9577553769 0901298148
1368832862 4816036689 9277390833 : 7074
9405147143 3848747692 0116426194 8192396449 4632004463 8474688479 1115170448
4391938676 5303110082 4238704981 : 7075
5245204055 7396205818 3162486945 0329535091 9858056458 9890874425 6831229254
3450747373 4151952742 3647641802 : 7076
8199548731 7377212901 2524214467 5404294585 5402423906 4717022570 0446424875
9363876636 7067747962 0244276809 : 7077
4377854825 7665128437 7690093142 0986660710 1870293385 9331043773 2853034371
6889518480 2547799127 3331396336 : 7078
G725069762 4004443999 8428715482 7354002136 2280036387 8357808619 3032813099
9490596589 1889375075 3834426925 : 7079
5703320890 3758324628 4679499477 2016161849 2448067072 6066423943 2321923207
1760037252 3136002600 7938117888 : 7080
5240250457 7314925350 2497399425 4405139909 9724277891 8511892389 5556724908
8332253210 7988627081 5640003225 : 7081
2780315109 7668662283 3467773834 9417461225 8945420980 0002909329 6907437726
0138690910 2791406259 8515250597 : 7082
7668184401 0781670669 3400075149 3482854055 5614304755 3911053314 3757464229
4862097446 7908447587 6468363178 : 7083

9277308598 8511135509 5794744954 2186298667 7496696159 4744440927 3113200669
7861138085 9054531762 6494590167 : 7084
8196971498 6139949226 9722338452 7051438093 8951350464 4375512254 8611170891
3966808936 6777973099 4384880861 : 7085
9331909104 3578607208 4967307382 5937923963 5993284751 0040727279 3862627167
0447540757 1920250393 4191972545 : 7086
1894831172 7902461404 9668955179 0799869102 5376489096 4845981670 9017139351
2067828395 7996122631 1973314913 : 7087
3778183433 8419318523 6753866290 1900463343 3285328343 4182492107 1916092767
3080132353 6947264848 9276442979 : 7088
8706811256 5462806117 2604607332 1917630474 1215064030 2017893205 7956876051
0277525053 0436122270 9604675845 : 7089
3127386616 5214241940 8683408375 8914009511 4133929545 7700947317 4811688536
4094183528 6109971034 1672355817 : 7090
8602823498 2020314998 6794017404 1719140283 9362105194 8060481901 8245180081
7013710421 9745921279 8404024268 : 7091
9630053355 3685494164 0086392177 4567695344 5865088328 6298216584 6016450780
0003598985 2129012395 9007042026 : 7092
6074498878 5062717607 7622255060 6390745319 4771889250 9905810093 6729891954
0995766335 3865837809 8044793537 : 7093
7991877121 4297746197 6409472721 2140235326 5517772361 6341966074 3763915386
6019450177 6914006240 6841273162 : 7094
9586080635 0676293775 1252934367 5960734271 9967513520 1580087373 9548389734
3990123825 6568290785 9114425882 : 7095
8521366169 2014029900 4131462136 8063235265 0865411218 0847499875 8441118362
9097908919 8341187537 6649604809 : 7096
3626489329 1043905977 2563295581 3887767446 9546181579 7870419159 7558335000
3772867246 0735185350 8854954642 : 7097
0293097158 5636949785 7640483741 5055580991 8840209343 3335128195 5145551857
3334888164 9167694427 7824055543 : 7098
3697731192 0143722541 5419274505 9760137984 1443735994 2028760525 5571893780
4114892612 5798303618 3801077110 : 7099
0500423923 9264415692 2779005702 7947401364 9422918479 3369253543 2484048780
2317774451 8417795835 5825754761 : 7100
8425495753 9587854945 8343084804 4173708798 4926741952 8932303148 9589916006
8542287957 8929481590 0068735373 : 7101
3333338638 1390842099 4872551172 6171087294 4088868447 8508116345 5852692141
5402996720 9886937158 8623222033 : 7102
1485581742 6826359506 8936234481 3422473786 9394609232 5660075472 1256741410
3421849013 1189639676 5473291674 : 7103
2471899342 4330280290 9269850752 2949079709 4430092386 7287743936 2551113136
2876159197 3832413491 6722682612 : 7104
2826312779 3811750807 4089589439 7190025426 4126649479 8017644201 6454158813
1760189726 5962461441 3913192067 : 7105
8503524638 3065776401 6522973204 7089201146 9171337228 7863037038 4531341942
6441424966 4984474404 8617738513 : 7106
5293731080 9985947281 5111736507 1913988126 9307482690 6140392864 2276627849
4329558828 5938372148 4750707783 : 7107
5511989622 4911955882 3704506458 2005610170 2053248444 9021502294 5617655618
5921457439 9009554822 9413751544 : 7108
1361329150 4304691417 9941224609 3381651362 6279028278 8421381220 7758942403
8840622943 3172798925 9418366829 : 7109
3679259604 2241845941 7706530195 4864394817 0335524102 8747044731 1715070347
7508343237 3331163258 7672763683 : 7110
4685844486 5625391872 3946082732 4713500448 9570868031 5032503867 6401735315
0794003455 7285503700 1845602769 : 7111

9615061568 2914161170 6104616174 0824846267 0335189152 5518824821 2726007259
5765678899 1256787020 1497866790 : 7112
8471066310 7487646730 9890914719 7987925985 9062573364 9734982350 3126098309
8734686162 7393580579 0735490824 : 7113
6830497284 0977323811 6708249151 6373468087 0510521919 1762054169 8826254760
5445817711 1799377677 8696542169 : 7114
9257855772 0426344244 3042074495 4897033904 3450720611 9007697363 5174019526
5632213928 2583120528 2400374669 : 7115
5459092804 5259687984 6081408707 1953425476 1388363353 5119112144 1431485055
2012358138 0626492313 5383387675 : 7116
8091889237 9758551573 2036588317 6242411691 6752381458 8592075164 0352366843
7267917590 6370539219 7981126597 : 7117
7081139473 5167196299 9970520901 7090758985 6916389066 4270423570 0744502775
1201049039 4814832945 7443809746 : 7118
2603510578 9819540763 9870607875 8177407249 1184506978 8429941383 4206081242
8390148814 8725998541 8148029294 : 7119
9227878324 3561055491 5641091748 8770670678 2011985910 9589098386 8839511718
3801349148 2549927491 4269855259 : 7120
5177653612 6242157266 2448896096 1482979703 0842021051 6027967147 8561594064
6136382477 5890201102 5199234215 : 7121
3210060175 2302574212 2375421074 9591872867 5189552155 3299453226 8942518840
9422826757 7442222755 2820761560 : 7122
7277103947 1825668024 7106606773 8312063031 4662847443 3862047432 5956850568
9287162653 2908327878 4399650716 : 7123
7242206138 2945329166 0046380872 5630595241 5328812093 7009922989 6006981396
2796862795 6763519874 1824018562 : 7124
9319849226 2333643189 9309017048 8199525882 7338807953 2658851449 3938124115
4327320589 1645786422 9452684117 : 7125
5188081840 5095690469 1381344307 7848902114 8797121162 8397734430 5484614980
7935624354 9671412947 3332666422 : 7126
4830500436 4454547027 4917232078 3995650868 0187617327 0331906565 7954753952
0692925201 5307064350 4713068812 : 7127
8903205385 4556398799 2101095667 2982873047 9466100543 1635846234 4874415554
0713381432 3441311788 4241960190 : 7128
3702868696 2422762465 0240042081 3713504640 1599347205 3275567950 3533906712
7216177906 0302162399 7804185763 : 7129
3481005448 3887031730 7162552564 7995999986 9353196813 0101562970 9787116730
6964940274 9166572674 9434320813 : 7130
2314613886 9255334949 2941183163 8778899032 5940109341 1157274298 0133181457
8281687913 5142439557 8734205915 : 7131
6145877368 9880807917 0711455115 4762661682 0817775747 2484278797 1298154823
8520368959 1652405437 3052466728 : 7132
7313030192 5829524987 6393209857 3120916105 6135722017 6392716995 9868610886
7606133843 6496169518 6548486046 : 7133
1684848240 7438381180 7417342224 4839478939 9827801473 4222305786 5410217729
2648209140 9485035013 2241515599 : 7134
9901630714 7814852705 1614432258 2547801434 4010953366 0239362642 4931385229
4054753641 6836467041 5743005583 : 7135
7298470401 8145571002 9696234698 5059977885 5736998021 8223035492 3831879762
9894156977 2123638440 8891789275 : 7136
2775284714 4776218920 7242545315 1037479977 7068623624 2189906303 0962822117
7321970823 0347745455 0602385859 : 7137
1140389896 9166622077 8768326012 3961741999 7623655063 6326601699 0661168908
8687741333 7809195161 4977054921 : 7138
4171181911 0149543478 6938206209 8082787895 7313278990 6002086339 1127847153
6843043814 9815097030 4770886881 : 7139

6359741925 3413412219 1396287457 8109934248 3271410917 8276317654 2096312507
1366926365 8821357511 9987540115 : 7140
7902802850 7806698333 8418338266 7394553683 1965657170 3194629336 4109272667
2742449927 4732291380 0983574450 : 7141
9134079930 8568360484 6778665354 2105866800 5211542648 6749723953 7936421566
5358720818 5541259819 4547427516 : 7142
1184169366 2556324193 5915856950 8890535314 1218336623 9987802211 5042456600
1502364691 0815997419 2025443787 : 7143
3610226712 9412282988 8805336794 3950329404 8049616771 1072878810 3161467730
3715021835 2890241464 8175430954 : 7144
9059448875 4104250168 0406999981 9671937982 0827733464 8558158591 0776178828
4975054685 6791399040 1138456243 : 7145
6040357440 2461912376 9440220076 0421433573 3872603925 3724664089 6649147146
5473007219 5292737417 6796135652 : 7146
3306780426 8200024230 7412186674 8669462805 8788787382 9952327805 0107066101
3854028801 1913855730 9113805727 : 7147
5589368194 0309380718 8637937582 1544435162 6559912030 8784820523 9374935189
5597158155 1828048148 0783131053 : 7148
5096823567 1612355096 3161528665 8578590195 2871775464 2506412984 9451057303
5332433489 0979972094 5729170095 : 7149
8784083221 4044050147 4114381709 5657712526 7157798170 9767638478 6420648761
3679394578 4423858603 5649664463 : 7150
1401310486 6705613462 7690174928 3229004327 1120819229 5641589481 5282655842
2192618902 5305138159 1180342310 : 7151
8851491226 6539179448 1743390432 2700506874 2888199670 6196068850 0673328353
6146804985 9608140287 9119631127 : 7152
4825466404 3703847828 8400798157 9134452321 0166867374 1121294496 8134510122
3798429402 1947946916 6481503124 : 7153
3187396911 9602156712 1142279430 6820104087 6756809373 8982054684 2147856014
2776882018 8071717414 1274936746 : 7154
7817513227 7109046730 0153094677 7306922329 0953495694 4916011690 4681818272
3477816197 5125727851 5355698616 : 7155
2922029188 4745302698 3015663874 9177021947 6050941748 6684119727 6604440854
8329750498 9078258061 5390200301 : 7156
0352180271 7920995081 7839121233 9700700278 3290619371 3368218336 9615408265
2039271615 1802279152 8640768315 : 7157
0329974060 9280483105 0625515722 6287131327 6818247801 1691549378 6248322401
7238370727 1608864192 9421181718 : 7158
5646748118 5694008705 2996113141 0594582938 7033052862 9791882649 3247420179
5512442329 4762903488 3495272292 : 7159
8308040122 2637227611 0275203800 1209375208 8562406453 1125037264 0153996437
1233707903 7439512032 0285350254 : 7160
7482779809 3030202196 6325093290 9448819057 4510298521 7283069962 2421954710
1651939329 7693706545 7633343243 : 7161
3256433136 3912210835 2913555770 0812650539 7931646814 9451341207 5121883505
4739685955 5387809194 7374656627 : 7162
9031868161 7136416152 1554077453 5438041479 8454584063 4474745085 6802808623
2412696939 9311229640 5603273109 : 7163
2793849169 1879333596 1485351700 1814969826 1648003440 2782281398 5879014862
8521286746 1753848503 4806838652 : 7164
0168074767 4402114766 5569643873 9675796644 2958864095 7755915419 2615071966
5373410817 0649748222 0419135035 : 7165
2239320369 2779239073 6955880577 9975526019 0308143550 8492475981 8577979802
9812641251 9269808310 7565746594 : 7166
6935112856 2797590578 0034193234 7600136961 1447290131 1372932651 8877314672
1410741275 2210105151 5586571349 : 7167

3912768855 7300645655 6663593525 3545094573 7896883580 0277072108 0754685197
9021567766 3559608589 9524492722 : 7168
0497752998 1545862535 9295568858 6552190862 0348857429 4435488340 1766168305
5326602458 3599445433 0952973620 : 7169
5642829474 5396641017 5938780787 5790401431 9567264450 2565389640 5200714870
6853706562 7117466761 5719181064 : 7170
2280464382 6867806417 8170171014 5579987859 4929810286 7487747167 9017435503
9931696835 2859211648 1487861286 : 7171
2891163759 9276847108 8743661617 7623930982 9498234064 1378280711 2174123783
4222414406 9984863288 2392406563 : 7172
7743898084 6317043398 1419017041 7704585646 7859934941 1376710185 1048288345
8475911298 5991132618 0631166520 : 7173
9771093260 2545777664 0478113979 8120273962 7905217807 9656593348 3031963863
7263605481 7261754402 6268435726 : 7174
4514741114 3619747488 5328135254 3231321040 6553747535 6091405133 5764484512
6503159949 9398475912 3386762589 : 7175
9886282674 2421410656 3170282684 2027427753 5478527849 8062554779 6730956213
5010751357 0108143767 1209736106 : 7176
5693389811 2877006439 5086822635 2419579899 8752445315 1326843577 1252695511
6568426062 4979530056 4134236794 : 7177
0326869432 0221277155 2845229455 9060131663 0023329786 1842729369 7054256408
9572247314 7960842279 9606431211 : 7178
5572289888 4134069015 7060791698 5539706270 6949869226 1315505954 5816240665
4276265360 9898824692 2842402217 : 7179
0510920884 5226419380 0405064723 5141065446 2939736295 0062687069 1857435034
6890787025 2668899031 0590307332 : 7180
0209977070 1955626708 0078421417 5004024877 6436938270 4966754069 1448956340
0092985235 4991296996 0465402832 : 7181
1299512881 7182931733 8025893433 6088863956 1881453054 9363321022 4670237123
4907974068 1256872488 3352801820 : 7182
8796868376 7489328659 5156350015 8437345427 1486153424 9712522615 1780864718
3501990499 3217819915 5353873785 : 7183
6234550871 4313905038 4328696513 8769032905 3401523842 8243963913 2568672927
2248708477 9696809273 6560421516 : 7184
9040230075 0173661949 9358544714 4041522141 0647497242 1615230372 4661255301
7965334543 5408898008 6901630718 : 7185
9158808455 5221984311 5742231294 5196703758 6260587762 6333668485 0752752847
9034425016 5376491633 1792832073 : 7186
5720436314 9629332621 7198412041 4969390021 5789875840 5506883631 3189782803
2706818206 2011470970 7824647760 : 7187
2762963513 2451352251 9224278604 4766017845 5463586895 6418344082 5955325433
4088710161 5107036746 8245292204 : 7188
6780848531 8529560302 6039336998 7912529602 3073278584 6937162703 7491448388
9163375005 2621697332 3321524007 : 7189
0875809848 5646289779 1809848253 4272120248 1272565551 5050305700 4181198951
8609216787 8092709372 4148047112 : 7190
0320927521 6382196930 0536677508 4683589386 5445281522 3534755755 9076380237
8521519460 1827245682 3639549056 : 7191
8125932747 2895456989 3447114180 9813153649 5974851541 0685460978 6330324864
2987770186 3553136745 0811076763 : 7192
8604733464 0295768821 1528898135 6644930949 3104358903 6451413175 5241478099
2535059148 1309531471 2262341354 : 7193
0646090598 0393007295 4096921369 4564498578 7813148498 3471061858 8580258335
3800866348 2207458754 7577692067 : 7194
4100675717 7870214802 4906809935 4034049780 8697475231 8213687886 7192069420
3441862152 9900468668 5357497787 : 7195

1014752845 8215215350 9827098635 2770357562 9384931393 1604238871 6899073982
3956189887 2218842061 1902770942 : 7196
9903227573 6687186079 7707077215 5923264181 3601697172 7774070288 4134571773
2062530298 8161726406 4980178019 : 7197
2696415503 8729328624 8745115669 4400901469 1037146658 9528076907 5063099165
8019040282 4912396345 6798392695 : 7198
6557159476 1366943873 9333006837 3830818123 0942044159 2959517187 1924926797
9297129502 9491581185 2694465448 : 7199
8816064489 0910857473 8497225266 2561359833 7391783194 9599150654 3998379398
3509005174 3387881718 2955161844 : 7200
7325706942 2937536525 5153471178 5761849774 7797331677 1607149942 1452638170
4752272729 5570500173 3744013560 : 7201
2978121687 1886130729 5720086827 0472815099 4493655761 3530068382 0300724872
3891860455 5823405676 3187626236 : 7202
2589043407 9602395529 1440818688 7350391637 2993471628 1576586285 3465444767
1897796640 0943420249 0164341084 : 7203
8630536408 4940912887 7119965069 8045788071 3797795159 1204253802 5709198604
5596738860 1901634414 8290099729 : 7204
7139228277 8638712630 8017669213 7049349769 0438081956 5618801297 9091465075
4406845318 0516940496 2484549525 : 7205
1863363242 2222614586 9950092979 0456268865 9703930983 9695602356 4388968632
9923323765 9081142486 0358035451 : 7206
6978206769 4889248663 8330687917 6594263820 0926377890 3165827050 2911378572
9750974004 9386753580 5163232169 : 7207
8473633858 4197352048 7633945539 4642739288 5249745167 4899046514 5947388497
7874365758 9470771779 4356270743 : 7208
2572285536 1014929296 7565026585 1006003218 1461536129 2859106934 3214878686
3743429770 2059825451 0958054174 : 7209
4556250811 6488195634 4323355550 9154070228 7345061766 3489800492 8111484860
3366735880 5224298027 6245065806 : 7210
4166164609 3309658322 7741201687 0046186470 8735460759 4600235537 3436366858
4630008298 9115932174 6543217351 : 7211
1755315755 5101863096 5885602744 7328817436 1764536559 2624617643 8777403799
7688608018 4145764476 4303571298 : 7212
3995556915 6559423117 4530659483 6250206036 4382217354 4096553539 5090457436
9386510431 2934271665 8105214647 : 7213
7362856041 0427152372 9397812487 1060598593 2017807577 3843670656 3113506859
3058886083 6202347780 6498797585 : 7214
3178401672 9059787657 0281387580 2603233337 9883203038 9685399914 5202912912
5226423489 6619763397 1826810602 : 7215
3709597596 6298091853 4819012504 0315899625 2047170767 6639986835 3153196608
6875291254 2311537503 4755005153 : 7216
6740686846 9956767733 0577510792 3504903457 7877826339 6473876429 0564016443
3316822456 5090395700 6685680098 : 7217
2518273305 8190424322 6112894851 8491882214 2539686131 0033713722 4301868395
5964417346 7084178320 4007571512 : 7218
4168083755 4131482479 2400452430 8125367729 6897044406 7964317974 2681593592
0313582151 4639678923 8533146916 : 7219
8480192378 5613665706 3609422132 9625671822 2140853765 8070683461 6858349637
5655648152 1327880020 8251696180 : 7220
7984840474 8675428648 1247189232 4572784880 9951784920 0238132149 5055976860
9780439227 2136566791 4325583872 : 7221
9396288555 7531898513 3904712378 0845594071 8448328361 1533615780 2570583643
8593371292 3478051058 0916071579 : 7222
2371735883 2712616456 0656794503 4809907997 2050842273 0547419077 8110649402
1601666662 9509459324 6968082788 : 7223

5873826896 0848774456 1183798448 6573045432 0536892684 0349142345 0881535101
6785757066 6874956376 0666307293 : 7224
6364979840 9105284008 4546233261 7818149797 6141218187 2861268653 4607116650
9551354778 3230627411 8114820717 : 7225
1646652654 2709914847 8807776892 8234579548 6110671504 9473647399 7364806652
9953736779 9934023532 7496480103 : 7226
4460008563 9769736337 3222947146 9730277640 9407753962 8839012582 4662939693
8987701586 2089379918 5516933084 : 7227
6381599653 9591339262 9245005644 0676097423 1690307802 7155970074 4418327838
6392005915 0419266647 2875278979 : 7228
5492770341 5082639154 7896926883 3081660479 5642091123 7931251245 2382628913
2391142592 5924887235 3135341403 : 7229
7167337993 0617119964 8037407664 4403610827 7815093719 8926678995 8837505255
4760929420 0184307444 3896493496 : 7230
8142535424 4228754826 3467259939 9787959797 2547224411 7287600771 2608840939
5048119215 9674887395 5586267106 : 7231
5490302132 4022477079 2652497769 4495520510 8056410482 2068021729 2763561522
3773851063 6020972031 3789483890 : 7232
8760877304 4443391070 3713871331 6374909282 1549212966 0943559957 6542505065
9702392398 0013397192 5439883533 : 7233
5819326042 3553346407 0907285804 8652141533 9035222970 6047122013 0494123455
3577895471 2486300047 0556288945 : 7234
5044608622 2664852384 7310733172 2303856418 6531180217 9351207883 6789316324
3192580644 8500511452 8227697470 : 7235
2881297556 7230419389 6966911516 8403425219 1266686056 1300384520 5389336911
2100812050 2524108795 2203462366 : 7236
0139062940 9586806381 0486156703 4918952889 9594434218 0753633129 5702241260
7656628378 5776310584 1332497008 : 7237
0210963951 3491608642 0519405513 7980373826 5147259899 9021247513 5972827648
7048030240 9576924624 3692551113 : 7238
9223531786 9482928421 8555724374 7361826577 9560035189 0158768759 8144266576
2658258377 7353798801 1685811161 : 7239
1885458719 9032823772 5765974792 5555015170 3522520432 9594566895 0459779859
3133546030 5804628712 1050992002 : 7240
6539199902 2032802985 5134990558 9529620350 8118519116 2393317684 1545106106
5516247882 1980694146 9531402163 : 7241
8542948465 0993057590 1842547725 8768576414 7409170800 5076586416 3376935814
6607818722 4450583794 2437576287 : 7242
0854540335 0110537092 6710216016 8897425190 1726647710 9118773401 9658761319
2382440167 7866944731 6036433235 : 7243
9105304639 2692548647 3405762211 6648426993 8330305048 1743973119 3500536761
6450508447 6304732563 7642166613 : 7244
9818172167 1531112891 0245585162 0731970070 0311254050 5208238290 8931287575
6421875416 7861569179 7009700938 : 7245
0501429846 6685212001 5915113272 0592321337 3624510092 7185727349 4135289481
2323115992 3461589690 9789674454 : 7246
1306276268 7256330468 2335313715 9651300040 6153320056 2642138432 2314827035
4735863509 1988446533 8356840048 : 7247
4168465383 5147529872 6685210475 9009105167 3159246264 8535707081 9430352359
8755049297 7333075317 2203557347 : 7248
0236517079 3159327874 4454981644 5932473938 3779436527 8003109277 5354373492
9701120308 8850811385 1103629856 : 7249
2059488346 9615636232 0828310842 5028737613 6412846726 4503679574 9821634469
4599244023 2038663609 5796527312 : 7250
5048912012 3599962848 0875170027 1355908821 2824459212 3659726362 3626193171
7499306027 8006727038 5204438694 : 7251

4202396265 9430717338 7544843964 9357394016 5446942543 8918315939 5967500993
2984755318 3201970958 6811147403 : 7252
4114643077 4907873234 7323831843 1697758006 9510475962 5347803929 0531227897
2028317103 0770441520 4096202753 : 7253
6066168775 0953874913 5648654899 3528510813 7546623023 0104416440 6078743379
9388659434 3197397276 2809887255 : 7254
3596421836 8862525286 8429209407 6693531017 4676061436 1504022410 6100657518
3857818459 4309738166 6527422822 : 7255
3595232656 3156004363 4606066179 9295434268 1278246980 2311078319 2217548828
2484434894 2155905096 2496809551 : 7256
6816141784 0810580681 7300900263 1527217960 7472345797 4475343900 8266238408
0704887676 6201256697 5422632476 : 7257
8414436029 9612706320 3642476725 5662395841 3633466963 4201833963 7825422544
7430632805 3514644920 8028401169 : 7258
3507171925 1343449175 9163446481 5473816005 8964567608 4057980165 9025574897
6711877225 3952700017 8383461807 : 7259
6833574250 6196377939 7290718664 1657756554 9223989511 6050443618 1020594168
5054187658 4715902962 5940307677 : 7260
1869386054 6243793895 2917738297 4148463926 9221118535 4790599095 0972997607
4738729626 4727605210 4595112827 : 7261
4356312709 6059502375 5731806195 4525987653 1988819158 3707806094 6073010461
5314079672 2525757786 0927309190 : 7262
2469192988 4517398716 3150768576 6351864937 9762973806 6031650319 8618172065
1187867335 9197652395 6744318887 : 7263
9533652290 8865200801 4049981149 5657664501 3571486997 3922194603 1665890119
2022613706 3954921500 0080968080 : 7264
4936515617 9044413993 6547321194 2270762716 8066232679 9317221812 5745760915
2455167050 9765492798 7579284966 : 7265
1673761936 8534354686 0426007927 8210505029 3510602699 1107729258 3322331389
4239808716 5405046314 6586178672 : 7266
1994997496 9729059791 5791846480 3703518738 8643820270 6098049340 0455678609
3076538261 9265263181 1202186385 : 7267
9251656925 8659483274 4831643461 4594036961 7235805971 7958882905 8381523697
8853316823 0133145229 2999154059 : 7268
2302274058 5503740628 5359045976 5785964573 8065216002 6315917742 8881248018
3076664637 7077199854 6947143246 : 7269
0415430871 0677882187 6907079694 1313678460 5770183828 3928309479 1609157762
2835129330 8159945269 7606840265 : 7270
9423392885 8711054408 6830636638 5810204736 3209449653 4988776275 7862699053
5391553279 1188119263 3510691931 : 7271
2733976667 4091486139 7456145254 5275688380 5182285266 0871349632 8388418368
4848013626 4582895400 7306507915 : 7272
6074102889 0834546596 2538086962 5526263711 5923594826 3845456087 2583686037
6198170672 6337448128 7637268299 : 7273
8877722944 5404077067 8994016438 1250027554 2016983023 0695451162 9983134310
7188990986 2541016002 9355244154 : 7274
8625913761 5974787091 8515340256 8888130198 9218844566 1116491063 2688832423
4483096288 8049657974 2710340822 : 7275
3736673828 2557725076 7715305186 0616483063 3559801934 0710989716 3098840874
6725910445 9888759596 0530329248 : 7276
8993638521 1939490968 2837108645 7135234560 1580438642 9710353846 6830596898
7968083587 3497857626 7855658883 : 7277
8364951627 2354520708 7795334676 5867204916 1720879813 1491416989 1375664645
4569742166 6338802887 0931159655 : 7278
2908909685 8324662094 2368725955 2095876586 7639645258 1062760953 9552914917
6518173767 2918987667 7984251192 : 7279

1680755258 1600469267 2892837696 5028649208 7987822244 2107140464 6730748519
7437139760 5307432234 5155253212 : 7280
1235870851 6162852145 3800836861 2462951548 0394533296 8377364248 6296457024
3581892316 8622649949 9218634983 : 7281
2468121494 6737511994 8195045074 4965970432 7176040133 8355222360 8153400047
9598193685 2230925881 4575459795 : 7282
6696947562 3894957247 3252484866 0706542882 5628767359 2883976141 9619059132
9083994691 6384100532 1889186943 : 7283
2565720667 7652323824 7316438079 7696044457 2969446502 8343061665 2216987080
2617575705 1241641110 9207357726 : 7284
2039704116 8885940376 1058433130 2931444829 0659556842 9648729637 7161321950
7321997164 1441702565 4689461012 : 7285
7030824597 2384182836 0403116603 4891537160 7623286396 6616561856 0009467155
4915197467 3084232558 6767410973 : 7286
3402984616 9675642107 6322315367 8979988580 6831989228 3085976337 5092946490
3879107437 0774008463 4936236240 : 7287
0830084950 0735581649 6691675977 7865808111 8230477074 2469643604 7327935182
0384719889 6117035900 8300455685 : 7288
1252805095 5106676996 0237387823 5376637278 2267740520 5106153201 8998380611
4919859528 7500266294 5542273526 : 7289
0581489488 0997940795 2381788622 4433013323 6455432574 1038837042 3239809214
1614963275 5395663617 6867524381 : 7290
7655662510 1337430464 8082075528 5156491167 1828345413 0472387959 8488899915
6275919082 1521959453 5894398739 : 7291
1987885447 8785588393 0195317034 7124025007 2072868196 6631889154 0923756298
2477373635 8962930373 0274864925 : 7292
1369197889 0676358246 1536907572 3811889000 3868340893 0397937519 9306538172
2873667753 8511417161 8214640630 : 7293
0539934079 3672109509 4123328350 5714265283 1949674694 8502122505 8627404548
1109395874 0537288809 6652279949 : 7294
1313504114 8573758189 9491522734 1352040220 1771742697 5610324053 9002593855
8957849154 9406875910 8510900445 : 7295
6698779029 6358999578 5804399594 7222843355 4403913845 5157545905 1012236770
9490042907 2516327250 6438911414 : 7296
9403143929 3381604921 4114829869 6151260756 8119300488 5160428925 6533437306
8623996218 1200837591 1431730894 : 7297
4198998029 3377230506 5225027098 4972059596 3536060693 0155405423 5809356222
1999017615 5133694756 8289739010 : 7298
0935189886 8910230562 0334560674 7815955637 3737243150 5131046388 4368461660
2221058076 6055016338 4439513392 : 7299
7539050471 9811126778 9378425274 2930717142 8737605543 7904153578 1461948559
8470604040 1016336531 1893068369 : 7300
6593975950 0845851332 7284730880 0484877403 9317183964 9232121257 5915598639
6831005033 4405605278 9887661147 : 7301
2430892073 9067717133 4489995904 9955652866 8704313897 4556386419 5065256263
3820072560 5325025449 7912589372 : 7302
8660693665 2747569545 8554937936 5494669059 5626482216 6011090722 2716643362
7147859946 4599992674 9049549615 : 7303
1264230852 9764040436 9503078388 7567665223 2278535937 5567721700 5441761522
7578900688 4922246472 5810068548 : 7304
3167616870 5370471402 9329379429 5759712945 6134524552 7545656824 9439714331
3858049509 7372060753 9412574156 : 7305
2910110232 6051852161 0876296113 4662379674 3611234251 7799340059 6014944329
4796416460 0310883776 6887059200 : 7306
4886201801 9633769761 4505829217 6891376685 4755419381 1607470643 6461835509
5042783676 3844727462 0716138197 : 7307

1191770444 4879751377 8892289958 8473820082 9650704622 3127280621 0512699103
9445893607 8904429066 1289121648 : 7308
8673293277 1245059557 6499953164 5688888539 3740496576 5716880913 0392423485
6937644501 9991202878 4002735463 : 7309
6310802804 8839039864 4628663159 4078401029 1779687774 1898286222 2702853291
9145503554 9524356674 4611953366 : 7310
8970392803 0763361342 2784813008 0590245232 2986453653 4759379247 0097712372
5148559789 6223350550 3714082373 : 7311
8855989996 3572576815 2824985732 3029102096 6336552980 9751869164 2928076192
7498322965 2194481696 0437154884 : 7312
7590855272 3586440480 2017581425 7540052934 3879461964 3196734902 9347026961
8697328305 6021266285 8359414291 : 7313
4176432704 3883997140 3798486296 5074407226 5264163463 4289829191 5406945822
3840053228 7197813012 0342846511 : 7314
5000214159 6938756059 8958467433 7883215104 4581733491 0293140849 2619625443
0106947558 6326736133 9601315493 : 7315
6002702882 2653755016 9131948172 1677272774 8879790713 0864591251 7689622702
3434826717 3744476254 0360443612 : 7316
3834266368 5290979347 3026221774 3440936100 4734383259 4705561796 2424701479
5775993839 6128282186 7040465947 : 7317
3671346740 0380288689 0673435056 0654196291 5439823233 1916634157 7275777262
5566712728 7567764740 3587825186 : 7318
3240142935 1230992652 4186105365 2623906898 0576795900 9378267212 5122055111
3158347823 9251417189 3266684869 : 7319
6741049338 6288058933 1412583648 7308559685 2364870957 3522735515 6417316000
5704356833 7512858648 6798777626 : 7320
8789119701 7419892629 7503739016 9668114553 1056396389 1334919478 4911260028
8727202243 6295541133 0210328883 : 7321
1252179038 6091534116 7857762338 8013856364 3471230253 1331275834 9214962681
6843461805 6550643768 6484432169 : 7322
1600993297 9358869713 2634594804 7658016730 2876236263 7540464177 3711712333
7555290761 6845760984 1203148149 : 7323
0671265447 8813087496 6926524950 7283763395 8248312795 5424169984 4491415609
0821234201 4466356115 1439878692 : 7324
8367644038 1999960736 1303565876 4068334110 2908782368 5230451771 6218148049
4326246784 2034037569 1218100204 : 7325
8571334683 8603164091 8904933187 0282565113 3422596513 9518362851 7923265340
0316253031 1776858304 3905855303 : 7326
1434700094 9954042899 3106200690 9384298598 4946257642 4364274755 0200929598
2199705371 3856754024 2399582249 : 7327
3614688183 5783905292 5627662578 1476282549 0521531188 4516726792 5851062996
4112189047 4290326898 2903001953 : 7328
9195564907 3481812466 8433899387 6522912442 3119814574 6620093780 8079651108
2080920250 0372217656 4662265462 : 7329
8516780697 5616512298 4690402876 3819653602 3135636136 4976638902 2197923617
2338854918 1980957352 2970142320 : 7330
4549139476 0020309831 1826513470 8319579082 7421779732 2911001498 1104209291
9972707939 1310541688 4305647668 : 7331
0682881432 6393122924 8475280351 5563282795 2732656083 4108816169 7961023963
0665231004 3702682309 1533999011 : 7332
0185178408 5347966645 7696399564 3862325907 2566307560 3947055135 8975851270
2593763532 5234545460 7115787656 : 7333
7775171323 1407198493 0870727417 2599038347 9908377522 2662385016 1890001369
6653310225 2957386519 3690993625 : 7334
8963337910 4117758605 1878192070 8777656099 9410980515 5176052433 8381498578
8441580214 8300377380 7294222057 : 7335

6114221873 4191201738 0974191603 9089694570 8128691961 8687718234 4133397780
5975939917 0408525740 2459527953 : 7336
5895516836 7104612241 4148488235 0704209894 9464053346 1029814818 1498384962
8748546950 0407432484 3010042370 : 7337
2377026935 9497780909 3900955516 4281254168 3795122263 4081034024 0060173185
6228367117 8791286350 5765122105 : 7338
2346487547 0892286547 5797991986 6349134336 7457317808 0001015922 8032458041
5606204058 8149733690 5468838305 : 7339
6831118856 4269274466 6825628260 5693182711 4971107908 0222167427 3725205608
5239809705 1928761728 1829491707 : 7340
9681084956 3420922156 8005222657 6590502708 1010596468 9600419159 4139713443
2079748752 0061970362 1436531076 : 7341
8194923146 6574683330 1942561772 5802965628 7105746356 6793619642 9335090417
5915476056 4372284823 1520544883 : 7342
2642676308 7111474606 3603473283 4323009419 0420867481 2321963355 9808929813
9013743217 2319029440 3380213835 : 7343
0378466582 4784928237 1602845709 0185093998 6308768974 6361213922 2874464372
8222306898 1442710808 7673984745 : 7344
1986859735 6568324758 5023790229 0438743386 5650163543 0920494975 1394651712
0423996380 4851524043 8271400499 : 7345
8736609221 5951679436 6552097162 4671902878 4972343068 7941462209 3306540351
5146533353 4381762121 4097499529 : 7346
0813343668 8742196085 4630056294 1871857198 9796549038 9226099400 6458134576
9898902937 5252603159 4793520517 : 7347
2873837166 7007215479 6195637574 0701285592 6075633358 0436117856 9570327151
0860973326 6002110088 2712319916 : 7348
3759340997 1230320261 3810988996 4222097317 5527414255 3634130615 9729484372
0485673053 6334535661 8321960256 : 7349
9723677437 6640419855 3908458665 4773133442 4306172126 0086297605 2967207859
3225622923 5584626284 6333189292 : 7350
8319836100 0259218711 3703997150 5577493204 8759783197 6713628632 0184428233
6539347163 6379457127 7158840984 : 7351
1768777705 5114469465 2404220885 3820282498 9296395084 6704657019 3475974601
0821090022 3798138939 3889173661 : 7352
9958702353 2221962941 4991396048 4299279341 1658670487 7857111337 2653492856
6536892887 5089626084 0586049730 : 7353
8118077965 0060100681 0979623045 5954801189 7685661967 6242389312 7565241655
9831945661 4716758220 5720515073 : 7354
3187438649 9592077292 1715411527 0702515737 4293274060 3038899701 8176392492
8444961449 8921199600 8741455920 : 7355
1197110619 1781356008 3117939437 0021856788 0801261181 1579946414 7173035479
0676703916 2644803691 5230016380 : 7356
9750954986 4785771530 1933207587 0767280738 2416272997 9856572198 1668444519
3115121472 1539462606 2415474788 : 7357
4492531474 5222342898 1444393566 8965208172 8352018491 0446850123 6353466594
4391017712 8605907892 3203698177 : 7358
8144140657 6579531416 9342756627 4427564792 4957225617 7112244150 8205905726
0848147140 4592395307 9597060427 : 7359
1724592752 1655060051 3715396777 2426318259 3431649599 9408488134 8962944398
0192805044 6990307433 2958696420 : 7360
2347799440 6085552980 2601687627 4594816524 7001320475 0956993158 2510565956
5545112468 1216741044 1892014971 : 7361
2917059113 2046486929 9521893607 8725719895 4332687027 6421120797 1112580634
3717907028 5365686818 8130649315 : 7362
2903586491 2313662571 9028116080 4499253317 3093318861 9738913721 1839634887
9203809214 8177397515 1215550607 : 7363

1421237847 8249514649 4932858500 0899517323 1334179449 1957195493 7381022516
2043355850 9640442999 5065494821 : 7364
8174918298 2098028501 6135764697 9930679132 2247849851 2958741989 7623402048
2642870620 2051130484 5432074564 : 7365
3630546829 8589979035 8840556461 0067458997 2300286489 5895241453 3037922186
0767572196 0300019984 0534610423 : 7366
6967394846 5981989033 2264471364 1778288583 1861564108 9981063332 5669588014
9087548911 8552944475 4877770703 : 7367
3254281654 0487011452 9025916253 5198012387 4140104360 3630920528 1399608185
7821750591 9920444071 6799015338 : 7368
4458640154 2054227660 3949509852 5317146476 5665886681 4588422462 7681547177
4897706020 1619396947 7385529779 : 7369
5179206393 4259381963 5523803884 2483958103 9275670564 0051774584 1545364779
2017073426 8207895378 4241931871 : 7370
8112513400 1913549075 2230188536 6424561518 8640302650 4152062258 7526974644
8305960600 0866480846 7397587469 : 7371
4031004430 5071100803 2857561852 6003370683 3217834100 6973073703 6403212982
5039545915 6645162518 0163585488 : 7372
9250188649 0417671308 6234741645 1200097805 2671659140 9458666952 9719432662
6136275546 6627639747 9883132420 : 7373
7660024708 2565551686 3465415732 2730379846 3349743541 7605339118 0554958485
1710030650 4714657091 7400082202 : 7374
2033342084 2716210255 0699829516 6962322792 0520679012 4999105979 0172386421
7585369064 2136187723 7109133881 : 7375
4995600185 2273629653 6092063731 6839784737 1498411623 1404795457 1631028508
3096492946 5614289077 7360309568 : 7376
9141375737 2422008431 9536767372 0751672340 2117432726 4670777570 7205664465
3817086104 3049130199 3304747985 : 7377
9635164887 9646194492 9528905825 7309790409 5062772066 6835694427 9880549619
8375127327 4650760182 1215205334 : 7378
9220808519 1762661708 3613572333 4251006968 3104178207 9718606502 7973257761
1571650724 0273651199 0861376703 : 7379
8988665032 7557743894 2275815661 7307697218 3643680194 2195847681 1417575765
7802259412 2936937628 5483761501 : 7380
1371209444 7772678839 0726376506 1893274494 7410926405 8693198959 0585599696
0097204635 6710955864 2594222507 : 7381
1342281083 1610787854 5620838655 2686224968 7755895674 2840096517 0789743998
3645219402 8789699296 7688552250 : 7382
9924548113 1679811960 9584499945 1283017439 5658645542 5349246733 1927699292
2114986442 8842742821 8962343777 : 7383
0714914385 7760805808 4856427303 7171353764 2453067937 8694579057 5233507643
8815218106 5660792707 5157252883 : 7384
9918515710 5260056188 3391085362 2127568842 6153686799 8180736017 0875673642
1701333242 6353061934 6354543601 : 7385
7260388552 5567745806 2131463820 5488997794 3799486592 5280732477 1277015335
6672430512 8745511130 5227076922 : 7386
6215106526 1479813961 3012054825 9553984982 9038221658 5400027871 8281794527
5934575981 6062682566 9353591091 : 7387
9702531758 7895850786 4251238169 0228684408 8555046233 8617680937 4387125500
3062767791 1446042717 5023690482 : 7388
6053496822 0034680627 9793975282 3723519731 0708566623 2473255481 5669120184
6615653934 4860475403 5740573532 : 7389
6497356594 4918990736 1608117631 3352491857 4724029491 4221910553 4669423028
6637390630 8379579526 5229165823 : 7390
0260997650 7409187334 0013305234 3101030377 7850787520 6351658346 2778448353
6836655902 7055834947 9612983505 : 7391

6123943390 2475739693 0505229513 0811046709 1324988204 8315378942 2613223556
6343098114 1830986684 7681028257 : 7392
1550286827 4838339757 1925873474 0115366467 1682629370 1552682244 2127711052
3909179693 1923912773 7977176880 : 7393
1542246883 4962014009 9732275759 3293917724 0864634892 8364644435 3876181138
3414398882 3534353062 3717836037 : 7394
7270922206 7901491315 7494612352 4864980368 0609448488 5719893841 1919684715
2483689960 8147222418 7543173353 : 7395
3374236949 0081762643 6893163678 8534545428 5713963603 2750992289 5712580324
1463056342 8729318699 7117594565 : 7396
8363103616 8351573519 2411529174 5856865483 9747092069 9815889999 7133603670
5125178322 1658704694 1565206271 : 7397
2941883036 7286169732 7759547489 8327009337 6170910380 5515938607 7975183164
7911315109 1832801275 2515936515 : 7398
0488607939 8848586099 3081193493 8012778967 0910847840 8470943154 8945577055
0455036182 1811546026 5023335986 : 7399
6262119075 4572843333 4065465592 3652677967 9009724022 5590376662 7701571787
4530106783 6845974624 2411941460 : 7400
9270255181 6535744964 0803707635 9218204807 0317078004 9503051527 9198046841
4802926237 5914410938 8211758454 : 7401
2230025871 0509897434 0538019608 0967060017 2944613159 4353931902 1552498639
1770306032 5109395966 0382062356 : 7402
4441379417 8842765240 6438997872 1528985487 3908254194 8112684974 0034138058
7702933160 0234975233 1437832506 : 7403
8346832390 2948629612 8917710704 4987706230 8584555830 0725385819 3699956274
9473165520 8956179446 2764776882 : 7404
7751115761 4704990194 8228432386 8027347074 6738628147 5267533372 0121010716
4668619616 0063377135 2137821138 : 7405
8572121554 7739721693 3044019471 6653483026 8285098055 3100367604 1798533591
0839912135 6009460421 1825482838 : 7406
2608059451 1633337459 9933098540 0909567553 1282250406 7646007341 2663544062
4826166056 3576912527 9981301615 : 7407
9259426864 8699500472 4775332770 6499938464 4411392955 8969146522 0429908122
4390628600 0465035738 1269521702 : 7408
2563921682 1773367320 2091586750 4500478898 1687036460 9615184600 4855741439
8817260973 8447212746 4419545019 : 7409
8335593231 7688295578 3064580208 9829527631 5490735500 5465411111 3719476304
3626173297 8590284651 3191614666 : 7410
5516766350 6411933457 5866713781 8109841564 6592962365 7690836271 6072233814
7885085506 3886317749 0787234919 : 7411
1987679896 7463625454 7126044095 4168197799 2207104276 4025154843 3196477390
5581049610 0016470598 9515785294 : 7412
3991917985 1978060893 9567864719 7363890098 5242234000 5442237631 1640315186
4279380217 9526568429 9781428838 : 7413
3213895982 4393958877 0155799078 3944892067 0917065520 3496769986 0541559021
0576514771 5787378675 1371254730 : 7414
4466033499 4239637877 9627117683 6236896244 5967764401 9969786930 2789570435
1674431509 8022898650 1287745403 : 7415
9340739724 2671605002 5587854988 8894099384 2091001681 7387489835 8456289352
7517079117 0549000543 6219621706 : 7416
4294043278 0627987378 6676730222 8520183704 1592135242 2427596283 7503729973
4959143125 1129239685 3190129753 : 7417
5611629672 3084861743 2317638929 8817540247 0391522806 2546847910 3097962829
3430421339 4022884129 3104759404 : 7418
3460835668 3432324956 1364754257 9862544554 4988963516 7136444593 7935735007
0277317673 4884517488 7099668250 : 7419

8104389915 5677099848 8740178829 9074948442 3316788301 8060597371 5188996429
5032419913 9880647782 1402001041 : 7420
1177747896 8839555108 0742111648 0422750798 7793103151 1511084338 3177288725
2558536680 1323192752 3396907869 : 7421
3015925808 5241915908 8376804536 2608975083 4211044419 1172855632 1464612372
0291167406 1466035739 6431901332 : 7422
6735113831 8086854427 5306115682 2006420507 7627624316 3090763395 7468027315
2917799081 0125354943 6125914719 : 7423
8875238666 3993748525 3979788879 3077692813 2087457027 2961219939 0812546255
4625410900 8343052040 3871083838 : 7424
5722642176 7563029049 3017692951 1528857908 5478954134 2729679084 5515266249
7710743422 1537656995 4921693119 : 7425
0035675588 7971969593 6089444056 0279853228 2974592300 5729290225 0191869049
9605151963 4665873845 7662861655 : 7426
8370926229 5490525587 1509887801 9153598544 1716721357 3070063569 7462366450
6999438595 8653045816 9557668141 : 7427
5188637516 7295828551 9470256814 8675216121 3458539326 2425217446 9525507007
6927008328 0535053695 1638370248 : 7428
2620563452 1190864360 9455087863 7455568841 6514747774 4660686802 7320905274
5681753951 5713203754 6217069826 : 7429
2614847590 7106925055 6887184036 3000530513 8681095753 8874806334 6329994876
9553743481 0824541310 9810837369 : 7430
0707510452 6381461571 3736203505 6374120611 5436983956 4311367318 5097174963
4340583229 0723583401 6182906087 : 7431
1658710160 1304215919 1713902364 5897059027 3481568107 2289867195 3025796837
6456398971 8769225175 9331558679 : 7432
7334077373 3043216755 6539107557 5423891616 0702122717 3759207005 6307136843
7478212135 2889897986 6821776283 : 7433
2572533345 1832303927 4761865403 9192087965 6525892860 8193070403 5739642080
9893210894 3220085199 4167611717 : 7434
5273539788 8202661874 4562282022 0531528314 2784830338 6860279965 1751797558
1988333102 5625438230 4160189423 : 7435
3917038633 7835792103 2037786682 1806189086 3264867419 6620083453 7726721380
3004835349 6696010662 2413864272 : 7436
5292530005 3729177370 6116234865 4387360333 5731512271 3930047525 7703092179
9026910568 3369868147 0070046680 : 7437
7131900900 6387690689 4203541867 6510749733 9382475859 8638628919 0118498516
0567627206 3568183641 2579224498 : 7438
2802899653 6186177766 8677814102 9039884734 7627179023 4863835367 1988418153
7764284357 9068363587 9056700585 : 7439
8121342784 8601734299 2731149852 5829426381 3065849793 2171960271 8883988881
3672103025 2937373068 7284284252 : 7440
6768184950 3279329574 7070453752 3032086408 8918820469 2025781143 7470677421
0439314645 2786589638 0704085307 : 7441
4482407805 3943147968 2354362054 0986524254 4613096095 6897299253 2000469372
2765192645 2523471043 8680616284 : 7442
4150263272 9061257192 7257767631 4022420185 6235148905 9030613122 4408194389
6347125981 0672223080 7362487403 : 7443
8823446428 0483917599 7118906930 4948686931 1881573358 9463337313 7830646307
5822060360 7272216512 0394083173 : 7444
6077127229 1089455309 9727713041 3106141567 7302487550 3030331197 4593678235
6969206929 0086840655 1784355069 : 7445
9097961301 2085913420 8252365383 1972164002 5784386531 5526851804 5595065276
2182978072 9270016262 3807539845 : 7446
3178749321 4571767442 8422404980 6304873363 4559557141 9216555990 6931601168
5456804757 8569445825 8146525410 : 7447

5690942130 4151860787 4243640505 0757001169 7845159928 5143636331 4191985622
0392836363 7698073696 4248023175 : 7448
0529550137 3358979603 5987444259 0363690717 2462752084 0312123803 0892613188
0232071030 1404611955 9039673478 : 7449
3221472762 1039428519 1545150346 9923928686 0608789482 7204013155 1785488924
2118587503 2601207662 2758794866 : 7450
1078061994 3166946023 1246670036 6406945698 8033789419 1692790930 5638007712
5569611138 5511106823 0718626350 : 7451
8781301659 1596657199 5616639269 5240132819 3712268673 8399710231 3023718606
1442402816 7500027852 3701329742 : 7452
8005731412 3377107673 0526300291 8854321849 5203772625 1725196654 8985863027
5201985805 5528655670 1741242478 : 7453
1398441147 9456140361 3372359291 0024028800 1695428252 7637001726 0558174084
4534308051 1456901070 9200685375 : 7454
1074980056 2299567939 3703609421 6558524012 6826086201 3989663997 9818266838
1464268876 2945498686 9869560183 : 7455
5872133246 8705175195 6171604085 0360702192 6593490727 0251099475 7252219108
4244559520 7830851434 2748979583 : 7456
1409061138 1368732186 5624780519 9733098940 1100475707 1818985229 3784438141
4345412750 8270984979 1964579292 : 7457
0802355036 4363535096 0125171771 6805655499 8278836687 2030579533 2335489227
3581439095 6051222929 4254511059 : 7458
6156659899 8015880640 0542294318 7694927076 2221110284 7618082615 9644660270
4309729054 9291809577 5775902696 : 7459
2478243427 1968425210 6673708953 9128792695 7103913170 1552419895 6659379628
8840942869 0519523491 9075493968 : 7460
3374338510 8678868311 2974840774 2561428880 2420254564 7075085740 3395398767
4644706472 4122440508 4157459877 : 7461
3169274280 6579384510 8082933471 3697457317 1707801215 0465560773 0879875787
0502442018 2513066251 3284579379 : 7462
6693426746 5917675447 3212872979 9255322939 5654158286 8358625639 2962270161
6958143610 4796463370 1681690383 : 7463
7002557364 9440139581 9022902590 4302917933 0141319851 9600605393 9593015118
0348550630 2148638173 9005927865 : 7464
9377962839 7460050166 0024561924 2505593516 5239389938 5785832922 5917116476
0183315867 3958922691 7986771992 : 7465
6426770722 8044516574 5545012821 7130748500 7367793445 4702487148 8371884766
8824378185 9855683230 0391829250 : 7466
7222124723 3954381450 8124959205 7273858216 3867174121 4554440720 0077462566
7799993033 8858143839 5224684186 : 7467
0609946505 5117494931 2624747545 4114900899 2988802447 5117143941 4717156204
8431661614 8349019046 0011909296 : 7468
2556851628 7760436851 2092217653 7252030663 2610279260 4571211234 6384308976
9100576076 0382050626 9489451831 : 7469
3368129757 0084946503 6278304450 3424298855 8033635619 8454450541 8523948398
9082514808 6679715953 0872371873 : 7470
2952461752 6196498905 9205469690 8404522519 7466547763 4065536705 0839612952
6943779829 7198225507 4008724675 : 7471
7960737559 2988511922 7004014089 9223099769 2507290824 3725293025 3655458496
2933437019 5169448316 0099816953 : 7472
8217539750 8939318308 8183490256 1945268977 2640110604 1349080453 1314501071
3937697105 4634766542 9387332788 : 7473
4965077889 1157044338 9987596862 0677669411 7023532572 1969700633 5598012767
9632173540 5701737387 3456002788 : 7474
4615557553 9188888190 1579378055 4417315212 4711048525 2795976660 8726189792
9914561575 5209797040 4808675605 : 7475

4469428312 2745402563 3219115710 4455429631 5225230436 0826363084 4221033568
3371003474 8286461973 4312032202 : 7476
4724439322 9338029788 3927316609 6657341966 4813971173 2905763186 0775941914
1898287478 3291136684 3302535292 : 7477
5249647921 1019646490 6521782428 0216584448 0119414837 5608706264 6822170527
8855586660 9397308492 1172488988 : 7478
0705529500 4138866907 6368399430 0818877934 8037755411 8045469519 3374369307
4015004386 9156290274 6936145881 : 7479
4570456637 2976279494 4061609311 9931117418 0452044928 3545614699 7128768235
0175144537 3892838376 8007204168 : 7480
0696395364 9250578693 0809325864 3237958397 9185083667 8526870939 4339609878
3481513142 3664526254 1524927558 : 7481
7799065325 6163320812 7186353644 0504910181 3148648797 2806483608 4966650248
9207356975 3621719047 5720554079 : 7482
2696638443 5426213099 4352432518 8309713530 8234078731 4154948096 6574879196
2070429813 2176243781 0817966000 : 7483
0362780596 1686458583 3110248343 3125102935 2663857765 3971473510 0522118161
6567263513 5384136775 5589213883 : 7484
3561375461 6690150611 7537764963 7297641275 7297961854 6006530588 1543374664
6375024676 6187368286 0135938997 : 7485
9661048870 3732129989 8807718930 3455828424 7986325865 0832971647 8813687836
9340431586 9944158405 6073105170 : 7486
0807409062 4232298342 1483645937 5406668988 7990245296 7750380630 8864246235
4000679205 0169402576 6842267333 : 7487
7762347007 3850861041 9471106994 3580453445 5704891685 7268425464 9000937125
6247610593 6696388997 3127846586 : 7488
2443799144 1399515694 6681083926 3252231819 1172864007 4558763410 4562885751
2550568081 5252293927 9259781486 : 7489
1747545269 4776380447 6995955050 6070304057 2307480234 7057423446 7241031229
6534950506 5170116543 1324285219 : 7490
7592232503 9634918024 6543163466 1291802225 6977140890 2123393287 8860417394
1013526311 5203691114 5392003875 : 7491
4109001217 4008003707 6406806582 4627805087 5572041054 2581570457 3154793095
8472208291 9046179453 7539554705 : 7492
5895534904 6238100816 6373194550 2358481019 9222719291 2016177520 2444668605
9009664014 2655724768 4365331867 : 7493
0352206558 0145891365 2142148842 7956558662 7059338874 9187863939 1231094985
6121262994 1292195705 5098215904 : 7494
1459138611 2526656218 7945691785 8641408418 4662918123 1277170185 6076429841
4034759248 5979536413 9295700295 : 7495
3996000447 6524174118 0636090891 0703299357 6012356745 0289489667 2436831132
3482735638 1370078481 8092204544 : 7496
4878697136 3944071406 8381085375 0237906792 5491990743 5453911187 5441697879
7744588070 6127946791 0261059726 : 7497
7850687549 6861910266 4289872660 4155400035 5937062096 1468777480 2119559012
9744347561 9036901503 2298765074 : 7498
4211659407 4948464211 6358360774 2142679643 9831027568 1555461210 5716791531
0722177793 0254573228 7453729298 : 7499
9194954644 4430214743 4944339915 2619976461 7560500446 8761414451 9713784817
6211789772 4143554673 5467024250 : 7500
7347836422 1897884302 4064895284 6314388163 4350752965 3333478207 4398744784
4243948343 7862158000 5295941201 : 7501
6958449975 9566219531 4594638517 0657449406 4454137708 8385322747 5395462267
4721781641 0785971506 2639482911 : 7502
7437331691 2308779475 5840162452 4346731607 6527923038 3361084033 9322785923
0437069614 2336185151 2020163003 : 7503

7106473392 3601594093 4876141393 1578014737 5520999305 7143969093 6141780868
6920081272 9995012689 4063381545 : 7504
9426180425 2487070552 9369512012 0933850061 8267008961 1255740262 9284289397
7981995365 8533467689 0135251331 : 7505
6544784862 1138975793 9533763847 7043789600 0543746315 2679226126 5540350013
2944795933 2038995044 0369123410 : 7506
0895651264 1833327189 6351165130 6511767120 2257937290 6101484235 2467437854
7404696125 8221106399 8326045158 : 7507
1254754677 9293221601 0039689183 5166867336 8303931362 9853299887 2877313365
1400992007 7181159495 8529964781 : 7508
8660678895 6873041297 9681933386 8640365429 9825024897 6083898727 8799683224
7994861290 6215381195 2781175065 : 7509
3500768127 0346380699 8532134523 9607040850 3822031138 3731246735 4408540035
4980559889 7628482182 5502984175 : 7510
1924213819 9529518355 3440312578 4310766698 1982362890 4985956993 9761472019
5040741534 1828849716 3679555729 : 7511
9811576928 9902945365 1754415265 3286097253 1124706097 7431006428 1021572319
9143928724 7182840939 9674138326 : 7512
9759137019 2060393452 4244628182 0941620536 6883589174 5861519946 1028929758
6116332625 3936348579 1650895497 : 7513
4755610887 0430826578 3353395487 2060436193 1381687421 1841752981 8004502462
4013213889 2154135406 3163302382 : 7514
1615654645 3399121918 8665880282 8303673099 4712897808 1654878897 2062587681
9014748657 6432647455 4358647966 : 7515
0505441931 4007833058 1576657539 3391766014 9690172511 0135193032 7641254379
0817246689 9998107870 7432988286 : 7516
0632105428 7292306846 1896542162 5785755601 6125523403 0058019732 9431255534
4214886520 6998097004 3014298345 : 7517
0790538092 3445824367 9438749162 8148340152 9428782991 9084621132 2390876235
0246378918 7701226754 7630879194 : 7518
5324149114 1091253487 7347045290 2228567697 3757021670 2723515203 6832281449
8653030193 3624735824 6400226530 : 7519
1781786230 6131826734 7574556271 9183138593 7038694232 2406078341 5856173750
1481032037 9510019132 6222685916 : 7520
2075839999 2841425836 0755023703 6751437347 2500610845 6521231782 0690269323
2717071701 8071750076 6710702438 : 7521
3008821349 5621142096 6662792576 2947535365 6122311963 0348729799 2396129561
0014411768 4309914435 9806556684 : 7522
7331837711 1148982516 6860528824 3158296681 5773674293 0175053096 1966351587
6755386412 8836864640 1680329010 : 7523
2098364145 3901361820 1231060000 0775796607 6771445323 7449036664 5381903303
5659640537 7484863770 5689195214 : 7524
3606796432 8702651141 9820587140 9962188495 4127590552 8966271743 7163787015
4631182349 9580805167 2638789869 : 7525
0116970574 3325252330 6688114102 3090637160 9653933829 1754247405 7228131824
8528656220 1408429483 5890305433 : 7526
1768669388 6272529901 3743012988 2853320414 9259647248 7084288708 6812156215
5659055520 1187669920 9953492885 : 7527
8695129349 2562058852 9975945981 4735055255 0331514573 4981302496 3566600571
5621294128 8201622999 8481452915 : 7528
8027274529 4336713461 3398954992 5197261429 5467134110 6550874868 7356789905
0526184186 4946700362 6154556517 : 7529
8590074743 4683022033 5306212633 1982820548 1038832943 7727848015 4352985423
4069099151 2207114081 9165328421 : 7530
6286828263 6635610834 3235666218 4583496584 2435114082 5271341844 6704388546
0564089153 3111831123 9118605398 : 7531

7811676276 7613703658 4284818073 1392005624 9047162328 2329910806 6060664706
2640354285 9190079697 3473018870 : 7532
5856542992 1291410650 8310502492 1181152561 0063742292 4329390473 2428585904
3691099780 8736870161 2715657608 : 7533
6252725630 4137946186 8532715908 9484682802 7643878196 8830346901 3986309935
3489043993 9614131385 9646288438 : 7534
5448545178 6146989483 6359302502 3338886138 6605788253 4938204297 5420534819
6605394372 6434779791 3552987090 : 7535
4409751671 4256549173 4186574028 3310563786 8993037656 1815662351 8204755499
4194349715 2154891214 2520647079 : 7536
2904376212 8454036356 5400426443 8858799154 4652650458 8441502208 3481899808
2523762637 8349391929 9087741863 : 7537
3119502333 5535079509 5501985044 2946460037 9207703264 7231653837 1677976322
5510461508 3447057278 6498431040 : 7538
6919552119 0290898190 9550664375 4003665989 1571296690 3738260586 8862209431
5856762141 6946277003 4461465588 : 7539
3709614541 8383678165 3264671798 8704751624 2002769841 0568104993 4797417474
2755566238 3448218723 9492559872 : 7540
4808179028 8697917808 2113078266 7388227175 6995368158 1006834944 1770141103
5314110709 8279679604 5742155734 : 7541
6773729313 1835742900 9881725107 0700102705 6816105909 2561977387 2526295496
7064382697 4876315270 9963323153 : 7542
8978021419 5846283702 3585522802 9775802140 6191250393 9000954972 0689692293
5154870995 7962168186 5172941430 : 7543
0236510071 4102755589 4312623499 9452621742 5183407314 9518226654 1367071120
5435950477 0529840551 6414408231 : 7544
6040094148 5932276718 3356104103 1952334848 4607952364 6610699869 6317658850
2819331709 0927754380 9750296912 : 7545
9282902871 8271868676 5605656010 3883625974 7689933191 8150868463 5102020444
3191415907 7236468302 5539168088 : 7546
9137742672 3321399471 2884977597 9539773479 2296789362 1930992511 2006630115
6617390657 5736837676 3659371892 : 7547
1142236849 9120214755 7197223055 7410057254 5257245385 5565056864 6053711226
8226795019 2963103945 8441745580 : 7548
6521223889 3467377781 1748771108 5358563651 0480400702 3513841408 3943449149
5873363170 3374524744 2076903104 : 7549
6589402807 0652040490 1026101806 3071440950 8901183652 8291001042 6311261217
2301607303 9142782998 2605482593 : 7550
7487197078 9455908633 4217056537 6911559913 3716964152 5291258655 0719274834
5937113075 6268223314 4193075059 : 7551
9400673536 3650372935 5007679804 2151214351 9725325605 6226439454 1411316747
2987783687 1409967589 6563070199 : 7552
6956082462 8014504911 9912407121 8574805370 6939640389 2123464697 8712255066
9606965161 5068130060 6294107400 : 7553
8047075701 3091617350 7733475564 8772547369 1225215098 1350331929 9338334382
1078855438 3323618735 6907160805 : 7554
4558098436 7005743508 5075319940 9765965336 8721043493 3222806188 3499200482
7873811252 7503049208 8881748464 : 7555
2831901653 9600630466 5029156208 9053229473 1999667891 0429991774 5341287689
1030909976 7188148030 9327162698 : 7556
1236572060 4331596496 4934205359 3097499554 6353415998 4382386204 9425069393
2014266637 8368948120 2199761417 : 7557
9860830589 8382939006 7395145517 5735499971 7075389248 0315294109 0118514926
1408759447 3721159393 2892637050 : 7558
6958329918 1008620840 4102864899 2346326034 5642370424 8242570639 0912938080
1704659653 6307241449 2818375533 : 7559

0479486233 5063366363 7588656105 8906603353 1629111765 9790023653 0121728575
3662018901 0164981597 7724729054 : 7560
0226707862 6833877730 1117009431 8514035776 2194548167 6206437531 9687841485
1842589344 8140529583 9963849702 : 7561
5395976212 6359785945 6691106370 6013322795 3341910530 3795404259 7830413766
7516749768 7874652996 4917834233 : 7562
6427000745 4754819494 7135986680 6915666645 8532914903 0823205989 0281205878
1268157674 3093072101 7613582263 : 7563
1899928879 7323093201 0149232126 3732671517 9649554896 8924776611 8431545671
6175749393 8401616522 9078440894 : 7564
2315087668 7054665275 7932388055 4912666169 3777598953 9255610804 0693811822
4800051350 8093720466 8602003905 : 7565
5009141165 3944819959 4173218719 0340971882 5590726295 8428371977 0085817941
8507010239 2475998828 9613573778 : 7566
7564358854 9634256114 6780783340 5187109513 1257599993 1805190921 9022665615
6950392826 1947901601 7023122433 : 7567
0575443126 0654464625 0086866227 4804386674 4194420153 9426038115 5782754000
0404020712 1819013157 4640109342 : 7568
7335748336 1014694036 8545125644 3203470752 8448389317 5456176463 1572209287
8102799120 2902189223 8282470350 : 7569
5463383094 1944779746 9130882924 5720502922 0624985552 3551056541 6430457629
8864176806 2017411348 0892342272 : 7570
8834745432 0119807266 0689569258 8229327024 5447753472 1765528390 8061743002
5428281598 2876747135 9831392486 : 7571
7754531846 8446083804 8150815663 5419462565 8132963595 5059482907 6330106815
6449665178 8032837772 3442649573 : 7572
4762093975 7595579306 3867100477 7439340064 9834055072 3621811989 4844821271
7277785138 9956849044 7627008613 : 7573
1269781577 5722954647 6033592735 5634301851 5256595829 2013820915 0226200880
0317497285 3899821503 9065355933 : 7574
1282828353 0229104248 4991010409 6008081292 2737134515 8145829596 2514381717
5466341676 3065800124 6761930273 : 7575
9397496828 2716096494 3723113557 5629078101 9612424029 8111767982 6186714048
3954668523 8558072759 4056093297 : 7576
3124055553 7304479929 3256493279 8114831833 2021373111 5623240409 2544419362
1712935732 1551955582 6930703620 : 7577
8693910309 5626257928 8494465718 1752916336 1050844013 8556367920 7115718885
7988513496 2980427561 3855008281 : 7578
6695303908 6989620802 9030355380 0625563214 0366817790 8180211648 4387794827
8512937817 1151953129 2060253499 : 7579
3978761754 6408001885 4911265094 7737794033 8081385165 8217736547 4212231932
8208269808 0401490534 8395503964 : 7580
3892782029 2472373482 7169643746 7813541562 3079293493 0724033675 1699252188
9224056696 1469968147 3878851860 : 7581
3142897635 3797624324 2698191015 9665618186 2939220488 0958915352 3367722568
7364384668 1697674177 3574492402 : 7582
7732442718 5217245824 9037944402 8830673844 5640018536 5607465417 5605371082
5468179256 9364696441 8020722076 : 7583
7950152974 6750838413 5231135616 2549952917 6351416883 8458188793 8427692077
0036330816 6744134057 1715043076 : 7584
4455055708 6201962187 5090213884 9321558849 3146361185 1668192193 3331068710
4157219222 5845136232 2199002670 : 7585
8380222348 7321575797 1191139682 6803848840 4028145926 9592328395 9641886626
7961493051 5982037118 6283109450 : 7586
1976913355 7938815895 1141532725 0422495358 8864505295 2518856976 6476775406
4198964612 5632782259 7017023337 : 7587

5585208862 2218260028 5359281446 3086109070 6741716126 1230022530 6229432694
3978032683 5730088162 4868449638 : 7588
0618338129 0631720666 9825392733 9740049475 8569587351 3934547695 1134818732
2641752631 1905776885 9219802373 : 7589
2093522988 1724959823 2180205141 6465423317 4602677104 7957385695 1746726028
0709681525 0437335898 2055047400 : 7590
8030133017 5581522716 9530967519 7201612009 2056623087 7542871069 6458634713
7428066751 6783193735 1325652214 : 7591
8383317367 2308119816 5342239802 6247437655 8947675692 1634368776 6915649479
9894490895 3335482608 6379823253 : 7592
9491667268 9941674998 3477046990 2146408996 8375824950 5829081453 3142200622
6370265889 0856758926 3050621772 : 7593
5045902749 9099932791 9762378666 6529191863 9558768793 5663877764 7427669516
0789608393 1623525292 7832503641 : 7594
5584678024 6181598805 1429269144 0698965247 1948193396 3231368546 3418650909
2841382717 2521695383 6200632300 : 7595
9210996206 2494250608 1118148675 1298160865 4863784916 8389142024 4074612537
3499118074 4446800456 5780723476 : 7596
2101130684 4607797942 2132044175 1848161601 0190843118 5778373692 3028533939
9275611116 0625500938 3801593115 : 7597
1113590785 2162560485 3869143238 1224590429 9772946964 3222737151 8952580297
3366045356 0070753438 0412669670 : 7598
5867914369 3280921830 4113925179 3786077259 0433010536 9386056453 1228257539
4317372233 5852211681 5430443635 : 7599
8497427720 8363422787 9617830153 6250280185 8578484425 9719867132 8342485127
6948108214 8289899874 5431722097 : 7600
7924036083 2619873253 6158595601 4193841365 1762588232 3166497136 6187149882
0913042815 5101622439 1160452496 : 7601
3384256540 7862054039 6847598413 7295093158 1487773712 4018717977 8380478989
9494365429 7767325701 5705381263 : 7602
7852227469 7246787424 1300793642 3284978184 7838418709 5000920272 3276546498
1769718563 1159468301 2099715472 : 7603
7353173557 0252640974 2948625381 4814007859 4209375626 3839468673 7163244650
0669475670 5315994726 8781125617 : 7604
0460064074 4455807429 0199970126 2105369442 8071409221 6615182130 7934869897
2837001195 2930810736 6672854879 : 7605
8578714822 3531687967 4787357526 6193854236 2400071391 1305675550 3388529014
2377185416 4892294155 6716384588 : 7606
6141110633 3383120411 0852764582 8402610255 5584547225 9375961872 3463643099
3963801234 4489255652 9898272029 : 7607
9003678439 1510418687 4958244295 4626212615 5525198674 5094446529 0221964963
2955400007 0385210563 2196765824 : 7608
8422650733 1253620546 2602686845 2266096888 0438380474 3726232331 6166601591
2793688199 5940502571 9999329176 : 7609
7339310271 0012595360 9469437966 3859680832 6643193164 9658496339 7729194414
5183731667 5736885364 5182023023 : 7610
8148263770 6530113911 2891369134 6249132791 2603253534 5919916323 4527758142
4766602795 4795407043 3050957305 : 7611
8571051219 8291209443 1653359432 4084681380 7488753872 9708537544 1682806609
8849350666 9963728979 4431656792 : 7612
6876367066 0592266495 5035210430 8343571403 3518387756 1742096308 6662250481
5251137848 5882570025 0405158386 : 7613
1836164507 6359197566 4564712150 6198985206 2746107207 7885340900 8974133378
8845290560 6432467837 2378442381 : 7614
1624539607 8973114333 3706096052 5973019965 0937439545 6246686628 1243275277
8817857645 0978686549 1389233961 : 7615

4539387201 6099547277 3168875997 3321677118 4419958948 4861426103 8915187575
3635313900 8447216715 9313610565 : 7616
9055230688 2270163900 6046465423 7193404309 8508250773 5015485199 3174718357
3730449150 6949757248 0079084269 : 7617
3598291438 3093138985 5487549423 2274494916 2792191174 4168176225 8515329230
9062702886 2627327172 7137367472 : 7618
8853638621 2322152156 5981401434 6174420864 1223824870 1842176211 3798480012
8918461150 2913407235 1040099682 : 7619
8163551558 8496259347 0228424529 6564463145 8122087796 4148139740 5213189828
7854284269 6578224348 9218621245 : 7620
3402142291 8738248798 3128365520 8388110223 5045871328 9651197529 7239395600
1142529640 2196067696 7957581479 : 7621
3597999314 1849812041 1115428221 6513239255 9070987481 6323371019 1611397919
8131133436 2875608519 1368235626 : 7622
3775811195 2015928301 0980245617 0110222573 6721118075 3783939970 5619783478
5899801039 5864273294 6843133101 : 7623
0946879646 3429787772 2682194448 0569992455 3685460167 5335399408 6181176225
9294277647 1534124000 0276901027 : 7624
4117619239 9224871721 2932198828 2132145815 5833081032 7676656821 5762232861
0235788886 6495575706 5519767142 : 7625
3095042057 6201061600 9270364200 2340450972 0835648577 1816682415 1192694485
0681518629 3519056117 1280365926 : 7626
7821369850 7508774982 2073193348 6197900725 8977582664 1428063197 5955986314
5370970654 7766476800 7258785006 : 7627
3099401087 8945547097 2063803894 8369303697 0026974582 9254188935 6827697675
9232867767 1016980350 9732769360 : 7628
2728831210 3987040472 9507335731 7572232713 9128868230 8969775881 2579145493
7934298535 9447300616 5593150559 : 7629
0245069290 1802949396 5319372764 1307597943 5030186195 1828494006 8279455659
2773381062 1490164498 8342847501 : 7630
5304083572 1432245892 6520002752 1484688613 5202683202 0123565045 1901911872
3235591670 3723297903 4597875506 : 7631
9469277215 7114718237 9966807463 9072294512 2990853465 3261746797 3382168394
0916461117 6212107568 2647338261 : 7632
4614878022 1208854610 9416072495 0146164718 0175574523 1575725649 1655201696
4627970618 3473704619 6119470719 : 7633
3166112753 7770751989 4464868916 1124640677 9934062221 9650686591 3705310362
3391875928 1965741610 7839829879 : 7634
5138647707 4064514224 4930292524 0762368664 3359498661 1393309682 0043150156
1171744028 4570589293 4250488997 : 7635
2302803590 2438855797 1513154588 3284619046 6314888130 0544616501 0619163392
5396324995 6689188558 1207683296 : 7636
1375023195 4026330963 0469405124 7986849565 8727645287 6970991565 7361774733
9190531248 6844946258 7226912095 : 7637
5020707217 0716218796 7918776070 5580481015 1922950743 2633782002 7918311553
6826437678 8757991198 1167968873 : 7638
9631778145 2995442566 5773092756 7143291248 4702302278 6475224533 5402514079
0949182546 2535291014 3935228966 : 7639
6482429845 5993425575 5917775857 5824235233 8973747484 4633400472 5931357292
6244839848 7770176548 1318806697 : 7640
0615022889 3096106568 8203005792 8247768456 5057591640 1812945586 9688032645
8117051128 8126050608 2220110294 : 7641
8137886762 3478883248 2528250394 6688656047 9147243495 9362408575 3393149412
8559506580 2576280008 9063568814 : 7642
9232095192 1640031880 9729991980 2546703286 3218567448 7704766797 4008052444
9412301925 7706168607 9333377261 : 7643

6679523723 1022560304 5607349457 2063560313 1707037596 2721464901 9425087954
2596683656 7349678477 9640178627 : 7644
7500941708 3925965179 2113718885 0269605808 5928715465 6418538911 5112448280
9575136933 2743829787 8402591292 : 7645
4252701523 8018394282 7764230778 6599800714 9900112718 6267256977 9749355858
5882762078 4419210115 0122755574 : 7646
1779763862 7058527065 4598893783 3296495187 7011526441 4216448752 0863294241
0854193186 1114096827 6285129014 : 7647
3784802343 9124804724 6668281527 7274311548 9646255670 8029122941 0470366544
1279619458 7245160059 1100008959 : 7648
9347648268 5734682460 4969511368 0191131432 1302307246 6341637342 5090192136
1990587934 8610301168 4913670312 : 7649
5159321122 3143289163 2315514863 9038920490 7304675379 0603338481 4622379047
6336372022 3176833541 1432973333 : 7650
1141893942 4737931987 5513336595 1923692060 5564181536 2927457172 4953820445
7474709743 3811199825 2069390559 : 7651
2573907077 7436069154 4834954645 4394555065 1751304659 3883516308 4824746341
0990519949 6236797103 5899271356 : 7652
2010978505 6162499523 1890502155 8146844677 2463615546 9783168324 8275696313
5717558314 7078970917 2877833718 : 7653
0485198954 5984382859 6699776869 2505943828 0995311638 0964438179 9776035011
3087073944 8519285162 5490589031 : 7654
1087233296 3148825592 0742740143 4402388147 1254559590 7001193665 4709673891
2587270273 5685273077 7519396886 : 7655
2038706430 5373419963 6785940852 1578977244 3560955383 2097371222 3499363623
4092989901 2531960339 9472142957 : 7656
8747514768 5543217321 6712746227 6803323300 0563823270 7227545294 9421725751
9412510539 1672186433 5859445349 : 7657
2687692208 2387331430 0392613546 6300572963 6733155649 0979598048 9924726693
8645643746 2022496833 8000050557 : 7658
8982685667 6967114280 1876726217 7865576649 1577003524 6110093279 4682563677
1615396243 7927712948 3813797234 : 7659
4729096852 2053123318 2015789584 4795748404 9665426250 2039567468 2354733532
5896674648 2118862207 7302441641 : 7660
3964005743 9589934451 2407081002 3107666728 3429860057 5549363747 4986490855
5630488045 7349808974 5784926319 : 7661
7751447359 5857773563 0772546785 2982333254 6870200957 1974896190 0232497446
8772535045 3317273092 4230889057 : 7662
8830620727 2855498567 4679048486 1180492665 3657381113 0031825472 9987787422
6322450523 5417218301 3256341795 : 7663
4396498386 7938294564 1196752277 2180797079 5055644450 8380435789 2004410159
9100871056 2086545753 5861988133 : 7664
7544255212 7302894241 6530758030 3308079636 0397960666 4242815793 2448605287
2495607487 4090361813 6406243020 : 7665
1765226239 5586836828 9209627492 4609118894 2919022126 9876819746 1064344788
9946335490 4936758430 4851434998 : 7666
0646095449 6732887730 0152215440 2956812345 3448946932 5923497985 5786079219
3479042487 3864420679 2284192513 : 7667
2730049639 0538838157 9374791162 9959337347 1044258657 3721918359 1342311892
4681215100 5641755378 3567312752 : 7668
7720339420 8457733093 2293933774 1474629120 4143364237 5845322780 0104181991
7548416469 0798896663 8034903140 : 7669
4208579751 2767023436 9730901782 0412023120 1633237068 1609809619 3762383531
6642814678 0856607220 8949378140 : 7670
5850826582 5612064156 6908039133 0145037873 7470086191 6342078689 6584131327
3363314363 3823725986 9247856701 : 7671

0819887206 1494310116 0093264302 0534519436 6867309888 9365873618 5274628304
6848650865 8931662844 1742815813 : 7672
4399120583 4318479331 4382513612 5768653093 7757477371 9660808241 3879789095
6863192427 8782136534 1453517151 : 7673
2012415632 1455772612 5717115423 7575230435 6111855891 9663144423 0869366705
1199138153 4032262106 2159439742 : 7674
7120765465 2317651989 6642447526 2047151909 6989174455 1184374331 2560411206
0004810628 3417744951 8999061022 : 7675
1341264453 4006534857 6158006336 4046688273 9220226192 1447571115 9414547570
5652435388 6708217499 5562889089 : 7676
0167708239 7310205194 8388718354 9834328808 8860711036 1769033231 0781379630
5572738981 1291277038 6827993216 : 7677
5904053146 8963259863 9194291520 7644121837 4053356689 5819942652 0915206072
2347870111 5078494599 2637942582 : 7678
7381004560 9373403738 4700526043 2400476515 1033974426 2915897859 4216190276
5461243710 0731415331 3395606701 : 7679
2099235705 5893555586 4373324662 3936827338 1376660988 5613860817 5608552575
1881822982 3656205930 3984802684 : 7680
6892564831 5727203819 6342750244 9053813871 2272836538 1381741189 0381862937
0668796552 7401835160 1106777215 : 7681
4427487631 8669385166 5268919096 6932412414 5765251754 7713861679 0707468769
0502863083 6414931896 5595623542 : 7682
7862455158 7599337404 8688880593 6350947640 4640422669 0237394346 9381319809
4553983059 6379527850 0402818801 : 7683
7151896873 1583106825 4147354750 4832093766 8798878682 0162497720 7965462296
5998697092 8634442711 8784043263 : 7684
4846582416 7244160379 0048629063 3522739626 3264368969 5218637458 5547737927
7081083209 9800656037 8549786179 : 7685
2816823801 8443737773 9659675829 1322060125 5392869706 1311830757 4973649821
0147313999 5137801195 8363046578 : 7686
4586218819 4414978493 7009840275 6006854980 2633573502 7690350133 4915999410
6340615470 7873329060 3707231161 : 7687
2403418705 5078837900 0898176947 0259740812 6366340233 2052347594 9462864772
0279625211 3686448377 8421277547 : 7688
0890607510 2553014546 4807254596 2761137431 6793083227 1444495182 5155320253
0688993058 5319483180 6190976281 : 7689
8016677230 3612166067 8136951717 6487597906 4176720782 7490044745 6671143903
0261848117 0988221888 9166496393 : 7690
8685131934 6911259894 9862477341 9222103929 3185653734 6282447189 8957875867
9611281511 0996593701 0037834603 : 7691
6699083660 5499539314 2099942349 2601951720 0560034985 9404096353 9910547277
3946979904 1357870226 3206920254 : 7692
5498416572 6774279463 9612974391 3685217948 5034061807 7285098742 8122769011
4136816402 8467528875 4276648347 : 7693
1728545123 8985915716 0621106210 3861150414 5324978791 3012138068 8937808878
5741135694 0236010861 1727474907 : 7694
6863458679 6259736100 1991170298 6963967536 0563303585 9850590240 4485705231
4430822798 5585804210 6676069784 : 7695
6685856139 9205325248 7035554585 3701010773 5008864951 9635411914 5399547557
0655262775 8435765745 8461258415 : 7696
6310747495 3677019045 0884604517 8961035630 0964498325 6798030526 3711009034
8336832738 0092585578 3963976978 : 7697
8757455040 9776166609 9027954291 8858930891 7950986812 3115630538 9211681423
3795813108 2941913018 5387664983 : 7698
8432666048 9096880186 0786989969 4639519523 5541223544 4495109149 0427932659
3168916538 4524647612 1029773238 : 7699

0494910269 7206319731 4457687223 0188864547 2310352033 6088032734 5189251460
4283782723 5737381379 9044513964 : 7700
6145877241 5780099182 9452769114 8925644955 4457820811 2585497462 9912272326
4649039750 7964230600 2619452609 : 7701
8123860389 4172852569 7644742742 9289378164 4102259782 7808080938 1188489828
6656450750 4699035503 6823640457 : 7702
0287861926 4007846083 1062566896 7210515107 8759980626 0533102100 1223580899
0878645255 2773927454 4838949424 : 7703
6097730976 8103288181 0483775584 9053575436 9854782487 2097962396 0285646337
0367224447 2560265313 9281324322 : 7704
6027749446 0178011800 6435380465 6177241828 8441537932 7973303165 6426254964
4575247852 2515140333 2921802994 : 7705
3636689200 4502808793 0145836265 5927453036 9320858199 2368582405 8244928679
7047004892 9341036743 2496807871 : 7706
6780528715 5715269795 3873176161 3935613259 3003098518 7458526074 6042807145
3029533444 2483635278 6309151458 : 7707
1116770338 4349810903 4713808687 9895128909 2704710360 3336771077 8140834844
6246860614 7254988365 4767143507 : 7708
8971650144 5276993860 0227922887 3124616118 8868651454 8757124587 5831948668
1960713512 9780973428 9009348299 : 7709
6091613721 7184080568 8912848320 3073780439 9029801197 7674459526 0872450178
7884827582 2031416079 6646845338 : 7710
4013730036 3393522391 3261746422 8397146920 1272934108 0322401486 2978907413
5636204355 1958301512 4060878235 : 7711
7990545993 7055983399 6396427254 2888444321 9350837396 0444126803 4988549867
8014241201 3984799469 4742672513 : 7712
4575043284 1611528318 9343362578 2555576248 6044698920 8108295158 1213080747
2484883173 7971440657 5523709296 : 7713
2046872292 2975057390 4325529358 4811980266 3290934039 8949735809 2902735650
2652096862 8278766926 5416618779 : 7714
3651464999 6335189185 9198281123 7177058085 1290889147 9895022969 8022775369
7818326436 5803065946 0809240080 : 7715
1768907232 5841441297 1928242472 0380048659 0762483298 4326661253 0268510334
9902657657 8579963305 7285781580 : 7716
4481506534 1472910340 1186510752 7575914621 3859575557 9846533174 6524212479
7035629657 6484971699 6877845984 : 7717
1432813641 5062658803 2373537267 2051002039 1872086548 8940491633 3923844805
7175638872 5093012313 7906973034 : 7718
4640283694 8914532187 4796890368 9080091910 7696487522 6793196883 9730831363
2855164428 4854543895 5119235162 : 7719
6705524166 1011619193 9562321215 4044745884 4931776033 2645864833 2699995327
6131156081 9867303179 8777096885 : 7720
4732069736 1110696873 5230086661 3257328235 5451328892 7073584038 1580895934
0954004776 6863373882 2098999477 : 7721
5888725251 3928947102 5761158113 0581373079 2569508911 1060363374 7143004818
0745447071 2358569670 1876163044 : 7722
0248592810 2941244816 5330430019 7678125184 8873242914 6546537476 5308407856
2990904537 3784763916 3410697134 : 7723
9483164149 4317725925 7359898774 1410425852 5650646992 2575333866 7022937999
2990248338 4497963616 5826017703 : 7724
7602404285 4352514689 2938266773 7884010050 4077814980 1065156552 9747650230
2530481848 2902235166 3490708944 : 7725
9876811196 1215106508 0708845289 4029830319 1343694788 6071972309 1087906545
4559290442 6962102412 6508962080 : 7726
0237754863 7921900582 2137541799 4513677375 5029344213 9478343845 4088249683
0055969807 2274271475 2346186384 : 7727

4963300372 0110584884 4426282284 7183566951 0578364180 7090871181 1976341346
7923048279 9919294233 4443433745 : 7728
2657118519 5827581724 2955697536 5547653558 5715371587 8867315582 3731791203
5819433685 9024611689 5493584548 : 7729
4381021820 9805645667 1152278111 1528663148 7979122410 4671503454 1226359174
0231050726 7578156516 9079497535 : 7730
4569546610 2343506351 7892628200 7387671578 4184485323 3126474816 9641045009
3295263939 1072551402 6380972345 : 7731
0742145266 3467912487 9921950424 9506398350 5756316700 2422209788 8845014235
4933264477 0854207329 5642006479 : 7732
9904567282 8827308973 6342401635 5981512716 5585708760 2631934760 3574797113
6732284425 4495461412 4103144112 : 7733
1365329590 7318446626 7811106274 0199008005 7408513360 2171132919 1019148172
3858110359 7632336215 9445906860 : 7734
6885817458 2721066360 7832472110 5539882285 3111623083 0772249320 7187603142
8355497239 9990954386 7479119041 : 7735
2064095816 3503069215 3653952559 3982910836 4440577894 7686559064 2695907855
5889278014 8115312960 5282973963 : 7736
7830522396 5687979329 1341582956 2315501056 1975005747 8825843506 8389480272
0131680054 4924176554 1341553672 : 7737
2967818672 2629752193 5729676215 9745729299 8278457699 9581857012 4704106527
5523470931 6760210887 4620189498 : 7738
3099905626 8035473239 1917438803 5285239451 9685590226 2991234055 6236681684
2026140659 4601661426 8981636489 : 7739
6322670567 1454527615 5208403199 7755218112 2228306946 4028275458 3090788009
5255382670 0117480894 4008582442 : 7740
2344840901 4439309950 7604587499 2960929194 4867872842 4465529426 0355040153
6630485727 8457506789 0334206375 : 7741
4856518026 0586465453 5049231546 3660672149 5597923199 6128389256 6068624538
2196621379 0956141809 4819626802 : 7742
3483737048 6445390985 7960411713 1070124331 4722770694 3816242616 0779174735
8930604032 2161173190 2362901081 : 7743
2414634682 3909101474 7845348126 4736939378 0345086690 2011785409 2269189072
1139084273 6264008402 2560952795 : 7744
3662226878 0363107499 2951896334 9372478078 4246207738 5456462974 6985678187
7946413327 7559594959 1985874191 : 7745
6684478301 8668875825 9850633580 9530473059 2627958788 2662813566 9574185663
5183662967 7963353661 6256900065 : 7746
8834528478 9326123079 4253332120 9343130986 1700139402 2451593986 3015330027
5885744826 1555201163 2165583054 : 7747
0110902479 9131744318 6094788894 4247191765 6585739383 3615293891 6463157877
7520692609 8992237784 2721076226 : 7748
7547136296 7726528338 2604580315 4881842918 3457905620 5585231384 6748688098
7475741829 9514360895 2757114896 : 7749
9698465913 3055157182 4901726406 3469473831 6224845702 9568821347 8321860050
2197579493 2795753714 0194691987 : 7750
1033919215 0346011748 9549350057 6483118480 0383210704 5510003197 2994606266
9081703284 6523263802 4224858174 : 7751
0354229851 0813693111 6248203781 5682402670 5402650609 2762957413 0039459067
4452791979 9664729933 2750032387 : 7752
8534455218 2030969527 7254118331 3194929837 4542996743 4508349456 9207945583
3089572194 1669742554 4682533864 : 7753
6561066514 1619474027 3233068918 0554887144 2397014824 8814130243 3322116028
2289288031 5913275460 4063931447 : 7754
5650451024 7078478270 0310083062 3983379035 0592085387 8236743848 7469071591
0716613835 2709755851 5843491321 : 7755

0350122255 8659285742 9605107691 4689252902 2359382791 2711007177 9878737897
9020601040 5379271265 5426027857 : 7756
2353850665 4446667038 6663145087 0991655715 3712069207 8442628725 8032345585
6531437854 3259039018 1683860053 : 7757
2754965792 5726069037 9957598881 3473068888 0747447547 1119322401 4850571325
8006452814 8510649450 6217879717 : 7758
3665367581 2418556178 1818650925 6591508967 8588385373 4600693514 9766940200
8541424119 5861577381 9902618019 : 7759
9575558106 2254836164 0410620763 0901758560 7457388473 2645071333 8575654604
1732533711 3028013789 4079385793 : 7760
6433995336 2780923121 0839700233 0283646942 3386202991 0331376974 5072519281
9304481060 5361488981 2723727553 : 7761
4536451517 9843869973 7572972344 7961754122 6385432501 5231780839 7255726878
5022568712 1681353425 3902834781 : 7762
0191915420 3432425934 6091331136 4251073193 9943370384 3238861375 8805682715
6164854936 5380237850 9039348336 : 7763
2248227318 2074099684 5329664200 6626803726 8782026486 7057829851 1928450302
2158150530 3564175689 3775421063 : 7764
1387943008 1690861925 8742869875 4675509234 9447396156 7075925019 1683691129
3037748049 5519909703 9823363826 : 7765
3648913505 8430399648 1957236211 5740772576 8233699070 2374463935 4292702053
4604480383 1023144481 1459137953 : 7766
8474923915 7294312780 7412044060 8030975872 9669731071 8198441066 9334082604
9698415157 7165929551 9486558164 : 7767
8675779423 3936898827 7590035789 4212761604 7047342116 7218792048 8769423319
6384054499 4751115932 5479617623 : 7768
9384985382 4019647708 1852094864 8559242287 7324966873 0030102462 6104466769
0852545609 7605252767 5426097221 : 7769
7580423153 2157673532 9075348153 6411137564 0334493764 4773157010 7441776235
8231865170 0354720537 5940431782 : 7770
4599133598 3391817819 0963298981 5139125754 3191389833 7254405281 5081545703
1022896802 9957722750 8331211659 : 7771
7704221998 2689658324 6941224511 2832949898 7297002528 8692782008 8561271599
4977513562 1524144621 2011857260 : 7772
1525246972 2909151404 6407531921 9928173428 0327618599 6457025802 1156017462
0906358907 5186175753 0000424919 : 7773
6720092148 5676401596 9761597040 5736942590 3568270576 2172698082 6125845805
9616706188 6633284026 3385648286 : 7774
4261038593 0749007326 1464768345 9041578797 9304998205 0590432130 0238863933
0297873873 6196275533 2185958012 : 7775
6622163258 4664077546 4765793633 8085449649 5709655491 9468161205 1155694391
0807968135 3938460597 5006717950 : 7776
6310256122 4795271996 0465708625 3843127219 1945200310 5093191236 9479545856
1223792958 4674483220 8038385192 : 7777
6631532382 8450706222 3554163557 5201604845 8595787727 9643566407 9289773151
7789254324 6016374195 1653324854 : 7778
8662125998 1179724635 7628056178 4916758392 3257722366 8439231390 4636250636
4023028317 2300137855 3839784405 : 7779
9603568332 6842799266 5461606721 1095960638 8342429504 3002336365 1087339459
1424660824 6878679142 2410972599 : 7780
5512135343 9193233952 0885700159 0380446982 6631106954 5853638464 2995474068
0406796096 3340089659 6476123350 : 7781
1181547182 2156520259 3800562590 6215027290 1222426040 4147261584 0342781626
0545987338 5724563714 7779216047 : 7782
4827443344 1112967312 3767611986 2978669863 0802417950 0473099054 8680786839
2900997525 7836056842 3252734512 : 7783

4938846464 2611743784 5901765446 9540965159 4708729976 5071127626 6611757869
6019032897 4286263834 8064602583 : 7784
5183651959 1472871025 4160903417 1076266775 8150465169 4625449579 9382899617
8666609857 2735726068 5506695869 : 7785
0375723054 9955720906 2621713947 0396850395 2363129448 7087415247 6422506065
2166970795 5283041281 0868604973 : 7786
9229536592 4315417259 9393270748 6079566741 4593870684 1232750041 6690860257
6148758879 8762359367 3475462930 : 7787
4596834705 7818243392 3747399924 9931821377 6303837529 9038515104 9213862685
5948622828 9802909860 4098401166 : 7788
5073222899 9532240562 2348477842 6908827773 3330025209 5707163878 9137572836
1113161508 2824058677 4806128378 : 7789
0997658812 2450097507 0984791439 9252401416 2240532859 8517840602 6775338192
2826784952 7886672456 8933375945 : 7790
1607319568 6216856063 5120350246 3099283741 8507807604 2047738430 5325483876
3592412557 5316364522 9316351178 : 7791
6522282328 9644483191 0269247416 3633999280 5353093665 9261798644 2495494884
6174813289 9568137313 9483095949 : 7792
9581124735 5548837387 1125219033 7005154680 4023677832 6154424065 6690622404
3635791995 8919161873 1985304828 : 7793
0061974222 3209883669 3647084018 1232141747 6276732239 4733060040 5172628593
5485411882 4613991225 4599360462 : 7794
8969714134 1652030935 0257355936 1261145139 6476466490 2106275445 7374977144
7155080588 8268603135 9661575978 : 7795
8880825134 0841751240 8423142188 7595702639 6166667684 2178850151 1665029595
5978618415 9605479201 7855481164 : 7796
6541858311 3141209127 8284596904 4480818199 8063143890 3805220749 7099964459
2680426847 3445415578 1033449320 : 7797
5950156320 1963053976 2141807399 0862708480 6430321780 0247609014 3977156423
1200722635 4349327379 9157315518 : 7798
5911006524 7277485199 7101984779 7665568944 9196716486 1686707903 7571706648
3562808032 5964867673 4045842056 : 7799
8650181937 0242695253 6481695581 4590909365 9078076868 1243863993 4256553533
0465056785 0655486555 7128211838 : 7800
1739659983 6335767803 0887104057 2972063848 1947048233 3079352209 0608439862
8018386708 9529794945 5903398039 : 7801
7505682344 9353678374 4414698853 8880452810 0813606255 8329519311 2119347517
7628452066 5192222527 3866969263 : 7802
0025665605 7137987467 7472263219 1103954463 9384612184 5585775692 6921127446
9656540715 7141418197 9492296144 : 7803
6403902152 3936521713 1970168237 9101625390 5630796286 7690370366 9998720101
0551972758 8390490266 6193971648 : 7804
3481149742 3911738704 2271429504 9803918440 9203535350 5646482647 0908831627
1727043914 1214238422 9216009271 : 7805
2912360067 0102952490 4826889528 5813608494 3203538041 3569645159 3092433827
7301069829 0740036378 1910721422 : 7806
0109190692 418/216027 7555804593 2649015215 0594142335 3135286277 8826912685
0570087709 4417670821 1117803391 : 7807
6132310778 8454769589 2356286062 6769063681 1548086194 4886505985 6349824207
8522482102 7355717116 8286043742 : 7808
7930019053 2421473269 8076035933 6634855924 6819120239 4714809212 3328418605
0062585379 1085553950 0693514325 : 7809
7214182173 4241696582 8916871047 7383110597 0509981349 4096939955 3351724534
6454725301 9452500219 7042367256 : 7810
3394199259 5913903004 3726038976 5339723331 0189273199 8036678696 6546139807
5380015007 4994058963 9656960444 : 7811

4634104888 9101877254 0175813648 2992701898 9318743514 7571951190 1643262456
5142006231 0747654521 9553062209 : 7812
4090762265 6396770318 2232859593 6090281625 2306279768 5291067138 8077127464
1902570560 2446123783 0789026485 : 7813
4860064900 8785877722 4582811024 3449387066 1477445356 7937320714 5334080832
4961036860 0316487116 0455763852 : 7814
9640009288 9905659038 8078720153 8168686057 8453602623 9537615843 6587641315
6895420184 5166804253 1125090612 : 7815
8057586051 8512612612 6372701960 4236219016 6991290755 2891484255 0002031663
9433680320 4057114116 0423757980 : 7816
3585659563 1247499469 5698495135 8676171716 1486134211 8764921998 5740236045
2581553140 0876092991 9647446288 : 7817
1338290732 1688226913 1573555149 0184217278 1677852130 3836603763 5612900768
5445315638 8109058718 0694081778 : 7818
1907895745 1433595938 3903513995 9232215815 4787478672 6203342273 0377952548
1013433488 7011269882 5270694572 : 7819
9704429254 7385863958 9403162524 1817680103 3883738925 9782856379 5133490722
5664398213 6040806076 9512299054 : 7820
7942207745 5869779933 0344439044 1999735760 7975028020 3525292863 6562271663
7530104348 1541454332 7065339165 : 7821
0995946735 8711349333 8216698948 5670557380 7822002173 1209391530 0782652612
8684514242 9072659990 5709582548 : 7822
6043035452 3299996515 5578146176 8952857780 5366548948 5926317291 1706262038
2979801608 4121593188 1886962902 : 7823
8287157978 7179368849 0402576691 9694789119 7016642307 9447116300 4796916602
9633665411 6567141292 6658386416 : 7824
7655842583 4662650669 0416995107 9819904156 3897096165 7836067803 4580385887
5490439836 6811079462 5922809484 : 7825
3989003473 5745765722 1278020507 6636124247 4530272832 7791975368 7094602692
1102641285 0520464771 5113195909 : 7826
7597847520 0954067737 2174118439 9590135072 7293059963 6253552751 8365125842
6047228081 3050170163 6458298879 : 7827
5296387473 5239442289 8404127423 0766926575 3891912937 9270227358 7158906780
0740485921 6483963090 8392340398 : 7828
8400951287 3222983061 8530714944 4124552897 4454318958 5611874000 1809527520
9628513711 7256845726 2195438787 : 7829
8592562737 2240011892 8592109735 9177445309 0991376018 5710513365 5268996097
9827966913 0166471364 5669707323 : 7830
7081484653 4938788898 1009948329 8222045461 0020172043 6127512031 5381165829
9101511869 4304911447 5937441511 : 7831
9921482776 9884667239 0919809215 0851582445 6105200887 1854606987 3013725536
3463299564 4554638726 4452326955 : 7832
7824381689 5531108965 3134069842 0412186385 0069053799 0290112965 5602973049
6460548219 6018497714 9958716016 : 7833
0186321879 2857765551 4826098688 9146641786 7435548639 8857672164 0798099307
8464470041 5892529812 1200807754 : 7834
6940444869 0228534770 9395086821 3234073387 3815211764 0444760483 4555293052
9995893020 7165021318 4557608516 : 7835
3782416290 7159794086 6179526322 6509087062 5000097852 9459880341 2110825577
6072014188 7727490710 1283893632 : 7836
4373447553 2766109946 2982029247 8457279595 9958669060 4600010182 5538486745
6565827575 0715858729 6867716751 : 7837
1148726521 2206589305 1470298169 9114593575 9706240523 8204272016 7609089719
3644031047 1427018901 1229197278 : 7838
3281602113 5597563255 4869999224 3388504519 8582826291 0381822612 2655992591
5970829197 8623319316 6881062975 : 7839

2541068411 7108625987 0307144086 0838890816 1734018394 5217867616 9102178400
0705721511 5331818346 3981090485 : 7840
5059841908 0653798403 9319676526 1825490144 9628263813 1368683019 0537277629
0221408228 2532050315 7581449110 : 7841
5214969340 0800591718 4288674742 3281889208 2210987075 1009609422 2717960611
8687523108 6513054879 4073032475 : 7842
5855158572 7129568516 6711502658 5753721524 9702073566 5542874804 8188381689
4380929471 9392497078 3426911981 : 7843
6210471308 3095702791 4404488752 9446522288 0916426932 2120891437 3532634347
0189414186 5498758085 4438978375 : 7844
4276111604 8219449857 3399194641 6345105882 7596415074 9708686870 6533308767
4685517086 3672200413 3550690898 : 7845
2270336380 8724832477 3623842767 1272998279 0004723750 3365691359 6485058948
7117971773 5895375235 1727045329 : 7846
8808976870 6037171689 3813114926 1179907638 6817572277 8250303799 4810530997
1413321067 9111769390 8249785759 : 7847
5017347343 2748633566 8431549974 2555434287 9813872888 5884082866 5550630311
6923034262 4006519246 1820851294 : 7848
0513841094 3833830994 3373235611 8408341561 0952547445 4266787441 5394527438
0854781574 8171805542 9230172577 : 7849
0825421551 4552311689 8159841576 3033104102 4867859344 5139563372 0568858890
6905711908 3999697995 2133448395 : 7850
2287592283 9778656931 6970545027 5524986037 5938138006 5895257056 5487153992
5942499971 7317167976 2518307807 : 7851
1800651730 5152956338 2632793212 7180626959 4806341649 6910211160 2364643235
8904772434 7766517250 4370188262 : 7852
1158437644 2266662047 5127825235 1887021789 8027267280 2771182277 3152103035
8726420824 5289884331 6831579166 : 7853
6499256821 6114499034 2466951532 4639135704 7807685268 9899848932 0525337210
1246744855 9451168698 2516508631 : 7854
5065590233 5060894613 3259647585 7404743071 6451021988 9646668576 5093083300
4568187275 7241680767 0592160435 : 7855
5468100929 2559395648 9533295027 9715672189 2027325708 1506277090 7173871031
3238424960 0991967666 5726212488 : 7856
7134992056 9723586933 2407381117 7055732346 0902157984 7223452138 2771579580
3472205402 5896644153 9053412137 : 7857
4470050890 3387153834 9390783651 1458079202 7210147906 2979899235 7303403249
9697624060 7889732184 3108824493 : 7858
0826619080 2622049990 1128597492 9556277217 4646114689 9392940105 9494209733
8175943828 7802504440 9968753401 : 7859
9038860177 1701245912 2767688467 5715236965 5212200577 2424503653 9737696902
8517858272 0108433598 3398454456 : 7860
8178406257 4904319170 8774382410 4031867141 4204172036 8114298950 3677256546
0710501286 0183188433 0590267041 : 7861
3591446899 9688471783 3470408291 4681241547 4309300998 1538425850 5632760473
9087689049 2793224924 0953499190 : 7862
3416211544 6082138983 6454362734 5255421371 5789152132 0603765643 4612877334
6423265279 4907883819 2386444755 : 7863
7705950091 1944414225 7604748727 3346424607 9592462981 7046415346 0651330840
9571476487 1552128608 6578470735 : 7864
2955176812 7582320953 3816373144 2904741798 4026893595 1138362336 1903652219
3686948953 9292986105 6065435057 : 7865
9702715511 2426086430 8592728205 7320382452 8633753600 4053715977 6632209913
4912226229 8215524644 1341529456 : 7866
5515770151 9189869240 7298625805 2706984831 2895480796 2543531974 1712858425
7661140282 0095349691 8021677875 : 7867

0906409638 8938910484 5046640865 7503729405 2210411280 0095473196 6004650043 9442168485 9422090144 5242927222 : 7868
5584905936 6382440277 3328362645 0303053353 9930969877 7034309071 2332576249 4149601610 7924287316 6852638845 : 7869
2751267060 3005707641 0952614298 9664881812 0396063873 4084985550 0941732198 7796675327 4996201186 9772569041 : 7870
4469020624 7765655350 6566423759 1469173330 7274891543 4058300807 0722148098 3167748193 0217368453 7002122864 : 7871
9151994480 8643918607 1365067385 2662802901 5686800738 6584826956 7625421052 9864116593 3869248253 8337878752 : 7872
1349222969 8895497703 3420478617 4098071621 0184415161 7722148191 1004373337 1169367430 8773266973 0859901037 : 7873
1973977949 6102842916 1868412757 5699139864 9211424787 8143531088 3070128710 3774766452 4240630432 8370837731 : 7874
5256119447 3055755237 9109846177 3531290644 2410447089 4516690488 0695091460 7318268944 9278788849 3093950009 : 7875
9507022903 5343539213 3255894765 2503280710 5285424124 2946878306 6310827995 1403992648 5381943346 1802956014 : 7876
3112432856 4846965317 1701739125 3878974666 0971453942 2772741011 4423938795 8959878910 5456641042 6814789116 : 7877
2685728035 9967830398 7097866634 4440474495 0237805374 9408916979 1738537209 7470734527 3979460572 4927590824 : 7878
9278258335 0682569083 7808354569 3636681739 5591500548 9117122945 8934250193 9702089639 8720423360 8131092995 : 7879
2781885040 2771287385 7443547058 1971449057 1304159192 5075571511 8568712451 3861708461 3762199481 3881555082 : 7880
9388483756 9145259023 5639700631 1126012211 8004370384 7770706721 5788291447 2504703857 1195085880 9347550637 : 7881
4838293531 7775380885 5894117419 2632937495 4300085072 2248392228 7543944226 2626959819 1189695630 5252790993 : 7882
4629461536 0936163549 4478791750 3777137415 9070623175 9672455836 8532599842 1558810316 2562444276 5549902532 : 7883
7501763136 2943127796 0119164305 0971695953 2112992045 2717744965 4633602651 5365658288 4193638793 3775355221 : 7884
0007523306 2499389170 8860305513 4982454203 3286014988 2251892444 9789713654 3887876073 2583968694 1239838808 : 7885
2043346266 1172204379 9813307423 5136784990 8458648166 6628620111 0167877285 4932272589 7317962900 8389380937 : 7886
8368862243 0928160362 6918073213 5646537153 5050488691 6477877918 5842773790 8188613147 5435737043 8396416108 : 7887
6684544778 6051856988 3643545539 4146744883 6225690758 2297656680 5177777484 8742973152 1740717222 4089962520 : 7888
2713446380 2893843314 5251698363 8073117992 1467836533 3421747888 0597728426 1602035843 0828924451 4390795487 : 7889
4192059946 7031093767 7692734600 1157263547 8653303787 0508825586 2616916954 7527360426 2765242628 0382477111 : 7890
2903565310 9206137550 1041535163 5002947736 0280304265 1560687044 5034565865 3273748883 1388713636 0128083391 : 7891
3126850405 6973058262 3675060661 8539761570 4174217427 5940940477 7662958560 9930464445 3696935713 1235331390 : 7892
1844731016 7028536689 4180350236 1632619326 7872577312 1497249824 5087303034 2047625852 2565871384 4171927986 : 7893
4813496990 0336823663 5925882060 1075142130 5993540511 8239822139 3595842421 2886801556 9496013163 4265883922 : 7894
9035170296 3854931431 2026857517 2092434167 7167595367 3854595891 5630415154 3408885004 1606705334 6654082813 : 7895

0700832897 7614882496 9628924016 0421141615 1036179777 9321029713 4762657997
9402154699 7803877847 9401598816 : 7896
0852142135 3530414979 4348531891 9528301157 5006555852 6423618771 6442360830
7897345825 0523293243 6402643759 : 7897
2459180056 3290872305 0762970364 8435508696 7621331082 8770177163 4937764001
5740770619 2909977311 8327157053 : 7898
7209205365 4029707767 8672665327 7933085996 5800109221 7362389041 3584279519
3350888221 4543651911 3434014342 : 7899
8094289321 4788848307 3872577049 7425332464 6609483535 1665781767 0744653783
6480284709 6926101318 6450293122 : 7900
9616599437 3555820292 8820234450 7282771381 3735310873 8681566684 6805841655
3952406315 1157367921 5491126324 : 7901
7750126529 8070868653 2078718046 1286792428 8077555706 5292782594 1943007588
4278012095 9101663775 0414014314 : 7902
4565160939 2042139955 0727498787 2445710941 3555504501 5722897094 7059287154
7857842375 4955553068 8277162786 : 7903
0681824647 6471999729 8003673328 7720747522 6535910397 5496408864 2547652414
3137886190 6622713920 3374830355 : 7904
5382843444 7644598243 6340653278 1363195780 8343899184 0207295633 1227424040
6329113697 6226856540 0010623898 : 7905
6047816686 2977778403 1279489991 7073967230 6752216295 0965287896 3150378284
6767916109 6745584887 3593475491 : 7906
0866919617 2246037943 3265367175 3266796427 5759439960 4997668066 1204001653
3028504949 9728607042 7542509372 : 7907
0773858637 0041544102 6059524493 7075850036 3784138186 0779567606 6806461723
4295007523 9776514594 3294896719 : 7908
0018394508 5350711525 0826284545 3452737577 9691224833 3719234276 1582030801
4628017675 6282502407 1460250971 : 7909
6056309317 6724593076 0486882487 9005384715 8052075074 4820305264 9668981999
4025157949 4232115902 7841043254 : 7910
2583533717 7788899239 5174636657 4326137285 1811005292 7573466418 4818518156
9944340408 8180376875 3519740663 : 7911
6304783372 8440558181 9299431249 2820699574 5614238266 5305098134 4664300969
8835798899 9333648249 6472266767 : 7912
8660948382 1820069176 4477995154 1006483149 5092340231 3298555020 7601297659
7654825143 2214277265 8066657557 : 7913
8245211435 1183573372 3773049243 7142595702 9585515871 1563888077 9956709924
0416783945 0785418474 5979089815 : 7914
8041308296 7546815203 4660396987 8439383081 8392374771 6278977138 2844434051
3409511684 2773409217 6907252476 : 7915
3108922444 6647891168 7943251322 2007365153 4562311855 0138742152 3354450308
9388539861 0146110690 7489109566 : 7916
3596629481 5046854675 6229289415 1089237294 3193195614 9182318256 1129060258
3642878172 1949291753 4538823527 : 7917
9936285889 6363679518 0669943553 9473796067 8838639432 9921827855 6841255867
7995497954 6133028786 2146465157 : 7918
1399165914 0284587348 7027052610 6981706256 8063586601 2816449797 2466681641
6196987988 6773684747 0963519634 : 7919
9675767724 1171938898 9116184101 6757249065 2812301639 6816931500 3082520148
9745797388 9001526824 0237779830 : 7920
9724004631 0850649974 1059120950 1570775841 4148918052 8324673061 8351409517
4913707885 1259753482 4769265004 : 7921
2368546235 8436015970 7296182804 4309168026 3700578592 1511372201 2165857223
3654137444 4765941395 4964395531 : 7922
3251616096 5801809920 8779083525 7903757726 7506223225 1858830607 5693231895
6337473318 0715366869 9884986386 : 7923

9143703937 9137969021 7509625869 2842571691 2905420020 8155469777 7269088469
1552481174 4999365707 2604850439 : 7924
5080918943 2853044749 3989630060 3185187727 4582105246 5577410812 2548587461
3984102455 2315469728 6056159900 : 7925
4919730133 8745304316 0232464499 2651250170 4259895981 8255652388 7587795429
1090967219 0883553577 7249868613 : 7926
6615280495 8762154417 3261930411 7924969971 4236480394 5636309930 7208351692
2495396202 0083881511 9183724445 : 7927
2275263668 8920992957 6677095667 3835188966 1959616053 7535008602 3233648287
2126727945 7182592717 8717732484 : 7928
3095828273 5739819410 7946656428 4153735209 7892288900 2103731720 2758664290
1183835024 0103826226 9418554517 : 7929
4623600553 9053368444 7990378023 1493500910 4335155598 3611865012 6049569602
5913457693 6587622265 5770632075 : 7930
1802591029 7097449295 1788854202 1192971691 2663104142 0914010786 4252386132
0813476345 4729775317 4408567322 : 7931
1032254607 3930167741 8956020272 9920316247 9753768627 8078459710 5213268572
6453736242 2553192509 5186171876 : 7932
0134511243 3762990535 5546195917 5254804303 8904417376 1791316962 7434487853
1155019525 3699780779 3447126384 : 7933
2488993435 8191719567 2961221652 0534622308 5534104546 1753516028 5303531706
0140881213 6373334586 3747449007 : 7934
1286834136 2130746643 1499196264 5876623832 4794568554 9526493664 6501737302
5269911908 7348893838 0822049031 : 7935
3999050562 7544398890 4463472232 5658585987 9118866887 4080697010 3202682039
9164596539 3982763139 7675758675 : 7936
6306611912 4852892485 6994955452 7475825958 3805926698 0566909315 2791187730
2059789706 3137870893 3883220709 : 7937
5089767335 3008555615 2945784936 4066391245 2296012669 5960096262 4923623543
5006717356 5750932462 1100978763 : 7938
8785940003 7811122554 8186286591 9130265466 1216074980 3532560234 4356367631
2411815896 4697385615 0829519318 : 7939
0065715412 6306228994 2986318094 4708829996 9778236008 3731932759 3124537302
9308031970 8391679211 2382203775 : 7940
9386912820 5704012577 2444861579 7828970323 7209824314 1995219135 6933990701
3764865723 7940901897 3102677314 : 7941
6477049154 3124633531 4923116498 2846119122 3204432979 8798865455 0488550281
7224127196 4129621425 5244081433 : 7942
8331848355 4853164891 5655594728 4549415951 8329931788 5179785723 3334982197
1674522093 9822794703 8948309387 : 7943
0068880950 0950017601 8651075682 6190182122 5987911067 3576574102 3718866272
4699351075 7585531487 5850606889 : 7944
2653011018 3871898352 1117215435 3512759382 6432969022 2339493804 5976378558
0907317163 4956875032 7089865704 : 7945
5091638820 5248079423 5222753568 7082560260 3182812925 7727003799 1853012635
2379901061 9425957035 2197978694 : 7946
6037934090 9076816903 9374240329 4222448564 9062429240 6904135555 7978176280
9939784678 0875087378 8927217350 : 7947
5415721686 1062524353 1297175287 0170021713 8926526862 6196686871 2382220018
0819981055 6711287217 8866928665 : 7948
1408685730 7314203026 0294175800 6147543396 9961445419 5780173471 3074532518
4665024873 9186631226 5912601566 : 7949
0236378626 6520161741 6091315672 1283704302 3831858011 8360359563 7611440219
6341988517 7735871538 0279808392 : 7950
3962306959 2894450288 1270771132 6597922053 8364752490 0638650538 4673526547
1279601392 7057425335 7239928439 : 7951

2566821190 0759207612 2827272386 0856213827 6534993392 1722883132 8929444455
4185591185 2332513183 0035127913 : 7952
7925207603 8822932907 6831775728 4347671056 8432799744 7620079687 8545960902
3037060154 7443724261 1467397637 : 7953
6141025089 1819342787 8230418831 1556795804 5412431192 1034413514 9026909550
1799996042 0723226099 4274936196 : 7954
2125944584 7015799426 7384082691 6807034200 3733428173 5882422048 0345776191
8055384418 6706109278 6475446871 : 7955
1090427656 4909613367 8966932061 8650990481 1679659860 5503404920 5961741584
0318373662 4449878891 0350470616 : 7956
5000925499 4233945662 1656246048 6362752367 5719584621 2709710103 5867837376
2859455190 3569480784 1844204611 : 7957
6427432686 0787480844 0542518712 2732222002 6214347989 5429528267 4932403815
0098184804 6570078416 1918386349 : 7958
2574776926 4605691767 5531836754 8227892033 1802726653 1093127116 4367933819
6228009595 4133047870 6842758615 : 7959
7839815966 7863531221 2649107416 2834036375 8489867323 8782661376 9035930136
2324296156 0311663457 9503348069 : 7960
6041468095 9998514167 3315641663 6610638135 9333702910 5580951861 0025965969
3358538133 7026672093 0385742574 : 7961
2167864290 6364254627 7737124516 1425008963 8653859527 5801322289 1865379060
7932844115 3123957194 4333059711 : 7962
1674626649 0238943266 7179611307 9458231044 1191900797 8388171522 3000670094
4002400638 7592269436 2285108398 : 7963
9082522875 5833967504 5354327424 1265645394 8010664801 8793370264 7796786435
6301541959 2850256243 9369802444 : 7964
0700414898 1012850384 7612557665 3219679989 4997511651 6655256803 5436426047
3149980694 1576711495 7179087061 : 7965
9144799725 2271479529 3279630735 3587611207 5829068056 4592707528 7717611731
2662945676 0366335713 3534836225 : 7966
7492316090 8849973395 0008239080 2270834018 4421860239 7156226894 6975901121
1164176352 8196178800 2013550898 : 7967
6069980355 5039550769 6017523557 1734913676 5805036634 8098917662 3747277794
4388920986 7951693893 6159532508 : 7968
1576104264 9807718788 5831916660 2587545585 5802071249 6020356901 4360241160
6765094417 5116399552 5117604635 : 7969
5458354558 6837333273 7835385586 6576517556 5323846658 8986398386 2957934979
8658078391 2028837322 1654180551 : 7970
0150451839 1046892095 0564429261 8713072718 8975971985 2246213142 9519367378
4547251284 1913917280 8431950208 : 7971
1487214486 8180912262 1419507962 4858367872 5120084959 0524925366 3525575397
1301282522 6663193126 7472711708 : 7972
7401687401 9818205300 9569012107 9949369324 0463863410 0522606283 3597847775
0993619741 1021609915 3341241745 : 7973
6322272914 2898453154 6044679324 6316585082 0599008444 3044529235 6835613059
5857276984 9765100357 6373809488 : 7974
9043789814 9713389441 1215534047 8285383410 2515862734 5220929813 7895801512
5267959050 1688381107 9779373785 : 7975
2314075453 6423832677 5372591139 7231685879 2687904119 5675255583 2558943262
4746316959 5932379328 0773972152 : 7976
3413165102 3326955510 0425671347 8165687894 4325901020 3076129931 8706117131
1147244684 7380987616 6132982158 : 7977
9523733673 2356716715 4561417555 1623880797 1200015234 5311199454 1783387246
0867591965 3156964945 9754343526 : 7978
2414717377 0273516235 2186908458 8169433264 9543609630 6903278636 7827217343
3787234212 2012779145 3073615284 : 7979

4291764593 7575452766 5511736166 2493776258 3412326694 7038661801 0995264161
4144694368 6665541214 6535572510 : 7980
8823132604 4430205947 8295706469 6420840506 6397206393 1736507351 3173003880
4452622596 0365484863 6596942695 : 7981
1483721919 4053748866 4822469551 3800141661 7534413109 4397669630 4892594972
3208317530 9962614307 8452993318 : 7982
0165084520 8032363582 6442101627 4310136612 9855870542 2918469185 8535800025
9657093610 6888670740 8368708490 : 7983
7612579947 1641309963 8885977280 6057254383 6777759459 4948790395 9265587631
9211032276 5058387304 3998704354 : 7984
1540772170 8153371228 1697253470 8480460678 4815303398 2742535460 6736133106
2553648078 2351719373 1976382923 : 7985
1976068793 1863997393 9709329235 8465259252 5487847560 6695873672 6025074964
6865932053 4760333026 7022587893 : 7986
9971559384 6707346382 2210871654 3115987110 8323635901 5069170462 5433514220
5615002339 9316362122 6931614164 : 7987
5163467224 8700890857 5597842867 2090870503 5059539110 2855612564 9913517596
4493878468 5362718801 6450215023 : 7988
4656608575 4400621293 2808424563 5478066713 7045656851 9259335922 4746434449
6813893964 0957288030 0774156674 : 7989
0902382614 2450163971 4899267150 5880621404 7363887717 7652091612 5774301134
7519545220 3749089616 3122104904 : 7990
3534651801 9000909935 2151964047 1458864773 5602146512 2479417805 9004235182
0766796507 8580026385 1082616667 : 7991
8848558950 7491586451 2661930983 8305834826 2969071317 0673107171 9728080776
8484805796 5875444137 9929449107 : 7992
0474718297 2801787259 0170340333 0009230392 5380610416 7548466988 9573135068
0386861742 6399244369 0005077832 : 7993
3469824069 2427033122 3442533816 8695545685 4241317436 8857672121 8523392455
1451954366 2928877490 4004394559 : 7994
4665507408 4270251388 1542258012 7799362494 8276125530 1095759410 7015575245
7017401978 8509769995 2595617208 : 7995
6894340915 9725896349 2835929491 7641877073 5118875573 3523909753 3861390461
1744197549 1085823789 4535948817 : 7996
3409788639 4506170092 2735583569 0040585622 7237806491 1621799257 5263587677
1733782545 0819281108 1021590238 : 7997
8311849529 5970302890 4546355239 2776037669 7180464550 0936638939 5271293756
6183649877 2135574322 7583521179 : 7998
6176974028 9374567771 1710209601 9708609988 8919748227 8183406451 4963240191
3419859546 6675186304 3821782298 : 7999
0536476808 5964134878 4882765133 5906070316 1660080738 5645023674 0863265104
7956555552 3905693241 3659061767 : 8000
3191577439 4061383800 8162286942 1478366058 1227966013 7868810364 2239344928
9901558277 5604767214 5288752696 : 8001
5184984097 5725311412 6969799968 3583131116 3623016679 6763740682 0847352207
6481987519 8822986305 2801822887 : 8002
3740084633 5839839835 1191797700 5782448199 5867123744 0728854255 4258820673
3038659351 2348849979 7026231342 : 8003
8093258461 2061873144 9057393342 9526185556 8315864723 4650823356 3111689903
9298074623 7301858961 7512746201 : 8004
1024135025 3016504735 9873912996 6877316578 8793314681 1462093006 8822230887
2061583309 6858364793 8117175872 : 8005
3221188044 0756360456 2665101608 3443677231 3414846187 1266939435 5809447299
2220580954 0034174771 1692494857 : 8006
2249761131 6668715094 9060130424 2787861170 0218095472 3913179444 1841677024
5763014022 9750183193 8044884481 : 8007

6047770140 4081374189 3954547596 5527353961 4603519113 3930122313 3299133292
5650798965 2308090598 5857406951 : 8008
0691920327 9334733972 6614955417 0561661608 1542952178 7924208501 8926158908
4277089990 8154102051 6351796459 : 8009
8157545596 8806942601 8001303557 8554702715 9876006807 3314518161 9407836722
1223851733 1773178365 8569999622 : 8010
1593848758 8392248992 9110687127 7741693137 4725695651 3343006306 5227379245
5105626692 1880807811 8325867313 : 8011
6960112090 7714665794 4366288330 0144081586 3098631726 4190633304 4044508058
2281128154 8891619832 2932731818 : 8012
7999588512 7378387293 8509699050 0690958150 5599686237 7712942317 6197294040
8139418136 0272405148 2082073795 : 8013
1363148433 5422793933 4277653686 1035254077 2583999654 8618746068 1108544159
9378263442 1838393844 8584852903 : 8014
5304569715 1099125044 2313825249 2529540089 6002776538 7386635741 5638827431
4110235995 3392227466 1428089503 : 8015
0235763018 4738996956 9687857964 7614813385 2746260001 7284007661 0353359970
0110742958 8617909344 8568823861 : 8016
4022120262 8100987841 9477855669 5767060208 2972972849 3217283594 7885074785
6268835413 9604520503 2733900115 : 8017
9768997592 3488237896 0037047435 0944012621 8101094815 7151330500 5269673475
0515957930 2414192467 7188377990 : 8018
7897096741 9856206206 0336577626 0662003934 7514589512 1231388967 6805215858
3214858756 2100493982 7433779291 : 8019
0008563459 6084787274 4334605561 8482163033 8722910556 4330934672 1926478665
6314689814 4200381872 1109343897 : 8020
3898630381 7212957909 2991127430 7671977729 1578376748 5462892852 1808203967
1437850793 3637369500 7727664266 : 8021
7558541999 5885697258 8894171885 4732457041 2654538727 9293835554 2304128413
7013213116 3168616764 9503182078 : 8022
1233528078 6703605753 1692611111 4131927428 7790446455 9074531083 0304595611
9947788929 7649507064 6397168145 : 8023
1462945312 4294363555 4886863612 4882656124 0064409327 0462720991 9380249935
3896059744 3351284906 5936304283 : 8024
0104058174 6061232043 9445967861 9943758818 2107887055 2909724031 5088518044
6976720704 9321932642 7327474947 : 8025
4657352706 3140084600 2270606955 9676968769 9358161220 8087614289 0824382822
5082716265 4510150765 0711225487 : 8026
7678312710 4898581258 9031480801 1532348223 5485756250 1636134324 4575548970
8401852562 4878558411 6155906291 : 8027
5706400186 1938267871 9167145422 9784270131 8556353007 5923391743 7150422523
6871430802 8697389723 0659408747 : 8028
2615583200 2885605643 0543675473 0126975992 3912140930 7820516347 3056839443
5766080505 4953912914 5864564189 : 8029
5198421266 7557032614 0401652646 4718191682 5561780500 0436626398 6812632605
1567703737 5763228255 7237224033 : 8030
7326563439 9261010727 7749134762 7176169780 0242024174 5421963354 2368342108
9982050983 9969309009 6683351729 : 8031
1556051800 3291237502 7974138336 3465570545 2783485374 7615770729 7860282642
3169895661 6215364150 6119169118 : 8032
6958507630 0887739841 3314433235 2216108648 0753762756 2795357326 1927680750
5917919753 8220805055 3608592143 : 8033
7064255882 0407949945 6805320998 4521619529 0348747172 9158788043 3789802719
9175755284 1249516455 3897366903 : 8034
6679774585 1060492373 1647493348 5138913107 6704781689 3150928005 6525080740
0336026255 8797465873 3895886504 : 8035

5549105944 1871151078 9894764277 2391413200 8590273134 8230514711 4710862568
4424430556 1264573131 7975946690 : 8036
8304134034 9698418952 6412808130 3923795354 2339551451 6436858329 7176727033
8799660705 0275380884 4666613249 : 8037
5208470595 6236035132 8876356928 3104042351 4230736898 3022576189 3982210451
4711649750 8649579313 3102305164 : 8038
0070274041 8722925525 3125550507 5196263090 1419042881 5837860801 8722688533
0334260324 1887419947 4309487098 : 8039
0900779689 6221521278 2040363871 8498691155 6892852678 8514369208 1119896699
1991759394 4222765985 8529607316 : 8040
1102727719 3039228750 3735276746 0313649872 8601485908 7801089027 6964810178
4192505137 6836394432 7798589783 : 8041
4996707510 6459160807 9978149867 4621295942 9929495103 4556123952 8518996648
3893151341 6634256253 8722904207 : 8042
8570031005 0576672387 8249942293 0433187629 3310759921 6334103488 1853078945
7862130438 4469902259 7151095378 : 8043
4248396404 5195613952 6499409105 4193480278 3338141050 5497503863 2521344166
5253257834 6241054313 7242051343 : 8044
6561007097 5026474975 0601879329 8970049526 3510803848 8001221708 1134423115
9233018250 6552329000 8053296095 : 8045
0529674761 7827764990 3380464927 5642460784 4006224775 1022984090 4464576088
2840084383 9653757251 6330908693 : 8046
6595011316 6805903176 0653994130 4678318604 5006298095 4641179322 2760643945
4509746403 1190004046 1013636375 : 8047
6652514763 3922056408 5394881156 5665156414 4995469997 9896155170 8519238964
1735048616 4401960282 2837023105 : 8048
3195494455 7157779195 6758859753 0366609882 2024968864 9624023737 6974149089
6227826267 1716470905 4545896434 : 8049
2513581072 1550993127 8700181094 7933780019 4288116971 3845597813 0343759117
4498305039 7272005088 3851002048 : 8050
6452766917 5972913171 5166904145 8699094829 9639788578 8471241582 1518191513
1193646592 6550224699 5225206586 : 8051
1237108617 0777529724 8009628557 3811172643 5043977239 9159135333 7279437121
4939748639 7926247924 5746309268 : 8052
8060161268 3328108013 7307606789 6388530352 4762479577 2962330363 2060797531
3117003215 6315773961 1414172670 : 8053
9684579198 8967823317 4980250090 4092769839 8894869465 8406197088 3711871716
2678950598 0379928186 8790835967 : 8054
6744386739 4110895371 9698618297 6738458954 3924411217 4432607281 4344156833
3778396580 0933241626 5311975886 : 8055
3863145759 6991628226 0607790333 7395378068 6154995433 4396662534 0413723933
6058376135 7136631484 9602284656 : 8056
7534731883 3395976940 6024258503 3382282696 7160251488 2648376703 1393661503
0465569440 3590997958 8086530647 : 8057
7527207618 1214153941 0681855315 1735880056 3179789508 1069982071 2326357146
7145311627 5825250080 6200983530 : 8058
2998598765 3456158772 6559057767 7705235068 0073472116 0449806488 3912344314
9699456629 6883810992 3179375961 : 8059
5151983265 0120598781 4416179328 9005863320 9563904496 7327409677 5615657921
6951256957 5734306501 1177279446 : 8060
7699754846 3963676474 0951978746 9857405002 5666304304 9693037828 6467308890
7406762172 0816291000 9270841088 : 8061
0219803664 6903200660 4898295265 4419736165 0808708999 1399726806 5256227964
1804946454 1564487670 1280058394 : 8062
4430038777 1355707804 6564329913 2187060632 2081514275 0622404635 7396332303
3417129205 3079551861 5942218913 : 8063

4174393464 2698600396 3040050800 2042870415 7295937353 0995771667 7615251624
5240929006 8764840944 5244870829 : 8064
6372347946 6340834676 7889242904 5159840508 6071695341 6892727685 7026100718
9617528775 2516051145 7579234875 : 8065
8289688272 5005829643 4832572501 9488704916 7329327414 4146938616 1100821207
3014894509 4220915611 3531966015 : 8066
0600951530 0421518796 8005009256 7027746169 7526917366 3358847895 3104875162
7925415229 1573647418 4880129370 : 8067
7600826420 0657418502 4170407054 3171728758 4945349675 5969256058 1068000917
3553233924 6553976580 5611506266 : 8068
8571458886 2313687782 5027075077 9389298522 1289837675 7339444061 2389601946
2548014968 6695371667 5692027945 : 8069
0041465728 4567068530 9943119591 0398723702 4927090104 3532552598 0501157121
2263284707 4507459034 0596438552 : 8070
8208729191 1849839551 6680440487 2154753233 9125619945 9302810597 5250334168
5465676518 4945399888 5243277578 : 8071
3503720834 7774127463 8187405458 6392925447 1150555432 2799818208 6818796986
4701046561 3411868005 3965131346 : 8072
7232381249 4414047414 5964713110 0723335520 2812338522 1405072386 1536931660
8946059307 5675862667 9075053701 : 8073
6110407034 0909262826 7488775028 5975237510 2385452677 9620553060 2752777436
7819405779 5338407659 3719123390 : 8074
4161229938 9658775405 0420933165 7806959653 4125892927 6900210583 5347234056
1065145554 2932009170 7450439470 : 8075
1995669836 6277022470 2288304266 5069045512 1943837056 8467083823 5730118163
2285434001 8430494979 8582621768 : 8076
4156199179 1993512415 3094915801 0079374774 0403927636 8882693841 4993758057
1751713447 5581593103 2745774187 : 8077
5763711877 5713224689 2272242493 3777711395 7558484346 9314320561 9352281759
9764563384 4635236679 3269416636 : 8078
5293829299 4731554160 2622781043 4045971282 5822426701 9760969254 2335680025
3834894656 3112495170 1468994004 : 8079
6400911880 7408707732 9964303094 4043581834 0841417580 1847899502 6247389569
0754781926 6522199349 6740541728 : 8080
8191105033 2527900215 8545943175 7345887327 6958639774 1338302654 2251574881
2728581889 2732849776 9691108525 : 8081
9385652172 3918404258 2582314470 3106476046 4814502022 2102865333 0865637440
2259588047 1321869450 1443668175 : 8082
3853073158 9200519065 7548015823 4491548443 1777081767 1252046119 6493945018
8625676374 1453768645 6006014215 : 8083
1261292424 6459673064 2846488510 4976111918 8467204986 6454759999 2680177566
4337900340 0568470375 4475628782 : 8084
4934898408 7140077023 6296504170 7501045403 0069683112 3531729340 3026317221
3170861762 0296043445 4820954348 : 8085
6763695249 3941538476 6753809577 3551214637 4990805038 2956467090 2052304673
9855643558 2380712001 3405179535 : 8086
8177592661 1792974601 4869213202 6306003931 7995648494 4444731611 9372536942
0623475637 4504782675 9397321270 : 8087
3372805781 3431266211 1738722249 2806333639 8345193036 8038925930 9688995209
2520352785 7198099238 0576236122 : 8088
8496065232 7697097076 4790391394 2314021333 6869039480 8687684743 3080122688
6994620347 0823716309 7247047250 : 8089
2810603314 1647014474 1205421382 4030812834 7191009308 6202697490 9153607225
2751286597 7495195070 1446835957 : 8090
0342661460 2530281533 8482727764 4979007768 8980744583 0108058076 2580282652
5398467121 8499044394 2589188021 : 8091

8806297599 5631428163 8485112094 7134995242 2729096625 6213105720 0278602521
3027165243 5730201321 9950531711 : 8092
2041933385 6243218115 9685353143 6428098866 0109585436 8526010852 9344637841
1885371826 2721650745 4141542909 : 8093
2412576342 8168346424 8035818333 9986607357 7309409498 6065706615 8440701673
5064684551 0834483041 0340714330 : 8094
6886135064 8161231335 0084423362 4141744252 2038471620 6858157780 0344074422
4089977379 5250772272 2425221632 : 8095
5270798286 4834239360 3199360077 0157814548 5499732791 4957152420 4777183259
8625574711 7217600047 5918686134 : 8096
6576680074 7913882388 8252605950 3696145637 5550948364 5524943318 4937579583
4470825651 1202977130 5489059858 : 8097
8936049041 9843661556 8028063910 1693150794 1239844261 4146314527 2074906342
1674666562 2082073387 4054410275 : 8098
5527732642 5726662603 2543414164 5180646462 0135910746 3690481409 1394881734
9150190905 8878658865 7560805463 : 8099
3191153957 9519922691 5101752611 3553536548 0449153805 2352430920 9568391339
2366418602 1848657405 6531386585 : 8100
8918038882 9010049544 1053950428 1908521709 5689709852 2266049149 7703442563
4152011335 4359560162 0804833457 : 8101
0567858284 7518288694 3264572847 0941550706 6390042021 8474284148 1507293211
4255525384 8768239840 5964036946 : 8102
7558558706 5077805288 8118437054 3707668539 4020436679 0879472757 6997674271
7439082411 4225151702 3195532601 : 8103
4929205324 8523316443 0336141058 1348081211 8784454016 8004984186 3719537750
7912250981 6883959439 8437434905 : 8104
0092769074 3049721973 2166826907 8681894298 8655432603 9347996027 9152866631
6742452049 9357683219 7598296140 : 8105
9060960027 7500754116 1182341365 9519543647 6826597435 3872503433 5575607603
0942645937 1768443525 6584561894 : 8106
1330436265 8549333964 4881366781 4189940353 7812518636 1809861431 0938046860
8423894176 5717091983 7530273529 : 8107
4605318217 7484980676 4100727806 8935394877 8562080766 0764628864 3497578138
4916754969 4801333616 4585535923 : 8108
4218744439 0487282202 3274827900 0438097437 2224090556 6579827925 4079201827
1917640731 5686878899 0869823443 : 8109
3324298461 7134807794 1579260789 4378693787 9321819276 2383208856 2198256262
0537061570 5336509986 6604373352 : 8110
7800430297 8538582577 7258208143 4930689509 9746902842 1232148054 3625214410
9926584513 2086983569 5037934121 : 8111
9278770154 8376684625 8848514803 5328210057 6225953123 8439549984 5597718366
1774696948 7713373602 9490243201 : 8112
4008936954 3524642846 7039062829 6129253331 7560905458 0152415336 2697046334
1560947714 7038783311 4080455327 : 8113
2026698516 9165505458 0885262347 2270091856 8381152913 2516075575 1473298310
4817451169 0118544738 9050027079 : 8114
6628135737 3328915219 4735131715 0714211819 6223950640 3606625079 3766101141
0909757198 7519506279 0767199208 : 8115
4156183831 2623278705 3677949587 2093480853 1033700370 7967698144 3339865319
4730055395 5036137371 6903540441 : 8116
2274418482 5089725543 4113291414 0992101050 2794060781 3434527454 5748910397
8703959255 7677197303 2662920585 : 8117
0021841243 0561101471 1726658858 8758581109 8013622427 1917836015 0563008450
6126095803 5114622186 8970392656 : 8118
0067677522 7127811508 6918240931 2639832388 9514330730 9208061818 3670533272
2219967122 1382439249 4133845030 : 8119

8553742913 9510060612 1406401527 7299204721 6247469060 1996875361 6129309594
4353196702 1387026224 2747134252 : 8120
9519834641 1502152585 2171907334 3528760508 9694898596 6187372464 7148445754
0042632111 6913066108 0007590199 : 8121
2768527723 1229433845 3753734455 6192707318 4207700352 5188819851 0000910588
6970246361 4133946373 6403629167 : 8122
6485024313 9592226108 9143124584 3108024918 5337663654 7379542810 1980063946
7549165673 8822093720 6795502245 : 8123
3974952793 6043218760 4167371252 9418689041 6787560507 1581918462 8397499517
7618984701 2014172849 2253165997 : 8124
6424913961 5534630455 9673700366 9827139244 7461722405 3775638845 2506057383
1172206330 9952246138 2751543842 : 8125
6248418305 4546161423 9767958075 7579289295 5358231583 7910599881 0001363558
3580253819 3049608748 4840623244 : 8126
2130196731 2857279756 9638089158 9114151062 6829367132 5339044322 0443935122
5627134258 3571937580 7555547099 : 8127
5280586509 2748914856 0615014905 8672218630 1814539552 7154337744 1157484301
4604542104 7237106590 9374589455 : 8128
6018507470 6555125049 6207976349 5262962858 5741910611 9868433743 5959902767
5135124284 2162787382 7040736179 : 8129
0182264211 6190565452 3559821346 5009844346 8474989342 5519419061 2568539471
2155093835 7783997339 8535979920 : 8130
8044091401 4013019692 0588481624 1657792074 3850616592 9884295428 7592676532
5764612786 5813653869 3023064324 : 8131
9514872291 5683419489 3397732577 2383110186 0728599213 8274399553 1670717977
8694631024 1209656299 2515636907 : 8132
0709797657 4622302248 1118369933 7954003728 9040035528 1383879166 9645146305
0174466783 0267733915 6584851313 : 8133
3733221345 5591201694 2819944635 5869119901 0467025724 8863039143 1318926902
7234278807 6559567143 5508100852 : 8134
3328736761 8883314330 6258402854 4613817909 1649151138 9868861020 4244109694
9303994618 8156434692 2592382271 : 8135
5423732556 1865763731 9111383935 0829844737 5787630908 8819066974 8750346261
6077362010 5614764919 5858889526 : 8136
1757302986 6000284492 0883309869 3563896669 3573658315 4321980511 4630291803
0353282391 2512265145 7960451129 : 8137
1362048141 6074263483 6890487253 4774146049 7928968665 4718043110 9637020693
6618003815 6492646017 6322823546 : 8138
7525772510 0634810833 2460496397 4581894856 2507159279 8514006882 3456587664
3286169649 9447028761 7278607401 : 8139
6278793760 0313403053 6537200163 3610673119 7354021657 4274552661 3772464146
7970163423 2210948392 7352053992 : 8140
1618466844 3165780514 8578748738 9956117356 1574229271 0799789378 3041174634
2331723123 6870629887 9939563796 : 8141
9323438140 9307730316 6738558758 3365894879 2192292991 7029253219 5313106313
7516995648 9555627920 7732633443 : 8142
9956069913 4012345566 1866002723 8644091295 7768174567 4556204269 6966397914
9248546317 6155558347 8849112031 : 8143
3298081093 8202001667 0821426195 3839391351 9494332455 8741577258 3694811634
9214048473 3546704852 5276266990 : 8144
9591444857 0908316042 3866343355 2347995241 9332174312 7082642757 1501537381
1447046489 6147792343 2391883659 : 8145
7604173276 0691839647 6067866322 6749291933 2316113187 7673919132 7131564491
3056937058 5515339505 8229226297 : 8146
9366928009 8890127374 4011072991 3075848391 9483725163 8752515268 1209356155
0689661281 5652785043 7743856737 : 8147

0659686571 2904074504 0213967864 0980501628 7163242664 2267337613 8215295623
6522040211 8843091692 4496201670 : 8148
3983772402 2900775191 1990173388 7256549451 6702488423 1446673301 6979564931
3895385861 2398111668 3082572022 : 8149
3337246985 7377872517 6730167468 8527011564 2775820059 3935709812 2586901258
8927727534 7751245969 5452503898 : 8150
2611668021 2877573805 6368563156 4442199458 1874028106 5680175318 5564565295
5822886169 5528627420 0281964030 : 8151
6143910590 0321537979 5396930218 5194232688 0794685279 1407258771 9484690411
7641691522 7472110682 4390965834 : 8152
9368174072 4391256726 0141392055 3750443877 8509718690 6128308954 2144509045
4534852381 5226121360 9136327962 : 8153
5618713641 4316494221 3935544220 6005338273 4515307986 7230668801 3562930131
6501765553 7704776300 9150624792 : 8154
1237391937 0575413872 6037784409 4217302591 1250492386 1549381900 7039732269
8270844593 3093837163 1480611283 : 8155
4117948631 3084619959 4307896849 6411168708 4337933440 5700795264 7802553242
9988270212 2395890715 7262154491 : 8156
8831198911 7964922556 8725795187 3764372652 1041961886 2359708076 3708251916
0197482234 4820994433 3236500401 : 8157
5103308033 4509873420 8212792141 2004204801 8050107979 8561723216 4735044013
6981148554 1058811972 6087395394 : 8158
2549086287 3783903746 8098832081 7221479683 0743130269 3815436368 4537845761
5575201994 7911774123 3670859357 : 8159
1769209592 8402888000 0629172087 1627379774 1535129580 5029709944 2419138076
2872308506 3578557022 0029013432 : 8160
7092777298 7375157616 6247484047 3928515508 6316421530 2820835265 0155756311
9590836823 0734030439 2715181050 : 8161
0275265003 3708689498 4288132356 8496500724 9398847401 0569473433 8056373384
0238243607 2538873309 1113738840 : 8162
7645000377 3447847091 0018648045 5411711002 5614054087 8369928869 5274139261
9379085138 5229297895 8106283980 : 8163
5904499241 3872737639 8571948291 2844834765 9740140414 3259818850 2453910670
5917683246 2269483973 6115718148 : 8164
9852877065 0237921321 7892695361 1637930446 4677102539 9165684431 4443202029
8116859366 1665925520 6559795684 : 8165
8269167629 9777340317 3737803087 4820811784 8767254586 8373342463 3452419941
4107801536 9212415987 7833997466 : 8166
9973954295 1868182009 0649697610 3678215280 9895298760 6995710356 9096028371
0085997898 9894633487 2095708021 : 8167
7923366574 8707146777 3673609724 6399221575 3219402369 8804256015 0000758404
1175731957 4984167403 3303526039 : 8168
4809737885 6539028866 5909901358 4031964803 7329389860 4594203986 8966162227
5489436550 7600023458 1410708961 : 8169
5093946926 4527739423 5114694011 1277191491 3692543785 8539483527 2648575521
6020872048 1249169101 8527179951 : 8170
8375607328 4426677184 7679540711 1621801535 3987918247 7498161858 3564645228
0308632399 7713849928 3592092705 : 8171
5530786651 5564527689 5916218692 0347015645 5573402876 6926384132 5770360160
5964596904 0322551231 0202955997 : 8172
0989164180 9071291457 4898624430 2977071138 9414628364 8484871576 7856779487
8143125124 3293550247 6872684689 : 8173
4693387893 0968619372 1724524216 8739271859 5032641171 8319813570 8927075601
0792991851 4562750286 6219243697 : 8174
9841781141 4045403196 0916424784 3921439027 8338683726 4171155494 3963041940
5888067523 8187742910 8551788002 : 8175

0788117679 1477877311 3638802772 8566969401 1827275547 5020009359 2740494837
6919641741 7474376430 9935882811 : 8176
4902415856 7172118654 5181954032 7446308508 4900186721 3652717887 4236275133
6854382584 3572609765 7633985935 : 8177
3648790620 2036888810 2701585005 8083028005 2905250059 2040995400 9559933828
1560551578 0805692775 0376405932 : 8178
4228538216 9458924730 8372693348 0345155440 8908180300 0097135903 1387603695
0646295255 8119323319 1042017397 : 8179
6968511078 5451020588 4117474295 6674264086 2762260667 7216076333 8326092443
1171632661 0442814595 8606250699 : 8180
8686074775 4290412444 6463286078 4627920804 1631617188 7736540720 5146931212
1432516923 4294535433 2472824172 : 8181
9604332479 0290635405 7442643551 0165761886 8875755289 2490583073 2266885794
0636107263 4713145128 0390967940 : 8182
9368497933 6814875713 2986978132 1192473059 9496014027 7818648319 5060983999
4902924848 2264519690 7669493668 : 8183
0918529246 5402548007 7219908917 7291960931 5660548678 0271484478 3766043906
0031471464 5229672337 3821018725 : 8184
0391731552 4586995542 3886136497 9298934057 3091186898 2296875692 2198783526
7888481656 2012179932 3557181193 : 8185
3947899729 4113162503 8237716074 7601225078 9091391360 0736981616 4496155077
1156247518 4867186427 4187522209 : 8186
8369926251 1079458767 4427102605 4910183771 4149395384 6017308993 3449360476
9733019287 7913802703 1350707321 : 8187
0981882099 0421774795 8244675990 2008357116 8178448830 4787391475 3449351938
1401175208 8137059843 9846549557 : 8188
0550101894 7463317851 3544780605 0024532228 9869595610 9812744537 2003405068
4331619519 8363031817 0374789862 : 8189
6632707244 0653794392 7775061173 8437739100 3701560450 8542441718 2946223230
9874159926 1379231030 6397975196 : 8190
2906214954 9367514934 8295534255 7326734057 3662545638 7820877247 8017012519
1080613807 6329419104 8376862061 : 8191
5503990171 0577549379 4371981498 6022745743 2768735866 5472119372 4736483688
8473863695 5047230458 6168757789 : 8192
9262417991 9242888086 8466327656 2154894537 7584642693 6601705490 4106967913
9658565047 2314269792 6688920856 : 8193
9296287842 3316515400 8797079484 0446062660 2059142071 5726414911 8427257725
4451851792 2522992289 9892552417 : 8194
9313295561 6293338761 0416084919 9666317442 3087794058 8735083947 8730728309
9169773983 4979436847 7463448015 : 8195
7906910420 8387495354 9261759171 8541229359 7518990089 2226217219 1445930678
8619760356 2218812780 6388132245 : 8196
6355560680 6941525529 7897496905 2102555871 1666880394 6253165815 5902647017
1799000399 0248334019 1847418611 : 8197
1779142508 7957832200 3743133899 9438878790 1749321783 2857091362 2593452398
7992119413 5829648604 2387182406 : 8198
7417098622 8121901857 3336815685 7032359430 9199084131 0032742416 3085607040
1839676454 2197565982 1525834228 : 8199
8679649061 7533288038 3375925188 0213557889 3511916333 6359125653 0761963446
8859555567 9031931342 6753383306 : 8200
6316434060 7697250337 4612045657 8575427494 4560577318 0947349044 6364211529
9853304353 6983846609 8461376693 : 8201
4308461700 0549083610 1239165487 4021552018 0743832046 3746783904 9999351856
7810792289 7258746903 6774957913 : 8202
3853503513 3607550732 1325600291 9103155131 0728317711 6250772551 5312573049
6232722086 0225214648 4022028427 : 8203

9859329283 6688799077 1408192398 8378503562 2052945336 1568028203 7531395403
3176431589 4056325311 2358682394 : 8204
7289210428 1477709045 2957049128 7593524120 5135868760 3284514258 2734694488
5456431933 4456061039 2771958212 : 8205
9219413108 7466667665 9245749138 5391579446 3552398869 0676799695 3556594034
0083926630 9489156370 5155183329 : 8206
3940005032 2641708848 8994601744 9665907668 6847228668 1018342347 3358074816
7860825699 2636145794 1434157737 : 8207
9697276195 3625148160 4750464457 1393528525 9217578395 2785644120 2769631859
5151992537 0647385437 5073484198 : 8208
0458523109 9524566402 6394525671 1483053596 8678311138 0040816192 3406923403
2609897041 0456803793 4836010544 : 8209
9292069273 1323910928 9489416122 5287725417 2588071576 9800290232 7692992653
9672222312 5954237897 8187961729 : 8210
7369231269 8629041329 4069304779 2690340796 0859369695 5308287063 3498853589
8063170379 0655102345 5503598110 : 8211
4722630784 3243788750 2680866528 6661970123 5843107548 7193446996 1351102465
3823076326 3859894685 7435306560 : 8212
5827530017 3591298306 9515639541 9433781212 1118044070 1536153143 8457987316
6672036184 0466592291 4000728615 : 8213
7047092318 3888837125 9011377511 0154672815 6831261731 3584544495 5569340401
6062515912 2144520263 0400731678 : 8214
6242123473 9841566360 6157002295 7579512596 0678909491 8071487219 9201151338
6033420124 9033432690 1903152471 : 8215
1111770537 6749037925 0704771080 0078072437 9721999748 6780512705 4386580977
3836608455 7141592231 3112507699 : 8216
4385057495 7208447061 6176464289 7016734248 5313072653 1131411329 6922449973
2589869379 3362053823 1043046881 : 8217
1672702967 8179353151 4792526227 7150873178 4134720117 5082919918 1400243851
6518729332 1922367139 3302183937 : 8218
1801828431 9234620686 1224877124 8161446437 2623846672 2738716927 0434322667
6282133595 7208657592 5299335704 : 8219
6930167137 7062937231 6184028128 1466540339 3929976479 9450835563 5293134044
8149974608 2157314367 8277060209 : 8220
0362000236 5614794817 9616695338 9820330364 3151976058 9388240313 1045864624
5390394575 6376887059 2794191446 : 8221
3513165656 3853034138 0551160738 3011523496 5016026289 8331175266 5408951427
6454873943 6429000310 4649756341 : 8222
8646706338 3937026404 3271999344 4501765362 1194183980 9140543543 1204661107
1022949147 9572822300 4888150989 : 8223
2880052029 8222195603 1371555319 1807236758 0873515739 4941586386 3465924264
9298271246 8742873788 7107211779 : 8224
6093428528 2218507517 6120316328 7865050956 7859784696 9439706685 7436394417
3341905638 3504407294 3862175261 : 8225
0060673994 4657144525 6508218511 0811234734 9770923533 0607315603 4383366681
2802897283 5529497834 6861207930 : 8226
6536662298 4886844589 7302368695 0573180717 0927104880 2093698860 0525942346
8558242146 6323714836 0334708879 : 8227
0508346751 4163185997 8731821377 1136093379 6792267130 4808400635 7127404476
1901699021 2805036249 5514649375 : 8228
6285725553 4722015529 1627360048 4207174507 8807950761 8459136069 5778824861
8747949602 7948214683 2553636964 : 8229
8055794751 9580422659 6181757545 5372575780 0638500788 2894288015 4045147286
6450769363 6429261656 1464092376 : 8230
0991944230 2530717760 6647646097 4890130071 2832067009 5834441486 7959642884
4265653348 0626985008 9001435859 : 8231

7934320495 4700739790 1976240218 3186450225 1873557681 9538466957 1307006288
8292637561 6032787642 1596559312 : 8232
5292492422 1613295745 0406538220 1672623986 1866164858 1434298883 1519203159
0579600733 6630592644 7801768242 : 8233
8057937751 9254991161 3874081655 5196180080 6191448772 9485116625 6997724515
0821371463 2819036518 0877334737 : 8234
5322139017 9569455469 8558946099 5099448197 6623160582 7389739243 0853104944
4078708472 6990640317 8359352904 : 8235
6138482240 6081401306 7392119403 5926589229 1196901683 0434362279 7153048445
9807315903 7135399085 2305554135 : 8236
8503033059 9426307530 0844749731 2132922813 9004149372 9257315810 3903671573
4392981372 3660967707 0372399756 : 8237
4711313836 5036061040 5488716098 6190396184 9936105390 3230560359 0895271652
1688244120 7600029377 7205420054 : 8238
9345059959 3284765791 4451389480 4179203585 0852529938 7445006490 7705032042
6246039291 3479186052 9641815140 : 8239
7354308345 8770362016 2613635647 5367166760 5974780085 9031429361 3337929331
9175008485 5096043897 1811040988 : 8240
7752004921 2223706955 8581249625 7133144760 6650693367 8827688873 1032445424
3982890773 5107267733 2667843932 : 8241
3322019262 1936564466 3220548552 1963628598 8698212459 0594841453 8983552431
7193424912 9900618146 3959235638 : 8242
2859406169 0297564763 1126229146 6746536626 5823575832 9722166170 9968921215
2632249540 9725300927 6090968909 : 8243
2981539854 7781047546 0397048056 4097019438 6112437832 1613301721 7352016953
8667789380 5184729999 6113301753 : 8244
0953639202 0079729438 4334427132 4196298010 7195951189 4080207607 8542247534
7076597267 9570860437 3801337156 : 8245
1484930232 5710198132 8749438961 4948548783 7329709427 8163362544 3633308269
3990557968 8104948920 3758750785 : 8246
4537640505 3657575719 5557194024 1392108268 7609088769 0522636865 4030187082
2701883547 5847239992 2894034079 : 8247
0795136796 3796350726 3442245541 9128095770 5903158772 5558922077 8969111186
8653692807 2383024038 9236271049 : 8248
8072888312 8755350717 5113342480 9692876972 0158667518 9142171434 2557639048
9594698580 7500609279 8100439092 : 8249
4945240817 6625095274 1727415601 6145431594 5185221718 5574955268 4752772716
7389139029 0147202350 5989549615 : 8250
7417316989 0428553994 4028302783 8377623650 1085909369 1920154027 6948681440
8826839215 9342531859 6862968867 : 8251
7073556817 8360341970 5184191167 9193966183 3729700019 3822980150 2724830835
0801097177 3091110587 0894894672 : 8252
3054289271 5118246415 4409106645 3295323935 4505193052 7899093172 8114996492
7253060347 0215987196 1345650020 : 8253
6310713361 3651786165 4416497035 7368209762 5837494386 3039224877 3435717559
6671101166 9312791920 4620889828 : 8254
7538871071 5363136881 5546679590 1553145462 3653372771 0941929274 8404169131
4541970017 7540191661 9419279523 : 8255
2601888253 7303377523 3601219617 1161255538 8978356176 7083773878 5260756313
4205865681 9701929804 2050503450 : 8256
9501358373 0270205832 4470696463 9692236906 3859331108 1122013967 7100067477
2455361451 7398246578 7334364145 : 8257
5185212468 5804072880 1878105976 4871776307 9789332546 4580093215 6135484194
8421779175 5933093578 5967194699 : 8258
1950561912 9316799344 1194597294 2434600010 2857975899 4049694365 6661909757
9329566805 4246701383 7449990948 : 8259

8656972175 2978624999 4440386325 6488733167 6004076776 9071769110 2171332461
6671369335 7767751796 6874823798 : 8260
1197416413 8932966510 9832191318 8912306128 8321306175 3473064594 3263123690
2194248765 0442656800 6403723735 : 8261
6520012431 9482373191 1164158611 9940162345 6107055588 0366626031 6221668784
7134896499 7725714436 3570318750 : 8262
0753298600 6333082899 7051279145 8197779206 0780694205 0549492682 0444046342
5629715753 4094927435 5797251601 : 8263
5720797972 6607019969 1870098612 2241889692 2478331223 0698835193 0268001333
5158234809 7469113783 6974486763 : 8264
2078154664 8812639240 8284186516 0249149549 7986442720 9866051820 3571764568
9499356020 4307104892 9581586506 : 8265
3951917305 6385593822 1751430131 4348075986 6933265048 7479903805 6468350956
5616398416 0231551047 9659885931 : 8266
6847452297 8453271156 3022567963 8245805708 7383359848 6159426999 2353031747
2056202203 7261527089 7606684602 : 8267
6088744711 9518675285 2500561782 9728887196 7524660489 5413107910 5130025731
0392626908 8847423715 6518309916 : 8268
3546737457 2399438350 0092516539 1918101842 3706418278 4463199649 0552888569
9393252826 5626282428 6907604695 : 8269
9122933888 8263678905 2588962979 9260043658 3512928591 6601681627 1158503859
5099200450 2382880525 7871607999 : 8270
1485779511 7410714587 8927892859 4455422975 7489266391 4906061225 5901846742
4204898306 9603260992 4160537319 : 8271
8039930958 0318458741 8756119655 1694103025 8842746132 6771528604 1675625997
1668900919 7446205707 8876061938 : 8272
2471443068 2007869995 1582186915 2348094620 5994733672 8416187438 0183762446
8431146227 1997183540 4199085721 : 8273
2670067522 0557081779 0802076422 3877008302 3145963243 7697248430 2281374748
0149945967 8297652451 1116475628 : 8274
4494882391 1265802232 0723393967 2487353483 7968347090 3721727807 1821993045
7961708748 4668322607 5483119464 : 8275
6363162955 0461428918 1870344025 5160661099 6043968227 9760085105 1090393629
1091994211 9388266551 3643110459 : 8276
3773982337 5470223489 7098938238 3349606222 4144588181 5717448698 5800768017
6809831354 4889080730 9801045982 : 8277
9884067101 2861381855 5977913112 6585794627 9763440209 3254046425 6523214487
8549983704 5212678648 6295967723 : 8278
5993867028 7589062826 9927949214 8880889025 9752971774 7890672029 9367123678
7634519865 9570801187 4771798648 : 8279
4510898251 9553391404 5264040275 7528622015 6098390974 3678839234 3368690293
7949023880 6297699255 6920231250 : 8280
4277051089 4350978320 2370260907 8772210288 8386917306 5202970742 6870592354
3037688984 7491331150 8457272408 : 8281
9272768529 3202568303 8229026985 4983092664 2798169621 5482643789 6461283683
8042073209 2446348106 2482376286 : 8282
7848195810 8547337889 1721650337 0931716230 0827835446 0953550001 5708732585
3718296069 7550817043 5834992204 : 8283
3478239737 2708582326 9636237017 6097674845 0030410906 0401168774 8131236415
7224925413 6735065969 9997435146 : 8284
8311300047 9043763389 4683811507 5044798362 2497756891 8706331215 9825736569
3430609810 1058107512 8320284644 : 8285
4444669225 8358776454 1076542454 4616023777 8278842301 4324148376 0775272286
6645863176 8687572843 6203463872 : 8286
6464703378 1045580384 0869900184 7001329490 6720162910 6732086656 0152720035
7037700877 2363933708 4619152832 : 8287

0488231140 3505825354 1286718497 6918987401 8370111971 1247408166 1540189701
4577602302 3745038112 3110997102 : 8288
6466141404 1857261089 5636960583 1244662510 3344217698 6955312306 9183533005
6188518487 1146756537 4712842537 : 8289
8375360270 0254047815 2676963868 0028149067 0826847143 6627044938 3420886829
7956055915 3091431959 2305379389 : 8290
7090912350 1683175231 5693892208 1723667794 7717971362 7192414555 8888086019
0380407494 6015109518 1573219926 : 8291
3163081536 7277863953 3982650376 9350931962 1749307103 6054682746 3851923840
1085893804 8215379605 7037554136 : 8292
3419453179 1021440277 2440022859 5051052507 8853005636 2548796203 9630514167
5053890281 5483989382 6056184605 : 8293
9692543092 3050118448 2024440573 5333994864 6232469861 6427152552 0414183945
4588339064 5070021120 2816269037 : 8294
6437236786 3270923848 0835704928 5566542256 5700417166 4346794270 5669316946
5903559002 5009811020 4615997069 : 8295
2226960430 3931341501 1238520208 4330783492 7685212802 3229115251 9737913751
7432840567 1719578654 8200968348 : 8296
3949354987 0633410451 0911558023 9978965553 7286716719 9806356218 8229089859
7135945659 9565839007 1990841135 : 8297
2900703212 7984868173 7876376919 7501596503 4762104926 7928002979 8828161442
6247045499 3187358737 9611446122 : 8298
0750417737 6803842330 8899512489 2738217191 1705995177 1345342994 5729725215
2140383430 4607340293 2129297183 : 8299
5990233671 6755190204 8367988934 8578542813 0740917112 1274913518 8966503880
5950364886 6001930517 7479737720 : 8300
0603859479 4431466501 0410771923 5172244725 8169876135 7315539473 6063610226
0817195494 7748733465 7530769762 : 8301
3829299706 6593811303 5566698283 8308327669 5476109668 8653181122 0411325508
8889820627 9798064803 0112172279 : 8302
2334181960 7984129997 2072008841 8393872211 3903472473 1085153277 4836698378
2486796544 8539605467 3324517828 : 8303
2183737129 0684884322 6097503190 6355301794 1264823895 3511473878 6399505486
0224600033 5769136031 8382569522 : 8304
3930439416 2767786502 2636715905 4260821794 1626062786 5341378178 7284203815
6593007446 3640673989 6675492687 : 8305
6493957184 9031321211 3656202390 2615229840 6287309564 8128169303 5018696850
3713109547 2329376747 2224785729 : 8306
6417081985 8940521696 5981052533 7889233503 1987258494 8893728640 7683296645
3384003413 9733684656 6429981296 : 8307
2074532556 5166107548 2537381673 9692277169 5636536682 5813785392 9592804863
9346240680 0456128973 6777656499 : 8308
2644452755 6162223994 3174891109 9786814130 0340878616 0964068909 1934466010
6857817399 6496691929 4071519797 : 8309
7061967355 6327808303 7486762428 1895329937 9935017437 7033041267 8463941074
2390004075 1986059181 5646594775 : 8310
8609699994 1255968962 2869881578 9024265778 9777920452 8945726859 7880151203
9187864497 1604999362 2264612495 : 8311
8687721434 7718177827 2154033157 9087438333 1809450293 5381915728 1454183268
2112481972 3259721432 2349402628 : 8312
9546957476 2101080787 4258476571 4780160883 3404962554 6506324147 4442371159
6789923705 8982776656 8589771945 : 8313
7925992692 9040833893 2723324793 0254203274 1208279443 6936354791 3979579720
3963663957 8210495841 6314403965 : 8314
4240938604 7528332592 4343500781 3065949179 4990770388 1670567144 8564759014
7081936288 4606343870 8308730139 : 8315

9925372529 8366891041 0331359439 4347713254 5251112552 1110927987 2778595138
2708916366 5909520741 7092752512 : 8316
9902620455 4103080348 1032270004 6208199313 4977410339 6993570520 0813490694
2080378722 0379046438 2890249902 : 8317
4012213804 6339798024 2182106839 3468508849 3016791828 9662130425 4933387996
5487438610 9320851491 0533275877 : 8318
3226902672 4129296625 6736459068 7559148963 1205017424 3945263893 9790244232
3703264903 5968525782 1378096515 : 8319
0454498931 8660685590 9756632394 4226182229 5196564215 7254180903 2099804496
6138139531 2098534796 7114699342 : 8320
6949382451 4966747515 9852900475 1805276612 2600719775 7224967081 5150580391
4346182401 2521809293 5634426976 : 8321
9077593665 4520908202 5060278046 7149336326 7787205816 8159702504 8184007645
4284365495 0033719423 3565531706 : 8322
1100284423 0300420530 5295293308 6763760286 4565461103 8275374154 7484910093
1287259564 4205697610 9993020881 : 8323
0403531866 9489262396 0995356572 2335747574 3158786184 5904831966 4295632227
2610527164 8198398546 3379437083 : 8324
6572964093 9488820397 4869606943 3331514579 1639720748 0723433442 7069762865
6850889473 0949815971 9068311061 : 8325
2001028675 2309520111 0639978597 0418819427 8438731917 9548374603 6719035569
3038399401 5483738186 2624892618 : 8326
1581754672 3094662836 2055651216 9174703278 8728457031 5345444858 5223696069
7986889224 9345328209 6934356761 : 8327
6792601084 7238262589 7599052637 9232591641 5032763625 5946077462 7417043325
4493457444 8894772168 7627182772 : 8328
7072947967 9929407037 2568210612 9508999370 2462171998 8989446787 6686273457
9413522640 3349817753 0833993906 : 8329
7029665213 3803469837 2728905324 7600661939 2545820659 2897014152 6129507274
2629227932 4657917043 8371626932 : 8330
0753650111 9601557525 9469406191 8181848737 7713426834 4442405293 0660057334
4588869058 8803931774 8451177410 : 8331
2397358775 2822843859 5867202823 9874374352 9592115562 4322438928 9630027291
0568728872 6816161177 3035695272 : 8332
3169774369 5929142484 4621189894 5750116903 1295742514 8284517441 9871337174
8657674635 3974745761 5954160878 : 8333
1521949380 3821906317 1978546364 8068772488 6181039189 4489750730 5385580490
9207963214 8308935231 8480379090 : 8334
6668134527 1782335322 4661252194 9926765291 4275908909 2262117510 8174670500
0595680931 3519528400 8043900757 : 8335
2617866577 5125745288 4330535531 7484142917 5337424877 5094899335 4373583595
5457882706 0373973912 9226937030 : 8336
1243965689 7123773945 1673185967 0419317393 0742310205 3944927937 2556695143
4978805545 7033059083 4011304552 : 8337
4208837745 3018234714 8368540570 3839803008 3490146661 5756278297 2454384494
7365389473 9985342875 4328227478 : 8338
5381373116 3246992938 3670295832 1529676293 1690157701 6376459707 3115455682
7663490194 8250420327 1623554376 : 8339
1606289610 1031778920 8130071331 3496833865 4865407252 6999424381 8874654827
2867272278 1054698999 2436385383 : 8340
8917100915 9271708230 4906672765 9616237816 8640441085 8757454793 6667543859
6986709554 7499965912 0236471863 : 8341
0251342342 8660312308 3288725426 1484650491 3339140845 5713489742 1213326279
5637514158 8593834370 2328836763 : 8342
6142721091 0916326438 1199307118 0581370532 0521818716 8903404084 2283314956
1397141000 9101715099 3735501625 : 8343

0498698021 2807755241 8620045870 8968444383 0634459894 9555144009 8765192200
4434826887 0127019830 3940692242 : 8344
8539144344 3769252569 0603785633 1636359169 5975636285 5116855631 6245250277
5453761962 9428904366 1945896802 : 8345
3918515806 7914386065 5457630666 5385508979 1671967227 7397520763 5829190576
6479671803 8856453881 8866035064 : 8346
5431683355 1245320523 8283278277 2109259629 7947436508 2733419069 4841174771
3586694162 3551515418 9766502736 : 8347
5243778292 7375010905 7898462586 7898462418 6849472218 7945092867 3056564729
3726572105 5693249737 5301720300 : 8348
3429846293 9903576040 5532694801 9752126003 0669803843 5039953622 4586496757
1735364486 6229608676 6502211476 : 8349
1971906683 6700078786 1952572772 4960774953 0158270401 9156303489 6276315535
9129352170 2092981509 9957118977 : 8350
7124690854 4676144850 8350541333 3978482613 4395314953 7191502413 2149252570
1145762701 0326835919 7885516410 : 8351
3614737647 5962509762 2302881118 8954047134 8042461791 1538541632 3284005437
1084642690 9503621868 3372877564 : 8352
4555584451 3212607093 6508889689 9626104066 0972714901 5825926516 8477635062
3657335276 7195383297 8920009067 : 8353
2985653235 4522274816 5410194800 7407498391 8823027932 6391449423 6957352889
9527704109 9533947552 8270108194 : 8354
3357336971 8431950816 5751817731 3617892037 2220046232 0225025710 1959975795
2402244477 7321462083 7600834855 : 8355
3877306273 9020918868 3525101963 9767078418 5032783930 3456164014 1876545693
8000416667 2187859830 1833249780 : 8356
4306684137 0878997780 6097087513 1224537331 7921047765 3219632269 2920624441
1860240382 3939582693 8487694386 : 8357
4799215827 5076780160 7506536356 0192478163 2884895067 3931704750 8196646271
9511896879 2595048559 8142253735 : 8358
3491882022 5223225452 7064031100 4505864934 0196268324 3996750270 9419772579
9996211263 1549818066 2935407155 : 8359
8361027497 1906518427 4065659372 5451257474 2135652740 6125514208 7368319535
8915340183 0056437614 7556000590 : 8360
1887559432 4899873423 5441853629 8897724641 4291129851 8495310609 0530703685
2909517474 7046621759 2782127028 : 8361
4427202763 2422188650 0362829327 3448127781 9082473719 7178733122 8262452933
9033105661 2313694376 7215970190 : 8362
5627862510 2314965083 8505178495 4746257928 6335484676 4750561938 9560487127
2476315354 2713060257 3246197073 : 8363
0588914499 5762866108 0519401608 7738399359 4759687934 2063061649 7610162893
8474378762 7083980936 5286890936 : 8364
2413539742 2309740440 1233773452 8350622583 0076819495 3505737271 2472914630
2429342011 8205594285 8754096729 : 8365
9824774332 9952532893 8891028826 2385002918 6860662230 6007695414 5344014154
3780274354 6527798112 4805950108 : 8366
8156886539 0958105517 9251789261 6859476198 9018512885 4853300197 1913658050
9343086513 7339156714 4253106933 : 8367
4558535936 9068057311 1213522090 1489843226 1639643263 0776114024 9595727575
5180179589 4194013197 7457342289 : 8368
2233099739 1962454237 8153163739 9205324766 4553480610 1436730683 2579576051
6674364736 6202346212 0548832579 : 8369
6206777946 5896153466 6628496225 1255998837 3663561545 7380994239 8223413977
8573181185 2669450921 9333400278 : 8370
3956605221 9043439079 5218769528 6295362583 4511428833 7418801389 7668334834
5199235437 2759509972 4884754998 : 8371

5348212875 4160212142 0007167425 2732281865 8471302437 4038012472 1275771551
7354380686 9321781709 8469304772 : 8372
1386934362 3935185177 2094380919 0247679191 2350163419 7498300194 3492514392
2732839989 5275284543 0980061397 : 8373
5570079141 7081678257 9339825803 4505303504 3559971630 1845528168 2926422796
3795173998 2625697213 9310348886 : 8374
9523650338 8767235345 9179213883 1157879766 2440444585 6862661187 6186607785
4423457825 5621751391 5151217506 : 8375
9970282671 2148235376 1675339029 9724794386 9400984398 0337239260 8257591497
1225249699 9091625168 2241883027 : 8376
7064831538 1122368712 7561226085 8402325217 7282389919 7546169668 7100468066
6839513940 5468301470 6632437280 : 8377
9717308526 1750040540 5846357996 4387130602 5046653245 0985137113 5047840666
9674081206 2280849524 7082736778 : 8378
4896750668 6806656952 0461593590 6403278260 2281023655 2083797774 9099988133
9305724930 6686654387 8693836289 : 8379
4312535175 1613853047 6569608483 4268921637 9531764454 1891627305 1752216789
7208041102 2372283886 2096566304 : 8380
3269375053 8126058074 3571556442 5203015360 6598273724 4631942002 7263684000
7290391352 3216097806 8208980025 : 8381
0397115413 5638074184 3338384377 5594568899 3432757328 7635899539 3433301321
5225900120 8386005125 2010931866 : 8382
8826735672 6049987995 3512265864 6076878454 4884183383 4136625422 1969714632
5189211728 2500974121 9838894379 : 8383
9647742461 1182875649 2740080109 5806810716 3190905554 4066376841 9924830303
8244538612 0476391807 8777478409 : 8384
5532936773 1266650623 0463491542 0945503013 1869928385 8704049776 9498762308
6816011982 2506037840 7782933137 : 8385
1486931955 7690412480 9600289428 5901471563 0035995211 8751134960 9284646438
8277636681 6444290872 3542365626 : 8386
2418491311 0970711588 1107599568 4882418627 6594293115 5326435533 6578107862
4936606809 7352567283 2824388047 : 8387
1495336516 3044632204 1993562377 6365923546 9498248612 2240436933 0506445470
8669838194 5613271731 6706287211 : 8388
2922330882 7822876856 6112936704 0431097366 8158215665 2530953192 7357606575
5366338130 8114126150 4182742591 : 8389
9791584686 0975661711 1553592650 4724528901 3979730748 3656845667 6376607503
0008038868 2744809256 0195250182 : 8390
2877867751 1683518513 9009239907 3510597032 7066961915 4073287289 1168466075
0520909200 6071456463 8393565915 : 8391
6554266871 1062586079 9966340457 7588827698 2303474499 1771278741 6589237797
9611704433 0665499088 1949703719 : 8392
9281218530 9204245501 0101872809 7074432904 3394827028 8632007292 9682300716
0613009672 6729562697 9183862541 : 8393
9239274660 3900071210 9973496105 3233558472 5675941583 3536503896 9578883612
2277161020 9908180784 9942356230 : 8394
9210965202 0074968190 9702336820 4647962109 3852321055 8762150886 5676768483
5432116346 9821573876 5508378320 : 8395
3733814319 9007420263 4978428101 1684895754 0102189754 5070983266 5427621469
3339080399 0467531115 2024150248 : 8396
3200566561 5806359792 3618193230 0762882729 4668878379 0788539740 4896953393
1470031131 3244930932 2770532613 : 8397
0028850543 4290778900 6340319100 2285599374 1995905453 4198294838 8657641910
8382166429 9146114191 0541040721 : 8398
7183757715 5061513515 3927714008 7755200122 8409781872 5816627089 3127396454
6477259808 8949046387 4441203383 : 8399

4039847054 7482648434 2166015060 4097870387 1976045332 9681594560 3974179277
0386611691 7540306056 0427594749 : 8400
2737317580 6087531129 6607988071 7230218830 9181631035 5469967899 9676814113
2238405848 7215045110 9777541309 : 8401
2733480856 1399431389 1956459179 1374612961 2274326490 2894505828 6960183976
6326676818 4863672978 4429610824 : 8402
5753273532 3785581012 7991695760 7536616328 4457154796 7757002225 9203947901
2456471885 9527332353 8013204986 : 8403
7071615509 1587828895 6727461343 9615495902 4810526757 8991639561 5629228002
4734147290 9294565424 1442384279 : 8404
7513489457 3060583395 5546620662 7021001410 2767079458 4352116489 0881686243
6979656823 4197708223 3313015802 : 8405
8218768411 6710285191 1373496255 5044156500 3220132187 8072083632 1567275832
8941194293 0094201762 7734310749 : 8406
3222163016 9690371102 1196817814 5961129850 8035678247 1755722595 2337646404
0239924499 4117133227 0648140922 : 8407
0890393406 7741659079 3358224796 1761271957 5790623216 0753334804 4259252472
1663765328 1249173787 9135545453 : 8408
1828388653 8707564763 9734081624 4449879336 1431231856 9653401386 4422093057
4391287276 3381581387 1255067371 : 8409
2242883009 9858186321 0156353494 0227831056 3117033107 6712499040 0513200129
3489702721 3099521549 2391507859 : 8410
0421402689 3130046098 6561523614 0525303927 2543131409 7867037236 7159813508
7041444155 6847409342 4285806826 : 8411
6918870587 0133146463 2050819150 5624847600 4435207080 7540878211 4949462115
0927923356 4167673683 3501642284 : 8412
2786529339 2832792845 3321515289 2040943012 0008170861 8584107504 4157621681
0260608335 6828369738 4319713651 : 8413
0829362124 6800257976 7991155399 9076484038 0499281718 0375653459 5183845950
9934009339 2603110508 7975376413 : 8414
3549052939 5708765991 3428997297 7018161429 4760801328 3728437159 0590628796
8664004706 1491784659 5143380897 : 8415
9790174722 8882213053 1415145267 5047969517 3436233472 6153303000 9304974265
6539457947 4740788563 6678194708 : 8416
7581203460 4862122119 7326839850 3198398806 7512355607 2123142248 3976820693
3579702545 4114267856 8828685762 : 8417
1814616682 4675502952 3775266140 8949262179 4102342151 6354117757 0269072944
3076957090 8960644149 6588167174 : 8418
2121668318 1149637091 4477913934 0867791720 3636047183 3759073820 0996909450
1230844029 7824319898 3074299124 : 8419
7475096595 0540243211 3462983344 1563938684 6663845113 0417168804 6800828350
1799096954 4557428581 3277440301 : 8420
4003363683 7871236116 2753223918 5200931510 8695478540 4060385142 9675337051
4492285816 8231754675 7859933248 : 8421
9704331947 4811631365 6876224420 9211961639 8478074939 9063255065 8610472649
9462785709 1184829307 6400523023 : 8422
9571694045 3022977484 3375344969 3479104278 8046497550 9168928481 1027335593
8094404693 4895784831 9661916191 : 8423
5687678664 7442091767 6961460159 6143007118 7637115981 8435709487 6341973991
3808562861 7818195168 3356605131 : 8424
9780945322 5854265525 1653405256 4189836041 9180987756 4754700933 3545646386
3745881837 0719308992 7774747776 : 8425
5194007121 0016021292 4290428843 7751885569 6937984137 4619594878 6640495288
5179702994 4034170922 5712698364 : 8426
3477923420 0044508972 4019564277 6843574000 4691806885 7882963825 5568576955
2433481059 2353696323 7766541213 : 8427

6136594165 8964893601 2690830391 2189079669 3463878269 9462568989 4323842694
7900195449 1764907992 5967283332 : 8428
0150204055 0563958228 8322986542 1520127390 3857125511 5833894601 4788267961
3070593684 4627140731 7663585074 : 8429
8773515360 8785471059 9450815737 5037687217 5758668974 7637142045 0858593475
5203715928 9414490384 5551888247 : 8430
7822488860 5676817948 4488505424 8271176560 2041275625 1081698730 2947899169
2904177807 3208202945 3912387288 : 8431
5057804711 5027943406 7819720679 8706677346 8991759685 7017096422 1498843862
1317233301 4036484090 6229633661 : 8432
3973120512 6785480197 5140106876 1497867822 3829515530 1447754388 0094209195
8111908455 9317284191 2844754245 : 8433
9302404344 1560468496 0365223220 7039181979 5473902374 7928894306 2875587989
5504346332 7292294265 8198189938 : 8434
4964339039 0174859190 0745498519 4324377468 8971516350 6178404476 5817263836
9808979750 9331606867 0902736290 : 8435
6796736528 2770315463 2011642375 5537799298 4746403323 2739853551 6109777781
0752126229 6894986051 3517165602 : 8436
4102871037 7241294082 7807555891 9925350758 4971547702 1490914683 3655432231
0865474877 7138622887 5760810079 : 8437
2717858979 0259875918 8635196306 0456666333 6319217407 9445303345 9277301240
4904323289 1698863107 2549085903 : 8438
9501306666 5927301170 2603766298 1068329188 8015400774 0068222930 2138595764
5423568417 2436497530 3391034247 : 8439
5954667977 6970800273 7594358006 4715248683 5066819946 2078500178 1035428128
2583528653 4039521232 7966035363 : 8440
2408223180 8982544771 0520475037 0425226479 7228699159 1452243000 7083320007
4295977322 7257950376 5299376768 : 8441
7202659189 3146678879 8396187665 0840897212 0716214708 0505329655 3068382337
5864780997 0173621775 2518266225 : 8442
9448897555 4791079002 9432807377 7695412037 8881938575 3362453555 7555386215
1372157904 8564519552 4782723839 : 8443
0439225555 8608545983 7832420422 4899605866 2215842368 8887828188 7503287720
5784097787 0899101239 7962235928 : 8444
1304154281 4620670946 9072943044 2763735707 9519463824 0638535397 5389325145
5320403986 5813187666 5067180128 : 8445
5529209290 2813884649 4499131489 6215110965 7353827367 1105194612 5607048321
1206288125 9687496905 3325465166 : 8446
0985515328 4705020721 8448979151 3038599618 2707552535 0830941788 1533073713
3334832472 8777479051 8106099400 : 8447
6506218469 5791416090 2586333765 3702695033 6325159012 4006107726 5518504085
7437205040 2869641903 4506015434 : 8448
1482587482 1359489668 0516971622 0412921890 9013651942 6616334910 1517770935
4187823411 5944342573 0184584604 : 8449
7967497734 1123967367 4607693758 4906354299 9397454530 0717430914 9674014521
8588375807 9084100939 5282512399 : 8450
3941887800 0980008529 8325011797 1552469662 9805239359 4260533425 6683484171
0659646899 6024067593 1818730076 : 8451
0771656964 6002749184 7845392837 5977395610 1054362297 2283307967 1242759581
9133817907 8340962140 8277302609 : 8452
4598302411 6813392425 4021024790 8295842719 2272091231 0498777743 6008228204
0479398238 3576317324 4317194831 : 8453
5697133001 0828525340 1780917584 6522941747 3591973493 7221873348 6503775663
7694545173 4805841274 1929648062 : 8454
3788474696 0032363296 4560718750 0856199400 6290196361 8143276961 0791010248
4774499481 7747303697 5189922813 : 8455

5576935750 1446845470 3817964356 0419274820 2966411484 2647553884 6160928431
7366732609 5117141455 1466420877 : 8456
5937211606 6400571318 7183194037 8249303566 0264211455 5465509638 1561717598
8326306606 3541502910 1213817574 : 8457
0705460347 5894377765 7343347443 3136145706 9549958552 0681596871 9207955264
6702350322 8932588692 6552115837 : 8458
4045717679 2697869830 9365844160 5217539839 7969141646 9213052887 1247348215
2684048633 5441603666 7164545205 : 8459
7287892065 3903968965 7009883303 9278228312 4639883225 9368184889 7300762029
5019139221 7469639190 0812982244 : 8460
7578103017 8407124137 1181333474 2069153806 3731963420 3722700135 3128231561
2820927307 8733360657 3188224330 : 8461
3526775361 6851440128 4814216046 9279280062 6149042372 6475295528 9806723868
9801124635 2617089223 6094195142 : 8462
9831850549 3877642205 5983978423 5496068384 3080444630 9187389828 1103232617
4942490245 9296870542 9095989327 : 8463
7188672788 1814902205 1859424964 9786437220 1915084587 2524151377 3308659016
3437389903 6909618394 1846604804 : 8464
7641285774 8573396024 3288848561 5948165391 3099507843 2615276124 2743730419
8132191310 9713823323 5365219625 : 8465
6565684132 1009977934 6586712530 9809163123 6945456552 4086709902 5795737378
6907357079 5762333041 5204577601 : 8466
5138834558 4741962374 7926673163 9431708110 4616149280 6358838918 9301292776
5043664289 2222952486 5961964742 : 8467
5015639365 1304555422 1841369811 5505602264 5692242688 4427092190 8249138797
4604688426 2153522223 2159695297 : 8468
2046003562 8448018051 4309235146 4906483155 8147073373 9099094033 0635162847
6364524307 0399022890 6910322690 : 8469
6335776036 4860551940 9027826803 1593780882 6592838678 8589283339 8144312107
4324210574 4407797255 3048758075 : 8470
4382718089 7381605829 4605104830 2938386321 1204406323 7985310181 2009680047
8401312104 1931723115 8801989412 : 8471
8999509449 0518235202 8550174784 5472762059 8636970700 9621505367 3678007104
0186618081 4138596278 0769153033 : 8472
0859735979 2227429779 6806443236 8930843822 2161613445 0290924444 2413428682
0459892391 4410058649 4855598206 : 8473
0284922716 2477870269 9558974228 1427014367 2583620201 9104692411 1432481136
5678238853 1661678230 5910130295 : 8474
7723739494 2182206285 3229253296 6281056278 9429374661 5051753207 1023254039
5606954202 4998214315 3977132554 : 8475
3297586855 2527248013 2525920496 2363918642 8240229505 6529171734 9820738772
7486453474 4992666383 3468080472 : 8476
8431021137 8092719503 6693983708 8898079287 3532815339 8474260645 0074080844
3294502104 8660235279 2585313312 : 8477
9653132245 3687733089 5416706614 8363110682 7794190105 2865443855 2547588213
8943087838 7554697438 9267645496 : 8478
6213807228 8425723934 5052083454 2156644577 3935903272 3197581717 6591609149
9223005364 7772838127 3341666225 : 8479
3384147224 2629942411 9246224009 7854472979 8291278440 3926399816 9698249831
9988102820 2402018949 6060671263 : 8480
6566100746 9397089206 4689403357 0492380927 0107051053 5093856117 9427302169
7988253541 6280152727 2038979683 : 8481
5160423690 2381883598 8721040292 0190710560 8751001679 0371111051 7939171375
4662368383 2541447178 5938653029 : 8482
7056462626 0948159605 9731112820 7255718281 1133246076 1042177477 5964548391
1179713618 8734748786 8253984586 : 8483

6897492106 1770350321 7366706570 8216985559 8660531527 2702364292 9621060332
7629512934 9217521429 7488361749 : 8484
8973053872 9795231771 3765169565 6076009410 2572096526 4136722804 7189019484
5365773052 3247184576 8564345313 : 8485
3418091260 2575401390 3941163886 1092776307 3561477104 2982037148 5588099882
8077011820 7688604358 1805556974 : 8486
2895134939 2085027036 0998529136 5667242000 4093406815 6266480004 7750369267
0100671875 6549830267 8403949779 : 8487
0284984080 4112864904 2737318787 3235749233 5157779266 5464058755 2701973317
4152553436 9343335878 1774394766 : 8488
7698653410 9034241828 8560046824 4271735589 1956299625 0797908273 7456067765
3849024226 2424341394 4410514766 : 8489
8328626098 1169869629 9573291894 8031309383 6578772030 5440635688 8087392173
3643168563 0197495882 4077856564 : 8490
6911005844 8506322174 8560278698 7082544923 4350094624 2878114247 9255709287
5881603713 3370498944 4793554135 : 8491
7861767774 2593003501 9948788189 3546457069 9898023884 2859434023 5293957359
0363877995 5485801844 4355982316 : 8492
9084248835 3550065678 4002486228 8539779059 1902101008 3266439140 4237858347
1415392112 5211966441 2719679850 : 8493
0140377737 8161139546 9455626393 4396372291 6983970344 3161802263 8018535077
2747903835 8387760381 3757838647 : 8494
0121363152 0655028504 3820926854 7160480494 4672487705 1115295639 9198461966
0704801990 9225438759 1936048994 : 8495
1776424323 7878245127 5097624575 2855490091 9496634300 5632504215 8247850394
3865657347 0302650698 0027224965 : 8496
3816227554 1258133830 0224668505 5927019280 9527663208 0553239136 0748859854
9525769899 5079276140 7664345764 : 8497
6429097181 8152704084 3411686487 5195291524 2506986869 7209121972 7276416639
9894803529 3815572061 0365285429 : 8498
9804227933 9909846309 2628786791 8884474582 2818384915 4137902575 7617305573
7219098917 3358087060 9521181391 : 8499
3922837017 3304768818 0850991710 9050444013 0249073627 2237452998 1247942216
5881185896 3808812967 8928602717 : 8500
3502479061 3642266669 7096556330 6010179050 5226554257 5044259197 9884309629
2810306578 1734716370 5682133316 : 8501
9546753852 5704127557 0407558625 8683249666 6639996077 1750714245 4243476380
7349935092 5572652501 9287640864 : 8502
9276184971 7304589762 5148716488 5915989512 5537522822 2955357809 2755157717
7349425449 4064636532 8643887343 : 8503
3421753070 2797210788 8439357840 8051947775 6982541739 3212928035 2819043283
0422660811 5277615033 7280932222 : 8504
1614627280 2255172759 4025891404 9567804027 6685368356 0064837512 1195655603
7929089017 4970568828 9249537680 : 8505
0055117044 5147090402 7647709526 6126178911 6435270849 4766333633 0375047626
6818134938 3168469838 6036717861 : 8506
8075709038 4099944166 8188850885 7156733750 8594523816 0582885059 4971914115
7735194691 6382663291 0009363196 : 8507
9376262656 3399714388 5908306240 5002210685 6248393754 5166453897 7952645501
4543848991 7421968312 1931401372 : 8508
9951184100 9750979419 9237468840 9542501321 2364767075 3289568716 4905935102
8968444317 0331513881 4848447542 : 8509
9115655149 5493239860 4734701430 9945309592 9657329964 0696179056 5718915571
3959225222 3779966163 3492924092 : 8510
6900212172 3513543080 9375801321 6123137754 5123487290 3356146091 4816427581
0499520023 6087135985 4964801275 : 8511

9693337261 1487822048 2714165428 6109728738 5314981069 7327536968 5591326889
3124658863 9737786052 7964731457 : 8512
7443870375 5129143836 3101480880 1025549750 1850563438 3742326494 2884038079
8143255578 0016392499 2969085284 : 8513
5898363913 2925179370 6254015022 6681735964 6575985941 6332244151 5757367266
2192304256 9556660186 2213826190 : 8514
8801925812 0372388154 6007760038 5904490388 6176642120 0157127662 4408763052
9283978600 2937708353 1090043918 : 8515
6588120008 7431950741 0289980656 5937988870 1230973100 6970179557 9926044738
6963611607 8651598376 4865016794 : 8516
3285933431 9628128655 5160845291 9059142677 9204939904 1189678877 8786674303
9299757616 7646554681 3424735632 : 8517
5818313412 2662690037 4833698503 1872001174 6032135725 1155487510 2276922014
3344417041 4759365038 9209545499 : 8518
7699002142 8049303569 5458920060 0890881228 9808480549 7614642398 1241706535
7454304376 3040631682 4995543589 : 8519
7677978729 0679404769 4811267909 5135574378 2017524164 6753461329 7580009812
9577909972 7071190393 6380944971 : 8520
3956296258 7268238358 9471854165 6384732924 2758695098 4276737772 1923321752
7653968660 6177282371 5965708211 : 8521
3035581036 3815182791 3256944611 9309158813 3531942734 0665428840 7415997081
9492402918 3884976257 4831533937 : 8522
5254667365 0828439655 4011389663 0167639046 3017835475 3694786523 9365547983
2094346814 7633789840 6578472456 : 8523
1420759944 9897741287 5174494581 4534657383 7899796025 5958626928 0739502699
2775699077 6084920221 5589743467 : 8524
3926377917 5303666438 7053071485 2519470220 9570892828 6369504935 8558489320
9095586866 7023345755 4319254514 : 8525
4251812880 0784991514 1850033013 8018283582 9191679510 1513506325 8915735687
9487674191 8382330615 9137685922 : 8526
3498332250 8319995527 8486255059 0848674550 4695160242 0521525235 6676382666
4320624401 1346246184 8355382319 : 8527
1903078979 0489156492 8816756949 5129760602 2618515957 5090914427 7224633135
6835722350 9941125528 0916228242 : 8528
4069411322 3256899532 0003903020 2196968217 3861309879 6980072131 1638620390
3272971919 9555770591 8946517770 : 8529
9310867334 3597019086 7546375077 7494181642 1366245871 2283625713 0969042522
1424092741 4444286294 6764111157 : 8530
3339026068 9928399591 4207344591 8210129878 9382713047 5037383316 6795787273
9006283488 1721234327 9316790078 : 8531
0179277820 5762794247 4196395117 7750945739 5274438958 6335334736 7966070550
5270121425 4429947908 0491346447 : 8532
3573810923 6893078603 5662366461 1507441594 1923079912 2635805375 2635962493
5883885499 4535786088 8234906479 : 8533
0166624557 2094788231 0387009991 7210865162 7001741277 6478931430 6360703172
6839611668 1796735997 4052437218 : 8534
0120623481 9404677351 5110115013 5753039355 3613503983 8765404636 3031729240
4004439345 4222886650 4375595216 : 8535
7963855990 4741481073 6635762226 9326643306 4224662251 3619647559 9479394475
1674283039 3884953878 6086663136 : 8536
5660876086 7401686825 4243859509 3098390095 0824741461 1518279715 1479253878
2363231160 3910159790 8370532676 : 8537
1335653050 1089255973 6947825669 3522289815 4208014466 2048550197 6532321021
8161662195 3034657184 1288016502 : 8538
6448317775 3037857572 1075721672 7037351922 2403141487 0533281255 6025237909
6528551710 6044744791 1806731970 : 8539

7100206860 3349095233 6928994351 7346016999 1071845070 4909528695 7774131794
1205305663 9315825980 8009454159 : 8540
4568474016 7337619934 4464418595 2484585780 6791846718 2672145793 3027162864
8423325497 0208681474 0691585705 : 8541
2483014262 5134913013 1793189738 3824525493 1717540343 5106305944 1518585179
9332289184 6398568766 1286780829 : 8542
8214112906 6850039256 2077476960 0566532434 8516251485 4282704856 1423339710
6339613269 3140524821 1840228020 : 8543
8764932824 6007079295 1867118770 7464164573 6456342222 6184718124 2843835488
2660565417 5590498795 3693629595 : 8544
6649725411 8371933569 7984699399 8266970823 2831209910 9341255994 8081987322
0386864574 9761500731 5013080359 : 8545
4050406734 0560123257 0978746962 9188299464 9670095532 2993288831 6237622770
2344608416 1786295841 8100330595 : 8546
1772290600 6608958130 3058313139 5588588048 2762259625 1755183942 6498063120
0451271810 0192221949 7057669748 : 8547
8445965926 9299769162 0797266423 4143396980 9608501454 5299116867 8452787722
5801508574 2859764318 0504071622 : 8548
5494651215 2689507976 1409835692 4309417467 6545181719 5667472044 4984286032
6968037184 1082593827 3355849743 : 8549
8685580513 5952284552 8759636138 6027589819 4531001707 6089442733 2474687242
9589116478 2188536209 8458294683 : 8550
0304075110 0830546676 6121316949 4465638662 3369731490 5363048789 0788328740
4207267338 3396925834 8281353332 : 8551
4626119663 9727672957 6987444036 5471360016 5916747714 2386181991 6453062722
8981557735 6622922661 0897177977 : 8552
1500834627 9464409360 5843157320 6378347619 5170000165 8106021009 2878404435
6820652019 4527028568 2643221760 : 8553
7135816017 5222197346 7223327780 2743989435 9711559780 3812762780 6526046703
8575085560 2556081051 6667781838 : 8554
2636916211 2027595476 3575092735 6103375651 7976994657 7949596114 4911621311
6791600460 7234256813 4822109174 : 8555
7041610254 0842483992 4042350962 1969126368 1209164349 0379266934 9222546351
0174034101 5465437528 3162070610 : 8556
5390821226 6935397141 4467016387 1339195724 5620135139 2040509186 1473232182
9361951891 2364549208 8239049722 : 8557
4882879142 5729913397 7822247818 6521013391 4143716010 7780810007 1612966209
3680467263 3770301940 5918478585 : 8558
9655730889 3645078579 3600087862 8686633807 9763689028 0780619857 0100232244
7725140393 3032211957 2056967186 : 8559
4238802321 8633477611 2599435486 4992447475 1661178360 3166952637 5416043600
6356632387 1059279357 9217568711 : 8560
9998228111 1089142464 6101546553 5659621270 4209994403 1975873515 3439833319
5609894173 0938654745 4651409993 : 8561
8939769355 3922458643 0358876231 5761562586 1587462872 8187813371 1235134587
8837855804 2169776439 8525992789 : 8562
0499624290 6538896215 8212221897 8878116258 3829659073 6324849697 7987612007
1306818833 7251900370 3774034872 : 8563
2504297349 6099347660 7284864000 1631992960 6695437071 4271183192 1409924423
9469582566 5420541914 5120455361 : 8564
4236488414 8160260537 7486614986 4110283759 5162909996 2209123298 1921678339
2396425613 9072775365 7077363937 : 8565
2511982297 3696786924 6891513716 5264869759 6513443576 7122826858 3753144012
6280418404 6228693588 9735824637 : 8566
3787848447 4816412107 3381675877 5922822307 9154982210 8047959043 4831933876
3433073399 4992519424 3372136321 : 8567

1915365810 8725527549 3903497011 7953105652 7436888894 4085474666 4727270780
9098080469 9730940520 2816129582 : 8568
4460656291 6550769680 2356145139 7859995356 1449185685 7949608990 2281540993
1896227352 1074758244 8567245261 : 8569
9510048267 2561527017 2300934394 3029068805 3019264553 5304297709 9978289188
7127297756 7312250801 2156908989 : 8570
2104620610 3032654195 3558830843 4633118423 2432667935 0246989405 7473910493
2355738728 6249786807 5144049873 : 8571
8143435118 3852558258 5965084861 9765348305 1655774653 5484925184 7062358463
6112896123 1063475613 4878291962 : 8572
0808029021 5418895671 6954705784 1263606512 8050970044 8653394569 2126761074
6489021826 0151360021 7640420759 : 8573
3430425154 9604712165 6063826907 2668810603 2814204741 2200888667 3494151799
7246444314 8785230281 9566773009 : 8574
2434525278 0354723713 3249811211 1447527196 0517285263 9093240007 4100850410
1353495340 4377507086 8269290958 : 8575
9645056777 6975505181 6997654774 2490749871 3768970805 4222310307 3699862144
2134449704 8083338862 9036192462 : 8576
0994170065 9527552349 9456084088 8352271199 5125641178 8748775542 6906953789
0166583717 0505976068 8177908491 : 8577
1618570218 7460703920 0411210127 3866342584 5845123645 9384881049 0498891712
4685739269 6221806481 1341034779 : 8578
9910285334 5043363529 3559946999 1079748386 0730938322 2418565543 6504976494
5838570957 5475892418 1079549377 : 8579
6431689626 0734364589 1301527446 5652114578 6141557709 9210493343 7132765485
5020781975 7756130190 2631627312 : 8580
4397224267 4146601694 0435362112 6647620668 9557181471 4997912220 3905936045
3171295309 5554012285 5108112121 : 8581
0265729697 5654697231 7828275358 9325263383 3111069657 6581556117 3594867954
7594241902 9558201244 5750907537 : 8582
3370460353 6226154971 0395443112 9334077832 3908570180 9046135796 2884855271
6340269965 0631291909 9426969596 : 8583
6225589497 9756287212 8775719854 2419673475 3015874634 7073305073 7845796145
3545425206 4423087824 0841408078 : 8584
4627967368 7388788958 5204470927 9081010712 1198779872 7835924233 9335575561
9371324997 9593876098 4147798982 : 8585
1381826503 4322401306 3857454488 5493589707 0486251427 8724296521 0065664978
7070156838 8386453995 9036508017 : 8586
1114364104 2662786967 2414199832 6547334111 2848793309 3439580866 7890754340
7020267937 8477434553 5265665129 : 8587
2901343372 3462897801 0782011552 2188723937 3120160937 0970406863 2553677092
6191900398 9418841145 9261537011 : 8588
6668658956 3673150522 1633041077 9848867721 1494160046 6211065860 5248504107
9547948381 4251404996 3682497073 : 8589
9689414424 6667174873 6746316851 7045635775 4242310515 0580986476 4641739106
2874599047 3792488810 7274308142 : 8590
6295248909 3198549576 2898324171 9089619983 8410018154 6644378082 3753328402
2544160414 8959931908 7300847247 : 8591
2855748818 3769758189 5347939088 0198458062 8942112361 1333546325 4004046653
5717316673 2386520784 5030801661 : 8592
9577080902 7132411891 6546201980 7089554609 1872385437 2611387496 6299766658
2523147148 1880032379 5003806670 : 8593
1185989421 9599482184 0489226045 8854876888 0233759401 4215791542 9528712356
8227193941 6595000624 3911835934 : 8594
2678164048 3541623849 6792420808 7709963757 3387339727 1385700510 1721639010
2814006199 6402955555 1092051482 : 8595

3727888939 5862250635 8200576235 5860460670 0333545148 8830880320 6990796123
1844923797 3339003311 5058829483 : 8596
8020687671 0240916485 9588883944 3246563446 6757463358 4953224141 6518803282
5488208991 8241029406 8318220651 : 8597
1712210563 5810950805 2676082430 0375506483 3828150365 8350774394 0163080243
7480979848 2956648727 7021341700 : 8598
5589113966 2180093523 0435395488 8255902667 0900065711 8853359523 3013070087
6190428286 5576790762 1690733940 : 8599
8560706158 8936193013 4878025952 4996703159 1436244219 9778519830 0437984441
4548655165 9082820414 1139376731 : 8600
8104087719 5965789360 8900600929 8571878482 4543162180 2453861168 3785858623
4510782849 1821691458 6433097247 : 8601
6771131495 9030827346 1852478922 1595763617 6194158034 1303131918 4160775944
1012157439 4177453130 0010795056 : 8602
4422525109 3067375232 2368854324 2791213530 5575926210 0311221186 2038297502
7538178425 4173117337 5517119681 : 8603
6524092843 6766301402 8433123610 3292345231 8918370223 0344454974 1418863973
9865083728 6068616943 6251316559 : 8604
8064294041 5960955828 1527947448 9407960067 0401560663 5423305660 8882241092
4622558187 3588042827 9346966420 : 8605
6274112459 7521522027 6131826209 6736092737 0341863068 0429620948 0151901121
5646323354 6199496049 7408344435 : 8606
4531421150 0672682516 6877950667 2541821696 4868333082 3372445854 9241292167
3190981802 2915171702 7639539076 : 8607
5960845015 9458745976 0737822363 7182049117 2490303721 7284752147 6818938282
8585241866 4014070914 0991778837 : 8608
1395692161 7076979010 0601923052 6629784219 5100992184 0123220629 1400604821
7819490156 1947902646 6313128750 : 8609
2908324448 3715546624 8494038615 2485353273 6635483810 2400072695 0375367248
1333156495 8131952983 2958248945 : 8610
6994701540 9838635883 7510527445 0387542374 1635053454 8439059129 8010160337
0240650914 1965440640 8930992874 : 8611
5635036122 4915848602 3751328733 7313061777 6438334911 4167974279 5635358264
3265650934 3897438769 7125404890 : 8612
9954397610 3094470122 3017049596 7791613454 6024092072 1494649710 4487480382
5509647557 9233784085 0157046965 : 8613
4092609937 5292975065 7563244547 8640178182 0868997603 5501204605 2898752372
2840965641 5401043048 7466276071 : 8614
9959290035 1813908226 4656278006 9174993818 8975755041 2455262811 8769537357
5052501272 8015922292 7782677371 : 8615
9461371921 6516400180 7111032606 6546576457 2568399041 1126551032 6989847762
0049452573 2076336611 8794668546 : 8616
8075655577 8816168336 5059804799 7528938859 3446218272 7482769757 3534811670
3851100088 1206002684 5121931613 : 8617
4987292279 7354351452 5162066264 4675504351 3009958875 9986591144 3487330276
0578327386 0610819603 2567833536 : 8618
2521398537 2146327137 1437336685 2594289784 0927984778 7866474664 4957883589
6680820451 4776727190 7031261880 : 8619
7710771780 8759446592 8794910582 8127513178 4511939021 9204362173 9992710417
4817473553 0055149680 7137461376 : 8620
6826102458 2297889026 8640706208 7169184080 5906684021 0711797550 7316360971
2321042997 9833192165 9946411876 : 8621
7390473835 8239727102 0669138686 7582223407 6837140251 2878602433 6013754569
5612101211 1866856952 7576098387 : 8622
6242013180 6009730015 1094877041 8647501460 3471909560 0164459132 5816011088
7003424095 1086006596 6824661861 : 8623

3700340454 0305650156 0321089611 1956432794 0113323206 2441296852 7189733900
2930438752 6826413252 3728118187 : 8624
3418377266 8317079822 3668198492 5517311140 4842922636 0049729836 6464717470
3203589251 1190314552 6378203697 : 8625
4827376444 7465796347 0343627745 2109556074 9209997068 5966310831 8811970526
7540760841 8520990745 2641268309 : 8626
0143476734 2985506555 5504958367 1068719038 4924438850 1716241895 9241597064
3895517919 7612480105 1183266210 : 8627
8395180919 1931686471 5007622854 9154633210 0294923138 4348040570 9777013982
7445193483 9221971306 0823701654 : 8628
6337256801 3935034810 1212266051 1497687829 4628586232 0643474121 1262663352
7821577473 2452483124 1354286604 : 8629
4140191905 6371445619 3361673409 9663782714 9600978057 6875416953 4454405994
7939618148 9149477288 3934449386 : 8630
2378544571 0715026668 2904961403 7984996997 9904177314 5349852052 5154680296
7974626481 9668702286 1642312147 : 8631
7629241242 9658312763 6615013215 9956875563 0321383495 7850419502 3669392820
4408381491 1150686568 4280960304 : 8632
4852969738 2538002417 2676987094 5805588238 7667508804 6999123811 3646498178
0323271388 6275399981 4306474612 : 8633
4047754171 7786963338 6139656178 8596483175 6351902365 3894286098 7981324310
5964107502 0220490603 9359870915 : 8634
9547794213 8098641791 3250939151 4302291823 5919452911 5029434493 1341236215
1981821583 1203202948 6003940276 : 8635
6901266322 0706625706 5165460052 7475224010 1992387346 9029975061 1631661928
1850976779 2260931005 1354456493 : 8636
6116459656 6028491128 4552622478 5726874706 1159462603 2730492603 8731909842
7270223022 7936371756 7119268583 : 8637
6471857815 1551335203 4020094255 7325702413 5674979298 3260661689 2374772379
0923156448 3699398221 9062239598 : 8638
3512927748 7404921884 8651486180 0676468364 7476634904 3546852270 4418150329
4734668805 9802599704 5147190867 : 8639
2697542090 9271463724 7939872429 0800146596 0306126644 1660222860 4050721331
8150829460 9888507098 0566227981 : 8640
5499892792 4314053239 9034861738 0802149334 1026331550 5110372233 8875148162
0439881629 6144993711 8414730654 : 8641
3972563580 3732060537 4193767157 2155162520 2628791243 4751769566 2745040518
3387160885 9843704646 7204976925 : 8642
7186263068 1746111178 9650712733 8941343126 4004219002 2842688463 2219249602
6992853768 0961589319 4890230425 : 8643
8339012860 8529021358 5525352287 2936927725 7311248398 0587096220 8930684664
6263634164 7431789867 0004740615 : 8644
6685763785 7107584274 9474296485 7966762548 7697959410 6944926811 6576569257
9106373912 8091743329 3427660082 : 8645
3474451226 4681724479 1345411283 2897445755 1465078695 6589905282 4665349909
3871511116 9679515364 3282615119 : 8646
8278986688 9710931016 9591041784 5024882873 5231923844 8624226362 9734975768
4392822631 1202000471 3218720170 : 8647
1783710975 7566855371 3938247585 3381189805 6805524561 7589253290 1241044606
1386262529 7201449551 8845924861 : 8648
0761575320 1952379210 6931822465 5177285864 2266047786 9383984421 5191391884
9477120401 2480322261 4617113859 : 8649
4797565685 9117457472 4441827556 4366755031 8617371053 1284061591 5488212497
1937173021 2674392008 2041117549 : 8650
8546836398 4135677460 8838167338 6741710047 0330451223 7269963296 7533503556
8374255932 7885052848 4397099534 : 8651

1517659032 9384028506 9219964260 9108990684 2903712845 2934549490 7934984037
0395015943 6563446313 9951298245 : 8652
8133353138 9648303954 6853777423 8675823879 9595073127 9166633912 1357622930
0823813749 9508804241 4367706341 : 8653
1867945790 2022252519 7702359954 4884292304 6328754923 4967220873 0074226185
3262394415 8468650826 1531865605 : 8654
7785769972 9253351807 4404638289 7304361213 1248741664 9557732058 3031985264
9228400338 2122961982 9400035721 : 8655
8896092227 6076892173 7321375612 8171401278 9476808601 8461734761 3358304779
9555570110 8465899184 6526471602 : 8656
9062432682 3097213791 9999502600 2007677928 4281280102 1890465503 6864494068
5163746947 4009796705 2287174665 : 8657
3346534766 8328529930 5839175291 9256842294 6133303350 2992661474 9035530997
0592944301 7566334383 4322304415 : 8658
4343703476 4664049273 9502658109 1264882415 1780638495 5847321292 5984863914
3378040535 6376028060 6861278168 : 8659
8892152483 5645436191 6584590511 2065370194 4847924254 6205587901 5583333543
2558659101 8391532755 6343253047 : 8660
9137407024 6658885855 1732641557 8510827116 2140919115 0201876161 7581702513
1170079441 4086348631 4831905529 : 8661
6155411318 6777476030 9553999986 0807938437 4976227303 7037169749 1622920021
8300013533 9181299918 2960402359 : 8662
2948162240 3757349964 7898156526 2268246922 4661822660 0334656315 5444069190
9194461335 9229476476 1753098401 : 8663
4456964949 8541787417 2123136077 9557004623 1575740701 6476128273 4096389797
6977401987 6160194157 6015291091 : 8664
0925312276 1838347867 9518524193 7060797916 5590705751 5180512954 2831018535
9173186365 2920291308 4205780739 : 8665
2675741156 3134564106 0048148590 6597772775 5892339730 4760176118 6109466683
9383170513 6767659806 0864546532 : 8666
7538441719 3324821035 0034614387 0064257499 9821742521 8212189204 2469834596
9460171454 1080967173 5478479028 : 8667
9649005709 3695658507 3602799626 6961684643 1118237198 1949954017 5555079372
4878019693 3765045134 7412370232 : 8668
7858171705 2728754067 6808677865 7331919180 6541465070 2346930410 7602604380
7649839841 8776243353 8838135818 : 8669
3139633229 7830451927 3542000124 4347701439 1410202805 8351376152 4868346540
0922414855 5589073720 2921946094 : 8670
9678309380 4725225427 1539715644 6139320717 5620510574 7947382556 3034044998
4090551893 1225258151 7064469135 : 8671
9945494107 8976605283 9379950219 1260261204 7720514588 3687727742 0393934927
4661742316 9661272448 8814202863 : 8672
9142022781 2419353329 7421694509 4679545206 7395697738 2806280091 5341955720
9296208702 1781623595 7315398580 : 8673
4940599130 9645978436 7461688033 2762471313 3218716363 9467180936 6747344033
5755526219 7725484420 2499963393 : 8674
1748616681 1685800241 6101935718 1587639139 3715915310 7634250433 8406774911
0062398885 9795461135 5539755839 : 8675
6530539242 5113385151 9729571507 2567194931 5914454388 6748094127 0092974257
7221170197 6780777554 1153146444 : 8676
1588881354 6404610034 3675463395 4381365737 9417552022 9837046337 8124045276
3115958787 4291541420 4162066248 : 8677
9126162385 0077703928 6348472623 3343506441 7462265548 8896432896 0847169212
3310844633 3350533714 7173330331 : 8678
9017211530 7748181597 5318740320 6520654666 3038340247 2404436419 2958515662
0772019573 1935194885 9162981533 : 8679

0055105279 9525400109 2334656798 5970645405 1429106571 0504302628 4793759359
3880541200 8468120716 5725958968 : 8680
2779436299 3111922904 6287498933 8151616124 1807541838 8765972717 0003193889
6653365652 7359657047 2983705556 : 8681
5100268027 9238516795 0336520665 3091784354 6800532221 1811784568 3243680044
3266590254 6285945991 0385475796 : 8682
5209123438 9503251059 5830284514 5019453098 9232771984 8928078784 5546749643
6275646169 6626183648 6662036715 : 8683
5784981383 9868252876 1956385736 0852041933 2025764108 6585308207 8346093735
4267441745 8791798165 0977606748 : 8684
0342358794 3788166111 9989595666 7944648382 1547715844 5223575130 9630861325
9832304456 6468192097 2502934490 : 8685
3578658988 8840520552 8878640659 3898270941 0986621527 3716175244 9269126472
2285626744 3179067066 0513250331 : 8686
4572067834 4046379514 2601733349 5920626461 8127328737 7940013015 3571573662
3761268528 3211039112 0166194811 : 8687
5587795504 0394086512 2360419497 5793571897 9740811556 3734672074 2742415773
7740444109 1238485558 4519673648 : 8688
3504888683 0993138982 3442124854 9562023391 9819006035 4898518044 0671358033
1408724132 6581558085 6335529235 : 8689
6505562434 0622173863 5871005910 9166690201 1060085192 1062061521 7298898708
3613379458 8258419892 9728135937 : 8690
8463864081 4620973215 7488764585 4545290564 8569347625 4092899106 6919225602
4642545291 0014982009 4514753886 : 8691
9058507821 5749901642 3832582661 2333030842 3601713313 3019740264 3659281051
6974600611 2974134995 4361141507 : 8692
7890155231 6163627582 1160734521 9385511127 2300896003 3993710873 6340084744
5858281169 2917284014 7159292712 : 8693
7197382253 5539639876 4592473836 9326112803 7429940138 7580817617 5069360473
8088259160 7654996602 8549415839 : 8694
7942913044 1789374134 9812851299 4331756758 2440767341 7695102424 4331173208
0541981225 0315547648 2578701650 : 8695
8657667023 0978527121 3432609608 2828084722 2735516127 9802179432 4863898906
0292519154 4808852944 9249132385 : 8696
7532148964 1284456505 8336515343 9839435370 6322369086 8184748916 3408293026
8567197857 9660416012 1522883409 : 8697
8644950300 6136379847 5278879019 5341774348 4446727480 0436348276 1808233991
6100874051 1223516596 7744517830 : 8698
8192167027 5125143549 4699668726 6873708278 4919199168 2025903711 4776598260
6846429716 8288715196 4827056579 : 8699
3794566304 8499534258 2827100202 8745751425 3134878869 8880801239 1825275474
8629359202 5797500281 8219751009 : 8700
9462917837 8511034667 1609157650 7689424057 6434882324 5410625517 1455768523
5574081545 5232660177 6743209479 : 8701
0156452054 6687532565 1052514797 6320111226 6026752483 3213955089 9312683263
4930136851 9242840309 4260238790 : 8702
5323204787 6793848815 7817991207 5883998951 1824201625 4924399375 2925029268
3380891296 5124723281 4990269823 : 8703
0237888614 4353189879 9215072001 7872694765 9321610518 5240507686 3648912351
7516719370 7377421624 3588562940 : 8704
6235947704 3120052660 6256976250 9684217812 1148882988 0026616044 0592223293
3162417612 2908743379 0222878045 : 8705
6170135772 3750619521 6034268628 0629053786 4968871393 3857125624 1696407932
4475831369 8859182729 9927578294 : 8706
9295751304 8250436660 2853237140 2054964473 3807382457 7555825709 2700759153
5821362248 7873951980 6447536509 : 8707

2283387321 9737894509 8948812224 3166650106 9873961667 2989920964 4205968175
6961923183 9866191793 4084742578 : 8708
6815461594 1458938640 6022961329 5003812003 8389767450 2086338557 8266798810
6569036999 0815678277 8516378293 : 8709
4599361943 3669806529 7922152215 3666288839 9402680386 2183878413 8954997920
0722893711 6950775706 1724002344 : 8710
8728986838 0888946932 5821863378 2343569120 7402895687 1885668709 6061273862
1934987326 3224096065 9606991760 : 8711
2005453603 8165896602 1438717712 8755309837 0990207133 0847123039 1797557448
3810050683 2809511893 2927219123 : 8712
1654940906 6402145683 5987446321 6265575739 7928373087 0286061293 9772368385
8149199392 5841574254 9146335154 : 8713
8204141285 0525611641 4384738621 5794850259 0959169401 6701922227 1520515924
4638478673 6844034101 9760225485 : 8714
0585962037 4520210340 1958672160 8171270192 6460704607 9599287131 2107480351
1882506823 3530449698 1265520956 : 8715
7080884541 9410225351 9913136835 2911597222 8197796519 1751091412 5749066752
7198799284 3990372741 0645886716 : 8716
9811835091 0120568349 8176773100 9548469910 0664217037 5101296402 7952668079
2620134649 0826570837 3127308877 : 8717
0349853816 8830180415 9107350878 0278978144 4525165407 7081274884 7379655033
1829893602 6105151709 0092011022 : 8718
2071001669 4799486064 9884160992 5577821033 2925422202 3824316169 3794552445
0771166127 8195720289 9490592371 : 8719
7876307137 9162077328 0519004359 5063027837 2052428607 1686319756 8319944953
5965461758 2379331954 9221835571 : 8720
4063821726 2170118990 6243630164 6834987994 3067324897 4840299006 2675636866
3934498668 7029574632 9559276358 : 8721
2274173904 7673664683 2709872540 0825658274 0793730134 2125015787 2246935933
6023202788 0213344754 7116925472 : 8722
4427882347 8857202047 8481096682 4957325946 5069381839 3289944085 6293295484
5234695473 2470720957 6815500031 : 8723
3628681830 5873597665 5244562923 3370980039 2024465397 0819808809 7516775900
0829452339 3382538737 9751663668 : 8724
4848199061 9171892305 3029327528 2052288738 9977798577 5746443306 6738428466
8342381977 7822394151 5224363898 : 8725
7424951701 0656678530 2676664370 8546240614 5075109823 8265082329 2192311694
6593605521 6243701274 3002394924 : 8726
6000846441 9137134537 3908817105 4329719962 5921785958 3689002275 3546934419
9270563544 9944646484 6356921147 : 8727
3954542342 0930350956 1912596294 2760332314 0283815641 9581239921 6843570592
6115521864 3672939081 1404988429 : 8728
9540135030 4582616685 6151191924 7042480677 7874883871 3189871896 7382619247
3974908922 1696564899 8157671704 : 8729
2890174496 6620596800 8681990919 5664839871 2799600060 6600983366 5085013172
6705066780 7381053625 3324043615 : 8730
6109801108 4767554948 7749423658 5163719465 2793284979 9057701845 1049091701
5335686136 3244389487 9659034367 : 8731
5903495600 2166426558 1524443928 2787227173 5945948307 8938720248 4734203222
9005213360 6846053129 2940974975 : 8732
9889320134 9050155466 3997880699 9170087371 5889175956 7768947270 6181150301
9647289132 5767848161 9091938845 : 8733
9773052888 1739137963 4191013912 2828818689 5766691581 7506640190 6642575911
8576388754 8293436299 2117191271 : 8734
0549773853 7301557783 8101884441 8606578305 9241045272 4319669227 6439468819
0230293660 3689392914 3527900678 : 8735

3454920522 8896117886 0518754083 1049180891 7759260962 5711828832 7086436346
7278181627 4725571485 0253575093 : 8736
5601944533 7057042972 7931651832 4369707363 8748560927 2821695707 5593521798
2829317630 4039885438 9505725017 : 8737
9465534119 6643840611 8328171225 8058093138 6536690163 2341550423 5539488039
7710071250 7041056787 7416202585 : 8738
9908400717 6820989414 4867462299 2276289202 5508580816 2173150838 9755388405
3494279190 5673448766 0483097079 : 8739
1086659225 2931017597 4785382557 1471946452 9629087845 1956517409 5947894396
3376856887 8413363340 7353907274 : 8740
3766372205 2802244591 6053457375 4066183716 9580524217 1800321860 2285837532
2597888350 1880423178 8756894023 : 8741
1975197437 4446133525 9745789740 0554662442 4324975934 4049537626 8236401505
7347269539 8011010025 6582513119 : 8742
5753891584 9382125129 6799677253 6212764607 6391067226 9184411105 9166671823
1748120661 9477228053 5025793861 : 8743
8987107329 3114319621 9558359007 5732545492 6544053448 5876237997 0486963982
1290623046 5260237154 5697463982 : 8744
1850402406 1606472122 4738628536 9145422758 2815893032 5786719200 3815231307
0301231450 0162038355 8559708463 : 8745
6288728568 6618295880 3815141259 7924271222 8006580721 7537599560 9753028168
1324319088 2675811213 9786458977 : 8746
9915670771 2334130061 4050720737 8747754227 1334777231 8791387780 6041160283
3892807322 9462986109 4389948042 : 8747
6877630904 1382820082 4932764378 4456946668 5596913509 7280929656 0296837842
4831906376 6489758940 2297476523 : 8748
3737070759 0295732296 7641074447 7902854220 5710833186 4160628348 3284039376
7133814809 4153081003 8363462098 : 8749
6740923141 6257725926 0164241310 7683838536 0967743938 9645388121 9871847087
8357602846 5758500662 6430131835 : 8750
6377598343 9423256319 5673889217 8647425115 3914648306 1056176185 2266148498
6211735299 3003941962 6797835115 : 8751
4324719792 1009990235 9950101850 4522633621 3662954175 8790411552 9116300595
0929887093 7205111995 3209519197 : 8752
6119111178 5656856314 5823742527 3363487442 6217894373 4255594438 8272109955
2583440480 7815336301 2318172504 : 8753
7611209858 6213911950 3812768576 8422270602 2880857922 8022789927 0135450268
9828981286 7084694808 5869077368 : 8754
7310488241 3520925337 7992815178 2807224730 4329560570 6622345618 9965692940
7990430103 1800558838 5154956003 : 8755
7101536288 3393296308 3886118375 7251344292 9623743625 6902862399 0818089667
4784072154 1648215364 6698525111 : 8756
8356093769 5388382477 9268205403 5562293103 3982346172 1774992876 1143107112
6186981767 1651010213 2817484322 : 8757
6860492899 9621374426 4891787470 8005217789 9145979368 3257690825 4470499557
3654607383 3295445503 7605455616 : 8758
9384629935 2695559825 4814369522 7451351596 3501284438 1657623878 2190283447
7841943484 9167543322 0898865725 : 8759
1072163801 2574559205 0062613832 3533310017 4635632696 7882997952 2312213359
2295598777 1478425621 5281966009 : 8760
5824803979 6074188068 6481462221 8467350238 4649962094 8290023721 6747151301
7616183486 6485690095 8044527129 : 8761
2413610774 4855016454 0165882100 9469531851 6709496532 0283685563 4394275525
8676230939 9026264688 0252100234 : 8762
8398810810 3139591567 2216752103 6403116827 6980204470 8468275102 1600765291
2859618123 2892391998 9837615465 : 8763

4015288476 4023895640 0800911170 7771686632 5847151888 6521341810 0963097892
4681427776 7444249098 4819462072 : 8764
0991186061 7837882725 6060277489 2025077565 4960922415 3721848981 9139994930
4356168693 6215964291 7710952569 : 8765
5097069900 6323100610 8564855544 8231768169 4918920335 3825938397 9779552017
8060026150 4538466022 3480584280 : 8766
6680800540 9727242488 7098899181 4030172108 3740851971 6844655506 8686682595
7621317618 5347414437 7640981168 : 8767
9746202037 1211318615 0318053481 6370992805 1005793939 5818396053 8157279905
3173564620 4725646756 4657337523 : 8768
2496044286 6627542283 3411947710 1158614822 5132905747 2336964545 9350778630
2813970350 2693355867 2025420653 : 8769
2019113645 6842785222 7113049930 8474015508 3205534502 2071115182 9250378245
8415159542 3857290920 5590931555 : 8770
2709371573 0436507139 1970662707 2083660506 5359257807 5387996624 2782962602
7190863678 5842034261 7942729278 : 8771
3872074227 0392586478 9969888520 1728543732 9638914179 8564954987 1231317421
0781117458 4783471481 0113022006 : 8772
6918811713 9063222377 4639144278 7135013391 0136146547 5823547313 2163877978
5942292590 9287326603 9806170514 : 8773
5105189352 6138463564 9007582348 6329152025 8655103110 3413814049 1015735161
7886075764 3964618834 1017565444 : 8774
1487477439 6958712593 1820683929 2181168088 3559712722 6537119526 7463913547
2889090102 5277774299 0668168005 : 8775
3198706232 4755696316 4794199431 8772899895 8907371744 7913822106 9168303144
3021088327 1873693722 3637159824 : 8776
8511277737 6585439522 0664987630 2769812346 1345369762 1047397548 4439806649
6855150018 2428724139 6293002078 : 8777
4239040770 1717530874 0062110388 6981275161 3311076710 4226560953 4206562796
4803091959 1712222568 6050866069 : 8778
7491958719 5285117201 8630145722 3111255958 0580300595 9045178016 2091560033
9271581356 1939615245 0582940942 : 8779
3830675223 1048760275 6833193139 5370615940 5690082546 7163407688 5187380620
2839376494 1416952247 8976448274 : 8780
1592439389 0738587033 6831106474 8295916563 6319407597 8797738936 8568253656
5679811940 5558294689 0756209748 : 8781
6397051586 2806160495 5380199298 7901072698 8526406869 4896100332 3272842034
0618845421 2897943218 8689707273 : 8782
8273627785 0137270114 9638708846 6407860794 2655552555 4856648253 6726238855
3019570089 9094418141 1968192782 : 8783
2521474367 3023757726 4790630213 6270092943 5193537520 2448246504 4502817747
3530313055 8784042890 5295653616 : 8784
6955757460 3044684002 0130258347 2857878603 4796429662 2856338390 9385040595
9963382052 0131345977 5949549591 : 8785
5282491367 5826557737 0863098534 3103165699 1864774935 2355876985 6167640256
9436794965 0826595164 9361856390 : 8786
6776134086 3449496109 2308528159 5759473244 6929978437 8750695779 6523406627
0348934399 2200393314 2022159640 : 8787
4753079877 2485471928 9903190331 1360675387 4099262658 7614582952 4546296274
4782530607 1370861009 3256561091 : 8788
0629015937 0323457846 5109425415 6584863367 5971714103 6824690682 1364595022
5938235868 9803420521 4584241562 : 8789
1097712594 1988605174 1846098105 2180233091 3491593055 3236402118 0135399382
7390760960 1881270857 0022161499 : 8790
4202352285 3011978515 1482540926 6951856567 6534104044 4548548817 7660629153
3342393558 8304097708 0382629431 : 8791

8944385077 0239040847 5017104093 2368683097 9044978493 2389215598 5002143587
0077134287 1584700693 0247167663 : 8792
1228530283 9112029615 8483322876 2187312702 5440275098 8697775241 8671972580
3984567834 5286723372 6268194259 : 8793
1376892237 3279869636 9957194750 8280574929 9080160929 6385648875 7743668130
5993300301 0651656716 8643311600 : 8794
3817843318 0947698249 2426608263 9256472210 8563028212 2858435912 9114203603
2720601825 2323794629 3109254102 : 8795
5124541700 5166491749 6978501765 8680012863 2544573787 2529326712 7745162433
4360127340 3978593022 2159961775 : 8796
1736148666 7939676563 3221951493 4298376790 3749308251 7016185284 9338344242
5041832303 0264100557 8188318544 : 8797
2893640874 0320396600 8892343871 0023409268 5238849673 2284456687 3657042343
1566989381 1311708549 8055633424 : 8798
1109039029 4020698788 3668650096 4163691705 2815658583 5647747550 4883191164
8430645989 9663270339 8001097062 : 8799
7143154871 7437048112 1700620918 6084162459 6321962758 1891687594 7150823636
8927617174 8516314584 5180705435 : 8800
6379707232 7895745053 8758447100 7564558747 3724567162 6075875582 6316241638
3017589481 2372734658 3284264334 : 8801
9842119906 7903327699 5061878866 7306344903 2827837648 5909168680 6539844039
3171382569 6659592365 7348223568 : 8802
7609504600 2069573673 6953943537 3448928789 4541429944 9242296541 9920687071
7179908202 7511232288 3020637409 : 8803
3324308284 0768023659 9622307450 7239548291 3251001462 3152283856 6996364648
1619930610 8035011934 0985512877 : 8804
3081595450 1549790983 2610070008 4351632209 1409713166 8390509307 7706782579
3848915921 3209928659 8075166477 : 8805
6274042102 2870695803 1691019769 0665849294 1163014490 4175524152 8407985584
1920459422 2472240857 9545248996 : 8806
6149963129 5674599317 8744978341 3501974760 0248558309 3556978815 3697313632
1445251140 8292841812 8045024919 : 8807
9295984561 7297886498 3652967173 7406535027 5757846534 1870784213 0980573575
8509870892 3218338602 7668096878 : 8808
6744587673 9370425061 0529455934 4800003379 4844118691 0343848981 9927217800
4569840882 5618002774 0246971556 : 8809
9634535370 5817713249 6654317079 5487952577 6642112068 5694340740 7369416521
1045301707 7751449502 9664201508 : 8810
5673418561 3308793690 7990859888 8195417742 6188031441 4174869352 9301286286
8769796349 7164412421 7738001969 : 8811
0974862799 6089460936 4253067910 4174593571 2831904029 8311315505 9303861120
6192754003 4742996012 9769845672 : 8812
8568007574 7868256852 6558880550 4465028247 2340621226 7230987650 9524679555
1167576075 5189736710 8186648733 : 8813
9135554730 3871771482 5992498209 0655636246 3688874428 1635475973 8020092703
3727972357 5620585201 9488731175 : 8814
7364152085 6887998396 2553950672 0457656370 8667868496 1673992899 0516639547
3480646884 1632161269 6232140430 : 8815
0430349787 9376589552 5591261273 3494431374 9318755851 5220350488 7715420612
8323215542 5010369584 2011777060 : 8816
5811310857 4067217688 4473924121 5189067429 7679952843 4604620851 0422989295
5901538861 7177785962 5965902453 : 8817
7479964480 5737542590 3395571736 9017939751 6001998758 3699094035 3460200600
6114570812 9727286492 4415558859 : 8818
7502427490 1019975278 5695834533 4494325002 5780204344 4408682890 7507743961
7367055383 7615787863 8538700095 : 8819

3573335902 5946681197 5123738983 8726653687 9554300184 1504480720 5276494457
0257994686 8034994929 4168674710 : 8820
4745236313 6504711526 9827810552 0599626500 2245440734 2871399194 9802538333
8505885393 6499417336 3696431899 : 8821
8036953211 4631746171 7020388070 8663490647 8634042245 8469135590 4242451402
8142597209 4336803940 4264695762 : 8822
2035197605 2537466919 6864684057 4852273222 1411263468 2007326809 9128359680
4337124898 6512484713 3386599958 : 8823
1557036241 2843119237 1380520698 5546305223 9628600169 3260924761 8752321257
0099594164 5450175979 1304831585 : 8824
2269009244 0553118658 1531978459 3140513549 6797501971 5913056364 0796787427
4388697431 8121596332 0242453695 : 8825
0908108540 1074867453 2233669488 7417447584 5601897763 9584490217 4934597104
7703154197 9472175590 3104895515 : 8826
0713033750 9226428947 4366150114 6171128540 4898362878 2321775540 3355815130
8900860023 1119089283 1719794615 : 8827
2733963981 4755795610 4816547218 2282092824 1262244086 6173161182 9531462701
1962136619 9594108793 5835643209 : 8828
3296418935 6289507521 8341609495 6286605476 0820233943 9036938294 4107069073
7842159371 1008435508 0993495125 : 8829
8480556142 6027948881 1735778231 4109215630 9775563343 5689280906 2401470430
4068096745 4142850010 5312911014 : 8830
4071931810 6005561952 9375994409 8161264543 7443677397 8928455823 6168065730
5686818893 2905552483 7773886978 : 8831
8338482126 9003385525 5772963294 8503625724 1617945606 8806250567 4398343584
0688586527 9847132056 8203326801 : 8832
5840118612 2537107299 4592972198 3140398249 5464301364 1412093794 4647846329
6730804124 0315791671 4681071721 : 8833
5465797595 0843790654 6392689441 6436702026 7173333214 2868727930 0067525680
8959472460 5400773921 4366237747 : 8834
0366937064 7989280683 4363066623 5735491883 6306740896 9053454196 2549218595
9482963529 9144250681 2421957849 : 8835
3976249362 6997668432 0117178309 4789764534 2821592110 5441925395 6738906802
5874295234 7024625362 7205862445 : 8836
2991614257 8749920154 7834925160 4342385349 2438434103 0380727377 0776570147
4373435807 9845112149 8902138772 : 8837
6113074931 2518484971 2899174590 9500393219 0625680772 3932545455 4691767350
0211427825 1591537139 2247521510 : 8838
2619575181 2555899192 2377560556 2518625777 8715204024 2356430080 1544064737
8686471774 5485337568 5133039577 : 8839
3050554298 4102745208 4882456380 1811743244 1150886669 4172029225 1387140525
9332921890 3930234849 5217837323 : 8840
5334465326 2693777473 2504109205 5082762701 3601076268 8057349283 4106150143
2117912584 1093281226 7491152949 : 8841
6919441405 7983354038 2007949205 2726207312 3858332785 8875647790 5672110416
5441661470 7612883610 0062438413 : 8842
0531050140 0810107987 5557731522 5042463587 2420813465 1707819681 3266473052
6526687009 7539010235 4840054319 : 8843
1030305850 5073284856 6621892307 3610160979 8730109604 5787862725 9697182149
7774593191 2172420855 2038322309 : 8844
7743733627 2600910791 7085415906 5400695404 7576946452 6935273895 8908946566
0935532169 4271294261 1401893368 : 8845
1755821233 6609288093 6868361084 1295931368 9766825463 4161807973 3799311434
9199450619 7674140383 6739591043 : 8846
3250837896 0976546321 6344296844 7504718087 7538796601 1594691058 4698569346
3154671519 6731054018 9434725327 : 8847

3510123305 5566492446 2530897985 5259889634 4445382941 4488382570 9671236053
3891982931 3649034991 3321228421 : 8848
9660873657 5713694328 6363383874 9656941544 7710706138 0343673939 5432994554
8894604432 8621170422 8102975764 : 8849
9010846130 3625809885 2044456528 8925386561 8355437466 9050757948 2110698111
6094362272 7817194688 4422390136 : 8850
4355582252 2433014093 7115484011 3662138841 0824798090 0542972492 8690877078
6394338356 9758089134 4855483753 : 8851
7677719658 9593658751 5432750294 9340116362 8628419330 6048171093 9239987919
0088420348 7294343721 4946123970 : 8852
1703430337 0791698316 2457660506 1362459054 9183588052 4520307312 9842425880
1870960784 9181635833 7632141776 : 8853
5526484660 8626744947 7751311633 7460985326 1567716821 6013147819 0445605770
8920308018 5215408812 6888224610 : 8854
8542068433 3127975848 0921954444 3889667113 1444617893 1474129366 5128198790
2593291945 6527368734 4836398893 : 8855
3899843612 1168068986 5797567488 5165488863 3769003537 5291987757 6930810573
5517395144 3795272703 8004470490 : 8856
0728573092 5262631673 0990740006 8490459975 8787132093 5334814799 8072978030
6859252749 2154032508 0620629679 : 8857
3680290963 6571196554 5474983265 7557609467 2472292408 9206137105 6217009793
3992793206 6567094589 2120839049 : 8858
8460475864 8011445523 2781360281 4534457954 3873365991 8540295506 0100187896
2582320644 6714509639 8089199675 : 8859
6614659823 7014128736 6468803859 4032665222 4087508860 5288410671 9799914085
4487007293 0220172202 6030478638 : 8860
0710886172 6314153139 2374899477 8191781040 7754525553 6936054590 3637816192
8639420022 6696480397 5868226345 : 8861
3758128535 5079620606 4636202276 3410015625 3911994632 5786788360 8702524372
5262830330 1021044894 3262262753 : 8862
2207366765 2910916281 9988871916 1678669769 8617210689 5009023640 5929175721
8594584763 0689212470 4365027536 : 8863
3283506500 4334618318 9703050828 3915358506 0525172234 4229331896 2943725776
1631522268 7395005813 5915637995 : 8864
0090704572 0072609689 8837387536 9824262186 3149512139 8835716735 6388063005
6290325475 1451996618 1726778207 : 8865
8962727991 6563774802 9910225094 7244094198 0146869018 8625285100 4233665906
6430765016 7100236787 3518980417 : 8866
5086476038 0556088271 1984788639 1169660571 2575811161 4332203162 3986195399
6081064489 1151299038 3218924496 : 8867
7115181985 7985027704 4697851841 6280732953 1875221707 3754278078 3674506060
8436887769 3049830230 4143646891 : 8868
3983700826 9396406069 4562881641 6952916654 4373908475 7281969614 6411957158
0663688131 2484878296 0051923653 : 8869
8166969144 1316437821 2808037745 8223191425 0665372731 8660158555 7631000545
8819146086 3415120134 0660568618 : 8870
3053991092 8422209722 2772766642 0070998758 2315907629 4391295156 3496720838
0989724723 0420387327 8350860147 : 8871
4116048285 2042007439 5960779396 7665745457 4157343814 3762929561 0953115848
2092000199 6834922274 6222332034 : 8872
9229787977 2093259373 4589318530 3632200218 5233292266 0432632777 3869929525
4400374606 5480447629 4982724042 : 8873
2914656005 2904598661 4905305330 3411842327 7471347528 7632417746 8600510351
9680259504 8933541617 7387650238 : 8874
9319108406 6521380466 7466529643 5771960522 8927287905 8233362671 7180104787
0741509786 5327441555 2281550979 : 8875

0153143256 9914109132 9952502769 9164121847 9890353417 3418028880 7859437047
4700374687 1669807291 3698781051 : 8876
9134823174 3199718175 6973247169 3411240421 9323832375 8158340750 5032512102
4232721599 9762255953 6081716396 : 8877
5955459215 2006263493 8327793871 7450987669 5534287878 9774664436 3855170650
2650448577 1475878951 6639052612 : 8878
6187671738 7540459387 7592449369 7246870221 9846805151 9182603434 1463515337
5173543983 4658403665 0078008251 : 8879
3376195811 1253960595 8941718804 7217874636 0466850775 5956087664 6149799759
0725725466 9309501812 2660597563 : 8880
3832045120 8463864464 9947755674 7110488258 1862451811 4821602417 3113511553
3763941801 6198608293 2584828502 : 8881
9721564124 9354354937 0147221837 1490931352 7465161404 2298840347 2587358480
4710030497 1036786199 6903966400 : 8882
3189027014 1021874714 3973930479 6671271469 6745248582 1961590443 5885750474
7116108355 8276186011 5988679925 : 8883
2327767007 7913488062 7067843058 2303768443 2240837285 5777585082 1627326777
5521575385 4931391889 4433113709 : 8884
7187976549 3099063043 7083812107 9727316041 4688741044 2732940307 2777436378
2884423977 5948723462 9173289643 : 8885
2364895042 3030339525 4772328529 2251820978 6329412792 2776106997 6483964461
5498030391 0368747636 7007442072 : 8886
0134868042 0978344656 7078085008 5124892609 1270815723 7896817953 8971471665
3163541792 7413249374 5551049067 : 8887
7088305382 9104660986 1801335492 7364715788 2317517022 9529255747 6729438071
8432352827 8938787305 8507172169 : 8888
8784087093 6084891275 8296731845 0352330091 0082450089 4600016837 2896985534
7815677089 8606637024 3799181871 : 8889
2713748359 4244296364 3209472757 2711041804 4334601305 1070221778 2472919884
0954442915 5924729679 3147664168 : 8890
6467990945 6377046036 9988700796 1285734670 5087176357 9926726419 0775864829
7958015096 1497179864 3932311870 : 8891
5902309745 1683435712 5233587442 5716502513 0783843967 8124895410 2878996867
2155583518 1821976729 2372675088 : 8892
2719132592 8904570539 2169962315 7341359810 1626063438 4197411159 5598712194
8557079154 0409912608 4315344947 : 8893
2936182597 1466635209 4030430179 4361263079 7077809538 7707949828 6453666763
2635334220 6894534303 0626974857 : 8894
2960188084 6564902097 4995525667 3401338028 2307886298 0681870520 5412520401
1999043042 8919912401 5463064896 : 8895
0755234800 1929454875 2880557065 0554523548 7917895598 7440250913 0074164191
8093997294 6822103570 1898118676 : 8896
7215490449 5028446259 9683767825 1688727079 2953349935 5139851172 3804911695
5666124418 8049351821 1954231454 : 8897
5793295497 3290115497 6327970042 5725292885 1676055670 6957888189 1666892649
6278266068 4281788855 1356822010 : 8898
5986486442 6897310404 2064203886 2021415931 3343356506 0797764837 2861174785
4101318195 3882096326 3739781867 : 8899
5015670201 0351613827 6223554990 7816708176 2580063265 1909072309 7311326126
4539480612 7461576397 4697290381 : 8900
9915758063 0745875385 1733483346 8607608896 4622701214 0401657955 9790815513
6431792697 1432781160 0039509295 : 8901
3053015664 0538544014 4681956741 6891439500 5060129899 5325206242 5640256999
7354056335 6851170626 3129378820 : 8902
9655763055 7832675616 1629221703 9451858995 9392779546 3337352050 1688984643
6488820731 4613992856 0157646190 : 8903

6088270052 1838829449 0528350185 6406504336 4175350139 3498570510 0635004423
2775532805 1663250155 5360016855 : 8904
8606321618 0167882288 9859277739 8070823443 0121876498 2988219507 6487937452
7324977571 6467943768 2597865380 : 8905
7770909315 8268698932 1856754102 2181137066 2891505071 9169240557 1751537297
9854679547 7169440460 8728583402 : 8906
0071682558 7850350258 0689802794 3094618467 2580186255 7176999143 6669256776
6247888256 3671023905 0892975798 : 8907
2489522270 9418467443 4146644119 1197624863 0886752269 1737956084 6434163676
3558830851 2954865371 1250744903 : 8908
7322882719 9257271519 9660002166 6938956050 3827951906 6233710710 2964526112
5354820180 8162340593 1612383383 : 8909
2787215450 9090544271 9803200644 2253236001 2498934484 3636759371 9142322778
5159626576 8452530758 4855373583 : 8910
0841651524 7778499835 5609967914 5290553712 8993380417 3580332233 1380434826
0191619102 8053475986 6238541512 : 8911
0389560611 3270055649 6689281316 7555129799 6763366535 5404709079 3988866895
3068578101 7330266056 8853689560 : 8912
2118077216 2258921919 9243117304 8925224971 2553122821 1758822528 2650923582
2912252041 3837008286 3879599408 : 8913
7533142292 0255378831 9259017888 1759789430 7727113160 4891567850 8678373881
2236288755 8552661186 5733674460 : 8914
2261173633 2880225562 0686149584 6722660537 7935752555 8360934098 9163829636
5998007307 8450013699 4588212051 : 8915
9742716629 5188936366 1788724519 3879883914 9850746367 0116146235 5909180891
4648782476 9832377597 8709634559 : 8916
3154570680 0552820706 2946431076 3848171836 4124284488 3222416345 3064177764
9280600316 7897001913 7344145995 : 8917
2908181300 8273571202 4454178146 0612372671 4401875382 2527453515 1552424500
7907975368 7991871156 7102845303 : 8918
1873265631 1281918735 8142050423 0774629722 5403696352 3357483065 6204856108
6409342738 3318334634 3273512715 : 8919
9264239390 0491279730 2888378346 3744236046 4419658158 4384053191 0829032323
9391500637 0525377701 0659429219 : 8920
0766393180 8108787655 7596907078 4073173239 7323550634 4135855674 6002928122
8194482626 3869185813 6721604139 : 8921
5463318790 7256169308 1540996943 7600214768 4823666895 9583386445 8433913973
4195773954 4589473799 6499396501 : 8922
9731801875 8215438818 5804924401 5386570718 8767878906 0589434397 0572439680
6766233077 7502154247 7708237790 : 8923
4412690412 0760661717 5145832906 5681240190 8880652059 2144597223 6876022617
3024555463 7405620748 0813993774 : 8924
6700941252 2215327344 8841706368 1524435825 6118696626 1363834029 2866449700
6603799675 0179376316 7639180694 : 8925
3767843386 2149089623 5618201056 4061401237 7885098835 6670847311 4371688891
3942684794 8538764665 0984117195 : 8926
4337028921 5848350725 8601976515 2460415435 6067627464 1178103795 8055110585
2750942447 2965571294 5889544502 : 8927
0468225672 0106209862 0677182166 7486885596 7781333673 0489413888 3039656612
1918933058 7140477845 5132876728 : 8928
0301432209 9270529021 0617713921 2773759052 6124680357 8323361223 1672451314
3018328278 9876952990 4655069884 : 8929
8503343483 9833599227 4164810683 3596317970 5040480171 6357581169 6110898875
2650503945 5450818904 5782033398 : 8930
8805527336 1740589066 0762567887 6023448996 5581821950 7130679827 9843747014
7130567036 7804002790 8882626096 : 8931

8751798433 0621616836 4975773394 8961813441 8824168865 0228996780 8162797562
5692274980 8740208876 8810359436 : 8932
9299203957 2656130506 8128838761 4411959186 2400223644 5248003947 9994244058
2531726824 6513520948 9596658726 : 8933
9366348840 1509928375 9846464753 4240305615 5906510544 9169124218 6078781176
8003899307 6090690483 5067279512 : 8934
1403003404 5948829208 4535672716 3295012007 1121468376 5449414070 6925962143
3285678745 7468380185 3930454624 : 8935
3136985598 7326420707 3373620952 6825330022 4659565421 1021883163 5555322102
2325835419 8699266649 1353192963 : 8936
1878234901 4915870014 7749921910 8946501280 1726186584 2411995774 7844638088
8179235896 7236549615 8227535354 : 8937
9969841302 9394932056 7962137577 6989665420 9615618393 5754851061 0187366314
0267325061 9956581438 4095884354 : 8938
5927104275 2474485532 0153629002 8879027363 7151170976 1157510447 4448575002
3258581485 6078898512 8350955612 : 8939
2124413532 3878162333 1816561192 9257620999 1851687924 2862423080 1705860076
5585346450 9921222138 6291932006 : 8940
2916671040 5344413253 0994050314 8420160032 8999231910 3284017248 0360326411
7395877373 6439315805 4756344611 : 8941
6674421959 0534169466 5636800499 7460891763 2616393699 7268005671 1919008111
6460002960 0999062976 6649508010 : 8942
8508005158 6703858197 1813055231 7324630173 5287693034 9789853304 6076126706
9151980521 0418792169 3861999113 : 8943
1368284102 5843874830 8631022755 6652408281 4123688895 1950624472 9375224036
6903159218 1864323402 6919293237 : 8944
6887151770 8076749523 8989921492 4574176299 1858043348 6296060893 6311062581
0014138063 0601231494 3627930687 : 8945
3326876877 1474454961 1824196603 7173011732 2672115488 9414471580 7643463644
7645759070 3340858879 3793885511 : 8946
7519467423 5340453941 2252406457 0712144659 0466567348 0924261538 8417502636
4996764039 7964039506 4536033058 : 8947
4671065916 0869493642 7670638418 7525076396 8961560317 3119292686 3383238674
4633518113 3083074391 3035433722 : 8948
7907141030 7952822716 5884110344 4348578821 8090802820 8295544281 2200380195
2660218635 9524610566 6063149576 : 8949
1270251582 3033352249 4707890466 1050788518 6132600708 7053125210 0924188140
3331109803 4325447218 7535643394 : 8950
3220404659 5402692850 4488556461 4251052654 7958472166 3057299456 3578827156
0772802148 2175044787 0011247793 : 8951
6570705673 0989211387 2193059290 5808649783 9863194346 6325792242 8340202752
0796201076 6746046940 7170560953 : 8952
5133339937 6074927130 1176506022 2240784478 0821493839 6398812005 4778938005
6657803599 0431148710 1637277035 : 8953
2144947284 4806598021 0246296286 3743293337 5005342109 8402485860 9771594603
4626507089 2757684803 2018361905 : 8954
4988522892 8095376821 3151504355 8251720372 8601695960 9587642513 9501382209
8404961222 6242281734 0434028089 : 8955
3997262257 7393303661 0298681992 2133775791 6373560345 3780755018 1325596156
9355010328 3299423849 9747515243 : 8956
3810011519 5012131780 5013879646 5628491543 3248919433 7183269470 1926816767
5960615918 7889763652 6032086512 : 8957
6585226424 5241199590 1898187884 5082876944 3767663184 9223848799 2413740076
7229406807 3105280395 3540236603 : 8958
5209984205 5043059212 0388275565 5930480839 1165973062 4501772525 2787907988
5468508425 8555173838 3833851994 : 8959

3442889159 1225364411 8669644171 2424001358 8796072191 0613494230 8330297896
6344308827 1107467000 5236299743 : 8960
2610231802 7142266222 6187505725 4396907738 1474263522 1552448324 0080437569
6699071029 4726405178 0301516189 : 8961
1026800926 3587698184 1801303452 6647105519 9507316064 3267550404 8745317728
1647974984 9316816351 8881132546 : 8962
1499640318 3140120849 9997545056 5440566511 4835838717 4381071044 4468199573
6346286893 0027137176 4306960414 : 8963
7832273275 6789030508 0957691434 7830867035 4016162028 1849114412 3200843999
2821318184 8433228813 4255124888 : 8964
6686544852 7084230428 4009883138 5549010037 9402648447 6236363753 6465110550
0810294066 0991524814 7926317308 : 8965
7744064207 0953919991 6755639316 7624758908 3224270729 4829544328 1512295290
3164750609 8107151709 4932166168 : 8966
1302220084 8999073351 9284840900 1433236988 6937917599 7792387280 5644858636
3560471696 4436602044 5259704868 : 8967
2215144197 8159123227 4757877216 3986575276 0984089988 9373777508 9374040658
4561154045 3448898263 5679449628 : 8968
6422470716 1326587499 5955834400 8024494005 2564753112 7582945824 5552383399
3886160216 7095409039 5096229843 : 8969
6975360579 4438167782 8156635171 7190156086 7850102297 1066937877 2149098891
9368454386 7259730079 7493811034 : 8970
2945358118 9213029104 8572049967 3562211673 3366350076 4326274058 8295448615
7569614320 4789633059 1725250029 : 8971
6559546804 8765362628 1549757676 5802778755 9023678734 8162504245 3157027418
3539064997 2371430862 3953642833 : 8972
3798525880 9365635808 7875872114 1673582000 2378585791 7146541612 1120260342
7255738422 1551801587 4630577677 : 8973
9395293678 9125329777 0050822625 0081937416 4511416847 3736572522 6777890874
5799682373 4527732736 0629946429 : 8974
2416996715 5086229280 6080316787 7150190204 1616630204 9350758837 7690661867
4616296470 1677056344 1830896762 : 8975
6618874403 2970517788 1452434012 5122267941 0412177617 2182888508 1597216420
8384793333 9829566994 0349593790 : 8976
7282033978 0087960497 0137807029 4014570618 3227529603 4853028371 9222610059
3956449912 4150927787 5461366823 : 8977
1461289469 9816725882 4711898544 7614664135 9747240400 1167264640 3383299904
0150263527 1218579913 8187518382 : 8978
1542253047 9921538890 2846165379 2947236379 6333479312 7084664272 2737043541
0768537912 1319034331 1924506345 : 8979
2076733343 8091681290 3926710929 8757177148 0282227330 7085859022 5913929052
5897400243 7571021699 5532657613 : 8980
3351851876 6386027619 9700398052 3930527428 9337090169 1023675207 4517701696
4047237538 6382876543 1904302903 : 8981
5798193044 6828632045 4301891421 6075051699 6685123364 4518831394 3158140465
2068503559 7675284602 0968648400 : 8982
1463298802 6383254956 2721325827 5734485355 8300022255 1331859622 8864977249
4481966641 5281904070 2879710950 : 8983
5677755838 3647075089 2928012992 1465508984 6527007269 6571688974 0132432879
5719821723 1190281099 0922494210 : 8984
6911519427 0447735875 2026602177 8729973938 0432917832 1634672128 8728433697
9031693485 9245577217 5986332169 : 8985
2291013129 9649345656 9456831267 2848095842 9250935515 6153586820 3373672201
3612851719 5799179067 8887948977 : 8986
8741557950 7858280400 5198795143 7931024097 3513754244 5229106658 7300786546
2514188208 0807307192 6898391350 : 8987

4925377543 7442026570 1651485490 3903784915 3357835239 1950918422 9410079581
7946261304 6216881844 1217468062 : 8988
2072287104 6251493876 4917833389 2585359415 4399135800 5859024298 5408557250
4489429103 1130668410 6105252152 : 8989
9436405894 2822561951 5090298853 4967011852 0896464332 0418793215 3336684750
0909379474 5862440500 9441979525 : 8990
9305808470 5730441714 2280778565 7037127947 5809345629 0877047988 3469716932
3551696059 1551290394 6546491946 : 8991
9769565801 0447721221 1529717885 4242063014 4935999036 4704881686 9639454598
7395664956 8446800827 9740648593 : 8992
9762888615 4206344959 5204778764 7960222248 1404518711 2205762128 2895120964
2426243976 9107779187 5989150916 : 8993
9674884969 0140417814 6248821899 2047215397 8970100410 0445191637 4635484937
7767240489 6305617608 5749019066 : 8994
4199208564 9882441665 9259136411 4979721105 7092004834 6356219112 5920531594
9520772857 2853502277 1786911343 : 8995
1709507474 1774046112 5977105440 6639288875 7183933236 0002445026 0387599951
7421359497 9764940400 0414409398 : 8996
6809319328 6423323138 0731072605 2347022269 9550297533 6413333363 7683830769
9122239114 7770558599 7784287425 : 8997
6964525973 0458979891 6184400911 8754738104 6980438055 9517006296 3032943375
0112437691 6592072295 3015125432 : 8998
1394054433 7789162781 9140621551 6820884736 3453419799 9887951611 7261028410
6323369853 4566227140 8982502069 : 8999
1286704441 1690258204 7965765068 0608338935 4490862114 3873825659 9464349788
0323272175 8292694516 9986312673 : 9000
5875109548 4558784631 4075971720 1962433708 5219967792 8830820417 0836282188
6710429402 4260058440 0437735875 : 9001
3310704188 8142219209 2460714913 3502963690 5846644883 2031947410 1734611287
8673517942 2094145466 0418534030 : 9002
1551815562 3214316574 7332666107 9898031090 6817008268 8732101936 4595617858
5173450547 2858980078 7287211541 : 9003
7256740244 1979202843 2253154101 9214013509 1238671110 3232137314 5940511561
4706721289 5932638196 7580376907 : 9004
2313032161 5824730407 0138858933 4636633597 6771547070 1977324954 8814517149
5615889159 7270403164 4349512185 : 9005
9747041467 1715097311 3294738480 8502107073 0048952123 7484215403 8998185951
3224901441 8572919357 0943752415 : 9006
9215545692 9631150144 9384703394 8930762435 5383423543 9507857917 7058758873
2868726361 3772313179 5763188119 : 9007
1749399736 4582955995 5961684714 4784415189 8543077414 5594300916 2727770640
0678452622 2188606338 1067248472 : 9008
6902440264 2674133907 2193530058 4244062259 4642539483 6856547845 0534349052
9674305897 4864956438 9293525069 : 9009
6872825573 0738865347 9795697379 6373941631 2512211357 2366124201 4026468319
8752349137 5325919651 5806193872 : 9010
6661939160 5104935926 5271321692 2096224639 6992453394 9416814876 9759450227
5693160173 7297825225 9321139227 : 9011
9726446990 7870797211 2927010072 8931641413 2897554051 1298607130 0454244972
1998255923 0173355939 9196662588 : 9012
6284890280 1610297741 4728147217 9960743046 8636839435 8376209663 7059217800
3581516991 2947673154 8326243472 : 9013
2529800380 0959587555 5451363524 8529233660 3666133452 1578492026 8506151949
2034529021 4617851420 3242331042 : 9014
2848635208 9687974218 4540038734 9417283201 1762737822 6479639784 6777136587
3511193020 7072225600 3750749407 : 9015

8103946338 9519984544 1663143229 7316080844 0498281354 3030383363 1635314540
5299148316 4256012510 6820856569 : 9016
0016030297 2916584678 9183221058 6994891004 0780107692 4778257280 6721865866
4493575923 7706601999 7260659525 : 9017
5433273364 2503894798 3366014319 9307308480 9345161508 8048076463 6667529086
6716936206 2492873981 4887990436 : 9018
5333871639 6911672736 9702731265 3742840860 9734869729 3255278854 1993019041
6842823213 9585796602 4873754065 : 9019
4392608495 3186341346 9468678923 5833606803 3944557618 5648701132 5964277558
2026319256 8099715894 4893454073 : 9020
5451669323 8449214991 1855493382 8244577076 6882305254 6979612822 4404159966
8923715929 5093923732 1195478945 : 9021
0740806774 4489003806 2443457522 4611555723 8942268385 9305152775 4976545431
8083490238 7291984674 8693162608 : 9022
8717921512 4829247615 8935141491 4158904235 1050735349 6796948749 1863344304
7936252036 5105567215 6988823952 : 9023
0349805230 1531223852 1251326166 4494737046 1248186099 0143956546 3727101755
6216112211 0472247926 5060881879 : 9024
2187856456 4770201918 7081740982 7426388517 8517823195 2934190481 9315715640
4001782600 8047464154 5364258579 : 9025
6882213147 1202195068 7073703931 2153332239 4296471014 3388176399 1811507421
5554226048 2199024500 8205203155 : 9026
1588031076 7656881219 8575038451 2044736027 9692388489 4398504077 6693919191
7803851311 7904637264 5787280056 : 9027
6499501595 7625302767 3424749035 5778730320 6946697620 6793710953 1408787466
0907190900 5478715022 7573861562 : 9028
2840311999 7936014817 4018140726 8559346424 7081865137 2676127973 4277641240
8940702412 2505759128 3320448767 : 9029
5083824823 3549006224 3196257292 8264805660 0967750928 5325730388 8341824250
4410194438 3749082928 9077044151 : 9030
8151343279 0126318627 0934410280 5833319718 3938084511 2478775779 0528799614
2480968537 5809766676 3701569484 : 9031
3487431747 5748991463 8891633504 3383627398 8511029559 0997268995 5904715112
9179455591 2698359429 3067385743 : 9032
0486989898 5594432619 8964253434 9217117176 1949868813 8115373601 1925283763
4812218777 1094392593 2205737095 : 9033
6269816464 5264593052 5413081768 0476849179 9670945909 7562709945 7464166873
1299851777 1315588620 7655433151 : 9034
0263023608 4922353201 8400246442 6949822200 9388561981 4174235294 2110120448
8878651762 0477231007 2355773711 : 9035
7569645402 6773786987 8293238488 4658685482 4307251322 4599718195 1763782065
1677017349 6390729119 7323152110 : 9036
4508388963 6900343634 5649771388 4180568029 8414053230 9783687878 8733235745
8437167785 9623193118 2129965442 : 9037
6422746033 1165621899 5807385709 1407481709 0777072060 1258255372 5598818255
4000170967 9090974133 8551791505 : 9038
0346241362 7962943375 2798039212 1612449422 8573480554 0929961742 2186755267
0663871540 1971649592 5804198284 : 9039
5727233943 5872738491 2980625052 2990823041 4417964201 8632393359 7564085626
4721140987 1027568423 2847105442 : 9040
0476927372 2795869343 2551623728 7061306248 9483176830 0595031627 3539272221
5559603719 1260927056 3209001688 : 9041
4464223997 4599076283 6038614515 6011467908 6719522744 2253415373 5630436368
0765820929 4481681575 6244075835 : 9042
4209445041 4818369400 7247871993 7160807471 4370480527 2412272057 6200148265
5673842585 2761520422 5756167756 : 9043

6344890835 5159040347 5597055278 1149851302 5087412165 5616058542 7292302899
3316547354 9907915612 1786647178 : 9044
1343392824 9941590501 4092363201 6984086805 9967723646 3118003230 9172314490
6596018394 4335732467 9947213636 : 9045
6714309332 2687259227 6995978663 4219848604 7640383312 1515982463 3481575389
1362137470 5062677609 4939156543 : 9046
4449665030 7157560190 5256149343 4123986500 8633497687 7258201426 1603587642
1886575309 1740518241 7491784121 : 9047
5303222383 0041880663 9385455889 1787620068 7881404876 6927605976 2638850841
8767172390 6882151375 3446907420 : 9048
5279687593 8629657498 6544177629 4251870300 9114961352 8443892051 4500715511
0873094664 9594990708 9979305234 : 9049
0129573493 8668817859 2724423081 5215906606 4996075502 7237608127 2387058512
1372745528 8861773544 5449593851 : 9050
5895687751 9518026877 9856482520 2662409444 8618828672 7054207475 0435367998
4584680211 8161245119 1791640838 : 9051
8220977886 4182756810 5850767756 5728648482 8360370249 3287158198 0604355587
9980375757 4763317200 0054495984 : 9052
9872516688 5657063033 5287606809 3081590181 4105937213 7856078810 3151292531
7504110509 6097516542 5371030855 : 9053
1748548992 8079279216 5082670247 7524637499 8378504723 4114872240 3887877968
5621658918 4157356593 9687030319 : 9054
3507502981 3828952996 8303573043 0607120754 6629980584 7951077322 9041914306
8162870295 0900718814 1342145828 : 9055
4156116327 6458979779 4318524467 0333572201 5183008067 7300984342 8145985559
4365738971 9903262861 0071674691 : 9056
1509026594 6427923755 6249374235 1217445080 3121349987 4102105040 2625411576
3114123064 0337384023 0248447393 : 9057
6132777143 1778326487 2278720000 3132437991 1584541073 2008325471 7655335778
8419738811 1987830811 6128253343 : 9058
5001379109 7326458045 6753562692 8483455102 5317569761 3783144368 2524778543
0693706314 3255096407 6224942709 : 9059
6972762106 1679816307 4586477313 6210291691 3190193505 3917363387 7209593077
2880211384 9522530852 3356420091 : 9060
4758211321 5081416345 5937327663 8164620996 4150418142 7926147848 5611225096
9744180739 9401218649 5761708774 : 9061
2985390839 4199011888 5877336373 1131301710 1357779033 4756204439 5262607677
9765685385 0415178002 8622026017 : 9062
3983153578 9490454442 7165705596 4920522231 8835447428 3111934696 0371194121
8609396474 3696835216 3008411309 : 9063
2122137612 3619315550 9118775346 4456042937 3792151668 9620242547 1680377818
2746385907 9682073564 0934299433 : 9064
4271792080 2887522112 5433179011 4149116004 7963896033 1877220471 4551925930
5894869335 0499223357 6520706393 : 9065
3665786108 0859200577 5957357706 0563469345 7603884910 8050669551 6093810694
3662128758 8273316132 2864831431 : 9066
4717672115 7046192356 1465003774 0538721762 7411136601 7823585584 5173100298
2077899936 4681776875 9805771969 : 9067
0442932656 4149288895 0616174327 3954534823 3166639979 1748498402 7478354053
5912002226 0943990531 2070766019 : 9068
6672743214 6673132505 9919615374 9191206109 2648781953 7779061425 3518922346
6139609531 9606252617 8425715869 : 9069
9243782660 9161717464 9716347204 7738961314 8671942948 2490291989 4191675830
8889233973 1174155541 7268094753 : 9070
3102737797 9970981756 5045054736 0227678621 0697540450 5926143883 7781516179
2537901060 6402291673 8026962573 : 9071

4343046453 0042110425 2766230305 5207247573 9306792726 3937131887 2288012695
8554904248 6632283070 2277401555 : 9072
2803422055 7317260915 9292751328 7204433777 2363815466 0224262722 7955242640
4790691285 3466474395 6703901536 : 9073
6644825118 6234027804 0253780886 6611353566 4410691376 9723882365 4053705720
3264851330 7118001886 2177768059 : 9074
7953218065 4367532102 2250428000 4399406185 1812889536 1407337239 5066311517
0700057138 6315302132 9368553801 : 9075
8489869696 3028510893 0120217950 6470724877 5032099948 3675687172 4700290558
1456984051 4467469450 7188717376 : 9076
3680287347 3556196853 1753075661 2015693057 0344309876 1497230689 5286644415
6407483458 8089865256 6166437972 : 9077
0289586844 2203921819 4317151275 6411177614 7563714059 3686400010 3588026389
1259692381 7062276371 6762874806 : 9078
2838160227 5941051146 2692288809 1294330277 6649594724 9738447309 3376327460
0371084359 0785997667 1800558687 : 9079
0287301832 2966729256 6511959261 0059415810 0365089290 6260399978 9107646931
0195227174 4645199443 6169991555 : 9080
6415641215 1087143820 8088680752 2978508148 0228623413 5318439205 6663971152
4608904813 1844519231 4929106328 : 9081
1540279224 8937822825 1545768271 6245961176 3956688646 1742395371 5865744626
6439961554 7890516373 2521825783 : 9082
3325356445 8989290595 1926058659 7986713448 2744782626 6678984191 9627360593
5202214966 8157043655 6904167082 : 9083
5752744588 1757281160 9561481857 2243695464 7505083028 4430753170 7792355713
2934876117 8390813029 1059918355 : 9084
2262237468 6711575705 9377490937 9757938195 2473316322 6623598269 5699804734
3344026168 7965475130 4293461624 : 9085
2661346074 7325269570 3114881469 6916429336 9071948154 5481790829 2910720694
2973187597 1973101542 6199335646 : 9086
1532836182 2870151559 0331070614 6530421700 6688253379 7013234495 0607141683
5268609881 3122722054 0903094664 : 9087
6066185857 9999141539 7814484774 1564082258 9035406449 0646351061 5433719400
4013861603 5071455973 6014278623 : 9088
4514865734 7962179784 6757021898 9951333364 4381929190 5300857739 9504523493
4957189684 6127113768 8957597933 : 9089
2349533208 9538145398 4677028512 4109139999 6240942861 5356154952 0156418899
6212593005 1264420968 6597252899 : 9090
4184350366 8188048075 2910597233 6008365482 3570191986 8550926035 0048765737
8829516292 3741832713 2367686584 : 9091
9464000596 7095067783 4536100367 4425949188 5819559592 6902512393 1107259512
1211563382 4158960673 7480071832 : 9092
4687784130 7809693824 1482915189 5604275501 7542065174 4208813401 4543607071
3556026763 4995975759 6004103616 : 9093
0961213773 6218202235 6398010145 5924936015 6897148979 3336585499 1863497304
1034950079 0550971037 3294892197 : 9094
6405886995 3201896649 3350820431 0048852305 9429848680 1785556564 5389715296
3868713982 3938927886 2831305388 : 9095
9870441633 8748532366 5505625430 2382861317 6831474399 3446156093 1076538494
7584648931 6215158358 8989339195 : 9096
6732944334 7903909096 4500201525 4529742236 0933487377 4857090601 8648070516
9912575593 3251820304 4120573389 : 9097
1169249497 9374444181 7210180048 6952748158 2486075771 2217241382 9852529767
0352688504 2133034637 0320590111 : 9098
2769270842 3122473740 3990344676 1895700102 5917858966 1470456118 8690554318
0001357411 4545384809 1623845601 : 9099

9398214576 9801540367 4473093324 2141647275 5521908773 9691741737 3506414595
1846078512 4018137745 4588376298 : 9100
5179066094 2517996950 3658723513 2911540694 1185580040 5756107804 3579191051
5438952930 1786070568 8578117217 : 9101
4213915509 5320721197 0898415221 5425316479 1930461604 9811760099 9404341319
0991189215 5136512261 1550181311 : 9102
0735194067 4896418609 4028486928 3055022119 9243438663 0966122998 3761658981
2747306690 0407133131 5325819303 : 9103
2028149674 5570289271 1980502308 3429490724 6105491095 7995507893 6602634698
4656628188 0554901043 8789895744 : 9104
0931415296 4143377690 2605064364 0983268217 6336287098 8262723974 3023005506
3851675289 2264837509 5088613721 : 9105
9833353460 6984890685 5685902444 6788863364 3960437818 2364931607 5069795253
6617770448 0786285210 4682093268 : 9106
2668289722 0715910980 0819778019 2649525383 0472463607 9589391737 0036932896
6358022050 6598028533 8702996809 : 9107
2228675427 1291338699 9402663357 7360863754 0472021149 9273339955 9638671394
1415995063 5550381622 7131799287 : 9108
6143298924 5958663210 2280507272 0176632829 0281395136 2463925987 9408411977
4242147849 7488792853 4813292261 : 9109
7580429695 3605684964 1633358836 1246477604 7176630339 8537726717 3732323243
5192979733 4237646067 0072590569 : 9110
7847782259 0102247186 1849551087 0041401552 7634922430 5850649791 7469994122
4701667003 1010443276 2653099301 : 9111
5284206842 4685952359 1053096968 1058431185 5103760808 5368103331 7095349081
3488313117 2235937743 8741462183 : 9112
9265017156 0903279402 8189935612 6944963967 1433207829 0473191666 6780851825
5197717288 0062773545 3991592789 : 9113
0340107862 8896366115 7080757926 3712515753 2125643458 7976758227 9860562178
5390463443 8782602247 6983164473 : 9114
0911677313 7698654394 4139748134 4800381829 8103754950 5885398354 2914632275
3291226062 3917829319 9621398691 : 9115
8817711118 4244196277 1878992305 7350447245 7753831194 3385179322 1285766035
2121687790 1140477658 9767784356 : 9116
9635136532 9151493096 3803910475 4501169967 5878027999 7955389558 5500590455
3329793563 7026407703 3348112055 : 9117
9679109660 8804654458 1991175696 7335381794 0202977420 8446714055 4762538001
6657961957 1991262008 0781668202 : 9118
8859158624 8572361559 9401625547 7707914111 6006764078 2608077107 8947343728
9911567613 0685073224 9631591231 : 9119
6341975884 6276472881 9202367627 1637519476 6953325420 4908916102 4916483733
4965917270 8001471152 7101290890 : 9120
2961211040 4724620656 2282096328 3626670888 9728464849 1945054852 4147558133
9237736269 2127662809 0107039603 : 9121
2994626272 5094714117 6912142913 3539751301 5143177467 1685840290 5968622217
0801110366 6071463020 6206422073 : 9122
9673674027 5444511153 1868035737 1197061263 2143552346 8555443824 5653255194
9622309244 2226276161 8107635327 : 9123
1218486711 0387486332 4167078904 6885223329 2111150197 9009872376 6740155479
1675344748 5891162812 6868673604 : 9124
2229943560 7682697833 1735176394 1137568187 3118531093 9147331613 4714642957
4802588661 2098433336 2644789232 : 9125
7799217189 3811049025 7508983329 5752311385 1163841181 0192449913 2930087784
7253627365 8801679273 2391195667 : 9126
7377346029 2311671472 5275438773 2395409644 0741744930 8810335690 1689944732
6506293568 1240746859 1689254650 : 9127

9211091423 1643396643 4965355399 9052260304 7114871175 0195108603 6214378879
2840744982 5270332425 1691779534 : 9128
3239338053 7534154286 3344920057 2757968191 8742184272 1885884666 2660281339
1591226508 7032955629 7431210060 : 9129
8476462382 4061202097 4088585109 7134502445 3456269674 8452174937 9519983660
1359599598 0442105533 9305799463 : 9130
5412156592 6037395454 8130709002 6816164735 8075309070 0579469859 5121857669
2820433593 1333658021 0439358016 : 9131
1079082794 2664487820 3528015749 8477771875 3666388687 1469284922 3359797020
1859216375 2637064707 2392328071 : 9132
1774975523 6536241706 2631546327 0059026630 4024739804 5335302040 9393130497
3971307917 1815148863 2385160351 : 9133
4091871517 2725963206 0397751818 9877379429 8335489621 4929883065 1687972617
3233429518 6029197912 3542091466 : 9134
1761858081 2065785097 5540518126 2454785358 7142349872 2824507628 0218555416
4393735572 8734131770 7953318264 : 9135
1069580231 8126782729 2621724790 4786733132 3026028790 1476485433 5809993244
3723491884 9958599486 2583067600 : 9136
0122047336 3446686800 3021774428 3089567321 2065731090 9298521268 5308293535
2033162609 6123871927 0474910316 : 9137
9411516483 8847479745 6771234335 5744298126 8446143275 3371060377 0238115873
0688628896 9394132363 0060605042 : 9138
8996520045 1060374867 6961364917 2511721417 1045397236 9837657482 5092862531
9917610379 6050507004 7452751987 : 9139
0692438307 9720813365 1074580862 5339870452 9503657739 4794375194 3255366001
4210556464 1482243606 1646770791 : 9140
7165851176 5610859235 6346094854 9764477962 1165511318 7009699029 1407315148
3903908991 8159185783 3265027795 : 9141
3957841825 1970561524 6751810745 6330457082 9594428891 5066671592 9760412803
3547451551 0043994933 9911357400 : 9142
3681082145 2010037166 3337695212 1337532395 9064451506 5233379074 7504285781
5969527569 6181784704 2381784203 : 9143
1599241711 2157281753 1382552899 0831722270 8031933401 8499746246 6150686413
7178679359 4805932728 5196433573 : 9144
6880274143 1586900765 2087234546 6373639831 8691202096 5620754134 8874115504
3517945705 2021920866 2862157046 : 9145
5012959513 1279374407 2467620419 2266556744 5333444729 6817148735 4493873384
8016654282 6423783384 8317565438 : 9146
3336174408 7321879219 9714309719 3907561528 9979919334 8168456648 6989431576
0143802862 6335331361 8572379316 : 9147
7236606367 5494380052 5296713997 4035099407 1219337375 8571204555 9496028444
5640461306 0336222636 2162934122 : 9148
4576151165 4193879168 4813280962 4695244456 9546212508 7911893539 8322196378
9994987057 5517487718 8610510452 : 9149
5870912001 5502718111 2140083303 3945999772 8658704523 4191667304 0685570047
1728611726 3358849682 7107174500 : 9150
3538903363 1066658091 1221611227 9535205973 5631542387 8627922117 4002792992
7660272309 1008788964 4867197751 : 9151
0644852854 2367606806 7832870271 6021491220 8907383598 6791677907 9846546847
6544328863 3275459268 9976471361 : 9152
1821919363 7197094309 1897609589 3307419509 1535789981 5945626817 4031091186
2136112387 0326632874 5925123801 : 9153
7221859237 5964203971 7801197330 1354548630 3115628764 5397333010 3535199368
9089171658 2118447202 5394047093 : 9154
1783306012 3964167270 9312163693 7919332391 8425977305 2761479229 3021230131
6365295613 7623330528 4546377449 : 9155

6678385572 4163055532 8610532755 2078438940 4424723308 7001494007 5648539493
8970856366 6247235115 5496842637 : 9156
0742241985 3407218843 3171180862 4785109999 8176232258 0581202049 0727023675
1559960385 5846672839 7347325959 : 9157
6127104496 9489969280 7040872355 6135501883 4860982733 4494211927 9511596389
1421701337 1362540595 9158400657 : 9158
6371033621 8594354090 7214950797 1926424741 6878866135 0962013130 3193981656
4431842319 1036741420 5125568633 : 9159
2809855207 7093239955 7422045837 2892438309 4811084233 0087641536 6308472416
8976375194 1939984808 6392769531 : 9160
7901643727 8029776888 0616249084 1933764103 6450961260 4065127369 4733432136
4751668674 5418754235 3324904525 : 9161
1400126199 1025504942 2060899086 5348912185 1977852080 3538297935 1647361636
3948528497 5628497148 8562703642 : 9162
5437615253 0348567914 2181383415 4676563036 2935943271 5688885113 9645341755
0113555234 2266095177 3817818038 : 9163
9386443090 8305399273 8653198839 2370825144 3497669579 5125406640 5582132495
3476082446 4237959520 4674037169 : 9164
1040228650 6016440118 8212816887 2783923427 3692926062 0640964091 9596145904
3145172341 6161791510 7061776717 : 9165
4151129700 9743626357 1691798097 9131076075 5444007274 8231658536 3917076912
5919005551 1285073280 8167705134 : 9166
7490741450 1195024810 8427677735 7730810360 8450037555 6502686582 7089490664
0961146299 6904292269 8380843496 : 9167
8138914924 7988622487 1671281240 8926279700 6509374129 1428012018 8192206542
1593897363 3819322591 2707130384 : 9168
8942162931 9110049071 4922536282 1862035617 6446854469 9594307641 9072713387
8182633847 9026905141 3488524088 : 9169
3415970409 3166717645 8485165390 4600109634 7293231702 4526860807 8649180077
0245426053 3859200916 6331507927 : 9170
7873248325 9016044217 1566874940 5791518967 7115913189 2750178044 5182499374
3874329932 9143554374 6809468340 : 9171
2608346425 2681707351 3602678441 1711754768 0302578284 3274127129 5550926710
8574023047 4696002644 5711893018 : 9172
0581121892 5757250024 1791066473 0201129469 3754953338 3927107678 3815855808
8756706132 9996499158 9394990408 : 9173
7497782355 0392105136 3016467163 4086226936 5394034567 6951865277 5268560312
8680881568 9169916046 0136793560 : 9174
0028878486 5017387036 1186136616 8233700637 6249017187 0354839165 3008880657
5237376799 0681554788 8893864623 : 9175
3804336788 1447386263 6975144463 5331513645 0336525098 7795413093 9941467601
1222285012 7827345575 5159561984 : 9176
4872672888 6216911391 2786444182 6501071593 4333181605 5288098093 1375760219
5448423668 9181404876 1296983574 : 9177
0368011755 1891330057 2269947591 9228724396 9471072449 7704047329 6751338485
3728989198 5144879126 9339956272 : 9178
7628630157 1782705735 5238450193 6652886942 5030157128 8649098993 0558977451
4806497400 7108137602 0676606100 : 9179
2833539832 0724359456 7205949451 2168440253 0561416115 0472376796 8712526931
5631930981 6082329795 0425898166 : 9180
7480087815 2648677364 1449356958 4287953879 5111120900 4138824350 6999888209
1565554032 8925022880 5141696787 : 9181
9299266268 6222467052 5490667495 3625013269 7003182451 0114073519 2981527091
1682876316 1525453362 3132422680 : 9182
4522288961 4970917397 1135352554 4012360861 8815454147 0853204672 2994693907
1488188603 3268282617 2282696478 : 9183

5169840975 5613280910 9049299420 5890209975 8680270118 2971438113 0616650165
6069405094 1744708413 6593172946 : 9184
0368323148 8678378340 1584666526 2779381103 4718565273 4290112646 9689951352
2043813883 5925408450 8757429340 : 9185
4830480525 7026367468 1999971113 9249943082 3809481473 1925760115 2853824735
7208314910 5271608169 9222814186 : 9186
7532991179 5524477487 9202469824 7835770179 0581768433 7666777689 0217764906
2193699589 6546765996 9428721801 : 9187
0978136921 3674462209 7478300409 2718190513 7635612325 4861272145 2226168051
8029325681 8310931413 9665924531 : 9188
0344236884 3397067352 8726638300 0454195146 4423032623 0190718975 9856124702
3586500542 0759825248 9819907503 : 9189
1653803249 5026016937 2305831481 7314752430 4359424989 1487918906 2802634091
2272673533 4485377798 5327688970 : 9190
4761672615 8528835140 6035252708 8519929217 1330705785 7638749393 7455594009
6761537521 7782801162 6903772652 : 9191
8989620344 1261598810 6321682532 0644381640 6129171172 1200955674 7383916722
2962355574 6124390155 9905448832 : 9192
2626441625 6871268704 8500344921 1415757614 3154878838 2262449382 5719072052
8224356540 3066864339 4952786639 : 9193
1978261966 2128890293 1708091506 9335476093 6306950387 7964838065 0097087712
5842074421 1499716985 5615899897 : 9194
4787651375 0578536272 4536521780 6628977750 7327157034 9854774716 7890295666
3958351111 9977254308 8210830083 : 9195
8719703001 6360375482 3203181103 4519634199 7195708016 2637542560 6969661834
3629726907 0662230614 3131863618 : 9196
1161133168 4184951612 9647994635 4081551662 8864531220 1056179623 8101443846
2014132524 6851026413 7934116621 : 9197
6666044355 5433967260 8390029334 2498560592 3047725430 1604859689 8781615324
2523488947 9927499568 0405750878 : 9198
5961584656 3996882770 5058248080 3752624440 9922842655 8107196531 3962147422
2234153507 7003136186 6522902424 : 9199
2427339752 2322011973 0089596891 0498540544 7427697563 8059626226 9087884764
3676551937 5681951996 3044228090 : 9200
2471965977 9814112299 7611309966 8948406547 0304306161 5428405289 8460555610
5277431670 9454797654 2569994432 : 9201
5615151270 4117768402 4726299051 8468739384 4031749092 2778671374 6504877565
4003526182 3361358220 9691595165 : 9202
3100302994 7026121379 8326995515 4794300452 8250404116 1789922994 7911176412
1739926937 7416582020 2835024261 : 9203
1557953577 1019286950 2646054359 2411800668 0782334174 9833422352 5119403957
8690357868 0997957355 5664634818 : 9204
4109235356 6380532162 5058733961 2730165179 2091526963 0774160353 9343614876
5086569589 4416687593 1028197227 : 9205
0842130060 6989032768 1248136434 0882914506 9353500784 2690028338 9692890036
7663065196 2125691137 0825149526 : 9206
4130730020 5723426006 1434794784 1846620763 3742474019 6523490639 3029662233
7730820640 2287040880 9540394489 : 9207
2602375593 0275783818 6727111955 5903626438 1803694410 2698956099 7022402685
1892905705 6341157634 5663453530 : 9208
9178364491 2706551465 2145274516 0957092696 0198193514 8250423083 0933240208
5693823257 3732465561 9783805079 : 9209
8236783914 8964413212 1190325383 7193051261 2143512054 3467213802 4917208445
7240675607 8389118361 4420617219 : 9210
6093241887 8715390653 1193456242 3143050595 9758138968 0014593272 6803699031
5314858981 7842184140 8627035413 : 9211

2340571406 3724233441 6230520114 6005372433 5454408580 4784915273 8356053700
8329841944 1940878577 2894289429 : 9212
8905564111 8489012798 8174242713 0941732502 2464998977 6184995844 4824319633
3877136064 1700505758 8112062601 : 9213
8903546125 8593451545 6181756840 9731473384 2014951893 7581589960 1208752575
6276033295 0030118318 8095642910 : 9214
8679299364 9140874263 2266721386 8491522412 9903291462 9320268237 3490956625
7903206428 0453385167 5572566335 : 9215
9643282983 6906797154 4894914414 4284457366 1312147165 2577292832 2838722519
1227818503 3318457537 5231181388 : 9216
9104687301 1202533293 4330332281 7674447909 2066563250 1883887499 1783124527
7956878032 5185708787 7108213218 : 9217
1754229913 7029990346 3408243198 2200181814 3016950158 6756477231 8455173516
0193539741 1806816255 4986334692 : 9218
9742793638 3683122862 0901500847 6329602715 4205540923 4721977487 5557737277
1253584379 2997336755 0413539009 : 9219
6260754601 7704783200 9209000043 7030477206 2396931123 6199692306 9451921228
0751280626 1090339608 0855119939 : 9220
3625766456 0584547489 2984566105 1643776323 0204762933 4883313664 5533457348
0473571567 4449977347 1782198157 : 9221
3926294356 6148533256 3525738007 5373424585 6962732264 4329253912 1854835008
4718726153 7611935992 1175544946 : 9222
8751722095 3402171496 7332300854 3031277343 0084421703 9223565805 2374699781
1952384744 4933383738 5774851142 : 9223
7462252203 9346757212 3278506610 5269132797 7306346288 7372622241 9584671667
2022151680 8291000526 7022364151 : 9224
2652274077 6004619794 9668504424 1492903303 7526153247 5565300931 5314557741
5607854888 4372041571 4060087651 : 9225
2807613311 4000215176 0928982489 8629450626 4798639727 8120873344 7929847854
5315123293 3405140684 7255746928 : 9226
4862631503 5477092571 9144201422 1858878025 7279128331 1779822123 3680779311
6875865477 7139994623 9543986001 : 9227
7821714044 5115877933 7645825217 5919910881 9238300516 6331028283 7236134127
2140722462 3795391293 3883641879 : 9228
3155329932 8948798748 6153861391 5230746891 7410066261 8607772267 9134871363
2214751656 8508441991 7806948619 : 9229
5460193408 9370819232 1419263827 7533759194 5703264502 3630434756 8717345295
8399553670 9739473113 7451394332 : 9230
8197791122 2269397254 5912493837 9823126607 0963822259 6701900838 1453286290
4610606586 8563209780 1508542233 : 9231
4848110590 6173852298 6205281789 6049500732 5704272220 2039361363 8247958310
3543259855 0726214034 0985962778 : 9232
6017216895 5987503032 8828176804 0946852093 8864033636 5236494428 5765333810
9795334202 5875230660 9947377791 : 9233
7483409964 0562083733 0431676710 8759298266 6684354670 0959970485 8953748415
1152214502 2499454415 2838657802 : 9234
9285301765 8562910138 8144172669 3837902070 5003419101 2138679134 6354652287
4814071533 8202901919 2351467212 : 9235
6838275100 0173948051 7922357591 0310629411 7826715838 1863781954 6488431229
7363020759 0729496131 3226423551 : 9236
0849102649 9847418870 1812740398 7203067935 8312315482 8787803868 6720763454
9849519911 3445099124 4247310505 : 9237
2272527668 3206603485 3805673485 1263693194 6652992516 2902626465 8941634139
6091509721 8723640275 5002697010 : 9238
8838683249 4142125712 0488696456 5829636160 9865368598 8378839028 0207060702
9639962089 2916924201 1756462921 : 9239

2717841443 8660944484 1530713275 3827418051 2475604700 8456141960 7860495448
5925581307 1615271768 1871096104 : 9240
1702864624 4510638699 2799031329 8023938322 9230786002 4611121256 2537492992
0696236055 4973977933 7090550915 : 9241
0615995807 4626476930 7061465473 3657295388 0108465930 7737092643 9327096173
3589798755 1332985173 5335805761 : 9242
9820375607 1739649512 1026056824 2153539432 2065787806 5433368166 8379183925
4310296299 7862558313 8150842902 : 9243
3460414642 8506331820 7802667408 5750429654 9353954494 8651852756 4708814351
3231959734 9789917141 5169373256 : 9244
8833893316 2833896451 8488703226 3989305568 9451839191 2430829325 1565402367
5385004309 4552275229 8621936349 : 9245
9930799560 6896844661 8745989474 8823413664 0851885321 9367311437 5894635657
0214222303 7174148120 1272628291 : 9246
0573318578 3922733479 5260680041 3122404444 6906957003 4326579109 5617342284
6551383028 7770817092 8004370327 : 9247
5264455762 0090294898 7017264718 2289327617 8823467995 9538966801 1402866870
5263367060 0630426129 9460849499 : 9248
5638275599 0602647776 5219702537 5830641181 4612875438 7609857828 9963422105
9502253415 0439826096 1876098352 : 9249
1652316543 3169772144 1251770038 0390215981 3797489132 0292927755 4387117033
9116322480 7524657249 7296231247 : 9250
6509351794 3567483811 4315286413 3302908912 3777146612 4690448645 5116492679
9346341556 2118822817 5642302405 : 9251
1694895444 2816831414 0490438057 8860590107 3700671829 8499365040 7494702785
5738627203 2710842602 7326956900 : 9252
6412015558 0946913710 1298425529 0544957645 0645756003 7403149458 7908210547
3559113639 9067278064 8145919170 : 9253
6433870697 1477366524 7784433863 0255698388 1025898793 0950197131 2840708918
7196967493 9400265719 4057221592 : 9254
9586883457 8669810318 1835949381 0271931161 5251530174 0904031945 1723832245
9633052678 6264210007 4573633679 : 9255
7264614352 9714988846 0552919078 2295721345 6926463834 7921759405 7805130367
3488795449 4733446456 0679667691 : 9256
2782679904 9420036288 0699002603 5221665252 6648809722 4672121294 6167822822
4742717834 1053585849 0938180843 : 9257
8207696712 2622155649 2524464101 1600663839 1181830873 0856354226 7215017218
8913491114 4340742316 7201858015 : 9258
4409683941 7218455292 4703066633 1743969920 3209991372 3079392087 0633268149
5027024183 6323739355 7565948355 : 9259
8643427585 2715303647 5346746011 8162312180 8611137993 2483545148 2289863062
5369332793 7473726404 6931267375 : 9260
6534019973 0090761426 2122865011 5856894482 0803714283 6120485831 6174750390
7712876046 5033612361 3522431214 : 9261
2049114096 2045858292 2554357490 0902717114 3100562027 7966427328 2036840883
5142189973 6766128515 4174170155 : 9262
0559669295 4335533849 8868702324 9020610644 5807169228 6334339185 5394434659
7418310331 5453291025 9130360646 : 9263
2266687977 9455734904 5467488232 7531737599 5937232273 1037104452 1133115338
2893042477 3972419572 7440116541 : 9264
8484315564 8940489213 5805570855 7627558495 5348891913 8564379163 8342408939
6022097880 1958750476 1416457873 : 9265
3843443198 0873515751 6674968200 3791537961 0297349443 2109476073 2700463633
4366125907 1179260382 9657765048 : 9266
9833996820 0528464234 2068544946 9930387124 9646642485 8116044200 0466693398
5741685551 7298369829 2635849104 : 9267

4717933844 6832504338 4471758725 2699366862 3375707985 8637995117 6474378774
2210295932 6217388171 7992112564 : 9268
9607665490 5036475301 1284605971 9986422397 2784339196 7774038958 2319175573
2599419379 0085492825 9806607678 : 9269
9498548433 3355330520 4429781468 6422621546 3907056678 0479389131 7765192204
9935761663 8821963223 5722413875 : 9270
8048818728 7554778343 0553371416 2429159181 4407249101 8337360725 8613130585
8393796369 1373160504 6386537876 : 9271
1619976568 3527896039 1654122119 7123163706 4638435087 5058804657 5531967200
8048106320 8311821537 9561380098 : 9272
3535595260 9363700064 5317080644 2028883772 6690826800 9424750615 7736530695
3699946473 4442641799 0880723658 : 9273
5691623899 6365175780 7623731861 3662803000 6775952545 6983035935 0209310340
1066548823 8760590630 9667152580 : 9274
3190270180 5651077417 9659964177 8895066406 0278847170 6807792755 5703510222
3714730679 5006509607 5380534263 : 9275
9820261540 7127213785 6032274328 8616802417 3389459790 5050321379 7484661490
3095301740 2300954957 5261795889 : 9276
6983609703 1429140848 0458384201 7705933308 7278988292 1065398608 5497841770
2268001994 3172312560 7279669350 : 9277
9378461673 8081453471 0813293763 4521964744 1631933117 8690649982 4823727616
2056150244 4394472323 3791069608 : 9278
3968856032 6743659447 6132436686 2391058343 5263725870 2655272723 5468109736
1367537998 8543402247 8297321958 : 9279
6473847079 8498514172 8538675277 9230658409 1743206050 1099102238 9298189386
4572160416 8949234020 8559404805 : 9280
9798887199 0753899448 3624575918 1795872647 8548243687 1784280511 8165701035
9994896167 5645814411 7743599941 : 9281
5574156405 4198094077 7060781817 8732780883 9235166527 2981172947 0451824894
8869402539 7849704040 1257850170 : 9282
8525229480 0326448553 9829339541 0250493410 5444614356 1304537123 6961682202
4270875468 0322577722 4676453869 : 9283
0691735846 3290996597 8927085724 1360685294 7228418998 8811197694 9257775673
4731492045 4188249935 3860754485 : 9284
3832734931 6024944583 0184005201 1005971211 2248818992 6014090339 0584301410
5055980718 8444154763 3560933892 : 9285
9558270335 6383918920 7244115662 4136346793 7554167389 0893091868 6080312637
8923091291 6607550098 9808404308 : 9286
7717386876 8493062385 3335091504 1060003830 6016394885 3687921061 2389410574
3940346062 4016371854 8425217716 : 9287
7545163976 0025505022 6439611525 9942943086 9349869074 6297837599 7016129503
0843803660 6600589226 5852930563 : 9288
7886695846 6784875726 0025329183 9307185472 6101201435 3181230082 6282453907
5652638481 6628430671 2414091535 : 9289
3230173735 7772231705 4545331857 3303986361 1629092807 9651400076 2580295868
3252113035 6252134998 5400678329 : 9290
0579810026 2663767805 1720624754 0163537025 2168218735 5287204019 9635961887
3606934730 6728409608 1288649892 : 9291
2816545218 5240832827 9128184938 6363527220 3008598275 4459989899 9583511157
4368787888 1270485571 7381485740 : 9292
3078036294 2048594206 4433415790 1693839596 8153358527 7508781574 3971924322
7798831706 0546340053 3096961159 : 9293
9543732039 4129955177 1974092493 7281938691 0424719168 0745755805 4131728168
3365537965 2759510402 5829376600 : 9294
6937948476 3050243686 6930874986 1291151557 9029908914 7511471436 1655097781
0891689159 3890432286 1163098089 : 9295

6160154365 4239707131 7339876255 6138393349 2789060574 7145381691 5692648820
1510262147 2183250340 9165624542 : 9296
9353117328 3968374175 5506978877 2460439855 6261085337 3740287709 9728804761
1491577857 6510475290 8911381780 : 9297
6546922207 2171325415 9467978055 5957405449 5325587792 8432324750 4820257296
1072119305 4272034454 3111901843 : 9298
2651599832 9511924254 9956886629 2451206155 4435485187 7843376022 8457318552
5530203857 8067996423 3347394328 : 9299
3255079768 1431749352 9036535523 5708336227 2954029760 3622459678 7022467961
0872900653 6915811032 9772411271 : 9300
9687184631 7120153108 7202282912 1678513683 2868288998 4100630830 5969973295
1240183437 9280807586 6877889849 : 9301
6043772754 0275895229 2936685392 2675139928 2371615964 4737329827 0175090837
5680274466 6915911114 9779944667 : 9302
1135691088 9243791993 0942472130 8073081984 2625931443 4796579085 6700825628
8588361144 6330706901 9106360685 : 9303
9951853870 4171062385 6804324112 2994069976 9765217189 4899349718 8045038643
2175982864 3313402327 3173503448 : 9304
7552793783 4864131041 9949659552 5770669045 6718435502 1562018967 2793973426
8216256860 8592248811 3166647341 : 9305
4298910138 7571275704 8305145943 6636099246 6106272011 2440987239 9971042075
6543915068 6310201357 5984601467 : 9306
3026511990 3429865063 9676000696 6879582828 9243397825 9058748567 8262692633
0346837233 2124066157 7602951565 : 9307
3722610682 2903836613 3683415049 9989593428 0932018654 2470360735 9076560816
2195997593 4382017246 1807695817 : 9308
8347221271 5039912393 7330805981 6434946231 3671749599 9946304211 7638181478
3019102133 4473569265 6258805710 : 9309
1644687984 5566172037 5874281490 9984339304 6523931220 0035644248 6502802002
2103872381 5085543608 0610859538 : 9310
1742785324 6549792311 0151018126 6741384662 9462674003 4072909243 0677564917
8579342775 1652950984 6000986282 : 9311
1986519350 1486314131 1338234081 8641810195 9888722959 3585603437 2242367239
6015146278 9656548533 5331740074 : 9312
1984242601 3606673552 9840754402 5731777149 5402192754 8762566366 3209795134
8389232398 4730934282 7299390954 : 9313
9226286852 5828037625 7104081404 1407092153 3779824712 1933464825 2071838785
3744500723 8525936056 7659576220 : 9314
4021945192 4792912413 0758546485 9181278455 5951253394 8537732743 9546532520
1686225053 7285001304 5372400046 : 9315
4744479074 5978251029 4447904759 7268994937 5374692808 9331155435 5051420516
1236368344 1007349842 9947070865 : 9316
3487282616 8261194995 4445988885 0359607914 3671119639 1393209112 0095413351
2885508992 4933928537 9476656164 : 9317
1592545275 8853479068 0348593042 1014317785 7711724511 3741846243 2155337224
0565412149 4232234673 4100328640 : 9318
9237227571 4731703809 3058466611 4105286653 4929215704 3843719387 5825489184
9894465897 4892112398 0435592536 : 9319
4919086589 0669173908 0886750091 3233054266 5482077157 3640252081 6248301658
9873036086 5980837961 5767364117 : 9320
7362734669 7615666539 2134829345 6423991292 8078599798 7881520429 2215190914
1690785497 3618725169 3099991322 : 9321
7000674997 2335155765 7951436647 4702374876 6149644049 2613486082 9976097836
2604927823 1738894979 2246852097 : 9322
7599508049 8826972392 4957659872 2306469511 8767799160 5495672699 6908515258
2265295227 3858854393 0217342747 : 9323

5574391874 4113766339 9412859483 1234384848 8127945760 1367100667 6165974395
8963255453 0670816842 9451211409 : 9324
1221200910 8666989989 1500102055 6924848523 7225542131 0716613919 8282765742
9818829175 1833720841 7523869676 : 9325
8280591023 1519925312 8014453772 2164743682 5950860888 6364434672 0804079957
4561042901 0196560880 8393098287 : 9326
1606160491 2163604586 9086222897 3756455741 3574307159 1089367242 3316644773
3282968241 8831492171 6494972514 : 9327
0119493690 5670952961 5270432919 6175641010 1851405960 8395422101 1253004320
3244772904 5095686728 6869283797 : 9328
8994453347 3254078320 0542835488 0451308874 1363193695 5816828746 0790465669
4590040744 2884187381 2325676996 : 9329
7164692679 9869558860 5208063729 8383211238 6246812028 8170434815 5814062949
8830626345 9334499888 4038652605 : 9330
7374223073 8578664000 2377415312 8859094551 2575353933 6940869444 2939407522
1828471001 0776658095 1275670201 : 9331
4775408298 2594365539 0077790618 0300371040 3019109218 5293284782 4116551890
9228702901 2404160045 2149317093 : 9332
5775360816 3420856552 3320144384 5388958342 2068417138 8239953227 0635638724
2611330217 2360887531 9692482011 : 9333
7906522275 0848154653 6063468432 0835251951 0833216068 4317334365 8468059130
1574087870 1822959876 5822830020 : 9334
4252835566 9320450198 1988171458 2611715984 0001172323 2484622368 0233784983
9057195842 0833941883 3025000410 : 9335
0260037883 4221148367 3054749609 6779242970 4499983799 4790440543 4971089626
5896766913 0028499096 0386063046 : 9336
2400933379 7909203575 5162551666 4005711218 7717239030 0150396095 4051845816
9993864304 4980401039 9161286593 : 9337
4744955827 6066834824 8909337338 6292669896 4697053174 1560892296 6624289143
8192737235 6726030305 0110341597 : 9338
0150390759 4115991561 7925116562 2892441767 5577202639 2710897852 6059947131
5335700459 0483012453 5860224570 : 9339
7716058212 3321852958 7582203051 9372890200 1773432061 9428734214 7523788608
3007029979 6531556810 3011287992 : 9340
5893918338 7796470067 5202703688 7724058406 6436919090 2874387633 8820971458
0101749510 1346458402 8127801131 : 9341
6813989780 6509007407 6746422096 3899804533 2620765149 6082597745 2275842390
4134502684 6186168145 7953371759 : 9342
4622683030 6366614536 5992028030 0843252851 4981788177 1272573867 5355028513
3836792305 6743243686 9620272756 : 9343
9049569472 1422424679 8843604119 2263169155 6788276484 2223911962 7403671459
8974144543 1800616886 2933763562 : 9344
3975254816 1092018062 8944206508 6508865174 3884451744 0293615708 9106653051
8191344083 5241738539 0895294733 : 9345
1269090022 8814761735 9240547275 5741008722 1186024807 0655273478 5464670810
0332528804 9487281884 6476645138 : 9346
7194846470 0273983663 9678691108 7224906894 4525449930 1361359823 0210096649
6626582497 9074179330 2604479616 : 9347
4678961217 6304735470 9410590547 6778743627 6981114648 1959467653 9533132602
1604518805 6852012381 8538359935 : 9348
2509055867 3048169589 3931266888 8710724516 3758078691 8529804644 3759849390
1498640886 7291215615 1469350544 : 9349
6800392775 3771628002 8444617088 7283461332 0160278469 3514171037 1898135928
5654440473 8893533643 4225299535 : 9350
6306714864 3575822661 5070847224 2121395749 0588123647 2608079566 5391821078
0697591919 6272996137 6825052719 : 9351

0168013501 8259365039 0431489237 4222182997 2943591050 4766510198 4354967715
8363903560 5090274409 4547376200 : 9352
0866255189 5379897398 6955249442 0945283689 3729162255 8644588185 7232200509
7340211242 0242701338 1380975063 : 9353
8707368786 2241346162 6607614186 5890367575 6728049468 5139294924 6947497670
4482862785 0379939428 3278791220 : 9354
3329713975 4384364472 2779419524 8300533133 0832659412 6816548143 1836724185
1907164537 1183945618 8537671861 : 9355
1463451009 8763556103 9688240346 9322743163 8685638936 6920782628 7866646316
2305865623 2080344670 3322414896 : 9356
5844290862 0119179775 1836078981 1784708762 6296153194 0034781546 4050634565
9858453959 3367839204 7177816196 : 9357
1151978159 9153348323 9756111621 2221045289 6830871383 5459988065 8578013548
5937490426 3956020172 9578681154 : 9358
9407988789 9895278594 4953129124 7582481713 7108859096 9140706193 3036180030
3323891913 2168402423 7117855941 : 9359
4793817512 2615373552 9282048462 0119087835 5782410767 9589872826 4838018836
3060516258 7458132380 7170212707 : 9360
0311601599 3195665321 1055908684 4637230112 1939352882 9932843569 5606597198
9301484189 4192469651 4195047131 : 9361
0036209138 4684087143 6786883238 1248718738 0582217796 9166872677 0528694917
2329692975 7412937157 6503104861 : 9362
4982649963 9425425153 5522789265 5817659328 1228135199 0499862833 8917695098
6498709388 5286522464 1624149800 : 9363
9133604809 4161672069 3342425001 7253335902 4122452069 6627428380 6079157097
4610193234 3274422842 7903009219 : 9364
7167819796 5979059549 1272105538 6447240760 0831000588 0818190724 4787034365
7454279475 0466602116 8615328207 : 9365
9367042283 1576774109 7870656528 8995809215 1207900910 6248893864 6517486033
6630488088 3858583654 7541959063 : 9366
5290169607 9598667197 9195154726 7539998847 7621888478 5107366055 6792237455
1580096127 3463703729 5470996414 : 9367
4895440357 0700509795 9712497079 1498940575 0420165030 7392100837 5739413281
6578085719 8028511304 2479613451 : 9368
4504277763 6660548705 7390164679 9638800674 9276335699 9017421424 7086042763
3687015388 9542554486 0519661560 : 9369
1145707431 1012678360 6189763376 5408595584 4167396998 9899171468 6664840902
4190014931 1117346206 2958230778 : 9370
7057948676 4638556758 7210995136 4683099777 1609454655 7201681228 5393776737
4099428303 9515574945 3169206582 : 9371
0371452045 8277535783 3798271161 5753554754 7595988090 2888250015 1326903062
1837535588 1522800804 9946219926 : 9372
3139514759 0076715044 4120102864 0422323467 5721465222 5543337464 5407695544
2963733651 8329440811 4816553112 : 9373
3178885685 3449362565 1092338223 2508875197 0064021785 6262405043 9203931155
1274241981 2786611804 5752037903 : 9374
1352263821 5002107721 3050240662 4183002862 4776559111 1413089474 9764170442
7632877747 3666951415 2962797298 : 9375
4736229019 6362231543 6319141841 7996968962 8035927750 6155139875 5785367266
3537814008 1715318318 7933147980 : 9376
3166307354 1838249854 7746143475 8212203503 3039491346 2672508437 3973133134
6134971878 6522154847 9521132806 : 9377
5589745100 2032487297 9223925689 2753749027 0425986685 6149040537 5284504662
6144264259 9122957699 4984595616 : 9378
8866129342 7415216860 4536950858 2771055380 6842470639 6860778132 8993106285
5112879954 3943367009 9191520888 : 9379

6144555645 7442279253 3094751032 7863999828 6086647782 4697766936 4695968209 3306304832 5230272861 6288409185 : 9380
4021075857 5059523354 9174517563 5055894316 7491291170 8620738469 6004878978 2639109056 2573959948 7492495979 : 9381
0110901916 1465908020 7484962639 3582792650 5836476767 0838301196 8855505051 8615798097 2185298078 2027546790 : 9382
7773528459 4885548642 0957184809 5773550264 1837966220 1056062017 6724101647 5961823177 1444198810 0961027947 : 9383
7760819625 6246820854 5749937593 9175577255 5043901644 2709090994 0333682021 0181891880 9431446872 5119448839 : 9384
7607261064 2894763770 5083816809 2847287045 3085471610 7278666310 1220366887 5829064624 9653294421 5040426089 : 9385
6079559602 8483768115 8331065639 8188715410 2229185636 3755468086 1467680606 2617591254 7513262658 9633416570 : 9386
6265117268 1705493010 9406586302 6692204229 8948302376 4432367142 1946364900 3200189102 9510553731 9742039399 : 9387
3038094287 0786655291 4769288138 5645878149 6647798082 3314826640 2166554846 7134062401 8597141217 5424409787 : 9388
1277182877 8534134383 7382580995 3777455566 8639030970 0061492840 7730247009 1720199524 6206245391 9858592371 : 9389
3450742399 9851718224 9542889513 2343503182 3344837883 1929595533 4879010992 1171899225 3652942936 5336258259 : 9390
5390946552 9635813744 9729369974 6511875338 5374870942 1770808174 5701742251 6040904388 4612157412 4452850202 : 9391
3798390369 9911389696 7347788049 7042088863 9312843891 5798686149 9535720637 6948214920 9306281251 3122808049 : 9392
9666262553 2242838399 1735202566 7452522990 8409946325 8646834111 3042084589 8803241428 7824148608 2621057749 : 9393
4753300377 0602151682 5542168588 8255205289 1371938772 4986908202 5393362940 4730475050 0997044070 9469358919 : 9394
6990053347 8304463581 1964910483 1638160680 7432397475 1887377450 4853932080 1189209217 6200325412 8590019212 : 9395
8507800878 0108096121 8699721567 2787878360 3783428505 0223359104 7386100379 0033681958 2153479531 2420332192 : 9396
3791186979 7381093285 0103678278 8060812745 2828831039 1831570441 4803727152 6911195891 3827350206 2661678813 : 9397
6738930058 9434987192 6275421758 6783775861 6929115641 9695497805 0050271744 1482142253 1771664564 8976255943 : 9398
7582676430 9491260552 9285775355 6527311492 9560879110 8215961940 1757024920 4462620779 4680537769 5544163790 : 9399
2384442862 7076001335 8829372290 5895613531 0992448792 3773970012 6380103906 2103629710 0540090233 2662052868 : 9400
5512923889 3654007663 9663983925 7245082449 6898926245 9924394384 5087655789 0902818985 6834334510 9961963840 : 9401
9116437670 5960548419 5253505687 8723052067 9162057973 9800869587 5004656119 6154504016 2976777009 6547018528 : 9402
4334977644 6794960280 3722422923 1209258155 2380645150 7317582639 7848971659 5610762944 9587379456 8519845906 : 9403
0936913497 3227024282 3829358483 4762009727 9573203973 9082440490 2652459373 9625729544 9117729608 5988591097 : 9404
9831916191 9237155743 1477730561 3275865030 1867128792 2333999409 2210626790 9462585081 1692827966 9238172379 : 9405
8551276299 3523860883 1771468972 8571559104 7969113529 0085367737 5489955124 7186890395 7446600048 4795401447 : 9406
8028822320 8242567018 4511475458 3859463997 7604121232 1829451207 2077908176 8233151618 4554219598 7497445589 : 9407

0151199270 6236051896 3407353934 8756409531 5603407069 3595825760 8166769558
7492098240 4806665275 6245519232 : 9408
2003251452 9417553964 6827190235 2195763796 3789759417 5052158675 4201928536
5596662014 5045633855 2783571256 : 9409
7120512822 7519609295 3909245784 1885540127 1282860622 0318403041 6252304573
3499198620 3368394185 4919174431 : 9410
2814170274 6834805331 2363654640 8009522798 6799808167 8614387670 2299176420
4219357703 0302889240 6338233711 : 9411
8831666892 3472566531 4715426197 3692488185 6774442368 3067628580 8035577667
0852493943 2726759182 7130580282 : 9412
0474498683 1092388451 8396569291 2826914135 8022850879 2026381557 5944588577
8391763041 5382477745 0273630047 : 9413
9676856895 9057429077 5328240318 3887419532 8472505758 2369989841 8557360973
7934792621 0756480012 7606600568 : 9414
4856289529 6270365033 4072849924 5707682166 0600430851 9214499399 4615927401
8679067024 9250918808 4254975382 : 9415
5000573641 1527656278 8206831683 7456136116 4661059452 9609876478 7617810657
3256994413 0069604044 1275748480 : 9416
5908154315 2521914759 3078104140 9918043677 0663956672 6344353562 4561935640
7513565467 2716790941 1879488480 : 9417
8680923093 8328738225 0042851308 6155646738 2313448911 9765303305 9034948850
7090436408 5511708968 1596818853 : 9418
5559683677 6708287592 6959001222 6813063407 9052180831 4175342883 8750131652
9218766333 3574314371 0101634779 : 9419
6658883618 8698834998 3289811640 0702596550 2047611990 6884593064 1188013743
3528093795 1098305220 5865755190 : 9420
2553313247 3935518421 5726088231 0168318176 4097241796 6428217055 3923573318
2359972105 5914675809 9096015712 : 9421
5453625914 6573583451 0808497696 7080717266 9437183081 2496641392 8529324205
8148352262 7446937585 8569571687 : 9422
0345022377 5777783598 1585809545 6674554627 2979473079 7569320508 5626405035
2830255672 2866775444 8284086209 : 9423
9813897930 3709253264 2596403853 4996170546 3860837230 3148948100 1371999013
1970126280 1084969868 3279080599 : 9424
5250749377 5423599997 8783744657 7837123753 5757852677 7106381435 6136789079
4737247924 2618159928 0191554236 : 9425
3584092241 0926951949 8675837014 0080691259 4594455592 5467047320 3928004701
1160015595 4170202879 5035929201 : 9426
7703968601 1345630444 9233397743 5455716545 7271495173 5345290462 2158776693
2103972565 4053386823 9130580966 : 9427
0021212324 8311387049 0726163498 8177752506 3398803735 2830441378 8504053529
1419573448 6262214803 3294954488 : 9428
8908545079 2480542683 7419406691 8855516757 2166221109 2089973185 2667678293
5266132904 2761712071 7343300234 : 9429
5258919537 3557475092 6874315293 6342844964 0771785673 9958438148 9421046285
1810384964 3427735519 2443854011 : 9430
0560036409 1860906598 3002337368 4591499584 0144499012 9369372589 3952889834
8115564558 1061030794 5458047566 : 9431
4650489357 6785275211 1834079945 9874420567 5506984616 2829748169 2743403195
7448112126 9219891324 3550319837 : 9432
0546644424 9609636337 4848655871 4369342400 0842683069 4664726867 8216054307
6055555157 1130105499 6369421412 : 9433
0145286056 7154928145 0356360885 7935420449 8831255719 5995587078 5833653305
5121083928 4998411268 4790571297 : 9434
2202465501 3870820524 4749272341 9195036030 3939460347 6770851534 7254338076
9135430210 3233118270 9941052543 : 9435

7316371791 8996081338 4440367350 9209111063 1733765874 0160001869 7304252984
2024488852 7033172514 6924974753 : 9436
9319823503 5252261761 6094384809 7053512457 0875131868 9267370050 7442774120
7097904073 4631226205 2900063918 : 9437
9290990331 9643533353 3778277303 3800956435 5202372811 8934142222 1316384124
6621876265 6292621316 5374744095 : 9438
2305454016 9190591210 2983325487 3844314996 9008179176 6624445625 7105500140
6036680013 3249580906 4102834877 : 9439
1641935647 1430409550 5763803857 3852209871 6167959104 8522208070 8395443816
6529535008 7746068113 6027308248 : 9440
8566261392 8667728371 7030323783 2334671640 6418651059 9162481767 0706345653
1467640744 9265899040 0249294338 : 9441
2034766548 2730341522 6579042351 0131092975 6783048481 3632976397 6322746732
7299098939 2744963857 2144129084 : 9442
8706448070 4661630671 8326269730 1895505226 1750367044 3765606386 0897547480
9199526394 4035360465 4399823773 : 9443
5619024650 2693129589 1402221059 8873408816 3222562586 8617445324 4115894930
3555099774 8247415150 7373411947 : 9444
5733373556 5276274185 1916424514 6201049878 2629453689 5088446232 1176358003
5118418359 2675986429 7128139853 : 9445
8414469096 9171907995 1674046781 8935875027 4435035511 8079967776 2498706284
7592913272 6168844888 4939141128 : 9446
7453205724 8359162360 6741616344 6388013123 5116126332 1415735073 5873789089
8646033191 0104790495 1088053276 : 9447
2436343801 9532053136 4134355459 3647041984 3997327319 2818302576 0085767530
2583216967 1464683435 8840039142 : 9448
0370724267 0826017616 2533902989 8671526756 2219813060 3529594844 6464040396
8856006481 7661090330 5607906503 : 9449
8768968446 7316948543 4389183689 3537933416 8987646104 0506024109 3598555326
6431299975 9390263679 6969251376 : 9450
9369261591 0228511721 6541488921 5726223597 7366770146 3984585521 0470747638
8972254902 2179557834 7385363008 : 9451
1933437898 8748156802 7697599949 5121254415 7027137649 0277515078 7799491095
6148957426 2273987940 3322128257 : 9452
5324578456 1955271497 1773748923 1107175800 4485261087 5339714409 3342364167
0007394747 6053387663 8025429586 : 9453
3552936913 0347698689 9061333624 5908473538 0529991695 2374065057 6570393490
1547390655 6528925808 1944696284 : 9454
0122139245 5111987603 8074056609 9410523116 4360192568 4722053285 1072587588
3708878783 5407461779 7601531730 : 9455
4950058293 8124750525 0302170023 6109573670 9078402235 1162225972 3813014440
7984791813 3032105443 2443110271 : 9456
9791005913 7380934837 7586361399 7719436720 0655924569 9381470276 1918466612
1180429617 1866528310 6957766092 : 9457
4377161598 3151245936 1728010390 1636552046 6193709252 5125363139 6559201177
8214276801 9428047658 6534858158 : 9458
7470311992 6770979133 5049110720 5165243246 2312538315 7448751295 4156035527
5026367961 4546109344 4168535781 : 9459
0273138996 9995468673 4351810344 3255259516 9300233005 0592572799 7815904820
2235890526 0479203482 5075042171 : 9460
7345546642 5312528525 9478440684 2133378979 7245599838 1452902491 3412772434
2450971795 1186446150 6281528392 : 9461
8650237212 9264368408 1326894231 6931518881 8953852681 8004763103 0778038992
4412151871 9858272225 4498999677 : 9462
8751458791 6023970766 6604413330 5940017725 1968288276 1813890254 0142141154
0322218807 5221493138 7303943894 : 9463

8386348804 8858725731 2960544022 6754011944 4344271239 5569023790 7150071386
1064198583 8887956880 5486335172 : 9464
8084446447 2481327723 8595610521 3013091015 1062642932 7631624722 2859852571
0263715946 2999401322 6550518804 : 9465
4657398980 0377544218 4629979193 3040060517 6161879860 2655576076 8914956796
2403302092 0353382300 6419857512 : 9466
1280549087 6812649998 6629682021 6008393577 4256923900 1450967655 1896830011
4985803950 0662463386 1680225506 : 9467
6870759127 4831975300 0145540601 5411980784 1134948294 6568060170 7418219448
2694202914 5919311797 2944253523 : 9468
5139331011 2186883167 8002326637 6919519389 8930495655 9636443049 9826362332
2221317379 5863375272 9015980267 : 9469
6243820849 0894364283 1249679621 6250016352 9966893044 6258458041 6724907109
0414279544 7432277640 5586064497 : 9470
9937748159 7906129222 0293061983 3526180042 6117966549 6758054829 0181068947
3722161202 6571626304 3635302282 : 9471
1293372450 8394353437 0967854324 3805831288 9505278663 5657171288 0369852845
4871495079 8562466555 7793170507 : 9472
9028998685 9714364593 0779735070 1401522754 4766934198 2392638989 2945353431
9021801693 8758502877 8687970204 : 9473
6168231973 5199280766 9758651276 0646823839 1696014867 1115009603 8459388200
5106152664 6256272708 6373994715 : 9474
0257718107 2308962043 6264155255 7152127904 5835429590 5910120518 5086051998
3358295202 7644524251 2357735153 : 9475
6145132912 2133578341 1967187075 8566063500 0297664587 2189965684 6809835422
5556797699 7861529583 1620657432 : 9476
0737210996 8440619460 8527521399 7207746028 3796182940 6778265980 9969835866
0897438651 0365624255 6250842354 : 9477
3545630151 2197111167 3283365507 0583247917 2735651427 6941098498 6357066618
8025284345 3611977423 4900016041 : 9478
0913598065 8253251047 7723487375 8588466697 8580299992 2419736650 4115511962
8160047321 5765070051 6628984539 : 9479
9637919414 4719612753 6849638484 1840783539 2194951607 6075298476 7084138274
6044030177 0757999666 7675686125 : 9480
3610514003 9171681725 6780501389 7837186583 7968976150 1720984806 0227215120
8763671116 3552935619 6130402092 : 9481
7396418528 6936047265 1396687556 3040087535 8568683131 4128686092 2825551242
2695956799 3035024901 1366770936 : 9482
4034990374 8458741989 1089018945 7051985781 2478440357 5786713931 9708554989
3809997065 4105791690 2095988974 : 9483
9873844727 2613724351 8065616499 3163897351 1197033103 9397804092 8094799734
3377265021 2249714340 9823782365 : 9484
1965892886 2792232779 9039418205 0685698376 7116623772 1514412022 7663379491
7374373422 8317972794 0119374539 : 9485
0490579714 4617183602 5673221405 5219462186 2859145438 9656234024 8945379811
9559649687 0707302186 0780513193 : 9486
0946785284 8444202128 4324993071 5413564423 0793862725 2658520849 2269484399
3948530535 2727068233 5863948608 : 9487
1705774075 1697385212 0210628959 4177160792 5413069913 4608143824 6328669352
3126590734 3036680953 8595608450 : 9488
1673932290 9654282885 4097863778 7221825907 2743419564 6611659593 4087134481
2029957960 4000576414 6848563841 : 9489
9208402583 2685521237 9549624891 1586260094 0098765413 5858706192 5113653719
4814860857 1077083760 2097237465 : 9490
9553114403 0733949423 4844861525 2372266625 3172090816 2269400021 1227591834
2552982816 8997196876 1014385191 : 9491

9012898807 4242805283 5555332523 2571548587 5614477520 1194680111 5509440542
9657310193 6215959187 5821948220 : 9492
7475615307 8334803008 5499433087 3982134027 0700031128 5887927967 3962736609
2071269711 5138195377 1554641063 : 9493
3745558491 9631691755 2555991892 2408979783 3124273545 3841796027 8459589864
7060095284 1808664677 1109464159 : 9494
8431500095 9749470220 7595874936 9609348923 5131553087 9522675319 2288751932
6905699269 5901124280 0843780897 : 9495
5802237272 1112814277 1580921578 6190314382 3282221584 3119736386 7972722768
4958136328 1470182765 2361699870 : 9496
3831654805 7146175201 7797801074 9005200986 2225386713 4378987984 8810445030
2717738606 8210671823 4856661032 : 9497
8159840918 5714908477 0741257737 2152962362 8951428193 9493449231 0024755294
6876881422 8268826338 1992107050 : 9498
0314822969 0781270685 0236096795 2456156317 6253784439 0868388545 4717662505
4868861453 8858940190 6509184101 : 9499
5885208833 6961298773 1672752691 9756238642 7609513694 5838426185 2183389570
5186413263 2025229313 9134838082 : 9500
1287964338 8136298455 3904237312 8573855900 6232871979 1590912181 7103349208
8287365725 9600675103 4516917303 : 9501
4840664773 1247285364 9869703225 5058028512 1031458139 7165500540 2199912584
6712475229 6233047947 6204174835 : 9502
7339651293 3884625986 0546690206 8727431079 8920093600 8629962585 2649556926
3422443490 7588987120 7540557239 : 9503
1778889837 4740062831 1035989363 9753683143 1637915535 0844599499 8151790357
1916645957 9726405356 3579622852 : 9504
2323209995 6252907295 6596862566 6476118217 4368893655 2658420973 5880483823
5632703846 2917410427 4634032764 : 9505
0442472179 1963389233 3004523529 2002875730 1356303821 1172897133 1694363722
0617508581 5202908497 2463690556 : 9506
6762892184 2626517819 7651953852 4643036425 6203212959 1608958533 5981540706
5024525216 5708822643 8896955320 : 9507
0330712386 0177199426 9799874271 1966035304 8528358184 4146085491 3450644431
7130866665 6732474479 4322805473 : 9508
3913760627 5818904283 6565865989 5466488561 7098590233 5718936471 1022191554
4801416081 0863286526 5720250373 : 9509
4779658698 0289697568 7596956655 1715729978 1741491255 1945083377 9497446698
0602686431 8152342292 4316776326 : 9510
5101550537 4677709204 1564764624 7512496723 0114288483 9539706060 5607246531
4227321889 5838355013 9850224798 : 9511
0616382349 4536612899 4040935591 7809265868 2360641984 9478639492 7392551467
5962185564 8434028716 3998424165 : 9512
6407922432 0499215193 5270942749 2550972098 3764055479 9507636962 3700896158
5314208285 4978064406 5756946107 : 9513
4012428301 1677091754 0880488062 6657504874 4970010064 4817281703 3718571687
6990520504 3126912675 8942354642 : 9514
4263219268 1612607135 2559377998 4687684876 6637463737 0848309130 2303187597
5519252439 9178262640 2799661636 : 9515
7094369112 5300862843 5029788667 1483877355 7009540850 9510925426 7237087162
8508720499 1001466660 6934352543 : 9516
9681324227 7505204120 8431177836 2086425913 7414037018 9390584913 0885307671
8033777597 7981545040 6004508428 : 9517
1692653954 9424241734 3968257979 4296332233 1213218107 8012932197 9360275038
8852631045 8725788790 4993301724 : 9518
9371699290 3363545290 7496514640 5609127548 7528815748 7485046656 4788157133
6432427015 7712065088 2647257091 : 9519

1545293554 1510645510 3509471072 0178800179 2464135972 1384299487 1045977552
7984274506 9774265641 4883406830 : 9520
0809230546 4626038948 3267223960 4062464659 4574202522 1084288128 2668896752
7898024346 8265578962 6562663797 : 9521
4536891092 9346892092 4841097023 5713162753 3023890207 6986731432 5842760948
1838812458 5758734014 0970686718 : 9522
9561813112 2294700219 7844824158 1544509819 8891529424 2890693466 0562607956
5664394339 3427998358 8235711675 : 9523
7511632925 5621499470 9518352233 1245810913 1586557735 9175847036 9414848815
0386326409 8660524416 0899442142 : 9524
3724467318 5768540526 1645960803 5904616045 9477472320 5628789108 8699234201
9575526455 5491518628 8103709855 : 9525
6289818492 0845362817 6074175578 2246663879 0726046775 5483451744 3481890496
8805637650 3250353592 6271374514 : 9526
9886240548 2956247369 3334496233 0902739113 7081058407 1237265540 3944406661
6546669812 7940161174 3803144254 : 9527
5836113138 7001800510 6422250474 2126749539 9707590971 5271031416 5536334773
2005496354 9146671949 9256643771 : 9528
1135196471 2248442133 8334482991 4740191692 9709782733 0482096570 2365744166
2164041385 9757199401 1075505481 : 9529
3835675097 5367702086 2148022476 9057593128 4579612052 0106065272 2159599745
5895639292 7416101354 4777148660 : 9530
2722228078 7891903104 9330486064 2348889412 6536509198 0468047266 7929079707
7022905020 2576713844 3684581563 : 9531
4829002226 6587224538 9242075867 8126480742 5984310372 3144835073 9349163285
6870879893 5308983035 5384383950 : 9532
0797078015 0899316677 2121535578 5047682448 0668120600 8160925249 0055065606
8820053943 9752940429 7779977739 : 9533
0954812182 3388281599 7862843934 4893791861 5556384584 7942592989 4905438450
3736976473 3322568524 2402160195 : 9534
9044724016 3994449258 9154522209 4738197285 7356081416 2944201208 2969234213
6751905692 1053373592 3778160066 : 9535
5668763641 4636657904 3573782443 6651668510 4935195776 5790791933 5490054245
3748362582 6410516843 7864450295 : 9536
9183589839 6440568593 6888263686 3710502768 7838531277 7705665995 6030015723
5823714715 4721905671 4260143368 : 9537
7966468436 4914394999 5727221505 8042899218 1009895079 9344944214 1990214468
9255392867 6917510398 2467582463 : 9538
7343805239 7350830280 6871873446 0641876299 3203121704 0704830461 2916426319
8208628507 8695172901 8997001434 : 9539
2896826244 1780781916 2644171950 4450757666 8563242456 7458175518 5913604359
9958574598 8250355384 6674132820 : 9540
4525370108 0509023511 4840481953 0588806416 0408235523 5129412814 0484544881
0370856942 6260002716 3923010086 : 9541
0372142424 8429126495 7195698732 1905242756 3394901622 4645462593 4745670300
5672837508 4672498252 7983673495 : 9542
6177089836 5165478278 8942618610 5183276920 8503603621 7800339152 4933714844
4550141579 1162505068 9091071382 : 9543
7580224465 0509860986 8832767797 1183257939 1871621676 7883562241 9886753868
3931575658 9776865201 6394528273 : 9544
8867440651 7875660068 9498217355 7480744432 5577590692 7363784818 0513357096
2701897152 0589097081 1986200522 : 9545
7679349913 3040582845 6720358566 4724105889 4553087522 3646183843 9640259601
1254852876 6088662848 3076360128 : 9546
7006667228 0267000361 4942302612 5416550458 2961699316 3843705829 6750705322
9269109161 1967493612 7372916312 : 9547

0586368847 9052509527 3515438062 2219327300 2895992407 9390744385 7366909352
9492586894 4010734221 7802437077 : 9548
1528161242 5319088227 1732715433 8237401474 5436218433 3256802959 9407713014
9833237045 1096996545 3202745335 : 9549
0702177037 0706113819 1051635880 3074747708 1980826531 9510552403 4346189080
4087728558 8710261912 9910922959 : 9550
5088225181 9205851887 4419863486 2518824566 5450780329 5363481026 3484330308
3724181136 5562783019 3910161835 : 9551
1740322384 5097914687 5216238771 4423922323 2455763641 0797470012 5539832471
2260190530 4886666493 3322848632 : 9552
9053808683 5170964354 0440256866 7911688444 3421694025 1772691672 0054236587
9524586487 2919319419 6398370591 : 9553
0834659556 5457374554 2747225256 3872049196 4846804561 2163467555 7800183911
4580507202 9104091774 6197882965 : 9554
0486123555 2872697552 0004238203 8183207642 5536409632 0893392454 4967598152
3092151894 7305019785 3510015299 : 9555
5735305428 1128364842 3659474359 5960959569 8016202753 0239594199 5334462820
8264979236 0794218868 0411060241 : 9556
5874150857 5194580615 6888083430 1854125378 1954596974 1423677874 1870667215
8427522319 3527707018 8776280303 : 9557
2337402862 6604207305 0523785203 5421042577 2442559140 4270087490 7643524826
9386810716 3766930730 3727237417 : 9558
1754245852 2477357582 7029599664 9854102311 5101387432 3704799159 1787902994
4855016825 5865451538 8125485742 : 9559
5414604291 2012322855 6148895327 1771724012 2627994410 8268691399 7298743825
5681581112 6238732626 1042642491 : 9560
9146730313 9324078996 7378614329 0414208441 1467415351 6742689733 7032190690
2874770602 0088422190 3212516552 : 9561
8911717385 6747773113 6615373439 1751962707 9549216170 9378005434 0457873789
6895794065 8402609222 6706663974 : 9562
7064746191 4851144652 4438074152 0552128686 2020067172 3263684717 9672315491
5335949245 3428928874 8593176642 : 9563
6993620967 3407497745 0530705684 3141013326 3287775921 3057623147 0087437384
5076260305 8757749787 3242071406 : 9564
6513617994 9569456010 8192843193 7367328417 9893518959 5423519702 2893472976
9771049753 5864995673 1850469509 : 9565
8739662801 3953152473 3606745957 4653673422 5096595010 9876696237 3414060683
9350348198 3827182118 6844060176 : 9566
1576560254 7011395373 5726783452 6945596079 1770944971 7272346947 3343678038
7223775747 9168689555 2041362153 : 9567
8014282548 6737769528 3703841427 9340044001 3056897835 5622798607 1360705966
0446853343 3322470819 9605172761 : 9568
5220108066 7106523795 0190709749 3741821633 5299386518 9170077889 8834763609
2361880526 9060064082 8079713434 : 9569
9789427595 9288720261 0785155541 1239737304 6097592317 8834688072 5383510590
8002186604 4902897911 8967259367 : 9570
5521396472 9068579735 0736757182 1697403667 0698961345 7874506109 7120350147
9565375164 1661531216 4732544167 : 9571
8927750724 5943628192 3825623362 9810375653 2891328239 2322272506 7941709469
1318566996 2267630084 7493329492 : 9572
3772482025 1680550666 3161990345 7800502962 1609509712 7831349754 9748497080
7501469286 1779720539 2087735573 : 9573
5426344654 0469175078 7836297830 3712221263 4499537604 5870682546 6567329277
5397228603 6781924732 6075875375 : 9574
0363960555 7551524704 4790468927 9007417440 8099162153 5352379551 5931682503
9170841157 3389472148 3377058789 : 9575

3695415348 2731607170 3023202492 9094546655 3120552501 2632253174 1427372940
8935823231 3041403596 7091049257 : 9576
1838352019 5357751110 3030189373 8736672956 8834759002 8046690150 2804156922
0627941078 9768280962 6966101213 : 9577
7488119556 3119677798 0468124930 6478374053 1627476062 8345868471 6459438243
6775342760 8269557653 2344276053 : 9578
3997129870 8052089898 5614420659 3472215753 5135731353 1509643256 8632699760
1181676887 2330904784 7387826108 : 9579
8300271876 5026082629 2523319447 6909940416 7002640065 5559713699 1814669791
1356778065 7451057296 4711963826 : 9580
4380698960 2338811353 0721498585 3687086285 8717892687 9629780160 8941639363
0120941635 5230277734 2996301526 : 9581
3435775125 0413519834 1187362058 3345473118 5380745726 8433720690 5220856255
0105061009 3794285614 0474184559 : 9582
2117933927 2708597251 1528800569 4028053614 1144924020 9246208794 8741836224
8544537016 7347935902 0150069908 : 9583
9471119500 9547716996 0645156934 0980957608 7230611679 8593054474 2494585592
7637542655 0790985078 2622745241 : 9584
4280564196 1957947016 1814101885 9396702928 8408817507 1326949126 4514792458
7138834722 0957012545 3762871154 : 9585
6135844710 1311323201 4954909446 4014760003 0237632857 1713953654 7149001355
5869633069 2581126404 7920053172 : 9586
8092117912 8700967881 3893732959 4906876916 2309178222 8643533340 5933967916
0242893274 8444663155 9457485611 : 9587
3204517830 6464916622 4181324629 5767509185 9029883332 3065514502 3629404347
4054925561 1764221609 3884711734 : 9588
1895740719 9850352736 6986933866 9851702573 9380660230 2791062808 5352549353
1661945852 9388540134 7619818297 : 9589
9019270269 9755397627 0972133207 7521428883 1363827940 3779548104 3639684621
6952494822 9844322968 9692085335 : 9590
5530853174 0953971002 7448732528 3527573624 7945801278 0445503610 6064558578
0357362625 2556360647 7349056863 : 9591
8324600588 2645729967 2867064706 8819718804 8995918209 5387698672 4126105812
3133718832 8153873053 2406351716 : 9592
0488373186 3483194487 8552453402 1310596054 3269787362 7899027362 3581526866
7728648413 7632175406 6899897348 : 9593
8261186018 0029360022 3626158849 5903893818 3834781502 1647310891 3836953738
0868316436 9908798085 9301283735 : 9594
2876220600 5362275872 8767946579 1680576358 1432409253 0550238865 4829492572
5127609771 0430841424 1327149223 : 9595
0145550249 1538011651 5701072599 1966088910 3344587780 2018420198 6872557983
4858927941 1579165489 8418079655 : 9596
9816529244 0028600089 2833089959 8461251541 3473641247 5537056580 7249607337
2896863956 5510344975 8583001718 : 9597
8013929340 8159346577 4074916873 1401990382 8427712262 3332446058 8756739838
5935007695 1311855631 6845738386 : 9598
5551229294 0803068422 0362567245 9181138606 3504801552 2616706356 4964286732
3459656693 7992435872 9329116688 : 9599
4983936420 6979703901 9159319455 9703612592 6270637083 7171360797 2229244838
9736599492 6322185943 0952934455 : 9600
1705400945 9274870328 4351993881 4026708591 5289496359 5076363807 3234705346
2309324415 0957569185 0480891957 : 9601
1739191653 1000241571 4293566869 0970675538 4850261080 4074340645 7426342832
5221102071 0345037453 8340721719 : 9602
2728609307 9709087864 0274037560 3419620326 0951802333 1944660470 4393480054
0635869102 9418314381 9807662636 : 9603

9229201519 6267454788 9054873008 5334220881 5974032892 5356782478 0457234485
5566388429 9365178593 8154287147 : 9604
3470540776 2504079807 1086832571 2720965952 4702809312 9849059790 3061967508
0599444217 9885069831 6109638043 : 9605
1857573493 2089702792 1443393913 4282900983 8902927600 9981034971 6753400553
5026657548 5135820698 1718943173 : 9606
6521873727 2703866524 3420592696 8399585877 1658075362 9304917458 2102753301
2670236222 7330521370 9274757549 : 9607
2754032248 6653632392 8428878807 1811943447 7544394315 7463373774 2190514462
6384014838 4522306013 2636502788 : 9608
4514717047 9058318058 3489408569 4942499441 5155483863 3423772040 6996019335
8031375344 9760284449 9515410901 : 9609
1381560664 1323294310 5354035663 3498250090 0534136214 9597475298 0282398461
9728367006 2105846139 7781582746 : 9610
7657982601 7847296576 4639589418 7742496331 6958842283 9119159056 5640228193
4968017581 6384013942 9208142088 : 9611
2045469029 9463765205 9988197831 7544801271 1996556221 3173244271 6080219316
6446071845 0670245160 4612011797 : 9612
6382723921 1348339438 7989629058 4017968636 0943255300 6506288897 3239251361
6327023907 5239598265 3489399468 : 9613
0658059487 6864627514 1094064993 1153419873 2172991431 2591009777 1886869455
7124444935 8286138125 9764637551 : 9614
1342798457 3762023435 6256898312 2530420590 2149070999 6741603215 4670753881
6502965863 9915531513 4290551333 : 9615
1653248308 8503470195 4905567448 4101032187 6589956758 3945582382 8688310814
1862835473 1947552527 4711575402 : 9616
5434804661 7748038786 8598378715 6949134278 5308294728 8735412043 1902320905
9529543286 1066132697 6076266926 : 9617
6113521146 6252769841 2773408524 1913826858 2809550758 3757875182 9441953816
6996475933 8058109744 1970408876 : 9618
8832525737 4385618259 1108975019 6431479372 5720708094 0589605539 7098285800
4455630785 9986110829 7845198039 : 9619
9882309416 2509868026 3516282280 7560816507 0483489644 1683618365 9463297691
9733266450 4433270265 2964407332 : 9620
6083594871 2972056368 0362999226 9220555502 1936131309 4392925616 8982589380
9531154348 1328951849 1654287254 : 9621
6286351978 1030237300 3490379917 9307688612 2045326513 1810138168 9879195678
4667866144 3310580143 8125991279 : 9622
1415887667 1702906759 9907122292 8172745278 5443191763 7751864885 4469055214
1829475460 7553734560 6085563464 : 9623
2039617667 0957528745 4949012046 6035196463 6873577292 9742823475 0549678654
4592736062 7608989246 9782418907 : 9624
1366910300 9266781913 0305591951 1695693178 3197540796 2403384211 0466446524
0458186863 9326096463 5303347112 : 9625
9213154369 5714422067 2372701903 2161283163 6606835353 5914027988 5260953147
4419767057 6401090753 0604721386 : 9626
5706766549 9726561399 5625908185 0853030455 5928407614 1246522180 9965435307
1631850748 8647893313 7159804019 : 9627
1058010242 5541713566 1896120601 1069761203 3871217695 3627748147 0240462879
5944796568 9291666561 5162911773 : 9628
6946184946 1776831663 5945285117 1641400879 6109655867 1942116381 6545789355
9437474165 9601910402 6506996537 : 9629
6089108849 0540908807 6624223624 4525313285 2178568721 1710750728 7258024370
2749583566 4624563513 9772059647 : 9630
7873476721 3709697787 4372222228 5444150516 2581475900 1600103498 7342164287
3794152091 7280874385 2870687452 : 9631

9967585063 4623615656 8380656846 8586659128 8392993987 4919280450 9759935762
1915303453 3964024162 8163756457 : 9632
3379859690 1282927418 7962762503 8064030579 9822393935 0958921995 2785102916
4638047832 9362091927 8040771504 : 9633
1873006891 7857381782 5379312653 2269564298 4805750570 4338593699 3433934560
4496232593 7243304357 6671477116 : 9634
6426205667 1619377737 5770182049 3615978206 5517759604 4557427140 1595850624
2086143202 2102794700 3286440974 : 9635
9853119493 3972205256 0172438067 8398080630 1980330387 1401082373 7020217802
3999196594 8474208100 4162463139 : 9636
9989387267 8129698371 3965334369 7806394667 6428600241 5282487378 5639412927
9353836077 0760830075 0085468836 : 9637
6849683447 3813800019 4809994639 2793972548 0926175571 2323072287 9914729499
6278293181 2117999531 1913933682 : 9638
9142170800 3917108392 0399625324 6571242671 1480762187 3238886630 2732632605
1022748558 7685824829 9627378563 : 9639
0375020526 2948831696 0894901291 6313726348 8580498192 4875455352 4887326239
4393675045 0165476893 4020682145 : 9640
8565566171 0507751194 3738058941 3425696041 3139475819 1970668230 2632423409
0244505409 5883410876 8898805836 : 9641
0019058008 5619949103 3238845013 3139604194 5468828359 0614806279 1702742550
5698363038 6819040807 6986074650 : 9642
8844236776 1705462260 8454397222 1940355420 2652033104 4552690070 4688277458
2145696636 6744699942 8841734711 : 9643
4807023479 5179430778 3615275740 0116754238 1049782265 1699679327 0179099132
5359692564 1310212031 7675960263 : 9644
6795510869 2598919360 5152931611 6399937906 0522162182 6257344736 3342026307
5057264252 2552550069 6250372133 : 9645
8053244844 6537714971 7057771217 3855502140 3641091178 3897214179 7635473176
4860997323 7020876571 6623286273 : 9646
6406666652 5224448442 3704743126 0270194052 9325392038 8124561167 8392260714
7800119571 8475045561 1615250384 : 9647
4822082691 8667452950 0194547955 4742670311 9533884633 6750475341 1924305172
8905583063 9606427300 9317899033 : 9648
9713439315 8405916100 8586393822 1382022717 0819247577 8210015039 1638486660
1709081390 9135953340 6190146269 : 9649
0424095680 5262401070 5647776618 4073651996 5983120159 1486191247 9104532820
8490037696 2357904520 4928147481 : 9650
4484658172 6874291601 1125678009 8117726912 2209070237 8514866112 4371445919
8466857630 9473751209 4243352004 : 9651
4650532973 9125167083 0182554297 1302260674 6609800526 0391962755 7983895090
6924319004 7375640183 6074549348 : 9652
5910179447 5577162863 1505548876 1028672918 1867586476 6444786596 2782729403
9993209905 3354969138 4757430422 : 9653
0358026682 0050183524 8565151703 4209943110 7260374750 8216434958 8541432104
5735574198 0118294006 5163845907 : 9654
7831309473 0099299395 4182718180 5997731995 5225376562 3522616879 8804828472
0314958906 2569689442 4278407617 : 9655
1974785212 1117086752 9446036705 3355570333 6195269940 6435233081 9095743708
0465607841 2350061934 1564951004 : 9656
9733173836 2040042273 4437891578 9653495861 1591918135 8972444956 0707032232
2782801579 1485580886 6326700924 : 9657
0423420313 9271646896 3013705622 1855039903 4622946791 7697832970 1524127573
5845801308 9759525863 9855021546 : 9658
3677050900 7033290797 5583265256 9733099251 9942335234 2674324526 2878343487
8039320990 1478697413 1728112549 : 9659

6445904277 7973691266 8770753379 3810529597 9219560094 5196472452 1456644781
1294088842 5973995022 8931567320 : 9660
4890035956 5314881818 1335248844 8698694800 6125364742 7755000504 2040462103
4425555880 8662144252 3793246458 : 9661
6130867091 5261439688 7816336773 4969512409 0077391672 6414094124 2164561853
6462085838 0521948089 8873774634 : 9662
2851400043 9792350670 2424670667 0697307923 5532297655 6684570476 2902563225
9783196183 3397249154 6952552514 : 9663
3514347973 0725058985 3935034414 3010372769 3308287010 7555261221 3237948915
3242485015 4784570097 7456836739 : 9664
5706462453 2316783427 1605995132 5335038447 6461804529 8891008925 7614236456
8920937121 6233577791 9010169852 : 9665
7465074331 3394038605 5838356051 2529911464 7525104974 0083923818 8040952466
5697821906 7500774512 7340413746 : 9666
9485989032 4303842177 3757608092 5883639849 4285249166 1239742134 0306639968
3994573153 1459218956 1854297369 : 9667
4163929127 4586602149 1932560629 0835478894 5387058909 9102877684 2634490924
2758195382 2771544550 7550828207 : 9668
8488360093 8857504071 3380643144 6435106635 7725420010 7215128214 8738012783
2552194772 6666964351 9639951388 : 9669
0976645324 3327235576 8426415313 8097822980 4632131537 7462523540 4290603580
1705619274 4688484775 9849178086 : 9670
7455498329 6585195346 1620010812 5422426766 7244659198 5003737246 7000945314
1832851380 2244338672 6425015956 : 9671
7592114147 4962451849 1204720646 7560940593 3598879179 7905900136 7488538645
0686962065 6149908349 6822578568 : 9672
1134556483 9572728890 3768564734 8587027874 7092443784 1701540033 7456982748
2324692217 7670738059 9850755861 : 9673
6687137871 8683096809 6765583702 1661114077 2207852210 6402682698 8873072896
0516843783 5203674022 5003301294 : 9674
2208799728 0733552033 2084982518 7947636617 4290552837 5912856416 4974137281
9365114332 5413215966 5447975088 : 9675
9663784405 8341384482 4557338593 1223675087 7569174580 7770666376 1431383703
3604334586 7465015719 9994935709 : 9676
2631498313 5394760109 9746000005 3765814494 5733267458 6106330493 2150297383
9393544273 7099386415 0604817313 : 9677
3810760582 5309439419 8785753236 5723323047 9592495750 5802346554 8576017407
8300407648 9467682586 4095103880 : 9678
4374399269 5534150692 6095520996 6849636219 6097541990 2566892730 0183039884
1155263548 7584101918 6258565195 : 9679
8949917543 6701377526 2962123556 2608907598 1447245106 3453168437 4203926963
4125122549 4255324466 0392238541 : 9680
4921802547 4882876575 3640669566 5685449904 9481491528 0503825352 8556467200
4425228943 1463574597 0560541321 : 9681
1134776183 4591435006 8040975165 7893967117 8548516522 9206778357 1075255139
5314528231 2224607743 2648578021 : 9682
4710696712 5824905495 3252002980 1187432980 3856098208 4749878105 1624257385
3621749468 3456079440 1386804272 : 9683
5973721779 9597613484 9268369259 0494544728 5453485554 7672855092 6269663335
1543953191 9926209022 2901179165 : 9684
0354053205 3541557323 9628658778 8501001210 5094483643 6935545656 5298702510
4273373112 7036864327 0185393046 : 9685
2298639018 4984977908 5270466381 0074106064 1993467683 4879214901 8237926231
5357767600 4525398309 4883499531 : 9686
2973156738 1103576713 8095060268 9881032606 6044317643 5502995330 7435121783
5363742263 0730894118 9098334119 : 9687

6386055152 2024968443 9394253053 1427282384 5197724460 1660403995 8354727913
7257994876 9056309963 8920470629 : 9688
3760505652 1820542134 5799773554 8935383680 3365282076 0812514894 3624690017
4250148369 4891032497 8639419114 : 9689
6005402306 2161855699 0861024673 1548646098 8020278107 3473737770 4918834398
1529606960 6171715477 0915580453 : 9690
9147213794 4886307027 5106259100 7953739995 1394861744 9030229789 2181523395
5882101576 4964020945 9194090016 : 9691
0611658741 1111980734 3351033819 0204012029 9293240314 4329877164 7219418758
9256763459 2321990922 9015572393 : 9692
3728886071 3694061776 1645093456 7152280499 8274750927 8302635993 1981639916
8556260449 5679935194 8320384470 : 9693
3965009899 8538011265 8613333794 7755213032 8891619787 9337327684 4741328316
2153821350 5022329824 7951985584 : 9694
2530644062 7327567040 4854215795 3717334383 0136579227 1697658152 5422725319
0269172158 3563479065 5014339322 : 9695
2118454577 4337318654 5271855941 0210622948 8252043473 0878381222 2983168723
5778373373 4451559948 2329233926 : 9696
9578929447 9982420094 9392664270 7394323274 7132009716 0347570744 1072850307
9630390757 2702838050 2095916175 : 9697
5070005016 6277924293 1812450523 8672399685 8519317795 9038840467 5565133255
7582149431 5351795949 8790175939 : 9698
9540186995 8616206697 1538933325 7560018092 0779578501 2715852501 3243702679
8381532168 3151058802 7374558939 : 9699
8225165079 6556806740 5529335804 1645869285 2316528925 0624410345 0725475174
6696957664 7599120655 6847983177 : 9700
8838624441 6469541919 1341166383 1359509426 9840789705 5494658072 9459831838
6546217752 1063114551 0433633536 : 9701
6177573050 3740634292 0891277502 2362094183 0938203794 2049430183 2564881514
6217473149 5312272496 6519403326 : 9702
6694654817 4825341725 2291247497 0511461616 0054281880 5403541989 4723457255
9079014198 5182981481 4599027140 : 9703
5143266071 0262296349 1948125934 2145337898 7245830976 0486188866 9585130390
7291733920 7722568960 9836520912 : 9704
7931148217 9747506446 5597613403 8757592240 2239603473 5708491833 7811214989
2582953233 5444378531 8333610174 : 9705
3721712707 5561619143 8053272660 2449486684 7503438075 2518392245 9239571783
0234714190 2235035038 0794112727 : 9706
7487289504 0023080474 0722455600 1247620731 5992125045 7886191338 6499033912
6851472433 0910608559 4366056248 : 9707
4867647533 0103363659 8577489631 5464134652 8176374126 1581100781 7367012495
4479654116 0122490093 9460349933 : 9708
0999402392 7494381350 4008029448 9916879393 6676108675 5035469238 6570894147
0394616477 8455893070 1189503601 : 9709
6968430078 1588665329 1356205568 9169725157 8127543955 4162151952 1053001313
3622187193 8145719744 3546784636 : 9710
7988318741 3333055129 2210015747 3047822274 7354362338 0608200983 3664536855
8689256331 5746920243 1516831234 : 9711
0672977314 1705079853 6830021146 5536273578 1206338697 3246879064 8595875816
7228676488 7437654650 4490483805 : 9712
6518022973 2111795405 6543794461 5042021563 4066456080 6217490834 4929657888
8295379303 5047260862 1682320154 : 9713
9849336065 8850369596 0666607613 6341254667 7142657669 6693825333 3538296866
7780185547 6375874137 5614974085 : 9714
1683629386 4444543728 2528212759 6573341880 6978040386 3756243213 5935333827
6914364031 8722051944 4882699899 : 9715

8678043790 5031726519 5333242884 7616800651 6298029907 5023247883 4434240656
7828812886 0769637406 4960256307 : 9716
1565676750 5203705308 3913616628 3921641845 0982844674 3051228444 2398334641
3015302739 2175042011 9266145727 : 9717
5082813903 7673363576 3925446052 9716017653 7577389894 6913977067 1718421230
2966812872 0497136453 2552899308 : 9718
6401233052 3226054081 6113878892 5319487861 7232763152 7480204769 9701084142
2239979291 1010289569 1963294280 : 9719
3327708353 5253890351 4318588286 3193742926 5422212927 6285260042 2545397998
1005955366 7839998923 9291809150 : 9720
3877736140 0928153081 9407860664 7484238225 9576352851 7849241474 4484368434
2520668215 6596691976 9728805736 : 9721
6703621563 5512499443 6122668933 0580394804 5462775097 8873567271 6123758257
1383913941 0872431955 1951605337 : 9722
6515555892 5355182697 9714756066 8931735310 5319152122 9835917302 3026758392
7416493142 2439389744 3100874491 : 9723
1962224480 7371585249 4755275828 4137168733 8856845139 7193571743 5137665104
8952390290 1706309271 2264684066 : 9724
9278348399 8862859594 2396793645 6417355871 8401990187 5725463460 6762070626
2789623220 3480537183 6351794691 : 9725
0877638519 9110783793 6690226478 9614265281 9899498944 0958364692 6514431656
5821247179 2078920335 1405936678 : 9726
2849401779 8965297981 5054754358 0317758562 2558690610 1230234010 9361295535
3576358432 9974630792 8840816702 : 9727
3366788970 1281459588 4742819942 8749866143 7711859701 0460786118 3122285674
9469131900 4482564302 8262007248 : 9728
7875253943 9790125220 3517990670 1088644441 7333294270 3850477762 9594843814
9909989126 2534888002 2470112685 : 9729
3866059437 6622270365 0826722123 2223515725 5037650385 4095253101 7575868733
8353119969 3602416600 3965092807 : 9730
2743807137 5470484844 5688684891 9687212310 9819909873 0747953205 4140103959
7361969675 2305244216 4056190052 : 9731
6742759939 7913726868 6278535430 5517912575 0471576601 8492371822 3934849391
9692953944 9988380196 5726625036 : 9732
5175749403 3119695979 4117212576 2371383165 1147962605 5791784402 7818812255
3834089639 8270497890 7257430443 : 9733
0334117910 8109059950 5417122077 3737974775 0303681258 2030995844 1194867999
8579401117 1393324239 5262703619 : 9734
9270637724 1703397811 1333238271 5682773420 1479054726 1654340193 2267144218
0196105337 0502633374 2731904551 : 9735
8794871348 5249862668 2222111113 1891445576 2210422898 3473903499 8125977811
0809013186 4256708891 0367429803 : 9736
0491363651 4213249820 3398923826 2331145775 4007631638 2512686668 4850315841
1111411558 8642679072 1346040922 : 9737
1705074598 2472070245 5243140352 0119565313 9245833100 9142536349 5878979074
3937136597 0955272555 6666007024 : 9738
2028391007 4594624663 1307454467 5113059937 7751204119 2805564972 9151234915
0553278102 2986430608 4053764409 : 9739
8917444310 7699871727 6003815163 4428606521 8306921009 1799279417 0931936294
2477458068 1733533597 0175989328 : 9740
6031485320 1568769791 5652124189 6700403095 9172775708 1603018694 5971407998
2443633328 7543192214 0735996952 : 9741
6268303856 5958452089 5655497781 0132745394 3408643865 2695414066 0420525015
1308378086 6457299749 2569037617 : 9742
0930028161 6225031870 0303273714 4860512516 4072390070 0882382390 6811473995
8043555903 5124122182 3298274366 : 9743

7778903853 8586239181 4781588353 2581378931 9130516472 3901590515 0073292582
7462042891 4392667534 9521549429 : 9744
2409278561 2941425869 3729514689 3142705444 2270009324 1610933444 7817626188
8491644169 2560813590 7757944738 : 9745
6206015924 0401206337 4998954252 9116340952 4485314231 8247458686 7921351026
9662875170 5210646078 4704974666 : 9746
1545188141 5103516734 9815783088 8050629025 2472784913 2883585857 1968770316
3300975539 0490488456 6448974688 : 9747
8824842250 4227410769 0691547824 1586198518 4219579094 5913926945 5393497074
1708260129 9136137293 3199089961 : 9748
2447611270 2770438892 7170172348 8617631963 6850246720 8266987608 4819752651
5117846839 7433083172 6048785403 : 9749
0332942786 4436091148 9762879741 3020336753 9268931859 4580161832 9179440100
9583249805 8750645836 6412769529 : 9750
2859826577 0333006234 5826549555 3231665323 0563737351 2195284921 4896392942
3810595598 2270927599 7303299473 : 9751
7505687449 8728129347 0260662447 7615834666 1704916269 7571797587 2429291141
8797507487 8217153341 9974526805 : 9752
5732256003 1417046342 2031897578 2077302373 8624697850 4165097975 8445271645
8522043551 3975928752 9508954652 : 9753
2806626969 4344990148 8020041811 8642039774 2204042702 6695544603 2990925495
9435202527 9648873458 0054345846 : 9754
8494529753 5391583792 9113057037 6177366337 5795239771 0873933795 4733211854
8790619268 5422400839 5361036877 : 9755
8991522104 5210650200 3800518083 4770931615 1054412972 6850899664 2282464489
7764232319 4767560243 8097694631 : 9756
0016887760 5725679893 6928008650 2487446760 8245459575 0132838100 0122974730
5653999137 6611276067 8583451295 : 9757
8030384050 2530416631 1739822211 3792207474 3939663003 4070496076 4328821987
3398773338 0285979360 9821535465 : 9758
5910072431 7097157060 9710596868 8690664790 6795150810 1151970513 6357516361
1207596373 8637573858 4999837864 : 9759
5305790304 4394301295 0410974378 3727473215 8210702226 7675703966 1418608774
4396909624 7776142982 6812572551 : 9760
8207382106 2914291792 8982577970 0233074929 8885823539 9319356394 2526806194
8642067083 3245104681 7067040554 : 9761
2134187651 6419219776 1886802958 9218724367 3912979296 7021700260 4707954075
9988069653 8294706826 2947500799 : 9762
2091780545 0108721318 3671030214 0341239939 8867410140 4729131764 4420908011
0919803524 4321133365 3582852718 : 9763
2843262367 2503700241 1625948225 5974488083 1760673701 5485628911 8665446550
4563052137 7904505732 8185120197 : 9764
9665409430 2700497446 3254612122 4141128329 3664379403 2198447927 6656109271
7076355940 1220553590 2446730707 : 9765
3784068108 9160484022 6316131653 7882261306 2347649322 8155229195 2234092518
3960171495 5706253393 0191906745 : 9766
9718990077 6543585394 5373057085 8767339377 5225561887 5908626655 7260714813
6026843048 0946337810 8948705332 : 9767
5334693152 8652281848 5015039938 0033663878 9733893411 2884345773 5233219997
6257543876 1947829060 8410492317 : 9768
0869792726 6856501771 7845357601 4440687171 6869009528 0680343189 3356304270
9727787065 0886333372 9703100159 : 9769
3202232479 7004101814 9467646133 6968844790 9453611490 1744629897 4932300375
8025319176 9550524162 5006552842 : 9770
6114373076 2265420826 8221345946 7537033662 1842181656 6443477572 0963001368
8515145967 9472036012 1393994632 : 9771

6114541444 6827526486 6158675668 1602323921 7047451257 0134865906 1643086005
8855687920 8478336062 4630419584 : 9772
6417470830 3366065330 0534206204 2836863188 8832426681 6035175214 0074640269
0075876108 9479463515 0844959617 : 9773
0005018277 0896682426 3274755293 9134464826 8756009627 6222425072 9627218837
8746955853 9680869873 1268923748 : 9774
1269813501 2870259485 2987093853 7227129950 5563015937 1662858218 8659160520
7403805726 0330451319 7216792914 : 9775
7718675630 5293557276 8823902922 6621973058 0487311401 1753081338 9217011861
8011733725 1436635308 7565734208 : 9776
9417048072 1933598874 7973642686 1987941028 5421295294 1043654806 1666466560
9535068126 7986083772 3426852206 : 9777
1587747745 0454408597 2412357281 3629395024 8772132291 8144746035 2824090500
4010177366 2698641702 1816703518 : 9778
1897467051 2042796054 3666279452 1749414856 4864034233 7595905490 6013609693
0916841062 9163679468 9232189120 : 9779
9127401951 7061803872 1228430708 7960773113 7254413060 5417298995 0548378827
7270466638 6419113779 8869740632 : 9780
7767999608 1032755656 2870906770 1614858118 5167152557 2067310925 9432660248
5559388718 4124304225 6167746151 : 9781
0838972883 4242258914 8505084729 0517606189 9777583007 0656525208 4740820884
2173373910 7673981148 8077333220 : 9782
1658891141 0051585422 3884063672 8656725089 7128850384 5294031629 1883714453
7878661305 4000917050 1111547185 : 9783
4363558310 3320721198 1334858631 1534236019 2029371830 4352616908 4401955004
8145065893 7687152384 7124185820 : 9784
7056441353 0487420515 6145120866 6809688655 3647307014 5171155583 9964583567
8003490949 0393274451 4411779916 : 9785
3796310338 2544506126 7296629820 7689283473 8348275637 6171383960 7939250893
8287519388 9083742482 8395334225 : 9786
6640130058 8877665791 7835977404 0670750177 1087193911 4578468254 2500121104
0115678138 1295662572 5510917646 : 9787
0365967780 5799613086 2820011501 2516792484 9476048498 8420373339 3954346866
8590235457 5230954073 8153041167 : 9788
5919196403 3950082322 1212203194 8582192443 4755293375 3017393518 1814669205
3006768354 9832608976 2673603011 : 9789
7845855487 5270307322 0035324122 3910296706 6481671955 9254532213 4782497402
5002702748 0599816837 6214118183 : 9790
3876083487 9258098138 1516616041 4642080752 0205374549 5801305135 5297538783
1795560660 9534527502 8999528430 : 9791
5259958631 4977990312 5995926852 3867599757 6441360225 7604706511 9871932352
6264910813 0193591599 6762477542 : 9792
0054683329 1360809332 3184530910 3266426957 0273636886 1586841986 3558986966
2161263694 2346269870 6529516460 : 9793
3659088977 6309436953 2892149718 0259712631 0791986342 3433683379 8542781592
4375361059 5231367670 5884251472 : 9794
6866925962 2556233388 2544491533 8951480780 3531600962 3627236262 9321093838
1213442925 9616897760 7129610965 : 9795
3285812638 5635284069 1870726909 5087199084 8758805976 8061543438 4983307862
2373299505 3859544652 8725800917 : 9796
3848216395 2476186691 9428440920 2032528586 3597342736 5210841779 2406533769
9487091904 0065362840 0265799102 : 9797
7808588169 4281124198 6780321267 6611908212 0687694390 2568983185 5029507358
1362833258 8132934978 7561199657 : 9798
0647703233 5460135933 0157371869 9852759527 8840155231 3044666467 0070445701
6737473029 4477842583 7971339579 : 9799

8102341927 4301141663 3563104720 0206230346 6720043473 6336209186 0740637938
7410837798 3726591022 6226628166 : 9800
8368174614 5088810586 7940209169 6237070267 2270785567 0246696621 5235923248
9065565411 4232165300 1230668315 : 9801
8137095011 7516497474 1770771047 8671144231 2703082596 4970280652 3097265225
9535026093 7603204641 8104012825 : 9802
7029715290 9966301797 2874967160 7818738643 4604365603 1260091308 0199055913
4496982430 5981044812 1422323919 : 9803
8832330517 4876177610 6038024224 8693360782 6983487990 5318712019 3656571881
1379882854 7000366180 3376461641 : 9804
8080056211 0656657135 4457380355 7216270206 6987066596 1163026692 8133512859
7234227347 4043550450 3018436615 : 9805
7059758602 5918972971 7629378207 5851521436 6300584413 7543530152 8736386393
7559202494 0199122961 4783120533 : 9806
9020402152 4957162375 1771392074 0481216320 6956168783 7440675142 7661181935
7040926225 4428125724 4674793566 : 9807
9902240001 6162793569 9977379362 2293288995 1096671881 4725472447 4423324508
3281611358 8506261778 1347522637 : 9808
7416689306 7961890698 1738542620 7116834520 2086617225 5402151315 2030142615
3635292417 6248872401 9384328470 : 9809
3145335685 5321163460 3241191096 9004980661 6365370483 0044010118 1291865610
8974698069 5576919135 8515559383 : 9810
3791589067 9818785736 9687334916 6531702934 8327448262 3496789367 1344072677
2268408403 9078504473 3709169016 : 9811
1948341749 2844768576 6055823894 9976266570 7260959172 8102611237 0882204240
4896741798 1761059712 1652434189 : 9812
8769732540 5183576390 6896752464 3049459814 0399601983 3681862821 7058607337
2014699356 9728500023 1851541357 : 9813
6999413002 8979845463 8720609259 9165675004 2574557721 3855821606 6238188184
3260855908 6488423057 7290245837 : 9814
5483327194 6592218906 0822557719 3031024428 4508802380 4118245874 5459540599
1187938986 6524346777 6067162411 : 9815
1861001010 4090734913 0306713696 9073215943 4848129745 4532146506 1611701587
0792378267 7675224366 3566391919 : 9816
6042731264 2890214401 8734759285 4707425670 3449969176 7523359847 8138865565
7058999833 1860123461 5036475938 : 9817
1711586038 5697047892 4993590444 1279760418 8982091303 4833021495 3067826196
9030240608 2509918409 4962411171 : 9818
4750193668 2547196684 4473398515 3038785005 1106979020 0649735354 5593857570
7883336768 8924611079 3462714441 : 9819
9802729030 5966709465 4269466800 0365572482 5053853726 5700346455 2984375485
6057664365 4846895919 8703255509 : 9820
0859093742 0980486241 3099267432 3728635611 7618097163 6873658852 5875429289
9868407066 7409131052 3331169139 : 9821
0175906117 7908405550 4104097301 2675087716 7686004326 4047173173 0874894457
9536828065 1668288418 8097638768 : 9822
7751677225 4015070039 3369379883 7135823136 7550158528 7524033755 3986870947
8397561579 4630852599 1462120723 : 9823
8560952292 2010241942 2643655009 6437281621 2365929564 9212034193 1085580480
5792520705 6097073311 0036108725 : 9824
7336551244 3639741726 8775153220 6542256393 3914249879 1922029924 3040151353
2618304251 6398217599 8799403271 : 9825
7706306529 6019659466 0331660919 3225213978 5174202756 0453282255 8009110705
5701608519 7780657101 4316301211 : 9826
8809125580 6498603094 8049146676 1077207455 0263450695 6158153283 0704419644
6264440197 8953041673 8083329238 : 9827

4654515372 5133316840 3315256389 6589471100 8798811829 5323646800 4613394537
6701491217 7042821942 8285050662 : 9828
2188463050 8804097835 7011532665 4916255249 5263850962 7579674947 5771603492
3473596280 1761556267 4390623303 : 9829
4700445381 1940952869 7550938580 6797026611 4568484484 6308991494 3151524881
7802441697 4099890603 9084394418 : 9830
5025304735 7246937305 6161853793 4068829460 1425402211 4233731397 2408574494
5862377619 3225518556 8911036463 : 9831
7684070516 0036061406 4357081184 7797770405 9956412001 0460141300 0890788089
3775952957 4504765503 9053183599 : 9832
1468545540 7570254194 4535881728 2302171840 8734015606 5947706698 2172966386
5913212771 6519251666 2912421600 : 9833
2608240455 6418182824 2071555930 4141284792 1238148595 1448399656 7150576460
2713610407 3454888170 7227180083 : 9834
2968140120 3966223752 4091762593 0501196457 4375389928 6461890218 7410750169
4155173074 8796555794 3412213401 : 9835
8619457111 1414514347 6791105087 3858437954 3228146239 2510321525 1780610220
0298506117 1511522924 6844062336 : 9836
7092708922 6453241078 1786234930 0203648615 2993070136 8469705066 3695876213
0387191849 6736262861 2877313530 : 9837
7721316622 0888069011 7845223199 4936532272 4431774790 4408052101 2426828275
7760576852 0721103235 3363551733 : 9838
2228465853 8557907259 2997658497 8688690119 3452927464 2275255585 0487824174
1596277439 2726950863 0037391972 : 9839
3243280964 2332098580 6746974551 1572367038 7955932445 7584531226 0023956451
8634241370 0109861502 6519509650 : 9840
1276333828 1967365976 4355940000 8983705072 1922683756 2534245796 4215208141
5381403528 3423274574 5282188653 : 9841
9984540277 4412225452 2942265414 5010246288 3885257064 6391990412 7385863747
2377197018 1661501559 3065525804 : 9842
0244909298 2449535013 2778095325 6426342626 3432798995 1686692296 9197876946
2307638213 4287918534 0166382658 : 9843
6138981055 6650674010 6342082853 9478180020 4536422696 7901634799 1779681426
9701962431 4183701793 3239208206 : 9844
9639114568 6534357517 8937024664 2695952596 0065432609 1604270903 2773341248
7937622089 7854569431 8474194345 : 9845
1062039746 6555147205 6170610194 6122268668 1596904838 3942904309 9283167735
4144429372 2192576392 1777923422 : 9846
3773397148 4881819101 6998201827 8621131812 8453632639 8438621565 5096776531
9885417831 8558674158 2390043053 : 9847
4511707373 7286728018 8233545785 3069599677 8806741796 4300979384 1325405448
1238083190 3058654085 1532278542 : 9848
3135428374 2353758721 6882363220 5507065270 6495251356 9636752121 0463206418
4322393563 7795209989 6545472001 : 9849
6968351074 3077019398 8419787070 9476949874 8642008554 7074572670 6021712740
9566926654 3143833776 9024013142 : 9850
7899977567 7872417957 1357223060 6323052385 6351476313 2557264467 5977337769
8628417509 4033938121 6921426735 : 9851
6386462354 3402066943 0635075139 4027442889 7658237003 0756493010 6573451869
8212694757 8850411919 6136104609 : 9852
3431644074 1533743957 7243588525 6502313780 9475431957 7305451878 5207209863
8611622730 4334532958 7547485512 : 9853
7338327972 2191850350 4787189788 4249287106 8961821089 9417158677 6120838015
1618886514 5400128585 9716463013 : 9854
6449651605 5149382279 2834449083 7674328198 9211429104 3106111086 8692165527
9207982016 4932937458 1342150765 : 9855

5118784892 7673825487 9351178323 5161120855 1784108376 2117528150 0785185477
8186076794 2864148432 3323319109 : 9856
7266850032 9341289140 4585820443 1557610778 7857625774 2383848993 1572373959
8381106078 1246778635 8556899652 : 9857
7368845240 9425287045 9643592076 0579346684 3241453255 9635624874 8821179120
8183970347 8464175492 9142248619 : 9858
8816983583 6819129231 9024071712 9570007687 4633450585 4605361897 4136529114
8748788266 7012766317 5041210072 : 9859
0315781088 9243766403 6773091175 5309040928 1711961362 1053923413 8733422593
7678768526 2571351224 3413482449 : 9860
2437863318 8527682753 7431049035 4455242143 5713693138 5640290400 0656993625
3681779918 7855871844 7307705825 : 9861
2974350574 8664127084 6544260384 7274453318 3529848936 8389889780 8289018623
0744840470 8449085284 9403039434 : 9862
2955463852 7408547590 1526881600 0374772522 8117811611 5742371398 1950740674
1753431841 4635054434 2438357451 : 9863
9248611580 8800396066 8343940190 8987378192 4404002198 3228452076 5154792113
6925507478 7416583660 2422185462 : 9864
1946430775 1521861355 8151988754 0463551761 4095905437 3857005250 1358093904
8596721620 2160698441 6156907877 : 9865
1684068127 4143766140 9111813963 1085599151 6614810795 4451532495 9921366825
4996327123 7356256841 4745414312 : 9866
3120604881 9565493502 8235797994 3767775718 3566590069 0204179933 6909172824
1049062313 2712442494 1601428516 : 9867
0962807790 6360383358 4141992709 1542276842 5817749922 1993057258 0325951237
7120129944 2484070122 7968679444 : 9868
7341806792 5826579349 5891476788 8378915440 9326316567 0060889473 1093256105
6198803086 0548652087 7456454524 : 9869
7485353154 0501116812 4284749579 0437215941 5539436702 9071257786 5368975422
8949640116 1850244371 3140843463 : 9870
0354358064 7721271143 5043136254 8438660924 3615500319 6510855005 0907158337
0415025568 8910245840 0418193352 : 9871
0252124498 4571676792 5568221802 3266396231 5709645890 7968699494 1373432621
1814954738 9785624882 1720105582 : 9872
7898453333 3136548489 4508887856 7519040445 9592653195 2158119244 8981135290
2213455038 3565982719 3694931765 : 9873
6578186760 0470077316 9181645503 4194766774 3801815903 3309930256 6595666806
0102509265 3011515016 0622468616 : 9874
3897193090 2143214170 4002191457 7268584078 6520972216 8098063403 4097430869
0729882230 9751951983 1002986126 : 9875
7048908177 8827687757 9611783302 2329018460 0190716087 4768423152 2405077436
0779330699 7142082661 8840794046 : 9876
9258063274 2472750415 6574254671 4723309962 6039572865 3859055288 8005911222
5774689762 1928238904 0655521371 : 9877
9749452232 1276833845 1551457684 6611277668 8207883475 8858600648 8657357631
6527556999 4119108346 6144561867 : 9878
0681274946 3392864273 6615115589 1224086859 7078927007 5033942198 7583653597
3430041069 5592267610 3553305993 : 9879
1059176312 7935116297 1407960650 1253293619 4013665063 0794157005 0211188043
5814945337 9132085873 2548430463 : 9880
6596357544 5347023689 5764075477 0441557149 3176867930 2277753119 8821848373
0191178628 9306940111 5893599512 : 9881
7482991205 3068129551 9298249382 7214779554 1012944377 3451756425 1171652621
6166999185 8356468746 4935792409 : 9882
7724551332 8023629366 7570990769 7852514226 6411911935 8543450137 4096079376
0050865528 3422806572 6478022042 : 9883

0613079516 9963953324 7616622891 5466846940 9691451525 2563415426 9767270965
4289236696 8403192535 0949067384 : 9884
5090202972 4318091231 2701825519 5138346002 9378821762 0776766147 0179105698
7095327706 6003084161 8105920658 : 9885
7486560051 4839521848 2499254325 4856132508 5851974364 1707255380 3196406928
0715632213 2217334545 1876615575 : 9886
5263903183 3939656842 0029460701 1191290037 4032234397 6701062591 8954941108
1661777562 1921623121 1370903139 : 9887
7199510930 0060103007 1533334529 0159788935 8798595884 7841800525 3796008798
3471109265 7875489541 9075386106 : 9888
6220189959 7895405934 3787394706 3842367677 5021833692 8872866052 7834544522
0240580571 1362960075 9746651977 : 9889
7111126309 6946950342 3846510531 4336670951 1076068629 0587829078 8220871334
6420036439 0366332980 9885008513 : 9890
3781496316 6188577108 0663125580 4432420769 8647512216 5823616208 3468148414
0831525275 3960506270 6669652630 : 9891
1902593848 7440341266 0635747377 1924725215 1652293940 6695524534 2248129991
8326565944 2375912055 9882004728 : 9892
6420420670 7422287275 9074097139 8215723796 3214549916 7296730802 8648644840
6832219033 6849026989 9297102009 : 9893
2041186578 7025178591 8357505527 0334768877 5638585462 7396075099 7614740057
2192898182 9667262012 0311458260 : 9894
8158553826 9222510138 2561102542 9443067962 4364008997 8100544006 7802134276
7076425499 2593671010 2284674866 : 9895
2259411752 9405166755 5818626941 7505939971 1537675399 6650983301 6157369709
2270055730 9695955649 2723851825 : 9896
7578755341 7888752639 7885559646 2447492326 4074821628 5423380363 3594937489
9526806016 7541421978 8350902373 : 9897
1057687503 2819182590 5212533184 9318306100 7022026785 8052785156 3024324195
5539385657 1133062252 2462341479 : 9898
7114479325 7899373626 5220222845 7992303460 1177104157 4409412468 1972950029
1141398676 1550352599 8194807352 : 9899
3915452858 1118222981 8156494479 2756355332 2350629049 6237602188 8577294081
8935145056 3943338935 3397765545 : 9900
6340866555 2826581360 0081646333 4525449484 5059555667 0516310770 8872505206
6260225605 7362921445 1674721138 : 9901
8601691629 3005420512 4206415555 2724019455 8735950975 2625533729 0938999075
2880852423 4255268789 6232612745 : 9902
3557578081 7153091024 5963535539 4814762771 9691641984 2611182758 8625890580
4366154875 6206816374 3408004760 : 9903
0164419843 8029373414 6828677717 6160414285 4126064256 4158093743 9615912924
2713631367 3177612445 8993385917 : 9904
7730611332 9884665524 5748283492 5231194587 0699891238 1060083820 2270265985
6257633391 9033672940 0012642906 : 9905
9966838423 8179436127 4698665931 8961904532 2223293603 1663296014 9436871828
0932743049 1523893225 6255191114 : 9906
7300753148 1288072064 2684222698 1136185520 1544798087 6010728760 9450269249
4540614879 5259778098 6360069669 : 9907
1013778141 2349002068 6919838926 4153767225 5176874951 5204014883 0312041130
2022780645 9645894805 8713779967 : 9908
6887349391 8703798214 3916720645 9086965189 7225691698 5099903102 0309665736
3301877215 7910787814 2644572561 : 9909
7411584185 8776996356 3529157661 1517161175 5619121463 7721380113 6522636271
2785035214 5333065842 4042719346 : 9910
5708056180 5642862699 8831932727 3724685080 8063725345 8677431435 6471343299
1361084587 1145670176 8060241563 : 9911

9874524038 5833637939 8356339349 4459503071 4427694213 9216593351 5161002806
8260018621 7108027589 9091634546 : 9912
2460226985 8671745423 1097729730 6019504557 5021217866 7292686336 6512580710
5050017472 1336920726 9807192779 : 9913
0633012919 0334291822 0130117130 2068612463 7361536176 3948055611 2267903595
9983458207 3680303215 6050836640 : 9914
1669154640 2608726050 6651984907 5749621296 3311920460 8642470105 9966275204
0679908521 1118873939 5262221717 : 9915
1839456743 5843662502 5940756778 7455206883 1770210182 2639289293 3319946189
5562143393 9875377741 8233490776 : 9916
3855993540 0871935321 5578106343 4926469216 6801973069 5877934717 2225448079
1181111963 9263927648 0012351772 : 9917
5719274788 3057839711 3669060645 5254331919 0322289193 6095497184 3249100904
5406627292 3850274056 3383854859 : 9918
0188268149 4358436458 4398026261 1116171765 8428160979 5765250678 7611795116
3239926351 7003261953 4150929750 : 9919
0553404886 6409658351 9165979491 9350863488 2192615981 4484369243 7545565112
2041948055 2682307533 2476974088 : 9920
4727787435 4523592721 0880405262 5835368689 1993198446 0989716084 4674263673
5916800552 8478634062 6912717217 : 9921
3171756061 6771714634 7556161980 7884390311 3584777164 2605104747 4576636143
8320854993 6721974573 9979786652 : 9922
2775035398 0618914088 8838590932 1374375272 0336230257 8779561047 2928563860
8513091577 8464960008 7363339203 : 9923
1048977816 9199045483 7211576932 6147221016 9373395677 0865691376 1110869153
2783540556 8948605071 0822954248 : 9924
0918055808 0958528406 6735287814 8038653802 1466467571 4389654758 0860434512
9553935513 0958693211 0862993111 : 9925
0605839939 4249657601 0657495240 2644946365 5244424073 0599036528 4908966648
0404679455 1760568902 7631717191 : 9926
8768727725 7489033656 7177856382 3216530569 2129115050 3264128157 3270750113
8355197893 0940891074 8803426109 : 9927
0882741413 7119409123 0943716786 9613636072 4776571046 2348615040 6068547046
4577187891 6603821401 4347509730 : 9928
5369103110 8440796955 0456237753 8119827552 1595213650 1877563397 0735439580
1204719660 1951288251 0545033173 : 9929
0516216341 0905181922 6052554631 2264355322 5929574757 2882001626 2708082360
4244459035 8136199045 9604491675 : 9930
4037553727 2061819889 9951477716 1493276079 7999354053 2317930374 3527584995
4268171872 1374730002 5933561536 : 9931
1921111292 6163891621 8469566956 2033564970 5965093323 7168855187 8420333041
8075030665 0556062517 4160505262 : 9932
3316640919 2522382588 7095521890 2881295750 5217116559 7917130825 3404608433
0797746547 6881666919 6814476897 : 9933
3284839179 2177677642 7103215174 5244795073 5880763289 0541673160 3181192610
2417003817 7565961182 5654152518 : 9934
0967987602 6163430172 7837032796 1732925081 3470476548 5765605901 0727673521
3854598276 8929873329 7583299446 : 9935
8536565991 9272027231 2841968966 6325935947 7266722350 0113719502 6467308449
2628609598 5262052240 9521822359 : 9936
2003147069 8269779266 7225042365 5929192492 0543503544 4239094088 0576201050
4653092697 7313494108 5727997638 : 9937
0113049279 7398655841 9898876583 3201593343 9610468750 7963520178 4729873173
0440842726 6965840609 6180546546 : 9938
5631930250 5014959888 4019250855 9631880923 3280147303 8791291957 9582510129
2043776533 4741089180 7580724712 : 9939

7240276166 6296862622 2223166070 4487522921 4715071461 5960773351 2382721669
1545527291 3078876136 7040033447 : 9940
7105207700 5942899727 1177365919 2429913212 0809706489 6312558843 9119442642
4834555020 7274615226 7209925644 : 9941
5652834679 8994906603 4174136847 6155773134 4073469800 3798042141 2267132037
2465321073 2173573760 4919320627 : 9942
5564676654 9039130286 8997801527 7912027247 5105329245 9552739864 2066245295
7180086809 1573355539 7019651293 : 9943
1004832314 7041350494 2935965118 2657240498 2044313975 6703147053 7098506131
4615599154 5967908038 2063327112 : 9944
7053976438 9461306833 5246691567 6444805847 9053185626 4957839354 5468362970
9750886407 2557823669 2990650812 : 9945
7064320767 4253904368 8571381094 0745855659 6741811348 1026729780 1287659770
5816267284 5756159328 1266064575 : 9946
3286983567 5416943733 5171869154 4942980395 3280956253 2967176474 2784192171
0549638534 1428321986 2148618952 : 9947
6791483040 0454230243 7244624942 7695881885 1304787948 0515092402 2184727268
7432602896 2048598568 3180743752 : 9948
1486290992 1339139921 8069538074 4363471106 2302102023 9908016642 8312197903
9310889890 2868127744 9139877816 : 9949
0123696349 3538379048 3373886164 9739886424 0856403886 0012173560 3712566302
8241932844 1393715260 3550255365 : 9950
0068469137 9951253301 5708816931 7619805957 6609611328 1453624863 8143747081
3760997339 2793407138 7102183560 : 9951
4437946559 5763210859 7263805611 3173862779 9618662828 1065800063 6060566965
1605002754 6320006428 3833990047 : 9952
0686106215 8978013591 8708023883 7576895579 1117132702 1871913861 2446092285
0946621788 0091235664 6714252845 : 9953
8131688323 3438617026 3454531063 5732686171 4173268252 1099271958 3249078323
2192898048 5122982303 3793785926 : 9954
9722693195 8950333541 2606776368 4519202799 0112943513 2900852589 6049061348
1768461244 8185863452 6732494412 : 9955
3950337024 2895528566 7455763776 5483130391 5449072417 4469903334 8963010326
9608512640 5368878222 1214621521 : 9956
9423825090 7818894026 4353675156 8910448856 8329212836 0258157480 0998458320
5487653084 0824256135 0893597229 : 9957
6031858838 8580383581 8566441215 3867087676 0124831010 4630844744 3218801447
9093367474 6884478564 1481744590 : 9958
9124539810 3230088592 0637101563 5875651643 0959611273 9640158617 6978131408
7950737142 8831776043 6689819626 : 9959
4134750069 7194056519 5045505167 7423979401 9689986252 7890085237 4458073288
6706974177 3963572545 0609854234 : 9960
5678985204 2507322860 7094202639 4841828669 2566061665 4440720457 7568393932
3122654078 1247831666 8801803018 : 9961
4225045200 5355186868 4850558453 0852538497 5426120579 4305353308 0747750826
4960885294 4557278500 3441395589 : 9962
3793350511 8403022529 8706162915 0596422599 9600585648 9572336531 7986913669
6994427866 5789125256 8462604147 : 9963
9811086568 5571739072 1330789331 0852590333 3113718587 7287034926 2790271567
3291686627 1667490819 9318258316 : 9964
6832828500 1575707801 6119316922 1931514754 9377551598 0465409283 9991094937
4201037170 8560860588 1785449005 : 9965
7041041360 4043513764 2468998152 6806092554 0112346532 9504349180 5374773561
6666710469 2983095678 9203164820 : 9966
6393173922 1248405512 0347806321 1131681337 3224063216 4554155882 3784609194
2738088502 8383123622 6654974430 : 9967

0558099898 2995742584 3235376429 8631465635 0552835604 7709075747 3282636770
4396309792 3465629794 9344566413 : 9968
9608514643 7130393213 6774212470 9044527215 4068792154 2630642597 2602920189
9465529811 5142612604 9007638671 : 9969
4173023572 7276839041 5597234506 6669386646 0588292012 4711441788 3178223482
1533891876 0583632761 8181943322 : 9970
7695553112 5819048475 1746290562 0013468996 4407161982 3205434718 4611020511
3155509302 2682510749 1990149608 : 9971
1785626051 0885903658 4745150376 3849151340 0032951639 9106219240 5572830081
0352176197 9616832213 9816924083 : 9972
5763955621 1657126112 1929050871 6325525855 8649616638 2541935914 8218187619
5923292056 9955063764 5818268557 : 9973
5221152787 0118029943 3546741536 2762077497 8540541333 0313633542 3682410108
4647637490 6388527984 1490076464 : 9974
6976485400 9479635895 4975461448 1376369705 9163569983 6811987525 0547930693
5320757076 6780148447 7014247162 : 9975
4190816682 2490074207 1118648815 4772891718 6535967765 3957993350 3342728214
6054169649 6009847069 7958559264 : 9976
3042870363 6647130713 1478233061 1576419913 2224206460 9989883076 2685836055
5274099047 8467610760 4241784215 : 9977
0628517557 3529996478 6255295428 3674298706 6457943375 8010140740 2116186144
8432976574 4263428528 7047785563 : 9978
0830963143 5278783041 9450197029 4657577773 2816746858 0874539316 0393725331
5899280579 4346314087 3586086177 : 9979
8826334927 7461511849 1165513068 1846713677 3488233410 8513640394 7939208876
8863363394 6138235834 4794081569 : 9980
6109142938 7734713893 4237736191 0964605642 4447477908 2076049660 2713561689
5410644483 2136598082 9389097296 : 9981
1891211834 2914906163 8963861069 3752089534 6883983344 4671898212 4347807238
7407457697 5545074368 4674713502 : 9982
4858818399 6655681963 4452881194 1833172636 8250506118 6490039412 5520574571
2036035578 0251419043 5267183721 : 9983
9213848299 0580322469 5842432315 8984432510 3965443535 0535432292 1674704077
8614684859 7625574461 5351188003 : 9984
1430569954 9278471674 5449726976 1283933251 8381972223 2836070752 2781292813
0106569412 6294873063 4268837338 : 9985
1817421706 0864754827 6394242391 4027532180 4295190341 1635170469 8074233515
5605785756 2450999253 2017874996 : 9986
3664047347 7038985587 3065076038 7099773184 3128109897 8988208543 5595509432
5390237189 5216820233 4424557257 : 9987
5307879263 3985509016 4559423733 9662522335 1648750589 5569421729 7244895998
8250892321 1203479589 4154654603 : 9988
0378786175 9157166139 8869326873 7496847305 4965329378 2147564810 5793808285
3005324470 8050656929 4223400109 : 9989
5934829461 4539078890 6616264021 5013073533 0033192074 5637263770 7709993999
2288621224 3248802062 6348508885 : 9990
3036010723 4368901360 6427581425 2839878594 9179979611 2196379757 6519245218
6709608809 2137111977 5000878159 : 9991
3043072934 4883930957 5741592413 7528597779 7291893453 8505080383 1986774590
0251865791 7237080857 4164297153 : 9992
8078840607 1306868036 1982419715 7747638950 7253468404 5691927595 3193722370
2229015580 0656076047 3854735990 : 9993
4477996748 7499697694 2713766869 5533195125 3377640985 8709668386 3263926164
9456086841 4037456842 0719405950 : 9994
7017430354 6918215090 0466493998 5517413893 8519757312 1568261622 8622318810
9672974760 6013028331 1937161140 : 9995

8747270676 2558567775 1199566674 8615196491 2970193318 0849941096 1813929649
2789360902 1253544332 7375064260 : 9996
6242994120 3273625582 4417498345 0947309453 4366159072 8416319368 3075719798
0682315357 3715557181 6122156787 : 9997
9364250138 8711702327 5555779302 2667858031 9993081083 0576307652 3320507400
1393909580 7901637717 6292592837 : 9998
6487479017 7274125678 1905555621 8050487674 6991140839 9779193765 4232062337
4717324703 3697633579 2589151526 : 9999
0315614033 3212728491 9441843715 0696552087 5424505989 5678796130 3311646283
9963464604 2209010610 5779458151 : 10000

Yes, you ARE a freak.

www.ingramcontent.com/pod-product-compliance
Lightning Source LLC
Chambersburg PA
CBHW061434180526
45170CB00004B/1407